# Likelihood Methods in Survival Analysis

Many conventional survival analysis methods, such as the Kaplan-Meier method for survival function estimation and the partial likelihood method for Cox model regression coefficients estimation, were developed under the assumption that survival times are subject to right censoring only. However, in practice, survival time observations may include interval-censored data, especially when the exact time of the event of interest cannot be observed. When interval-censored observations are present in a survival dataset, one generally needs to consider likelihood-based methods for inference. If the survival model under consideration is fully parametric, then likelihood-based methods impose neither theoretical nor computational challenges. However, if the model is semi-parametric, there will be difficulties in both theoretical and computational aspects.

*Likelihood Methods in Survival Analysis: With R Examples* explores these challenges and provides practical solutions. It not only covers conventional Cox models where survival times are subject to interval censoring, but also extends to more complicated models, such as stratified Cox models, extended Cox models where time-varying covariates are present, mixture cure Cox models, and Cox models with dependent right censoring. The book also discusses non-Cox models, particularly the additive hazards model and parametric log-linear models for bivariate survival times where there is dependence among competing outcomes.

**Features**

- Provides a broad and accessible overview of likelihood methods in survival analysis
- Covers a wide range of data types and models, from the semi-parametric Cox model with interval censoring through to parametric survival models for competing risks
- Includes many examples using real data to illustrate the methods
- Includes integrated R code for implementation of the methods
- Supplemented by a GitHub repository with datasets and R code

The book will make an ideal reference for researchers and graduate students of biostatistics, statistics, and data science, whose interests in survival analysis extend beyond applications. It offers useful and solid training to those who wish to enhance their knowledge in the methodology and computational aspects of biostatistics.

**Chapman & Hall/CRC Biostatistics Series**
**Series Editors: Mark Chang,** *Boston University, USA*

**Case Studies in Bayesian Methods for Biopharmaceutical CMC**
*Edited by Paul Faya and Tony Pourmohamad*

**Statistical Analytics for Health Data Science with SAS and R**
*Jeffrey Wilson, Ding-Geng Chen and Karl E. Peace*

**Design and Analysis of Pragmatic Trials**
*Song Zhang, Chul Ahn and Hong Zhu*

**ROC Analysis for Classification and Prediction in Practice**
*Christos Nakas, Leonidas Bantis, and Constantine Gatsonis*

**Controlled Epidemiological Studies**
*Marie Reilly*

**Statistical Methods in Health Disparity Research**
*J. Sunil Rao*

**Case Studies in Innovative Clinical Trials**
*Edited by Binbing Yu and Kristine Broglio*

**Value of Information for Healthcare Decision Making**
*Edited by Anna Heath, Natalia Kunst, and Christopher Jackson*

**Probability Modeling and Statistical Inference in Cancer Screening**
*Dongfeng Wu*

**Development of Gene Therapies**
Strategic, Scientific, Regulatory, and Access Considerations
*Edited by Avery McIntosh and Oleksandr Sverdlov*

**Bayesian Precision Medicine**
*Peter F. Thall*

**Statistical Methods for Dynamic Disease Screening and Spatio-Temporal Disease Surveillance**
*Peihua Qiu*

**Causal Inference in Pharmaceutical Statistics**
*Yixin Fang*

**Applied Microbiome Statistics**
Correlation, Association, Interaction and Composition
*Yinglin Xia and Jun Sun*

**Association Models in Epidemiology**
Study Design, Modeling Strategies, and Analytic Methods
*Hongjie Liu*

**Likelihood Methods in Survival Analysis**
With R Examples
*Jun Ma, Annabel Webb, and Harold Malcolm Hudson*

**Biostatistics for Bioassay**
*Ann Yellowlees and Matthew Stephenson*

**Power and Sample Size in R**
*Catherine M. Crespi*

For more information about this series, please visit: www.routledge.com/Chapman--Hall-CRC-Biostatistics-Series/book-series/CHBIOSTATIS

# Likelihood Methods in Survival Analysis

## With R Examples

Jun Ma, Annabel Webb, and
Harold Malcolm Hudson

CRC Press
Taylor & Francis Group
Boca Raton London New York

CRC Press is an imprint of the
Taylor & Francis Group, an **informa** business
A CHAPMAN & HALL BOOK

First edition published 2025
by CRC Press
2385 Executive Center Drive, Suite 320, Boca Raton, FL 33431, U.S.A.

and by CRC Press
4 Park Square, Milton Park, Abingdon, Oxon, OX14 4RN

*CRC Press is an imprint of Taylor & Francis Group, LLC*



ISBN: 978-0-815-36284-5 (hbk)
ISBN: 978-1-032-46814-3 (pbk)
ISBN: 978-1-351-10971-0 (ebk)

DOI: 10.1201/9781351109710

Typeset in Nimbus Roman
by KnowledgeWorks Global Ltd.

*Publisher's note*: This book has been prepared from camera-ready copy provided by the authors.

*To our families:*

- *To my parents and Mary, Jiawen and Jiani.*

*J. M.*

- *To Gaby, Michael, Jeanette and Grandpa John.*

*A. W.*

- *To Helen and our family.*

*H. M. H.*

# Contents

# Preface

This book is about likelihood-based survival analysis methods, including partly interval censoring in general, but also providing a few chapters that solely focus on right-censored observations.

The focus of this book is on methodology and computation, which is why there are many R examples included throughout the book to help the reader understand the concepts and methodology. In particular, examples of how to conduct simulation studies are included as this skill is fundamental for preparing journal articles.

Many conventional survival analysis methods, such as the Kaplan-Meier method for survival function estimation and the partial likelihood method for Cox model regression coefficients estimation, were developed under the assumption that survival times are subject to right censoring only. However, in practice, survival time observations may include interval-censored data, especially when the exact time of the event of interest cannot be observed. Numerous examples illustrate this scenario. For instance, cancer recurrence time and dementia onset time cannot be precisely observed; instead, the physician only knows that it falls between two examination time points. When interval-censored observations (including left-censored ones) are present in a survival dataset, one generally needs to consider likelihood-based methods for inference. If the survival model under consideration is fully parametric, then likelihood-based methods impose neither theoretical nor computational challenges. However, if the model is semi-parametric, there will be difficulties in both theoretical and computational aspects. This book explores these challenges and provides practical solutions. This book covers not only conventional Cox models where survival times are subject to interval censoring but also extends to more complicated models, such as stratified Cox models, extended Cox models where time-varying covariates are present, mixture cure Cox models and Cox models with dependent right censoring. The book also discusses non-Cox models, particularly the additive hazards model and parametric log-linear models for bivariate survival times where there is dependence among competing outcomes.

The potential readers of this book include graduates or researchers in statistics, mathematics, finance, insurance and data science whose interest in survival analysis extends beyond applications. This book offers useful and solid training to students and researchers who wish to enhance their knowledge in the methodology and computational aspects of biostatistics.

This book is divided into 9 chapters to cover fundamental knowledge and gradually introduce the reader to more advanced topics in survival analysis. Below are brief descriptions of these chapters.

Chapter 1 of this book introduces some basic concepts and methods that are either required by the later chapters or useful for understanding them. It covers the concept of censoring, including interval and right censoring, and summarizes parametric and semi-parametric regression models for survival data analysis. It briefly discusses three semi-parametric models, the additive hazards model, the Cox (proportional hazards) model and the accelerated failure time model. It also introduces the penalized likelihood method for estimating regression model coefficients.

Chapter 2 discusses the details of the maximum penalized likelihood method for Cox models, where the observed survival times are subject to interval censoring (including left censoring). This method starts with an approximation to the non-parametric baseline hazard function, such as M- or B-spline approximations. The penalty function not only serves the purpose of providing a smoothed baseline hazard estimate but also helps to reduce the impact of the number and location of knots required to define an approximation to the baseline hazard. There are two computational issues when fitting a semi-parametric Cox model: (i) the baseline is required to be non-negative, and (ii) some of the spline coefficients are often actively constrained. Ignoring the active constraints can lead to unpleasant consequences, particularly negative variances for some estimated parameters. Chapter 2 provides a simple solution to fix this issue. Additionally, Chapter 2 explains why reparameterization by transformation, such as square or exponential, may not be ideal when imposing non-negativity constraints on spline coefficients.

In Chapter 3, the Cox model with interval censoring is extended to accommodate double truncation, where once again, penalized likelihood is adopted to estimate the model parameters, including the baseline hazard. For the most common situation of left truncation and right censoring, when the baseline hazard is approximated as piecewise constant the Cox model parameters can be estimated by a profile likelihood approach, resulting in a procedure similar to the partial likelihood method. For the general double truncation situation, a conditional likelihood can be adopted for inferences, and details of model estimation of regression parameters and covariance matrix are provided.

Chapter 4 considers how to address the issue of a cure fraction in survival data. The cure fraction refers to a sub-population where individuals in this group will never experience the event of interest. Ignoring the cure fraction can lead to biased parameter estimates and hence false conclusions regarding the questions under investigation. In this chapter, the mixture cure Cox model is introduced and its parameters, including the baseline hazard, are estimated by maximizing a penalized likelihood function. The non-parametric baseline hazard is approximated using, for example, spline functions. In practice, classifying a right-censored individual into the cured group is often an important consideration. This chapter contains a section explaining how to calculate the probability, based on the fitted mixture cure Cox Model, that an individual belongs to the cured group.

Chapter 5 introduces the stratified Cox model, but again in the general context of partly interval censoring. A stratified Cox model can be useful when the proportional hazard assumption of the Cox model is violated, or when there is interest in investigating the effects of covariates within a specific subgroup. In stratified Cox models, the regression coefficients are identical across strata, but the baseline hazards are stratum-specific. After approximating all the baseline hazards using, for example, M-spline functions, the regression coefficients and spline coefficients can be estimated using penalized likelihood. The asymptotic covariance parameters of all these parameters can be derived after addressing the active constraints of the spline coefficients properly. Hence, inferences on quantities of interest, such as the probability of survival time beyond a specific value, can be calculated with the corresponding confidence interval provided.

Chapter 6 extends the Cox model introduced in Chapter 2, where the covariates are time-fixed, to include time-varying covariates. This chapter is limited to right censoring, but the methods discussed in this paper can also be easily applied to survival data with partly interval-censoring. Time-varying covariates pose computational challenges for likelihood-based estimation methods. This is because repeated integration tasks are required in each iteration of the algorithm for maximum penalized likelihood estimation and this can be extremely time demanding. Given that time-varying covariates, such as biomarkers collected at different time points, are mostly piecewise constant, this fact substantially reduces the computational burden associated with time-varying covariates. In fact, with a long format of the survival data and covariates, the computational burden of Cox models with time-varying covariates is similar to that with time-fixed covariates.

Chapter 7 of this book explores dependent right censoring in Cox models, focusing solely on right censoring. Dependent censoring poses a challenge in survival analysis; however, disregarding this phenomenon can lead to biased estimates. The complexity of dependent censoring arises from the fact that only one of the pair of event and censoring times is observable for each individual. Hence, analyzing survival data with dependent censoring is feasible but necessitates heavy assumptions. Existing dependent censoring methods include the frailty method and the copula method. This chapter focuses on the copula method. It first summarizes copula functions and then explains likelihood-based estimation methods, where the constraint of non-negative baseline hazard is imposed. The asymptotic covariance matrix of the estimated parameters is developed, considering active constraints.

Chapter 8 shifts the focus from the Cox model to the additive hazards model. It starts with a summary of the existing counting process-based estimation methods for additive hazards model, and then describes a primal-dual interior point method for maximum penalized likelihood estimation for the additive hazards model, where survival data are partly interval-censored. An efficient interior point algorithm is described, and the asymptotic covariance matrix is derived, taking active constraints into consideration. Due to its complex constraints, fitting the additive hazards model is generally challenging. Existing count process methods largely ignore these constraints, potentially resulting in less accurate estimates of the model parameters.

Finally, Chapter 9 of this book discusses parametric bivariate survival data, where survival times are assumed to be log-linear in covariates, and the error distributions are assumed to be log-normal. These bivariate survival times are "competing" in the sense that only the smallest of the two is observed. Under these assumptions and the assumption on the bivariate distribution, an efficient EM algorithm is discussed and evaluated in a simulation study.

The authors have also prepared a GitHub website to store the datasets and R programs for this book. The GitHub web address is: https://github.com/MPL-book. Other R programs from this book are available on R CRAN, particularly the "`survivalMPL`" and "`survivalMPLdc`" R packages.

Jun Ma
Annabel Webb
Harold Malcolm Hudson

Sydney, Australia

# *Authors*

**Jun Ma**
School of Mathematical and Physical Sciences, Macquarie University
North Ryde, Australia

**Annabel Webb**
School of Mathematical and Physical Sciences, Macquarie University
North Ryde, Australia

**Harold Malcolm Hudson**
School of Mathematical and Physical Sciences, Macquarie University
& NHMRC Clinical Trial Centre, University of Sydney
Sydney, Australia

# 1

# *Introduction*

In this book we provide an overview of the mathematical and algorithmic approaches used in our approach to penalized likelihood estimation methods for survival data subject to censoring. Our methods apply to right-censored, left-censored and general interval-censored data, and estimation is readily available in the R-package `survivalMPL` available on CRAN.

Survival analysis involves the follow-up of subjects from an entry time (and state) to an event of interest, or in the case of non-occurrence of such an event, the time follow-up ended ("censored follow-up"). While such analysis is very widely used in medicine, other applications abound, such as in systems reliability. In this introductory chapter we present some selected components, deriving our presentation from standard texts (Kalbfleisch & Prentice 2002, Lawless 2003) and course notes readily available online (Rodrıguez 2005) to which the reader should refer for definitive treatment of topics, notation and further clarification. In each section we shall reference the text used for the topic.

## 1.1 Survival data and censoring

We begin by introducing single event analysis, where the event survival and hazard functions will be defined. We will also briefly describe two models for composite-event or multiple-event outcomes, the competing risks model and illness-death model.

### 1.1.1 Single event and right censoring

Basic methods in survival analysis provide the estimation of the distribution of *survival time* from a start-point to the first occurrence of an event of interest. In medical applications the event, or outcome, being followed may be death, disease progression or a surrogate outcome, such as a biomarker predictive of the desired outcome. To simplify the descriptions presented below, we assume the survival (or event) time is a continuous random variable. Let us denote this survival time by $Y$ and the corresponding distribution function will be denoted by $F_Y(t)$, for $t \geq 0$. We may drop the

DOI: 10.1201/9781351109710-1

subscript $Y$ when there is no confusion. The *survival function*,

$$S_Y(t) = P(Y > t) = 1 - F_Y(t), \tag{1.1}$$

represents the probability of survival beyond time $t$. We comment that $F_Y(t)$ is also known as the cumulative incidence function function (CIF) in the survival analysis literature, as it measures the proportion of events occurring before time $t$.

We use the term survival analysis to refer to methods for analysis of survival data. Survival *data* refers to data on subjects recording a time from a start point to first occurrence of an event of interest, with an indicator that an event occurred, or otherwise an event-free interval of follow-up, when no occurrence was observed for a certain length of time. Non-occurrence of the event during follow-up is referred to as censoring. The variable $T$, `time`, then denotes the value of time to event or duration of follow-up for each subject, and variable $\Delta$, `status`, is 1 if the event was observed, or 0 when censored. We call the random variable $T$ the *observable* survival time. Survival data then comprises $n$ observations on the pair $(T, \Delta)$, where $T$ is time and $\Delta$ is status. An observed realization of the $(T, \Delta)$ pair is usually denoted by lower-case letters: $(t, \delta)$.

The relationship between $Y$ and $T$ is specified below. Let $C$ be the censoring time (time to end of follow-up without an event), the observable survival time random variable is then

$$T = \min(Y, C).$$

This type of censoring is termed *right* censoring as if $t$ is a censoring time then event time $Y$ is located on the right of $t$: $Y > t$. Other censoring types also exist, such as *left* or *interval* censoring; see Chapter 2 for other censoring types and the penalized likelihood-based estimation of the associated Cox model. Correspondingly, $T$ can also be referred to as the right-censored, left-censored or interval-censored survival (or event) time, respectively.

In survival analysis, apart from the survival function $S_Y(t)$, the hazard function (or hazard rate of the event), denoted by $h_Y(t)$, also plays an important role. It is defined as

$$h_Y(t) = \lim_{\delta t \to 0} \frac{P(t < Y \leq t + \delta t | Y > t)}{\delta t} = \frac{f_Y(t)}{S_Y(t)}, \tag{1.2}$$

where $f_Y(t)$ represents the density function of $Y$. The hazard function gives the instantaneous failure rate at time $t$ given the subject has survived up to $t$. From the hazard function, the cumulative hazard is defined as

$$H_Y(t) = \int_0^t h_Y(s) ds. \tag{1.3}$$

From the cumulative hazard, the survival function may also be calculated via

$$S_Y(t) = \exp\{-H_Y(t)\}. \tag{1.4}$$

On the other hand, when the survival function $S_Y(t)$ is known then the hazard function can be obtained through

$$h_Y(t) = -\frac{d \log S_Y(t)}{dt}. \tag{1.5}$$

The censoring mechanism is termed *non-informative* or *ignorable* if it plays no part in likelihood optimizations. In other words, its distribution provides no information about the distribution of $Y$. This is most clearly seen in the way censored observations enter the likelihood function. Under the non-informative censoring assumption, if an observation is right censored at time $t$, its contribution to the likelihood is just the probability that lifetime $Y$ exceeds $t$. If $Y$ and $C$ are dependent, then the censoring time $C$ must be informative.

When the survival distribution is estimated under non-informative and parametric assumptions, it is appropriate to use the maximum likelihood estimator to provide fitted survival probabilities. Well-known univariate parametric distributions used in survival data include exponential, Weibull, gamma, log-normal, log-logistic, inverse-Gaussian, etc.

**Example 1.1** (Exponential survival).
Here, the survival function

$$S_Y(t) = \exp(-ht),$$

where the parameter $h > 0$ is the constant *hazard rate*. Thus, the hazard function $h_Y(t) = h$ for $t > 0$. □

**Example 1.2** (Weibull survival).
The probability density function of a Weibull random variable is given by:

$$f_Y(t; h, k) = \begin{cases} \frac{k}{h}\left(\frac{t}{h}\right)^{k-1} \exp\{-(t/h)^k\} & \text{for } t \geq 0, \\ 0 & \text{for } t < 0, \end{cases}$$

where $k > 0$ is the *shape* parameter and $h > 0$ is the *scale* parameter of the distribution. The corresponding survival function is given by:

$$S_Y(t) = \exp\{-(t/h)^k\},$$

and the hazard function is given by:

$$h_Y(t) = \frac{k}{h}\left(\frac{t}{h}\right)^{k-1}.$$

The shape parameter $k$ controls the hazard function of this distribution: constant hazard when $k = 1$; an increasing hazard when $k > 1$; or decreasing hazard when $k < 1$. Thus, the Weibull distribution generalizes the Exponential distribution (the case $k = 1$) when modelling an event time. □

**Example 1.3** (Lognormal survival).
For this model, $\log Y$ follows a normal distribution with mean $\mu$ and variance $\sigma^2$. Hence, the survival function is given by:

$$S_Y(t) = 1 - F_N\left(\frac{\log t - \mu}{\sigma}\right).$$

Here, and subsequently, the function $F_N$ denotes the distribution function of the standard normal distribution $N(0, 1)$. The corresponding density function of a lognormal distribution is:

$$f_Y(t) = \frac{1}{\sigma t} f_N\left(\frac{\log t - \mu}{\sigma}\right),$$

for $t \geq 0$, where $f_N$ denotes the density function of $N(0, 1)$. □

A detailed introduction to these and many other parametric models is contained in Chapter 1.3 of Kalbfleisch & Prentice (2002). Their Section 1.3.8 includes another class, the models with piecewise constant or polynomial (`spline`) hazard functions.

In general, a parametric survival distribution will be specified by density $f_Y(t|\boldsymbol{\theta})$, for a *finite dimensional* parameter vector $\boldsymbol{\theta}$. We denote the dimension of $\boldsymbol{\theta}$ by $p$.

The specification of a parametric distribution, as above, makes available the classical estimation and inference methods in survival analysis, in particular those based on the method of Maximum Likelihood (ML). However, in practice, it can be a difficult task to locate an accurate parametric distribution for the survival data under consideration, and therefore distribution free survival estimates based on semi-parametric or non-parametric methods are then required. These methods do not make specific assumptions about distributions for event times. The Kaplan-Meier (and Nelson-Aalen) non-parametric survival (and cumulative hazard) estimates and Cox model semi-parametric estimates of treatment benefits are introduced in later sections. Semi-parametric and non-parametric methods have been widely adopted for survival analysis.

### 1.1.2 Competing risks

Before considering non-parametric and semi-parametric methods, we first consider a generalization of survival data, competing risks data. Competing risks (CR) data, a subcase of multi-state data, occurs with two or more known potential end-points (competing *events*) when the occurrence of *either* event immediately precludes further observation of the *other* event(s) from this individual. That is, there will exist multiple *modes* of failure (Lawless 2002, Chapter 9). Observed time to the *first occurring* event is coupled with the type (i.e. mode) of that event, requiring complete observation up to that time. As before, the observation history ends at the time of the event or censoring, assumed to be independent of the event.

For simplicity, without essential loss, we describe the situation of *two* competing risks associated with two event types. We shall term an event of type 1 ( event 1) as the *event of interest* and the other (event 2) the *competing event*.

An effective probability model of such data is a random variable $Y$ specifying *time to first event*, accompanied by a label $\Delta$ describing the event type, with $\Delta \in \{0, 1, 2\}$, with $\Delta = 0$ indicating censoring ($Y > C$).

The joint distribution of $(Y, \Delta)$ may be specified in various ways (Nicolaie et al. 2010). Prentice et al. (1978) provide a model specifying hazards of transition to each endpoint. Alternatively, Larson & Dinse (1985) provide a mixture model that factors the joint distribution of $(Y, \Delta)$ using the marginal probabilities of $\Delta$ and conditional distribution $Y|\Delta$. Specifying the full parametric form of the joint distribution is a strong assumption, as CR data is uninformative about this distribution (Tsiatis 1975).

### 1.1.3 The Fix-Neyman illness-death model

A further development of univariate survival models considers a series of health states, such as relapse free survival and relapse or death, entered in sequence which may depend on the individual subject, concluded upon either end of follow-up or occurrence of a terminal event such as death in the context of cancer treatment (Fix & Neyman 1951). The sequence of states entered is modelled as a Markov process with time-dependent intensities (or hazards) of transition. Generalizations of this approach have led to very general multi-state processes, with Markov and semi-Markov assumptions.

Thus, the illness-death model provides the time to death – a survival outcome – but also provides models for other outcomes (e.g. time to relapse). Alternative pathways to an outcome may be considered to provide different modes of the outcome, as in CR.

It is worth noting that when the event of interest is non-terminal and the competing risk is terminal, the situation is sometimes termed *semi*-competing risks. Special methods are needed to analyze semi-competing risks data; see Fine et al. (2001).

This illness-death model was later generalized by Lagakos et al. (1978) and by Dinse & Larson (1986) using a semi-Markov model with non-parametric assumptions for state transition probabilities and transition hazards.

## 1.2 Maximum likelihood estimation for parametric survival models

In this section, we briefly explain the maximum likelihood method for estimation of parameters of *parametric models* for survival time.

### 1.2.1 Maximum likelihood estimation and properties

For $n$ subjects, their event times (random variables) are denoted by $Y_1, \ldots, Y_n$ and assume these random event times are independent and identically distributed. Let

$C_1, \ldots, C_n$ be the corresponding censoring times and assume these random censoring times are independent and identically distributed. We further assume $C_i$ and $Y_i$ are independent for all $i$. Let $T_i = \min\{Y_i, C_i\}$.

Recall, as was explained previously, $T_i$ represents the observable event time for subject $i$. We use $t_i$ to denote an observed value of $T_i$ and $\delta_i$ to denote an observed value of $\Delta_i$, which represents the indicator for the event, that is $\delta_i = 1$ if $t_i$ is an event time (so that $Y_i \le C_i$). Then, based on the observations $(t_1, \delta_1), \ldots, (t_n, \delta_n)$, the likelihood function is:

$$L(\boldsymbol{\theta}) = \prod_{i=1}^{n} \left(G_C(t_i) f_Y(t_i|\boldsymbol{\theta})\right)^{\delta_i} \left(S_Y(t_i|\boldsymbol{\theta}) g_C(t_i)\right)^{1-\delta_i},$$

where $g_C(t)$ and $G_C(t)$ denote respectively the density and distribution function of the right censoring time $C$. Under the condition of independent (so non-informative) censoring, functions $g_C(t)$ or $G_C(t)$ do not involve any elements of the vector parameter $\boldsymbol{\theta}$. Therefore the log-likelihood function, omitting the contribution from the censoring mechanism, is then:

$$l(\boldsymbol{\theta}) = \sum_{i=1}^{n} \left(\delta_i \log f_Y(t_i|\boldsymbol{\theta}) + (1-\delta_i) \log S_Y(t_i|\boldsymbol{\theta})\right). \tag{1.6}$$

Recognizing the relationship between the hazard and the density function, namely, $h_Y(t) = f_Y(t)/S_Y(t)$, the above log-likelihood function (1.6) can be more conveniently expressed by

$$l(\boldsymbol{\theta}) = \sum_{i=1}^{n} \left(\delta_i \log h_Y(t_i|\boldsymbol{\theta}) - H_Y(t_i|\boldsymbol{\theta})\right), \tag{1.7}$$

where $H_Y(t) = \int_0^t h_Y(s) ds$ is the cumulative hazard function.

The parameter $\boldsymbol{\theta}$ can be estimated by maximizing the log-likelihood $l(\boldsymbol{\theta})$ in (1.7):

$$\widehat{\boldsymbol{\theta}} = \underset{\boldsymbol{\theta}}{\operatorname{argmax}}\, l(\boldsymbol{\theta}).$$

Since $\widehat{\boldsymbol{\theta}}$ is the maximum likelihood estimate (MLE), the general maximum likelihood asymptotic distribution results also apply to $\widehat{\boldsymbol{\theta}}$. Particularly, consistency and asymptotic normality results hold for $\widehat{\boldsymbol{\theta}}$ under certain conditions.

**Theorem 1.1.** Let $\boldsymbol{\theta}_0$ be the true parameter that generates the observed survival data. Under certain regularity conditions for maximum likelihood asymptotic properties, such as those specified in LeCam (1970), we have

1. $\widehat{\boldsymbol{\theta}}$ is a consistent estimator of $\boldsymbol{\theta}_0$.
2. $\sqrt{n}(\widehat{\boldsymbol{\theta}} - \boldsymbol{\theta}_0)$ converges in distribution to a multivariate normal distribution $N(\mathbf{0}_{p\times 1}, \mathbf{F}(\boldsymbol{\theta}_0)^{-1})$, where $\mathbf{F}(\boldsymbol{\theta}_0) = n^{-1} E\left(-\partial^2 l(\boldsymbol{\theta}_0)/\partial\boldsymbol{\theta}\partial\boldsymbol{\theta}^{\mathsf{T}}\right)$, with $\partial^2 l(\boldsymbol{\theta})/\partial\boldsymbol{\theta}\partial\boldsymbol{\theta}^{\mathsf{T}}$ being the second derivative of log-likelihood $l$ with respect to $\boldsymbol{\theta}$.

□

The above results demand the expected information matrix $\mathcal{I}(\boldsymbol{\theta}) = E(-\partial^2 l(\boldsymbol{\theta})/\partial\boldsymbol{\theta}\partial\boldsymbol{\theta}^\top)$ at $\boldsymbol{\theta}_0$. It is common that neither $\boldsymbol{\theta}_0$ nor the expectation operator in $\mathcal{I}(\boldsymbol{\theta})$ are available (in a closed form for the latter) in practice. Usually, we need to adopt an approximation to $\mathcal{I}(\boldsymbol{\theta}_0)$. In fact, when the sample size $n$ is large, the $\mathcal{I}(\boldsymbol{\theta}_0)$ matrix can be approximated by the negative Hessian matrix at $\widehat{\boldsymbol{\theta}}$, namely:

$$\mathcal{I}(\boldsymbol{\theta}_0) \approx -\frac{\partial^2 l(\widehat{\boldsymbol{\theta}})}{\partial\boldsymbol{\theta}\partial\boldsymbol{\theta}^\top}, \tag{1.8}$$

where the unknown $\boldsymbol{\theta}_0$ is replaced with the MLE $\widehat{\boldsymbol{\theta}}$. The negative Hessian is also known as the *observed information matrix*. From Theorem 1.1, when $n$ is large, the variance of $\widehat{\boldsymbol{\theta}}$ is approximately

$$\mathrm{Var}(\widehat{\boldsymbol{\theta}}) \approx \left(-\frac{\partial^2 l(\widehat{\boldsymbol{\theta}})}{\partial\boldsymbol{\theta}\partial\boldsymbol{\theta}^\top}\right)^{-1}.$$

Now consider a function of $\boldsymbol{\theta}$, represented as $g(\boldsymbol{\theta})$, where $g$ is assumed smooth with at least the first derivative. Due to the invariance property of maximum likelihood estimators, the MLE of $g(\boldsymbol{\theta})$ is simply given by $g(\widehat{\boldsymbol{\theta}})$. The covariance matrix of $g(\widehat{\boldsymbol{\theta}})$ can be *approximated*, when $n$ large, by the following sandwich formula (the delta method)

$$\mathrm{Var}(g(\widehat{\boldsymbol{\theta}})) \approx \left(\frac{\partial g(\widehat{\boldsymbol{\theta}})}{\partial\boldsymbol{\theta}}\right)^\top \left(-\frac{\partial^2 l(\widehat{\boldsymbol{\theta}})}{\partial\boldsymbol{\theta}\partial\boldsymbol{\theta}^\top}\right)^{-1} \frac{\partial g(\widehat{\boldsymbol{\theta}})}{\partial\boldsymbol{\theta}}. \tag{1.9}$$

The distribution of $g(\widehat{\boldsymbol{\theta}}) - g(\boldsymbol{\theta}_0)$, for a large sample size, can be approximately $N(\mathbf{0}, g'(\widehat{\boldsymbol{\theta}})^\top(-l''(\widehat{\boldsymbol{\theta}}))^{-1}g'(\widehat{\boldsymbol{\theta}}))$, where, for a function $g(\boldsymbol{\theta})$, the following notations are standard: $g'(\boldsymbol{\theta}) = \partial g(\boldsymbol{\theta})/\partial\boldsymbol{\theta}$ and $g''(\boldsymbol{\theta}) = \partial^2 g(\boldsymbol{\theta})/\partial\boldsymbol{\theta}\partial\boldsymbol{\theta}^\top$. Such an approximate distribution can be used to compute confidence intervals for hazard or survival function at selected time points.

Once the MLE of $\boldsymbol{\theta}$ is obtained, parametric estimation of the hazard or the survival function is simply achieved by plugging in the MLE $\widehat{\boldsymbol{\theta}}$, giving:

$$\widehat{h}_Y(t|\boldsymbol{\theta}) = h_Y(t|\widehat{\boldsymbol{\theta}}), \tag{1.10}$$

$$\widehat{S}_Y(t|\boldsymbol{\theta}) = S_Y(t|\widehat{\boldsymbol{\theta}}). \tag{1.11}$$

Let $S'_Y(t|\boldsymbol{\theta})$ denote the derivative of $S_Y(t|\boldsymbol{\theta})$ with respect to $\boldsymbol{\theta}$. At a given time $t$, an approximate $100(1-\alpha)\%$ confidence interval (CI) of $S_Y(t|\boldsymbol{\theta})$ is:

$$S_Y(t|\widehat{\boldsymbol{\theta}}) \pm z_{\alpha/2}\sqrt{S'_Y(t|\widehat{\boldsymbol{\theta}})^\top(-l''(\widehat{\boldsymbol{\theta}}))^{-1}S'_Y(t|\widehat{\boldsymbol{\theta}})}, \tag{1.12}$$

where $z_{\alpha/2}$ denotes the critical value corresponding to the upper $\alpha/2$ tail of the standard normal distribution, i.e. $F_N(z_{\alpha/2}) = 1-\alpha/2$, where $F_N$ denotes the distribution function of $N(0,1)$ distribution.

We comment that the CI given by (1.12) is not ideal as it cannot guarantee the range of this CI is limited between 0 to 1 (since $S_Y(t)$ is a survival function). A simple solution is to apply the logarithm function to the odds ratio $S_Y(t|\boldsymbol{\theta})/(1 - S_Y(t|\boldsymbol{\theta}))$ (equivalently to a logistic function of $S_Y(t|\boldsymbol{\theta})$). Define $\eta(t|\boldsymbol{\theta})$ to be this logistic function, so that

$$\eta(t|\boldsymbol{\theta}) = \log \frac{S_Y(t|\boldsymbol{\theta})}{1 - S_Y(t|\boldsymbol{\theta})}.$$

Now, as

$$\frac{\partial \eta(t)}{\partial S_Y(t)} = \frac{1}{S_Y(t)(1 - S_Y(t))},$$

we have the following approximation to the variance of $\eta(t|\widehat{\boldsymbol{\theta}})$:

$$\text{Var}(\eta(t|\widehat{\boldsymbol{\theta}})) \approx \frac{1}{S_Y(t|\widehat{\boldsymbol{\theta}})^2(1 - S_Y(t|\widehat{\boldsymbol{\theta}}))^2} S'_Y(t|\widehat{\boldsymbol{\theta}})^\mathsf{T}(-l''(\widehat{\boldsymbol{\theta}}))^{-1} S'_Y(t|\widehat{\boldsymbol{\theta}}).$$

We can construct a CI for $S_Y(t|\boldsymbol{\theta})$ in two steps: (i) compute a CI for $\eta(t|\widehat{\boldsymbol{\theta}})$, and (ii) then transform the CI limits of $\eta(t|\widehat{\boldsymbol{\theta}})$ to the CI limits of $S_Y(t|\widehat{\boldsymbol{\theta}})$ using the relationship that $S_Y(t|\widehat{\boldsymbol{\theta}}) = \exp(\eta(t|\widehat{\boldsymbol{\theta}}))/(1 + \exp(\eta(t|\widehat{\boldsymbol{\theta}})))$. Therefore, the CI for $S_Y(t|\boldsymbol{\theta})$ at $t$ is

$$\frac{e^{\eta(t|\widehat{\boldsymbol{\theta}}) \pm z_{\alpha/2} \text{sd}(\eta(t|\widehat{\boldsymbol{\theta}}))}}{1 + e^{\eta(t|\widehat{\boldsymbol{\theta}}) \pm z_{\alpha/2} \text{sd}(\eta(t|\widehat{\boldsymbol{\theta}}))}}, \tag{1.13}$$

where sd$(\cdot)$ stands for the standard deviation.

When constructing confidence intervals for other functions of interest, such as the hazard or cumulative hazard function, transformations are also preferred. Further examples of this can be found in Chapters 2 and 6.

**Example 1.4** (Exponential survival model).
The Exponential survival model has been introduced in Example 1.1. Here $h_Y(t) = h$ and $H_Y(t) = ht$, where we assume that event rate parameter $h$ is to be estimated.

For a sample of survival data $(t_1, \delta_1), \ldots, (t_n, \delta_n)$, the corresponding log-likelihood function is given by

$$l(h) = M \log(h) - h \sum_{i=1}^{n} t_i,$$

where $M = \sum_{i=1}^{n} \delta_i$ is the number of uncensored observations. The ML estimation of $h$ is readily obtained, given by $\widehat{h} = M / \sum_{i=1}^{n} t_i$. Thus, the survival function estimator is $\widehat{S}(t) = \exp(-\widehat{h}\, t)$. This estimate of $h$ can be interpreted as a rate, representing the ratio of the number of observed events $M$ to the total exposure $\sum_i t_i$. □

**Example 1.5** (Weibull survival model).
The Weibull model discussed in Example 1.2 has hazard $h_Y(t) = kt^{k-1}/h^k$ and cumulative hazard $H_Y(t) = t^k/h^k$.

Thus, from (1.7), the log-likelihood becomes:

$$l(k,h) = M(\log k - k\log h) + (k-1)\sum_{i=1}^{n}\delta_i \log t_i - \frac{1}{h^k}\sum_{i=1}^{n} t_i^k,$$

where $M = \sum_{i=1}^{n}\delta_i$. The likelihood equations for estimating $k$ and $h$ can be established by setting the derivatives of $l(k,h)$ to zero:

$$\frac{\partial l}{\partial k} = M\left(\frac{1}{k} - \log h\right) + \sum_i \delta_i \log t_i + \frac{\log h}{h^k}\sum_i t_i^k - \frac{1}{h^k}\sum_i t_i^k \log t_i = 0, \tag{1.14}$$

$$\frac{\partial l}{\partial h} = -\frac{Mk}{h} + \frac{k}{h^{k+1}}\sum_i t_i^k = 0. \tag{1.15}$$

These equations are nonlinear in $k$ and $h$. A numerical method, such as the Newton method, can be used to solve these equations iteratively.

In a simpler case where $k$ is known, we can easily solve equation (1.15) to find the MLE of $h$ as: $\widehat{h} = \sum_i t_i^k / M$. □

**Example 1.6** (Lognormal survival model).
In Example 1.3, we explained that for this model, the density and survival functions are given by:

$$f_Y(t) = \frac{1}{\sigma t} f_N\left(\frac{\log t - \mu}{\sigma}\right) \quad \text{and} \quad S_Y(t) = 1 - F_N\left(\frac{\log t - \mu}{\sigma}\right),$$

where $\mu, \sigma$ are the mean and standard deviation of $\log Y$. To simplify the expressions below, we let $u_i = (\log t_i - \mu)/\sigma$. The corresponding log-likelihood for parameters $(\mu, \sigma)$, using (1.6), is given by:

$$l(\mu,\sigma) = \sum_{i=1}^{n}\left\{-\delta_i \log(\sigma t_i) + \delta_i \log f_N(u_i) + (1-\delta_i)\log\big(1 - F_N(u_i)\big)\right\}.$$

Since the derivatives of $l(\mu,\sigma)$ with respect to $\mu$ and $\sigma$ are:

$$\frac{\partial l}{\partial \mu} = \sum_i \left\{-\delta_i \frac{f_N'(u_i)}{f_N(u_i)}\frac{1}{\sigma} + (1-\delta_i)\frac{f_N(u_i)}{1-F_N(u_i)}\frac{1}{\sigma}\right\},$$

$$\frac{\partial l}{\partial \sigma} = \sum_i \left\{-\delta_i \frac{1}{\sigma} - \delta_i \frac{f_N'(u_i)}{f_N(u_i)}\frac{u_i}{\sigma} + (1-\delta_i)\frac{f_N(u_i)}{1-F_N(u_i)}\frac{u_i}{\sigma}\right\},$$

where $f'(\cdot)$ denotes the derivative of $f$ with respect to the argument in the bracket. The likelihood equations for the MLEs of $\mu$ and $\sigma$ are given by:

$$\sum_i \left\{-\delta_i \frac{f_N'(u_i)}{f_N(u_i)} + (1-\delta_i)\frac{f_N(u_i)}{1-F_N(u_i)}\right\} = 0,$$

$$\sum_i \left\{-\delta_i - \delta_i \frac{f_N'(u_i)}{f_N(u_i)}u_i + (1-\delta_i)\frac{f_N(u_i)}{1-F_N(u_i)}u_i\right\} = 0.$$

Note that

$$\frac{f'_N(u)}{f_N(u)} = -u,$$

we can simplify the above estimation equations to:

$$\sum_i \left\{ \delta_i u_i + (1-\delta_i)\frac{f_N(u_i)}{1-F_N(u_i)} \right\} = 0, \tag{1.16}$$

$$\sum_i \left\{ \delta_i (u_i^2 - 1) + (1-\delta_i)\frac{f_N(u_i)}{1-F_N(u_i)} u_i \right\} = 0. \tag{1.17}$$

Again, these equations are difficult to solve exactly for $\mu$ and $\sigma$; a numerical method is required here to find the MLEs. □

In the next example, we will use the R package to solve the likelihood estimating equations introduced in the last example.

**Example 1.7** (R Example: Lognormal survival model)**.**
In this example, we will perform the following activities:

1. Generate survival times from a log-normal distribution where $\mu = 3, \sigma = 0.5$, namely $Y_i \sim lognorm(3, 0.5^2)$.
2. Randomly replace ~ 30% of these times by right censoring times. This can be achieved by generating censoring times also from a lognormal with mean 3.4 and std 0.5, i.e. $C_i \sim lognorm(3.4, 0.5^2)$. The simulated survival times are: $T_i = \min(Y_i, C_i)$.
3. Solve equations (1.16) and (1.17) by an R equation solver.
4. Then plot (see Figure 1.1) the estimated hazard and survival functions and compare with the true hazard and survival functions.

The R commands are given below.

```
# Generate event times
t <- rlnorm(n = 500, mean = 3, sd = 0.5)
# Generate censoring times
c <- rlnorm(n = 500, mean = 3.4, sd = 0.5)
# Identify censored individuals
delta <- as.numeric(t < c)
# Save minimum of event and censoring time
y <- t
y[which(delta == 0)] <- c[which(delta == 0)]
# Check censored proportion
sum(delta)/500

# 1. contribution to the estimation equation
# for mu
mu_MLE <- function(mu, sd, t, event) {
    u <- (log(t) -
        mu)/sd
```

```
    phi_u <- dnorm(u)
    sf = pnorm(u, lower.tail = FALSE)
    sf[sf < 1e-30] = 1e-30
    contr <- sum(event * u + (1 - event) * phi_u/sf)
    return(contr)
}

# 2. contribution to the estimation equation
# for sd
sd_MLE <- function(sd, mu, t, event) {
    u <- (log(t) -
        mu)/sd
    phi_u <- dnorm(u)
    sf_u = pnorm(u, lower.tail = FALSE)
    sf_u[sf_u < 1e-30] = 1e-20
    contr <- sum(
        event * (u^2 - 1) + (1 - event) * u *
            phi_u/sf_u
    )
    return(contr)
}

# Solve two equations iteratively for mu and
# sd initial values
mu.old = 0
sd.old = 1
max.iter = 100

for (k in 1:max.iter) {
    # solve for mu
    solve_mu <- uniroot(
        mu_MLE, interval = c(-5, 5),
        sd = sd.old, t = y, event = delta
    )
    mu.new <- solve_mu$root

    # Solve for sd
    solve_sd <- uniroot(
        sd_MLE, interval = c(0.01, 5),
        mu = mu.new, t = y, event = delta
    )
    sd.new <- solve_sd$root
    if (abs(mu.old - mu.new) <
        1e-05 & abs(sd.old - sd.new) <
        1e-05) {
        break
    } else {
        mu.old = mu.new
```

```
        sd.old = sd.new
    }
    print(c(mu.new, sd.new))
}
mu_est = mu.new
sd_est = sd.new

# Estimated vs. true hazard and survival
# functions
v = seq(
    from = 0, to = max(t),
    by = 0.01
)
u_est <- (log(v) -
    mu_est)/sd_est
u_true <- (log(v) -
    3)/0.5
# Survival function
est_St <- pnorm(u_est, lower.tail = FALSE)
true_St <- pnorm(u_true, lower.tail = FALSE)
plot(est_St ~ v, type = "l", ylim = c(0, 1))
lines(true_St ~ v, type = "l", lty = 2)
legend(
    50, 1, legend = c("Esti", "True"),
    lty = 1:2, cex = 0.9
)
# Density function
est_f <- 1/(sd_est * v) * dnorm(u_est)
true_f <- 1/(0.5 * v) * dnorm(u_true)
# Hazard function
est_ht <- est_f/est_St
true_ht <- true_f/true_St
plot(est_ht ~ v, type = "l", ylim = c(0, 0.095))
lines(true_ht ~ v, type = "l", lty = 2)
legend(
    -2, 0.09, legend = c("Esti", "True"),
    lty = 1:2, cex = 0.9
)
```

Plots of the estimated and true survival and hazard functions are given in Figure 1.1 below.

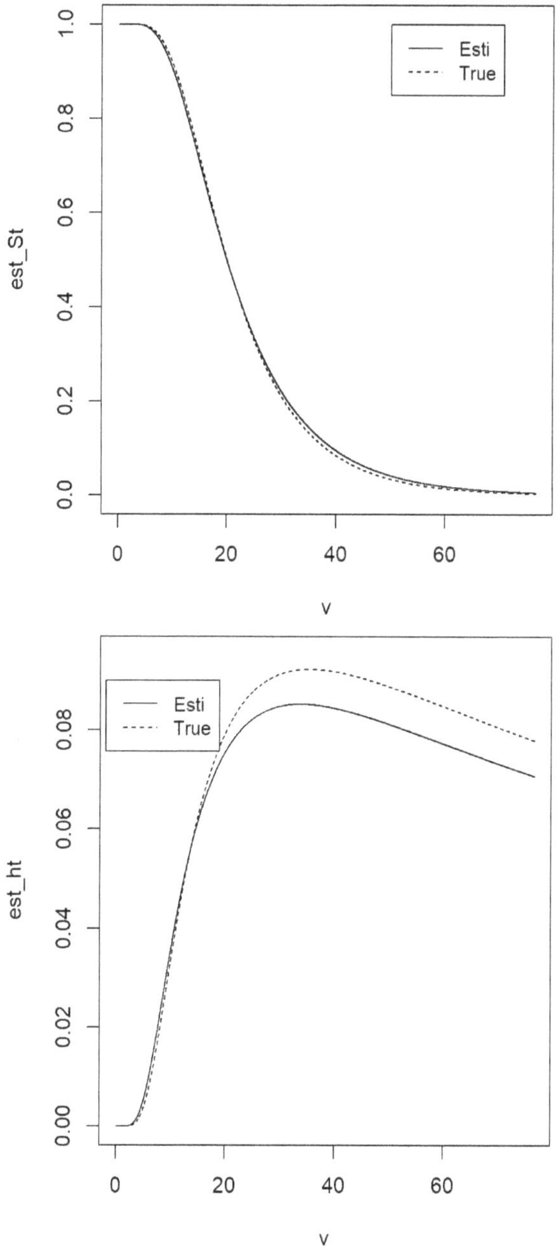


FIGURE 1.1: True (dashed line) and estimated (solid line) survival (top panel) and hazard (bottom panel) functions of the lognormal model for Example 1.7.

□

Examples 1.5 and 1.6 demonstrate that closed-form solutions for the parameter vector $\boldsymbol{\theta}$ that maximizes the likelihood in (1.6) are generally unavailable. While numerical procedures such as the Newton algorithm can be used to solve the likelihood equations, an alternative approach that can be used to obtain the parametric MLEs is through the Expectation-Maximization (EM) algorithm.

### 1.2.2 EM algorithm for a parametric MLE

The EM algorithm of Dempster et al. (1977) is readily applicable to compute the ML estimations in common parametric survival models. We first briefly summarize the EM algorithm below.

Recall $l(\boldsymbol{\theta})$ denotes the log-likelihood function. Let $l_c(\boldsymbol{\theta})$ represent the complete data log-likelihood function. For right-censored survival times (including both event and censoring times), the complete data log-likelihood is simply obtained as follows. We know that when $\delta_i = 0$, the corresponding $t_i$ is a right-censored event time and we also know that the missing event time $Y_i > t_i$ in this case. The complete data log-likelihood is then given by

$$l_c(\boldsymbol{\theta}) = \sum_{i=1}^{n} \big(\delta_i \log f_Y(t_i) + (1 - \delta_i) \log f_Y(Y_i)\big), \tag{1.18}$$

where $Y_i$ in the second term is a missing value but we know that this $Y_i > t_i$.

The EM algorithm of Dempster et al. (1977) is an iterative procedure that can be employed for obtaining the maximum likelihood estimate of $\boldsymbol{\theta}$ in parametric survival models. The algorithm is based on the idea of using the conditional expectation of the complete data log-likelihood in (1.18), given the observed data and the current estimate of $\boldsymbol{\theta}$, to update the estimate of $\boldsymbol{\theta}$ in a way that increases the likelihood. The algorithm proceeds in two steps:

**E-step:** Compute the conditional expectation of the complete data log-likelihood given the *observed data* and the current estimate $\boldsymbol{\theta}^{(k)}$ of $\boldsymbol{\theta}$. This function is denoted by $Q\big(\boldsymbol{\theta}|\boldsymbol{\theta}^{(k)}\big) = E\big(l_c(\boldsymbol{\theta})|\text{ observed data }, \boldsymbol{\theta}^{(k)}\big)$.

**M-step:** Update the estimate of $\boldsymbol{\theta}$ based on the expected complete data log-likelihood. Specifically,

$$\boldsymbol{\theta}^{(k+1)} = \underset{\boldsymbol{\theta}}{\operatorname{argmax}}\, Q\big(\boldsymbol{\theta}|\boldsymbol{\theta}^{(k)}\big).$$

When the optimization problem in the M-step itself demands an iterative procedure, we can replace it with a simpler task of just increasing the $Q$ function value; that is the updated $\boldsymbol{\theta}^{(k+1)}$ is required just to satisfy

$$Q\big(\boldsymbol{\theta}^{(k+1)}|\boldsymbol{\theta}^{(k)}\big) \geq Q\big(\boldsymbol{\theta}^{(k)}|\boldsymbol{\theta}^{(k)}\big),$$

where equality occurs only when the algorithm is converged. With this modification to the EM algorithm, it is called the generalized EM (GEM) algorithm.

For the survival data of our consideration, the observed data consists of event times $t_i$ where $\delta_i = 1$ and right censoring times $t_i$ where $\delta_i = 0$, and for these censoring times, the corresponding missing event times $Y_i$ satisfying $Y_i > t_i$. Given that the information $Y_i > t_i$ is part of the observed information even when $Y_i$ is missing, it is necessary to incorporate this fact when computing the conditional expectations.

Hence, corresponding to the complete data log-likelihood given in (1.18), the $Q$ function in the E-step can be expressed as follows:

$$Q(\boldsymbol{\theta}|\boldsymbol{\theta}^{(k)}) = \sum_{i=1}^{n} \Big(\delta_i \log f_Y(t_i|\boldsymbol{\theta}) + (1-\delta_i) E\big(\log f_Y(Y_i|\boldsymbol{\theta})|Y_i > t_i, \boldsymbol{\theta}^{(k)}\big)\Big). \tag{1.19}$$

Calculation of the second term in (1.19) requires the conditional distribution of $Y|Y > t_i$. It can be verified that this conditional density is given by:

$$f_{Y|Y>t_i}(t|\boldsymbol{\theta}) = \frac{f_Y(t|\boldsymbol{\theta})}{S_Y(t_i|\boldsymbol{\theta})}, \quad \text{for } t > t_i. \tag{1.20}$$

Therefore, the conditional expectation

$$E\big(\log f_Y(Y_i|\boldsymbol{\theta})|Y_i > t_i, \boldsymbol{\theta}^{(k)}\big) = \frac{\int_{t_i}^{\infty} \log f_Y(t|\boldsymbol{\theta}) dF_Y(t|\boldsymbol{\theta}^{(k)})}{S_Y(t_i|\boldsymbol{\theta}^{(k)})}. \tag{1.21}$$

The E and M steps are repeated until convergence is achieved. The EM algorithm guarantees that the log-likelihood $l(\boldsymbol{\theta})$ increases at each iteration before convergence, which in turn assures convergence of an EM algorithm (under certain regular conditions); see Wu (1983).

**Example 1.8** (Exponential survival model).
We have explained the ML estimation of this model parameter in Example 1.5. Now, we are using the EM algorithm, even though an iterative procedure is not necessary for this example. Our purpose here is to demonstrate how to apply the EM algorithm.

Given the conditional density $f_{Y|Y>t_i}(t) = he^{-h(t-t_i)}$ for $t \geq t_i$, we have:

$$E(Y_i|Y_i > t_i, h^{(k)}) = \int_{t_i}^{\infty} t h^{(k)} e^{-h^{(k)}(t-t_i)} dt = \frac{1}{h^{(k)}} + t_i \triangleq \tilde{t}_i^{(k)}. \tag{1.22}$$

It is worth noting that this conditional expectation result can also be seen as deriving from the lack of memory property of the exponential distribution.

Thus, as $E(\log f_{Y_i}(Y_i|h)|Y_i > t_i; h^{(k)}) = \log h - h\tilde{t}_i^{(k)}$, the $Q$ function becomes:

$$Q(h|h^{(k)}) = \sum_{i=1}^{n} \big(\log h - h(\delta_i t_i + (1-\delta_i)\tilde{t}_i^{(k)})\big).$$

The M-step updates the $h$ estimate by:

$$h^{(k+1)} = \frac{n}{\sum_{i=1}^{n}(\delta_i t_i + (1-\delta_i)\tilde{t}_i^{(k)})} = \frac{n}{\sum_{i=1}^{n} t_i + (n-M)/h^{(k)}},$$

where $M = \sum_i \delta_i$, the number of fully observed lifetimes as defined in Example 1.4.

The reader may confirm that $h^{(k)}$ converges to $\widehat{h} = M / \sum_{i=1}^{n} t_i$, the ML estimator given in Example 1.4. □

**Example 1.9** (Lognormal survival model).
When survival times $Y_i$ follow a lognormal distribution with parameters $\mu$ and $\sigma^2$, its survival and density functions have been given in Examples 1.3 and 1.6.

From these functions, the conditional density function is:

$$f_{Y|Y>t_i}(t) = \frac{1}{\sigma t} \frac{f_N\left(\frac{\log t - \mu}{\sigma}\right)}{1 - F_N\left(\frac{\log t_i - \mu}{\sigma}\right)}.$$

Let $U_i = \log Y_i$ and $D_i = U_i - \mu$. Since

$$\log f_{Y_i}(Y_i) = -\frac{1}{2}\log(2\pi\sigma^2) - U_i - \frac{1}{2\sigma^2}(U_i - \mu)^2,$$

the conditional expectation is:

$$E(\log f_{Y_i}(Y_i)|Y_i > t_i, \mu^{(k)}, \sigma^{(k)}) = -\frac{1}{2}\log(2\pi\sigma^2) - \tilde{u}_i^{(k)} - \frac{1}{2\sigma^2}E\big((U_i - \mu)^2|Y_i > t_i, \mu^{(k)}, \sigma^{(k)}\big),$$

where $\tilde{u}_i^{(k)} = E\big(U_i|Y_i > t_i, \mu^{(k)}, \sigma^{(k)}\big)$. Now, we can write the $Q$ function as:

$$Q(\mu, \sigma|\mu^{(k)}, \sigma^{(k)}) = -\frac{n}{2}\log(2\pi\sigma^2) - \sum_{i=1}^{n}\left(\delta_i u_i + (1-\delta_i)\tilde{u}_i^{(k)}\right) - \frac{1}{2\sigma^2}\sum_{i=1}^{n}\Big(\delta_i(u_i - \mu)^2 + (1-\delta_i)E((U_i - \mu)^2|Y_i > t_i, \mu^{(k)}, \sigma^{(k)})\Big),$$

where $u_i = \log t_i$.

At the M-step, we solve:

$$\frac{\partial Q}{\partial \mu} = \frac{1}{\sigma^2}\sum_{i=1}^{n}\left(\delta_i u_i + (1-\delta_i)\tilde{u}_i^{(k)} - \mu\right) = 0,$$

$$\frac{\partial Q}{\partial \sigma^2} = -\frac{n}{2\sigma^2} + \frac{1}{2\sigma^4}\sum_{i=1}^{n}\Big(\delta_i(u_i - \mu)^2 + (1-\delta_i)E\big((U_i - \mu)^2|Y_i > t_i, \mu^{(k)}, \sigma^{(k)}\big)\Big) = 0.$$

Solutions from these two equations are simply:

$$\mu^{(k+1)} = \frac{1}{n}\sum_i\left(\delta_i u_i + (1-\delta_i)\tilde{u}_i^{(k)}\right),$$

$$(\sigma^{(k+1)})^2 = \frac{1}{n}\sum_i\left(\delta_i(u_i - \mu^{(k+1)})^2 + (1-\delta_i)\tilde{d_i^2}^{(k+1)}\right),$$

where $(\tilde{d}_i^{(k+1)})^2 = E\big((U_i - \mu^{(k+1)})^2|Y_i > t_i, \mu^{(k)}, \sigma^{(k)}\big)$.

Finally, we briefly explain how to compute the conditional expectations $\tilde{u}_i^{(k)}$ and $(\tilde{d}_i^{(k+1)})^2$. Let $z = (\log t - \mu^{(k)})/\sigma^{(k)}$, $z_i^{(k)} = (\log t_i - \mu^{(k)})/\sigma^{(k)}$, and $\tilde{f}(z) = f_N(z)/(1 - F_N(z_i^{(k)}))$, then:

$$\tilde{u}_i^{(k)} = \mu^{(k)} + \sigma^{(k)} \int_{z_i^{(k)}}^{\infty} z\tilde{f}(z)dz,$$

and

$$\begin{aligned}(\tilde{d}_i^{(k+1)})^2 =& (\sigma^{(k)})^2 \int_{z_i^{(k)}}^{\infty} z^2\tilde{f}(z)dz + 2\sigma^{(k)}(\mu^{(k)} - \mu^{(k+1)}) \int_{z_i^{(k)}}^{\infty} z\tilde{f}(z)dz \\ &+ (\mu^{(k)} - \mu^{(k+1)})^2.\end{aligned}$$

□

Parametric inference for exponential and other parametric families, including mixture models, is discussed in Chapter 4 of Kalbfleisch & Prentice (2002). In their Chapter 5, detailed likelihood-based inference and $m$-sample comparison procedures are provided for Weibull, log-logistic and log-normal families.

## 1.3 Non-parametric methods for survival data analysis

In this section, we will discuss several non-parametric methods for estimating the survival or hazard/cumulative hazard function. These methods include the Kaplan-Meier method for estimation of the survival function, the Nelson-Aalen method for estimation of the cumulative hazard function and a likelihood-based non-parametric method for estimation of the hazard function.

### 1.3.1 The Kaplan-Meier estimator of survival functions

We start our discussions with a simple case where there are no right-censored survival times. Non-parametric estimation of survival probabilities is straightforward in the absence of censoring as this is basically equivalent to deriving the empirical distribution function.

Recall that we have denoted the time to the event of interest by a random variable $Y$ with distribution function $F_Y(t)$. First, suppose there are no right-censored event times. Consider a random sample of $n$ individuals whose event times are $t_1, \ldots, t_n$ . In this context, the empirical distribution function from $t_1, \ldots, t_n$ is simply

$$\widehat{F}_Y(t) = \frac{1}{n} \sum_{i=1}^{n} \Omega(t - t_i). \tag{1.23}$$

Here $\Omega(t)$ is the Heavyside function:

$$\Omega(t) = \begin{cases} 1 & \text{if } t \geq 0, \\ 0 & \text{otherwise;} \end{cases}$$

i.e. this function takes the value 0 for $t < 0$, otherwise, 1. It is clear that $\sum_{i=1}^{n} \Omega(t - t_i)$ represents the number of events that occurred up to (and include) the time $t$. The empirical distribution in (1.23) is in fact an unbiased, non-parametric maximum likelihood estimator (NPMLE) of $F_Y(t)$ (e.g. Owen 2001).

The corresponding estimator of the survival function at $t$ is then

$$\begin{aligned} \widehat{S}_Y(t) &= 1 - \widehat{F}_Y(t) \\ &= \frac{n - \sum_{i=1}^{n} \Omega(t - t_i)}{n} \\ &= \frac{N(t)}{n}, \end{aligned}$$

where

$$N(t) = n - \sum_{i=1}^{n} \Omega(t - t_i), \tag{1.24}$$

representing the number of subjects who are *at risk* for the event (so that their events have not occurred yet) at $t$. $\widehat{S}_Y(t)$ is called the empirical survival function. Clearly, $\widehat{S}_Y(t)$ is a function continuous from the right.

Assume, without loss of generality, $a = t_1 < \ldots < t_n = b$. Then an alternative expression for the empirical survival function is

$$\begin{aligned} \widehat{S}_Y(t) &= \widehat{P}(Y > t) \\ &= \prod_{\substack{i=1 \\ t_i \leq t}}^{n} \widehat{P}(Y > t_i | Y > t_{i-1}) \\ &= \prod_{\substack{i=1 \\ t_i \leq t}}^{n} \left[ \frac{N_i}{N_{i-1}} \right], \end{aligned} \tag{1.25}$$

where $N_i = N(t_i)$, $N_{i-1} = N(t_{i-1})$, with $N(t)$ being defined in (1.24), and the limits for the products are $i$ from 1 to $n$ but any term in these products must also satisfy $t_i \leq t$.

In (1.25), we estimated the conditional probability $P(Y > t_i | Y > t_{i-1})$ simply (and naturally) by $N_i/N_{i-1}$, the ratio of the number of subjects who are at risk (referred to as the "number at risk" hereafter) at $t_i$ over the number at risk at $t_{i-1}$.

It is worth noting that when there are no censoring or ties, $N_i = N_{i-1} - 1$, resulting in $N_i/N_{i-1} = 1 - 1/N_{i-1}$. When there are tied observations, adjustments to the formula for $N_i$ can be made accordingly.

The above approach of estimating the survival function can be extended easily to the situation when there are right-censored and tied survival times. The observed survival times set $\{t_1, \ldots, t_n\}$ we consider here may contain right censoring times and tied observations. We suppose $\mathtt{e}_1 < \mathtt{e}_2 < \ldots < \mathtt{e}_m$ are ordered distinct event times from $\{t_1, \ldots, t_n\}$. Now we allow ties; this means we may have more than one event times at a $\mathtt{e}_i$. Applying the idea of (1.25) we can estimate the survival function in this more general case by

$$\widehat{S}_Y(t) = \prod_{\substack{i=1 \\ \mathtt{e}_i \le t}}^{m} \left[1 - \frac{n_i + c_i}{N_{i-1}}\right], \tag{1.26}$$

where $n_i$ and $c_i$ denote respectively the number of events and the number of censoring in $(\mathtt{e}_{i-1}, \mathtt{e}_i]$ and $N_{i-1} = N(\mathtt{e}_{i-1})$, and here $N_0 = n$. When $n_i > 1$ we know that there are tied event times at $\mathtt{e}_i$. It is conventional to define $\widehat{S}_Y(t) = 1$ for $t < \mathtt{e}_1$. Formula (1.26) is called the product limit estimator of the survival function. This method is credited to Kaplan and Meier (Kaplan & Meier 1958) and is known as the Kaplan-Meier (KM) method.

In Section 1.3.3, we will explain that the KM result is also a maximum likelihood estimator of the survival function.

There exist a few approximate variance formula for the KM estimator, but the most common approach is Greenwood's formula (Greenwood 1926):

$$\widehat{\mathrm{Var}}\left(\widehat{S}(t)\right) = \widehat{S}(t)^2 \sum_{i:\ t_i \le t} \frac{n_i}{N_i(N_i - n_i)} \tag{1.27}$$

where $n_i$ is the number of events and $N_i$ is the number at risk at $t_i$ ($\le t$).

**Example 1.10** (R Example: KM method).
In this example we will use the "`lung`" dataset, which comes with the R package "`survival`". This dataset contains subjects with advanced lung cancer from the North Central Cancer Treatment Group in U.S. You may view the dataset by the following R codes:

```
> data(lung)
> head(lung, n=10)
```

The variable `time` contains the survival time in days which is the survival data of our interest. The variable `status` contains censoring status: 1=censored, 2=dead. We will conduct the following exercises:

1. Using R "`survfit`" function to find the KM estimate.
3. Plot the estimated survival function.
4. Using the log-rank method to test if two survival functions corresponding to `Male` (where `sex` $= 1$) and `Female` (where `sex` $= 0$) are significantly different.
5. Explain how to derive the hazard function estimate from the KM survival function and then plot the estimated hazard function.

```
library(survival)

# Find the KM estimate using the
# survfit() function
KM_est <- summary(survfit(Surv(time, status) ~
    1, data = lung))

# Plot the estimated survival function
plot(survfit(Surv(time, status) ~ 1, data = lung))
# Test for significant difference
# between two categories
survdiff(Surv(time, status) ~ sex, data = lung)

## Call:
## survdiff(formula = Surv(time, status) ~ sex, data = lung)
##
##          N Observed Expected (O-E)^2/E (O-E)^2/V
## sex=1 138      112     91.6      4.55      10.3
## sex=2  90       53     73.4      5.68      10.3
##
##  Chisq= 10.3  on 1 degrees of freedom, p= 0.001

# Derive and plot hazard function
# estimate from the KM estimate
diff <- -(log(KM_est$surv)[2:139] + log(KM_est$surv)[1:138])
t <- KM_est$time[1:138]
delta_t <- KM_est$time[2:139] - KM_est$time[1:138]
h_t <- diff/delta_t
plot(h_t ~ t, type = "l")
lo <- loess(h_t ~ t)
lines(lo$fitted ~ t, col = "red", lwd = 2)
```

These plots clearly indicate the KM survival function estimate is "self regularized" while the hazard function estimate from the KM method is extremely noisy. □

### 1.3.2 Cumulative hazard and cumulative incidence functions and their estimation

The hazard function plays a crucial role in the interpretation of survival analysis. However, similar to density estimation, its non-parametric estimate only places mass at observed event times, resulting in a noisy estimate which is difficult to interpret; e,g, see the hazard function plot in Figure 1.2. On the other hand, empirical estimators of the *cumulative* hazard function (CHF) – which represents the integral of the hazard function $h(t)$ and is related with the cumulative incidence function (CIF) – are more readily interpretable.

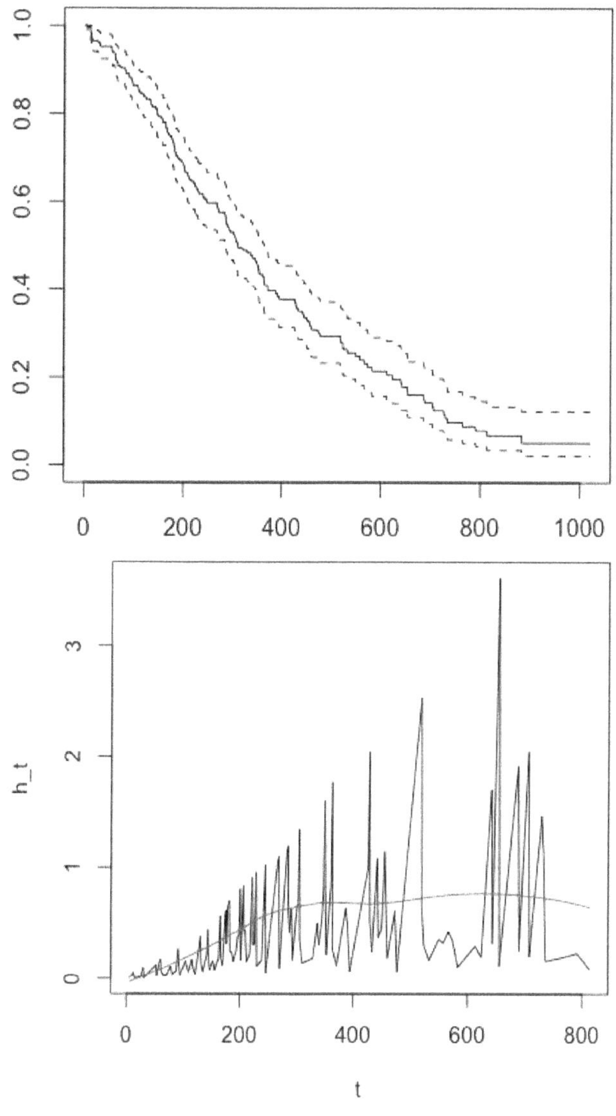


FIGURE 1.2: KM estimate (left) and the corresponding estimated hazard function (right) for Example 1.10, and the latter is superimposed with a LOESS smooth curve.

For outcome of a single event type, the cumulative incidence function (CIF) is the complement of the survival function, i.e. $F(t) = P(T \leq t)$, the cumulative probability of the event occurring up to time t. Therefore, in this context, an estimate of $F(t)$ is simply given by $\widehat{F}(t) = 1 - \widehat{S}(t)$.

Non-parametric estimation of the cumulative hazard function in survival analysis (and more general counting processes) is described by Aalen (1978). Empirical counts of events provide the fraction of individuals observed to experience an event ("death"), so that the cumulative hazard of death can be estimated as

$$\widehat{H}(t) = \sum_{i:\, t_i \le t}^{m} \frac{n_i}{N_i}, \tag{1.28}$$

where $m$, as defined in the last section, is the number of distinct event times and $n_i$ and $N_i$ denote respectively the number of the event and the number at risk at $t_i$. We refer to (1.28) as the Nelson-Aalen estimator of the cumulative hazard.

It is clear from equation (1.28) that at each time of event occurrence, the hazard estimate is simply the number of deaths relative to the total exposure (number of persons in the risk set) at that time. That is, there is a close parallel of this result to a non-stationary Poisson marked event process (with events at times of deaths among the $n$ individuals) and an intensity function proportional to the number remaining at risk and proportional to the hazard of a death of any cause at that time.

The Nelson-Aalen estimator is available as an estimator of the cause-specific CHF (Jeong & Fine 2007). From the CHF a survival estimator $\widehat{S}(t) = \exp[-\widehat{H}(t)]$ is available, but *only* in the case of a single event type.

For extensive bibliographic notes related to Kaplan-Meier, Nelson-Aalen and related procedures see Kalbfleisch & Prentice (2002, Chapter 3).

**Example 1.11** (R Example: Nelson–Aalen method).
This example explains how to compute a Nelson–Aalen cumulative hazard estimate using the "`lung`" data in Example 1.10.

The R function to compute the Nelson–Aalen estimator is again "`survfit`"; see the R code below. Additionally, we compute a cumulative hazard curve from the fact that $H_Y(t) = -\log S_Y(t)$. Therefore, once an estimate of $S_Y(t)$, such as the Kaplan-Meier estimate, is obtained, we can then compute its cumulative hazard using this relationship.

We plot the above two estimates in one plot window to observe similarities or differences between these two approaches.

```
# Compute Nelson-Aalen estimate of cumulative
# hazard via R
library(survival)
nelson_aalen = survfit(Surv(time, status) ~ 1, data = lung,
    ctype = 1)
nelson_aalen_cumhaz = nelson_aalen$cumhaz

# Find cumulative hazard using H_Y(t) =
# -logS_Y(t)
derived_cumhaz = -log(nelson_aalen$surv)

# Plot the above two functions and comment on
# similarities/differences
plot(nelson_aalen_cumhaz ~ nelson_aalen$time, type = "l")
lines(derived_cumhaz ~ nelson_aalen$time, type = "l",
    lty = 2)
```

These two plots are almost identical when $t < 500$, and gradually become slightly different afterwards.

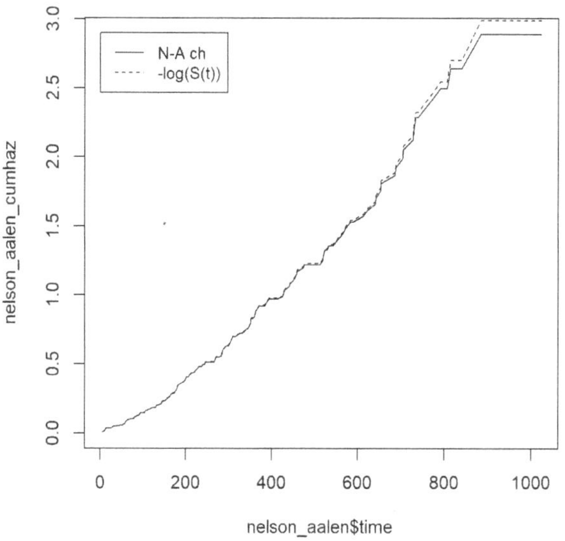


FIGURE 1.3: Plot of the cumulative hazard by -log-KM (solid curve) and Nelson-Aalen (broken curve).

□

### 1.3.3 Likelihood-based hazard function estimation

Here, we will briefly discuss likelihood-based estimation of the hazard function $h_Y(t)$. This approach uses the log-likelihood function with $h_Y(t)$ approximated by a discrete (i.e. piecewise constant) function.

Recall that $t_1, \ldots, t_n$ represent the observed right-censored event times. Recall that some $t_i$ are right-censored event times and the rest are fully observed event times. Let $a = \min_i\{t_i\}$ and $b = \max_i\{t_i\}$. We are interested in estimating the hazard $h_Y(t)$ over the interval $[a, b]$.

We partition the interval $[a, b]$ using a set of distinct points $w_1 < \cdots < w_m < w_{m+1}$, where $w_1 = a$, and $w_{m+1} = b$. Over the interval $(w_u, w_{u+1}]$ (where $u = 1, \ldots, m$), the hazard function is approximated by a non-negative constant:

$$h_Y(t) = \theta_u,$$

where $w_u < t \leq w_{u+1}$ and $u = 1, \ldots, m$. The problem of estimating the function $h_Y(t)$ now becomes a simpler problem of estimating $\theta_1, \ldots, \theta_m$. The precise choice of partition of interval $[a, b]$ is somewhat arbitrary, with an extreme case where the partition points coincide with the fully observed event times.

There are three commonly used strategies to partition the interval $[a,\ b]$. We briefly describe them below.

**Event times-based partition:** The partition points $w_2, \ldots, w_m$ can be selected as the observed distinctive event times (if some event times are overlapped we then select one of them). This approach gives the so-called *non-parametric* hazard function estimation as explained in Section 1.3.2.

**Event counts-based partition:** In this approach, the partition time points $w_2, \ldots, w_m$ are selected so that the numbers of fully observed event times in each sub-interval $(w_u,\ w_{u+1}]$ are roughly the same. This approach is also known as a quantile-based partition.

**Equal interval length partition:** This partition method simply divides $[a, b]$ into intervals of equal length so that the partition points $w_2, \ldots, w_m$ satisfy $w_2 - w_1 = w_3 - w_2 = \cdots = w_{m+1} - w_m = (b - a)/m$.

Let a piecewise constant approximation to $h_Y(t)$ be denoted by a vector $\boldsymbol{\theta} = (\theta_1, \ldots, \theta_m)^{\mathsf{T}}$. Since $\boldsymbol{\theta}$ has $m$ elements, this representation of $h_Y(t)$ by $\boldsymbol{\theta}$ means that $h_Y(t)$ is approximated using $m$ intervals (forming "bins"). Over interval $u$ for $u = 1, \ldots, m$, $h_Y(t)$ is approximated by a constant denoted by $\theta_u$.

Another way to express this piecewise constant approximate hazard function is by indicator basis functions, namely:

$$h_Y(t) = \sum_{u=1}^{m} \theta_u I(w_u < t \leq w_{u+1}), \tag{1.29}$$

where $I(\mathcal{A})$ denotes an indicator function for an event $\mathcal{A}$. Note that we shall demand all $\theta_u \geq 0$ so the approximated $h_Y(t)$ is a valid hazard function.

Next, we will explain how to find the maximum likelihood estimation of $\boldsymbol{\theta}$. Let $n_u$ denote the number of events in the interval $(w_u, w_{u+1}]$, and let $d_u$ represent the number of observations in the interval $(w_u, w_{u+1}]$ (including both event times and right censoring times so that $d_u \geq n_u$). The log-likelihood, as shown in expression (1.7), can now be expressed as:

$$l(\boldsymbol{\theta}) = \sum_{u=1}^{m} n_u \log \theta_u - \sum_{u=1}^{m} d_u \sum_{v=1}^{u} \theta_v \mathcal{W}_v. \tag{1.30}$$

Here, $\mathcal{W}_v = w_{v+1} - w_v$ represents the width of the interval $(w_v, w_{v+1}]$. The second term of (1.30) comes from the corresponding approximation to the cumulative hazard. In fact, when $t$ belongs to $(w_u, w_{u+1}]$, the corresponding cumulative hazard is:

$$H_Y(t) = \sum_{v=1}^{u} \theta_v \mathcal{W}_v.$$

Therefore, $\sum_{i=1}^{n} H_Y(t_i) = \sum_{u=1}^{m} d_u \sum_{v=1}^{u} \theta_v \mathcal{W}_v$.

The MLE of the $\theta_u$'s is given by the maximum of the log-likelihood (1.30), subject to $\theta_u \geq 0$ for all $u$. Let us first ignore these non-negativity constraints. The unconstrained MLE is obtained by solving the following equations:

$$\frac{\partial l}{\partial \theta_u} = \frac{n_u}{\theta_u} - \mathcal{W}_u \sum_{v=u}^{m} d_v = 0, \text{ for all } u, \tag{1.31}$$

which leads to

$$\widehat{\theta}_u = \frac{n_u}{\mathcal{W}_u \sum_{v=u}^{m} d_v} = \frac{\text{number of events in} \in (w_u, w_{u+1}]}{(\text{number at risk at } w_u) * \mathcal{W}_u}. \tag{1.32}$$

Note that this estimator automatically satisfies $\theta_u \geq 0$, eliminating the need to consider non-negativity constraints in this context. Based on the hazard estimate in (1.32), the cumulative hazard estimate, when $t \in (w_u, w_{u+1}]$, is given by:

$$\widehat{H}_Y(t) = \sum_{v=1}^{u} \frac{n_v}{\sum_{q=v}^{m} d_q} = \sum_{v=1}^{u} \frac{n_v}{N_v}, \tag{1.33}$$

where $N_v = \sum_{q=v}^{m} d_q$ = number at risk at $w_v$. This estimator is more general than the previously discussed Nelson–Aalen estimator in (1.28), as the Nelson–Aalen estimator is a special case that relies on an *event times based partition* of $[a, b]$.

The MLE of $\boldsymbol{\theta}$ can become extremely non-smooth depending on the partition used to define $\theta_u$. In cases where the MLE $\widehat{\boldsymbol{\theta}}$ is non-smooth, it is preferable to obtain a smooth estimate of $h_Y(t)$ for better interpretations. One common approach to achieve this is by using a maximum penalized likelihood (MPL) estimator, which we will introduce next.

Before introducing the MPL method, let us explain the penalty functions that can be used to enforce local smoothness of $\boldsymbol{\theta}$. Smoothness among the $\theta_u$'s can be conceptualized as a constraint where neighboring $\theta_u$'s are similar. One way to impose this constraint is through a quadratic penalty function that penalizes the differences between neighbouring $\theta_u$'s:

$$J(\boldsymbol{\theta}) = \sum_{u} (\mathbf{D}_u \boldsymbol{\theta})^2, \tag{1.34}$$

where $\mathbf{D}_u$ is typically a row vector that performs a "local difference" operation. For instance, if $\mathbf{D}_u$ is a row vector with its elements at positions $u-1$ and $u$ equal to $-1$ and 1 respectively, then $\mathbf{D}_u \boldsymbol{\theta} = \theta_u - \theta_{u-1}$. This represents a first-order difference between the two neighbouring $\theta_u$'s. The corresponding penalty function is:

$$J(\boldsymbol{\theta}) = \sum_{u=2}^{m} (\theta_u - \theta_{u-1})^2. \tag{1.35}$$

Another example involves a second-order difference: $(\theta_{u+1} - \theta_u) - (\theta_u - \theta_{u-1}) = \theta_{u+1} - 2\theta_u + \theta_{u-1}$, in which case the corresponding $\mathbf{D}_u$ is defined as $(0, \ldots, 0, 1, -2, 1, 0, \ldots, 0)$. The penalty function then becomes:

$$J(\boldsymbol{\theta}) = \sum_{u=2}^{m-1} (\theta_{u+1} - 2\theta_u + \theta_{u-1})^2. \tag{1.36}$$

It is more convenient to express the penalty function in (1.34) using matrices:

$$J(\boldsymbol{\theta}) = \boldsymbol{\theta}^{\mathsf{T}} \mathbf{R} \boldsymbol{\theta}, \tag{1.37}$$

where $\mathbf{R} = \mathbf{D}^{\mathsf{T}}\mathbf{D}$. Here, for example, corresponding to (1.35) $\mathbf{D} = (\mathbf{D}_2^{\mathsf{T}}, \ldots, \mathbf{D}_m^{\mathsf{T}})^{\mathsf{T}}$ and corresponding to (1.36) $\mathbf{D} = (\mathbf{D}_2^{\mathsf{T}}, \ldots, \mathbf{D}_{m-1}^{\mathsf{T}})^{\mathsf{T}}$. The rows of the matrix $\mathbf{D}$ are the corresponding $\mathbf{D}_u$'s above.

With a selected penalty function $J(\boldsymbol{\theta})$, the maximum penalized likelihood estimator of $\boldsymbol{\theta}$ is given by:

$$\widehat{\boldsymbol{\theta}}_{\text{MPL}} = \underset{\boldsymbol{\theta} \geq 0}{\operatorname{argmax}} \{\Phi(\boldsymbol{\theta}) = l(\boldsymbol{\theta}) - \lambda J(\boldsymbol{\theta})\}, \tag{1.38}$$

where the quantity $\lambda > 0$ is called the smoothing (or regularization) parameter and its value is named a smoothing value. The penalized log-likelihood $\Phi(\boldsymbol{\theta})$ discourages estimates that result in large $J(\boldsymbol{\theta})$ values. A large $J(\boldsymbol{\theta})$ value indicates dissimilarity among neighboring $\theta_u$'s, implying lack of smoothness.

The MPL method also has a Bayesian interpretation. By assuming a prior distribution for $\boldsymbol{\theta}$ with the density function $\exp\{-\lambda J(\boldsymbol{\theta})\}$, $\Phi(\boldsymbol{\theta})$ represents the log-posterior density. Consequently, $\widehat{\boldsymbol{\theta}}_{\text{MPL}}$ becomes a maximum a posteriori (MAP) estimator.

If all $\theta_u > 0$ (i.e. no active constraints), and when the penalty function is given by (1.37), the MPL estimate is given by the following equation:

$$\frac{\partial \Phi}{\partial \theta_u} = \frac{n_u}{\theta_u} - \mathcal{W}_u \sum_{v=u}^{m} d_v - 2\lambda \mathbf{R} \boldsymbol{\theta} = 0, \tag{1.39}$$

for all $u$. However, these equations are non-linear in $\theta_u$, and they typically do not give a closed-form solution.

Another complication is that if some of the $\theta_u \geq 0$ constraints are active (i.e. some $\theta_u = 0$), then a *constrained optimization* algorithm needs to be applied to obtain the solution. In this context, the Karush-Kuhn-Tucker (KKT) conditions require, for these $u$, that

$$\frac{\partial \Phi}{\partial \theta_u} < 0.$$

We will introduce an efficient multiplicative iterative (MI) algorithm in Chapter 2 to address the non-negativity constraints.

**Example 1.12** (R Example: MPL estimate of hazard function).

In this example, we will explain how to find the maximum penalized likelihood estimate of $\boldsymbol{\theta}$ by solving equation (1.39).
1. Generate $n = 500$ survival times from a log-logistic distribution, and then replace 30% of them with right censoring times.

A log-logistic distribution with scale parameter $\gamma$ and shape parameter $\kappa$ has the hazard function

$$h(t) = \frac{\kappa t^{\kappa-1}}{\gamma^{\kappa} + t^{\kappa}},$$

and the survival function

$$S(t) = \left(1 + (t/\gamma)^{\kappa}\right)^{-1}.$$

Thus, a data $Y$ from this distribution can be generated in two steps:

(i) Generate $U \sim unif(0, 1)$.
(ii) Then, a data $Y$ from this distribution can be generated simply by

$$Y = \gamma\left(e^{-\log U} - 1\right)^{\frac{1}{\kappa}}.$$

The required 30% censoring can be achieved approximately by generating censoring times from exponential distribution with mean 3.
2. For each of $m = 8, 20, 40, 60$ and with $\lambda = 0$, compute the ML estimate of $\boldsymbol{\theta}$ from the equation (1.32). Plot the obtained $\boldsymbol{\theta}$.
4. Next, for $m = 60$, first create the corresponding penalty matrix $\mathbf{R}$ using the second-order difference. Then solve (1.39) for different smoothing values: $\lambda = 10^{-6}, 1, 50, 100$ using R constrained optimization package `'nloptr'`. Plot the obtained MPL estimates of $\boldsymbol{\theta}$. Comment on how different $m$ and $\lambda$ values affect hazard function estimates.

```
# Generate data with 30%
# censoring
kappa <- 5
gamma <- 1
neglogU <- -log(runif(500))
t <- gamma * (exp(neglogU) - 1)^(1/kappa)
c <- rexp(500, rate = 1/3)
event <- as.numeric(t < c)
y <- t
y[which(event == 0)] <- c[which(event ==
    0)]
sum(event)/500

# For each of m=8,20,40,60
# and with \lambda = 0,
# compute ML estimate of
# theta from (1.57)
```

```
estimate_theta <- function(m, y,
    event) {
    w <- as.numeric(quantile(y,
        probs = seq(0, 1, length.out = (m +
            1))))
    delta_v <- w[2:(m + 1)] - w[1:m]
    theta = rep(0, m)

    for (v in 1:m) {
        n_u <- sum(as.numeric(w[v] <=
            y & y < w[(v + 1)]))
        N_u <- sum(as.numeric(w[v] <=
            t))
        theta[v] <- n_u/(N_u *
            delta_v[(v)])
    }
    out = list(theta = theta, w = w)
    return(out)
}

theta_m8 <- estimate_theta(m = 8,
    y, event)
theta_m20 <- estimate_theta(m = 20,
    y, event)
theta_m40 <- estimate_theta(m = 40,
    y, event)
theta_m60 <- estimate_theta(m = 60,
    y, event)

# Plot obtained theta i.e.
# discrete hazard function
# values

par(mfrow = c(2, 2))
m = 8
plot(c(theta_m8$theta, theta_m8$theta[m]) ~
    theta_m8$w, type = "s", col = "black",
    ylab = "hazard est", xlab = "Time",
    main = "MLE of Theta; m = 8")
m = 20
plot(c(theta_m20$theta, theta_m20$theta[m]) ~
    theta_m20$w, type = "s", col = "black",
    ylab = "hazard est", xlab = "Time",
    main = "MLE of Theta; m = 20")
m = 40
plot(c(theta_m40$theta, theta_m40$theta[m]) ~
    theta_m40$w, type = "s", col = "black",
    ylab = "hazard est", xlab = "Time",
```

```
    main = "MLE of Theta; m = 40")
m = 60
plot(c(theta_m60$theta, theta_m60$theta[m]) ~
    theta_m60$w, type = "s", col = "black",
    ylab = "hazard est", xlab = "Time",
    main = "MLE of Theta; m = 60")

# compute MPL estimate with m
# = 60 and smoothing values
# lambda = 1e-6, 1,10,50

# For m=60, create penalty
# matrix R
m <- 60
D <- matrix(0, (m), (m))
for (u in 1:(m - 1)) {
    D[u, (u + 1)] = 1
    D[u, u] = -1
}
R <- t(D) %*% D

# Solve equation 1.63 for
# different smoothing values
# using nloptr

library(nloptr)
eval_f0 <- function(theta, y, w,
    delta_v, R, lambda) {
    m = length(theta)
    alld_u <- alln_u <- rep(0,
        m)
    for (v in 1:m) {
        n_u <- sum(as.numeric(w[v] <=
            y & y < w[(v + 1)]))
        d_u <- sum(as.numeric(w[v] <=
            t & t < w[(v + 1)]))
        alln_u[v] = n_u
        alld_u[v] = d_u
    }
    theta[theta < 1e-30] = 1e-30
    cumhaz = cumsum(theta * delta_v)
    p_log_lik <- sum(alln_u * log(theta))
    -sum(alld_u * cumhaz)
    -lambda * t(theta) %*% R %*%
        theta
    np_log_lik <- -p_log_lik
    return(np_log_lik)
}
```

```
w <- quantile(y, probs = seq(0,
    1, length.out = (m + 1)))
delta_v <- w[2:(m + 1)] - w[1:m]

opts <- list(algorithm = "NLOPT_LN_COBYLA",
    xtol_rel = 1e-06, maxeval = 10000)



smtheta_1e6 <- nloptr(x0 = rep(1,
    m), eval_f = eval_f0, lb = (rep(0,
    m)), ub = rep(Inf, m), opts = opts,
    y = y, w = w, delta_v = delta_v,
    R = R, lambda = 1e-06)$solution


smtheta_1 <- nloptr(x0 = rep(1,
    m), eval_f = eval_f0, lb = (rep(0,
    m)), ub = rep(Inf, m), opts = opts,
    y = y, w = w, delta_v = delta_v,
    R = R, lambda = 1)$solution

smtheta_10 <- nloptr(x0 = rep(1,
    m), eval_f = eval_f0, lb = (rep(0,
    m)), ub = rep(Inf, m), opts = opts,
    y = y, w = w, delta_v = delta_v,
    R = R, lambda = 10)$solution

smtheta_50 <- nloptr(x0 = rep(1,
    m), eval_f = eval_f0, lb = (rep(0,
    m)), ub = rep(Inf, m), opts = opts,
    y = y, w = w, delta_v = delta_v,
    R = R, lambda = 50)$solution


par(mfrow = c(2, 2))
plot(c(smtheta_1e6, smtheta_1e6[m]) ~
    w, type = "s", col = "black",
    ylab = "hazard est", xlab = "Time",
    main = "Theta, lambda = 1e-6")
plot(c(smtheta_1, smtheta_1[m]) ~
    w, type = "s", col = "black",
    ylab = "hazard est", xlab = "Time",
    main = "Theta, lambda = 1")
plot(c(smtheta_10, smtheta_10[m]) ~
    w, type = "s", col = "black",
    ylab = "hazard est", xlab = "Time",
    main = "Theta, lambda = 10")
```

```
plot(c(smtheta_50, smtheta_50[m]) ~
    w, type = "s", col = "black",
    ylab = "hazard est", xlab = "Time",
    main = "Theta, lambda = 50")
```

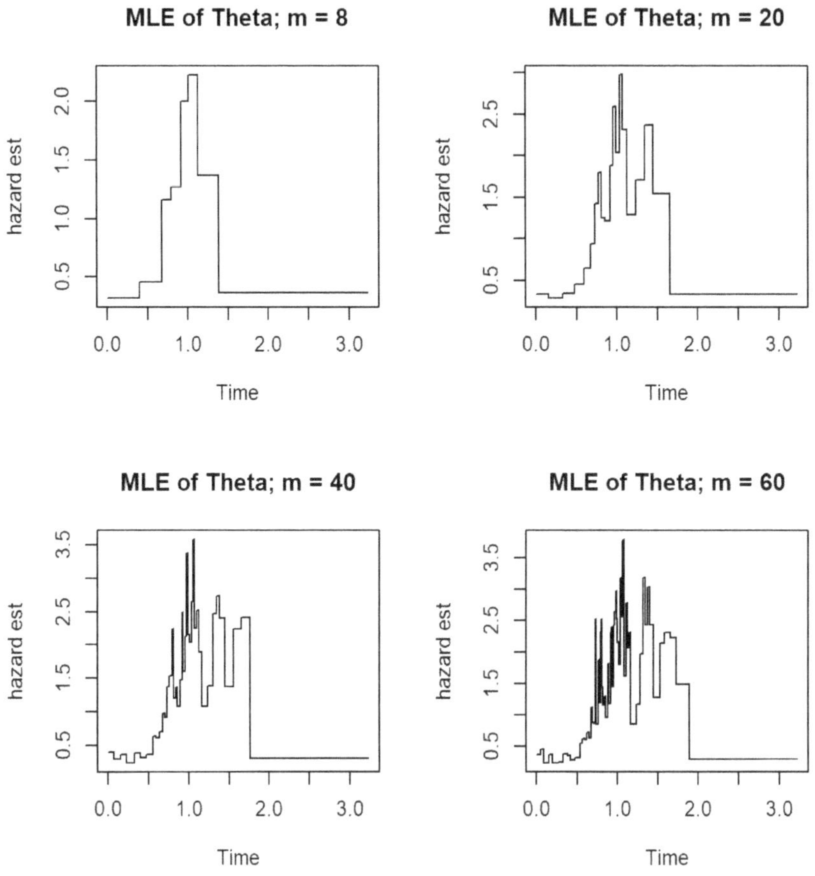


FIGURE 1.4: Plots of ML hazard estimates, with the number of bins 8, 20, 40 and 60.

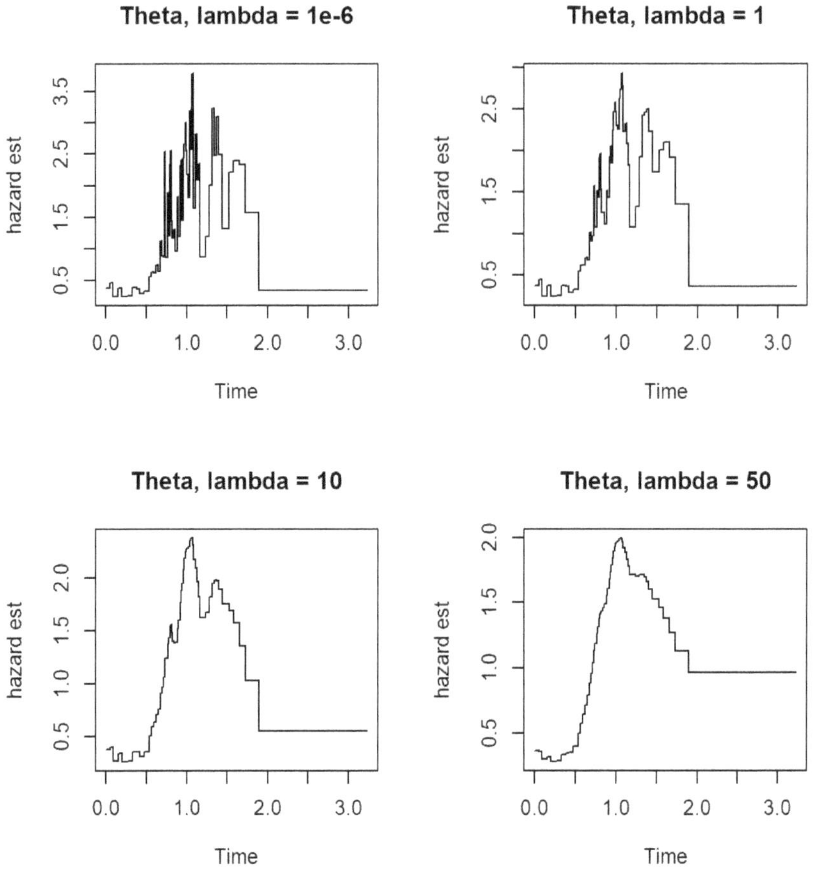


FIGURE 1.5: Plots of MPL hazard estimates, with m = 60 bins and the smoothing parameter $10^{-6}, 1, 10$ and $50$.

□

## 1.4 Some survival regression models

We now consider several well-known regression models for survival analysis. In this context, the survival times are still assumed independent but they are non-identically distributed, and they are affected by covariates, denoted by $x_1, \ldots, x_p$, with corresponding data matrix for these covariates denoted by $\mathbf{X}$. The dimension of $\mathbf{X}$ is $n \times p$ as we have $n$ subjects and $p$ covariates. Here, covariates can be time-fixed or may be even time-varying. Time-fixed covariates have values obtained at the baseline time and their influence on the hazard function remains constant over time. On the other hand, time-varying covariates have an impact on the hazard function that varies with time.

We will briefly discuss three commonly used regression models for examining covariate effects on survival outcomes, and they are: the accelerated failure time model (also known as the log-linear model), the Cox's proportional hazards model and the additive hazards model.

### 1.4.1 Accelerated failure time models

We now introduce the accelerated failure time (AFT) model for survival analysis. The AFT model is attractive to data analysts as it is basically the classical log-linear regression approach, and this model links the failure time directly to covariates, giving easy interpretations than hazard-based regression models. However, *due to censoring* and *non-parametric error term*, the estimation method for a *semi-parametric* AFT model in survival analysis is very different from the conventional log-linear model where there is no censoring. We will use the terms "log-linear model" and "AFT model" exchangeably in the following discussions.

Corresponding to the event time $Y_i$ of subject $i$, we use $\mathbf{x}_i^\top$ to represent the $i$-th row of the matrix $\mathbf{X}$. An AFT model can be written as

$$\log Y_i = \mathbf{x}_i^\top \boldsymbol{\beta} + \sigma\varepsilon_i, \tag{1.40}$$

where $\sigma$ is an unknown (usually) scaling parameter, $\varepsilon_i$ denotes the error term and $\boldsymbol{\beta}$ is a vector for the regression coefficients. The representation (1.40) for an AFT model involves log-transformation of the survival time $Y_i$. An alternative expression of this model, which is useful when studying full-likelihood estimation methods for an AFT model (Li & Ma 2020) or when incorporating time-varying covariates into an AFT model (Cox & Oakes 1984, Ma et al. 2023), is obtained by using the survival function of $Y_0 \equiv e^{\sigma\varepsilon_i}$, denoted by $S_0(t)$. In fact, it is very easy to verify that (1.40) is equivalent to

$$S_{Y_i}(t|\mathbf{x}_i) = S_0\big(te^{-\mathbf{x}_i^\top \boldsymbol{\beta}}\big), \tag{1.41}$$

or alternatively, if using the hazard function

$$h_{Y_i}(t|\mathbf{x}_i) = h_0\big(te^{-\mathbf{x}_i^\top \boldsymbol{\beta}}\big)e^{-\mathbf{x}_i^\top \boldsymbol{\beta}}, \tag{1.42}$$

where $h_0(t)$ ($\geq 0$) represents the hazard function of $Y_0$. The function $h_0(t)$ is usually referred to as the baseline hazard function. It is clear from (1.42) that the AFT model adjusts the rate at which time passes.

One of the reasons for AFT to be a popular model in survival analysis is that its regression coefficients have simple interpretations. In fact, the coefficient $\beta_j$ itself reflects an average change in logarithm of the survival time when the covariate $x_j$ increases by one unit. In the aspect of survival time, since $Y_i = e^{\mathbf{x}_i^\top \boldsymbol{\beta}} Y_0$, $e^{\beta_j}$ evidences the average change of $(Y_i - Y_0)/Y_0$, which depicts the percentage change of $Y_i$ when compared with the baseline survival time $Y_0$, when $x_j$ increases by one unit.

AFT models can be divided into two categories: parametric and semi-parametric models. For parametric models, the distribution of $\varepsilon_i$, and thus also the distribution of $\log Y_i$, is assumed known except a finite number of parameters, and this number does not change with the sample size. For semi-parametric AFT models, the distribution of $\varepsilon_i$ is assumed unknown.

For parametric AFT models, examples of popular distributions for $Y_i$ include, log-normal, Weibull, log-logistic, flexible F distribution, etc. In this context, the parametric log-likelihood can be easily constructed, where even left, right and interval censoring can all be included so that partly interval-censored survival times (Kim 2003) will not be an issue. The unknown parameters $\boldsymbol{\beta}$ and $\sigma$ can be estimated by maximizing this log-likelihood function.

**Example 1.13** (The log-normal AFT model).
Assuming right-censored survival times are available for $n$ individuals, where individual $i$ has the data $(t_i, \delta_i, \mathbf{x}_i)$, with $\delta_i = 1$ if $t_i$ is an event time and $\delta_i = 0$ if $t_i$ is a right-censoring time. Now, let us consider a parametric log-normal AFT model for the survival time of individual $i$, where $\varepsilon_i$ follows the standard normal distribution $N(0,1)$. The parameters $\boldsymbol{\beta}$ and $\sigma$ can be estimated by maximizing the log-likelihood, expressed as:

$$l(\boldsymbol{\beta}, \sigma) = \sum_{i=1}^{n} \left\{ \delta_i \log f_N \left( \frac{\log t_i - \mathbf{x}_i\boldsymbol{\beta}}{\sigma} \right) + (1 - \delta_i) \log \left( 1 - F_N \left( \frac{\log t_i - \mathbf{x}_i\boldsymbol{\beta}}{\sigma} \right) \right) \right\},$$

where $f_N$ and $F_N$ denote, respectively, the density and the cumulative distribution function of the standard normal distribution $N(0,1)$. The MLE of $\boldsymbol{\beta}$ and $\sigma$ can be obtained numerically using, for example, the Newton algorithm. Alternatively, the augmented scoring algorithm (a modification of the EM algorithm) proposed by Ma & Hudson (1998), which combines the complete data log-likelihood and the Fisher scoring algorithm, can be easily implemented to estimate these unknown parameters. □

For semi-parametric AFT models, regression coefficient estimation is much more complicated than for the parametric counterpart. The most widely used estimation methods, only applicable when survival times are *right-censored*, are rank-based method. These are motivated by inverting the weighted log-rank test; see Prentice (1978), Jin et al. (2003) and also Kalbfleisch & Prentice (2002). However, the parametric AFT methods are competitive with these methods as the parametric AFT form is more amenable. Wei (1992) provides a general review of the rank methods. Chiou, Kang, Yan et al. (2014) also contains a nice summary of various estimation methods used to fit AFT models.

**Example 1.14** (Semi-parametric AFT model with R).
In this example, we will demonstrate how to implement the `aftgee` R package to fit an AFT model using a simulated sample, where the AFT model is

$$\log Y_i = X_1 - 0.5X_2 + \varepsilon_i.$$

Here, values of $X_1$ and $X_2$ are generated from $X_1 \sim Binomial(1, 0.5)$ and $X_2 \sim Unif(0, 1)$, and the distribution for $\varepsilon_i$ is a standard normal and the simulated sample has its sample size $n = 500$. The censoring data distribution is exponential $exp(0.2)$. Using the R code given below, we can perform the following activities:

(i) Generate right-censored survival times from an AFT model with right censoring.
(ii) Explain how to fit this model using R package `aftgee`.

```
# Generate right-censored survival times from
# AFT model
epsilon_i <- rnorm(500)
sigma <- 1
x1 <- rbinom(500, 1, 0.5)
x2 <- runif(500, 0, 1)
X <- cbind(x1, x2)
beta_true = c(1, -0.5)

log_t <- X %*% beta_true + sigma * epsilon_i
t <- exp(log_t)
c <- rexp(500, 0.2)
delta <- as.numeric(t < c)
y <- t
y[which(delta == 0)] <- c[which(delta == 0)]

df <- data.frame(y, delta)
df <- cbind(df, X)

# Fit this model using R package aftgee
library(aftgee)
aft_fit <- aftgee(Surv(y, delta) ~ -1 + x1 + x2,
    data = df, corstr = "independence")
summary(aft_fit)
```

TABLE 1.1: Results of fitting AFT model using `aftgee` package.

| | Estimate | SE | Z-value | p-value |
|---|---|---|---|---|
| X1 | 0.981 | 0.092 | 10.694 | <0.001 |
| X2 | -0.471 | 0.097 | -4.876 | <0.001 |

□

### 1.4.2 Cox regression models

Another commonly used survival regression model is the Cox model. This model specifies the hazard function of an individual as a product of non-negative functions: (i) an hazard that is common to all individuals; and (ii) exponential of a linear predictor of individual $i$. That is:

$$h_{Y_i}(t|\mathbf{x}_i) = h_0(t)\, e^{\mathbf{x}_i^\mathsf{T}\boldsymbol{\beta}}, \tag{1.43}$$

where $\mathbf{x}_i$ is a vector of time-fixed (or time-varying) covariates and $h_0(t) \geq 0$ is called the baseline hazard function, as $h_0(t)$ is the hazard function corresponding to $\mathbf{x}_i = \mathbf{0}$. Here $\mathbf{0}$ denotes a vector of zeros with the same length as $\mathbf{x}_i$.

The survival function corresponding to model (1.43) is:

$$S_{Y_i}(t|\mathbf{x}_i) = S_0(t)^{\exp\{\mathbf{x}_i^\mathsf{T}\boldsymbol{\beta}\}}, \tag{1.44}$$

where $S_0(t)$ is the survival function corresponding to $h_0(t)$, namely $S_0(t) = e^{-H_0(t)}$, and here $H_0(t) = \int_0^t h_0(w)dw$, the cumulative baseline hazard function. The function $S_0(t)$ is referred to as the *baseline survival function*.

From (1.44), the 'log(–log)' transformation of $S_{Y_i}(t)$ becomes a linear function of $\mathbf{x}_i$:

$$\log(-\log S_{Y_i}(t|\mathbf{x}_i)) = \alpha_0(t) + \mathbf{x}_i^\mathsf{T}\boldsymbol{\beta}, \tag{1.45}$$

where $\alpha_0(t) = \log(-\log S_0(t))$.

In the case where $h_0(t)$ (or $S_0(t)$) is a fully parametric function, (1.43) is called the *parametric* Cox model. When $h_0(t)$ is non-parametric (so its functional form is unspecified), model (1.43) is called the *semi-parametric Cox* model. In practice, since an accurate parametric function for $h_0(t)$ is difficult to determine, and a mis-specified $h_0(t)$ can cause biases in $\boldsymbol{\beta}$ estimates, semi-parametric Cox models are more commonly used than their parametric counterpart.

In the next example, we will run a simple simulation study, demonstrating the bias induced by mis-specification of $h_0(t)$ in a parametric Cox model. We will also fit the model using semi-parametric Cox where regression coefficients are estimated by maximizing the partial likelihood, and hence evidencing the benefit of semi-parametric models.

**Example 1.15** (Effects of miss-specification of the baseline hazard).
In this example, we conduct a simulation study to illustrate the benefits of the semi-parametric Cox model. Specifically, we wish to demonstrate that an incorrectly specified parametric Cox model, where the baseline hazard is inaccurately specified, can result in biased estimates.

We generate data from a parametric Cox model:

$$h_i(t) = h_0(t)e^{0.5X_1+0.5X_2},$$

where the baseline hazard $h_0(t)$ is taken from a log-logistic distribution:

$$h_0(t) = \frac{\kappa t^{\kappa-1}}{\gamma^\kappa + t^\kappa},$$

and the corresponding baseline survival is:

$$S_0(t) = \left(1 + (t/\gamma)^\kappa\right)^{-1}.$$

Survival data from this distribution can be generated simply by

$$Y_i = S_0^{-1}(\exp\{\log U_i / \exp(X_i \boldsymbol{\beta})\}),$$

where $U_i \sim unif(0, 1)$. In this example we take $\gamma = 1$ and $\kappa = 3$.

We fit the Cox model in two ways:

(a) using a parametric Cox model, where an incorrect Weibull distribution is assumed, and
(b) we also fit the simulated data using a semi-parametric Cox model by the partial likelihood method (this method will be explained in the next section).

The steps for this simulation are specified below.

(i) Generate right-censored survival times using a Cox regression with a log-logistic $h_0(t)$. Here, the sample size is $n = 50$, and the number of repetitions is set to 300.
(ii) Fit these data using (a) a parametric Cox model with a Weibull distribution and (b) a semi-parametric Cox model. Both can be fitted using, for example, the `survival` package, or you can choose different packages as you wish.
(iii) Build a table to compare the mean, standard deviation and coverage probability of the estimates from these two approaches.

```
# Write simulate() function to repeatedly
# generate a sample
simulate <- function() {
    kappa <- 3
    gamma <- 1
    neglogU <- -log(runif(50))

    x1 <- rbinom(50, 1, 0.5)
    x2 <- runif(50, 0, 1)
    X <- cbind(x1, x2)
    beta_true = c(0.5, 0.5)
    eXtB = exp(X %*% beta_true)

    y <- gamma * (exp(neglogU/eXtB) - 1)^(1/kappa)
    event <- rep(1, 50)
    t <- y

    df = data.frame(t, event)
    df = cbind(df, X)
    return(df)
}

# Load flexsurv package for Weibull PH model
library(flexsurv)

# Run simulation
save <- matrix(0, nrow = 300, ncol = 12)
```

```
for (s in 1:300) {
    df <- simulate()

    weib_fit <- flexsurvreg(Surv(t, event) ~ x1 +
        x2, data = df, dist = "weibullPH")
    cox_fit <- coxph(Surv(t, event) ~ x1 + x2, data = df)

    save[s, 1] <- weib_fit$coefficients[3]
    save[s, 2] <- sqrt(weib_fit$cov[3, 3])
    save[s, 3] <- as.numeric(weib_fit$coefficients[3] -
        1.96 * sqrt(weib_fit$cov[3, 3]) < (0.5) &
        weib_fit$coefficients[3] + 1.96 * sqrt(weib_fit$cov[3,
            3]) > (0.5))

    save[s, 4] <- weib_fit$coefficients[4]
    save[s, 5] <- sqrt(weib_fit$cov[4, 4])
    save[s, 6] <- as.numeric(weib_fit$coefficients[4] -
        1.96 * sqrt(weib_fit$cov[4, 4]) < (0.5) &
        weib_fit$coefficients[4] + 1.96 * sqrt(weib_fit$cov[4,
            4]) > (0.5))

    save[s, 7] <- cox_fit$coefficients[1]
    save[s, 8] <- sqrt(cox_fit$var[1, 1])
    save[s, 9] <- as.numeric(cox_fit$coefficients[1] -
        1.96 * sqrt(cox_fit$var[1, 1]) < (0.5) &
        cox_fit$coefficients[1] + 1.96 * sqrt(cox_fit$var[1,
            1]) > (0.5))

    save[s, 10] <- cox_fit$coefficients[2]
    save[s, 11] <- sqrt(cox_fit$var[2, 2])
    save[s, 12] <- as.numeric(cox_fit$coefficients[2] -
        1.96 * sqrt(cox_fit$var[2, 2]) < (0.5) &
        cox_fit$coefficients[2] + 1.96 * sqrt(cox_fit$var[2,
            2]) > (0.5))
}


results <- apply(save, 2, mean)
results <- round(results, 4)
results <- matrix(results, nrow = 4, byrow = TRUE)
```

The semi-parametric model clearly provides coefficient estimates that are more accurate, along with better coverage probabilities. □

The above model (1.43) implicitly implies the *proportional hazards* (PH) assumption described below. Consider two individuals $i$ and $q$ and their associated covariate vectors $\mathbf{x}_i$ and $\mathbf{x}_q$. Assume the Cox model (1.43) contains only time-fixed covariates.

TABLE 1.2: A simulation study comparing a parametric Cox model against a semi-parametric Cox model.

| Model | Parameter | Mean Estimate | SE | 95% CP |
|---|---|---|---|---|
| Weibull | $\beta_1$ | 0.6330 | 0.3090 | 0.8767 |
| | $\beta_2$ | 0.6448 | 0.5314 | 0.8633 |
| Semi-para Cox | $\beta_1$ | 0.5254 | 0.3104 | 0.9533 |
| | $\beta_2$ | 0.5283 | 0.5316 | 0.9433 |

The corresponding hazards ratio from the Cox model (1.43) does not depend on time $t$ since it is easy to obtain that

$$\frac{h_{Y_i}(t|\mathbf{x}_i)}{h_{Y_q}(t|\mathbf{x}_q)} = \exp\{(\mathbf{x}_i - \mathbf{x}_q)^\top \boldsymbol{\beta}\}. \tag{1.46}$$

This property is known as the PH property, and model (1.43) is often referred to as the Cox PH model for this reason. However, in this chapter, we use the terms Cox model or Cox PH model interchangeably when referring to (1.43).

For a covariate value $x_j$ and its coefficient $\beta_j$, $e^{\beta_j}$ is the multiplication factor to the baseline hazard when $x_j$ is increased by 1 unit, and therefore, $e^{\beta_j} - 1$ can be interpreted as the average relative change in the hazard, defined as $(h_{Y_i}(t) - h_0(t))/h_0(t)$, after $x_j$ is increased by one unit. This relative change in hazard compares the hazard of individual $i$ with the baseline hazard.

- It is clear that AFT and Cox are different models in general. However, under a very specific condition, these two models become identical. In fact, by comparing (1.43) with (1.42), we can see that for the AFT model, if its baseline hazard $h_0(t)$ comes from the Weibull family of distributions, then such an AFT model is also a Cox regression model (Kalbfleisch & Prentice 2002, Chapter 7).

- Chapter 6 of Lawless (2002) offers the experience that AFT models are particularly applicable when covariate effects are on orders of magnitude. Such large effects are more often to occur, for example, in physical stress models, but are less common in biomedical studies.

### 1.4.3 Additive hazards regression models

The additive hazards (AH) model (e.g., Aalen (1989), and Lin & Ying (1994)) provides an important alternative to the Cox and AFT models for analyzing time-to-event data. The AH model specifies the hazard of $Y_i$ as a summation of a baseline hazard and a linear predictor from the covariates, namely:

$$h_{Y_i}(t|\mathbf{x}_i) = h_0(t) + \mathbf{x}_i^\top \boldsymbol{\beta}. \tag{1.47}$$

Note that for this to be a valid model, one needs to require $h_0(t) \geq 0$ and $h_{Y_i}(t) \geq 0$. These constraints, particularly the latter one, make estimation of the AH model parameters a complicated constrained optimization problem.

Function $h_0(t)$ is referred to as the baseline hazard as it is the hazard when $\mathbf{x}_i = \mathbf{0}$. Here, to simplify discussions we assume $\mathbf{x}_i$ is time-fixed, but this assumption is not necessary for the AH models. The survival function corresponding to (1.47) is:

$$S_{Y_i}(t|\mathbf{x}_i) = S_0(t)e^{-\mathbf{x}_i^\top \boldsymbol{\beta}\, t}, \tag{1.48}$$

where $S_0(t)$ is the survival function associated with $h_0(t)$, and hence is called the baseline survival function.

If $h_0(t)$ is a parametric function then (1.47) is a fully parametric model. On the other hand, if $h_0(t)$ is non-parametric then (1.47) is a semi-parametric model. The semi-parametric model is more useful in practice as it does not rely on any distributional assumptions of the baseline hazard.

For the AH model, a regression parameter $\beta_j$ represents the risk difference rather than risk ratio of the Cox regression model. In some applications, people may be interested in risk differences, where selection of an AH model becomes appropriate.

## 1.5 Semi-parametric Cox model fitting using partial likelihood

The method of maximum *partial likelihood* (Cox 1972, 1975) is the most widely used approach to estimate model regression coefficients of a semi-parametric Cox model when the survival data are subject to right censoring. We will first provide a brief summary of this method and then explain this method from the perspective of profile likelihood.

### 1.5.1 The method of partial likelihood

For right-censored survival data, assuming non-informative censoring, we can define a partial likelihood function. Recall the Cox model (1.43), and consider estimation of $\boldsymbol{\beta}$ without making any assumptions about the baseline hazard $h_0(t)$.

In his 1972 paper (Cox 1972), Cox proposed the proportional hazards model (the Cox model) and suggested a special likelihood function, called condition likelihood in his paper, to estimate the regression coefficient vector $\boldsymbol{\beta}$. However, Kalbfleisch & Prentice (1973) showed that this was instead a marginal likelihood of ranks under some restrictive assumptions. In the Cox (1975) paper, this likelihood is finally named the "partial likelihood".

We assume $\mathtt{e}_1 < \mathtt{e}_2 < \cdots < \mathtt{e}_m$ to be ordered distinct event times observed from $n$ different subjects. Let $\mathcal{R}_u$ denote the risk set at $\mathtt{e}_u$, defined as the set of indices of the subjects that are event-free just before $\mathtt{e}_u$ (or at $\mathtt{e}_u^-$). We adopt the convention that $\mathcal{R}_0 = \{1, 2, \ldots, n\}$, which represents the risk set at time $t = 0$.

Suppose first that there are no ties in the observation times, so each $\mathtt{e}_u$ only associated with one subject. The conditional probability that this particular subject would fail at $\mathtt{e}_u$, given the risk set $\mathcal{R}_u$ and the fact that exactly one subject has the event at that time, is

$$\frac{h(\mathsf{e}_u|\mathbf{x}_u)}{\sum_{j\in\mathcal{R}_u} h(\mathsf{e}_u|\mathbf{x}_j)}.$$

We can write this probability in terms of the Cox model and having

$$\frac{h_0(\mathsf{e}_u)\exp\{\mathbf{x}_u^\top\boldsymbol{\beta}\}}{\sum_{j\in\mathcal{R}_u} h_0(\mathsf{e}_u)\exp\{\mathbf{x}_j^\top\boldsymbol{\beta}\}},$$

and we notice that the terms depending on the baseline hazard $h_0(\mathsf{e}_u)$ cancel, so the probability in question is

$$\frac{\exp\{\mathbf{x}_u^\top\boldsymbol{\beta}\}}{\sum_{j\in\mathcal{R}_u}\exp\{\mathbf{x}_j^\top\boldsymbol{\beta}\}}.$$

Multiplying these probabilities together over all distinct failure times and treating the resulting product

$$\mathcal{L}(\boldsymbol{\beta}) = \prod_{u=1}^{m}\left(\frac{\exp\{\mathbf{x}_u^\top\boldsymbol{\beta}\}}{\sum_{j\in\mathcal{R}_u}\exp\{\mathbf{x}_j^\top\boldsymbol{\beta}\}}\right) = \prod_{i=1}^{n}\left(\frac{\exp\{\mathbf{x}_i^\top\boldsymbol{\beta}\}}{\sum_{j\in\mathcal{R}_i}\exp\{\mathbf{x}_j^\top\boldsymbol{\beta}\}}\right)^{\delta_i} \tag{1.49}$$

as if it was an ordinary likelihood. Estimation and other inferences (such as likelihood ratio test) on $\boldsymbol{\beta}$ can be made using $\mathcal{L}(\boldsymbol{\beta})$.

Cox (1975) provided a more general justification of $\mathcal{L}$ as part of the full likelihood – in fact, a part that happens to contain most of the information about $\boldsymbol{\beta}$ – and therefore proposed calling $\mathcal{L}$ a partial likelihood. This justification is valid even with time-varying covariates. A more rigorous justification of the partial likelihood in terms of the theory of counting processes can be found in Andersen et al. (1993).

The maximum partial likelihood estimate of the regression coefficient vector, denoted by $\widehat{\boldsymbol{\beta}}$, is obtained by maximization of the log-partial likelihood $\log\mathcal{L}(\boldsymbol{\beta})$. Cox has demonstrated very similar asymptotic properties as in an ordinary likelihood.

With either full or partial likelihoods, we have three approaches to testing hypotheses about $H_0 : \boldsymbol{\beta} = \boldsymbol{\beta}_0$:

- **Likelihood Ratio Test**: Given two nested models, we treat twice the difference in partial log-likelihoods as a $\chi^2$ statistic with degrees of freedom equal to the difference in the number of parameters.

- **Wald Test**: We use the fact that approximately in large samples $\widehat{\boldsymbol{\beta}}$ has a multivariate normal distribution with mean $\boldsymbol{\beta}$ and variance-covariance matrix $\text{Var}(\widehat{\boldsymbol{\beta}}) = \mathcal{I}^{-1}(\boldsymbol{\beta})$, where $\mathcal{I}(\boldsymbol{\beta})$ denotes the information matrix from the log-partial likelihood $\log\mathcal{L}$. Thus, under $H_0 : \boldsymbol{\beta} = \boldsymbol{\beta}_0$, the quadratic form $(\widehat{\boldsymbol{\beta}} - \boldsymbol{\beta}_0)^\top\, \mathcal{I}^{-1}(\boldsymbol{\beta}_0)\, (\widehat{\boldsymbol{\beta}} - \boldsymbol{\beta}_0) \sim \chi_p^2$, where $p$ is the dimension of $\boldsymbol{\beta}$. This test can be extended to test a subset of $\boldsymbol{\beta}$.

- **Score Test**: We use the fact that approximately in large samples the score function $\mathbf{u}(\boldsymbol{\beta})$ has a multivariate normal distribution with mean $\mathbf{0}_{p\times 1}$ and variance-covariance matrix equal to the information matrix $\mathcal{I}(\boldsymbol{\beta}_0)$. Thus, under $H_0 : \boldsymbol{\beta} =$

$\boldsymbol{\beta}_0$, the quadratic form

$$\mathbf{u}(\boldsymbol{\beta}_0)^\mathsf{T}\mathcal{I}^{-1}(\boldsymbol{\beta}_0)\mathbf{u}(\boldsymbol{\beta}_0) \sim \chi^2_p.$$

Note that this test does not require calculating the maximum partial likelihood estimate $\widehat{\boldsymbol{\beta}}$.

One reason for bringing up the score test is that in the $k$-sample case, the score test of $H_0 : \boldsymbol{\beta} = \mathbf{0}$, based on Cox model, happens to be the same as the Mantel-Haenszel log-rank test.

**Example 1.16** (Partial likelihood example with R).
The R package `survival` is included in R's standard library. This package can define and fit survival models involving event times and status, including date, state and censoring status of entry and exit. In this example, we just practice how to fit a Cox model using R "`coxph`" funcation.

The reader is referred to its manual `help(survival::survival)`, and R help on functiona `survival::Surv()` and `survival::coxph()`.

```
library(survival)
summary(leukemia)

##       time           status                  x
##  Min.   :  5.00   Min.   :0.0000   Maintained   :11
##  1st Qu.: 12.50   1st Qu.:1.0000   Nonmaintained:12
##  Median : 23.00   Median :1.0000
##  Mean   : 29.48   Mean   :0.7826
##  3rd Qu.: 33.50   3rd Qu.:1.0000
##  Max.   :161.00   Max.   :1.0000

leukemia.sva <- coxph(Surv(time, status) ~ x,
    data = leukemia)
summary(leukemia.sva)

## Call:
## coxph(formula = Surv(time, status) ~ x, data = leukemia)
##
##   n= 23, number of events= 18
##
##                  coef exp(coef) se(coef)     z Pr(>|z|)
## xNonmaintained 0.9155    2.4981   0.5119 1.788   0.0737 .
## ---
## Signif. codes:  0 '***' 0.001 '**' 0.01 '*' 0.05 '.' 0.1 ' ' 1
##
##                exp(coef) exp(-coef) lower .95 upper .95
## xNonmaintained     2.498     0.4003    0.9159     6.813
##
## Concordance= 0.619  (se = 0.063 )
## Likelihood ratio test= 3.38  on 1 df,   p=0.07
## Wald test            = 3.2  on 1 df,   p=0.07
## Score (logrank) test = 3.42  on 1 df,   p=0.06
```

□

### 1.5.2 Partial likelihood as a profile likelihood

The partial likelihood function displayed above can also be derived as a profile likelihood when only right censoring is present; see, for example, Murphy & van der Vaart (2000). We briefly summarize this finding below.

When there is only the right censoring in the survival data, and under the semi-parametric Cox model, the full log-likelihood becomes:

$$\begin{aligned} l(\boldsymbol{\beta}, h_0(t)) &= \sum_{i=1}^{n} \left\{ \delta_i \log h_{Y_i}(t_i) + \log S_{Y_i}(t_i) \right\} \\ &= \sum_{i=1}^{n} \left\{ \delta_i (\log h_0(t_i) + \mathbf{x}_i^\mathsf{T} \boldsymbol{\beta}) - H_{Y_i}(t_i) \right\}, \end{aligned} \tag{1.50}$$

where $H_{Y_i}(t_i) = H_0(t_i) \exp\{\mathbf{x}_i^\mathsf{T} \boldsymbol{\beta}\}$.

Define $a = \min_i\{t_i\}$ and $b = \max_i\{t_i\}$. Let $\mathsf{e}_1 < \cdots < \mathsf{e}_m$ be distinct and ordered event times. We will use $\mathsf{e}_1, \ldots, \mathsf{e}_m$ to define points (knots) $w_1, \ldots, w_{m+1}$ for partitioning the interval $[a, b]$ into $m$ sub-intervals $[w_1, w_2], \ldots, (w_m, w_{m+1}]$, each containing exactly one such $\mathsf{e}_u$. For example, when $a$ and $b$ are not event but censoring times, then we can use: $w_1 = a, w_2 = \mathsf{e}_1, \ldots, w_m = \mathsf{e}_{m-1}, w_{m+1} = b$. The key point here is that there is only one $\mathsf{e}_u$ in each $(w_u, w_{u+1}]$ interval for $u = 1, \ldots, m$.

We approximate $h_0(t)$ using a piecewise constant function similar to Section 1.3.3. Specifically, we adopt

$$h_0(t) = \theta_u, \quad \text{when } t \in (w_u, w_{u+1}].$$

Let $\boldsymbol{\theta} = (\theta_1, \ldots, \theta_m)^\mathsf{T}$ represent the unknown vector for $h_0(t)$. The corresponding $H_0(t)$ becomes:

$$H_0(t) = \sum_{u=1}^{m} \theta_u I(w_u \leq t) \mathcal{W}_u, \tag{1.51}$$

where $\mathcal{W}_u = w_{u+1} - w_u$.

Substituting these approximations into (1.50) we have

$$l(\boldsymbol{\beta}, \boldsymbol{\theta}) = \sum_{i=1}^{n} \delta_i \mathbf{x}_i^\mathsf{T} \boldsymbol{\beta} + \sum_{u=1}^{m} n_u \log \theta_u - \sum_{i=1}^{n} e^{\mathbf{x}_i^\mathsf{T} \boldsymbol{\beta}} \sum_{u=1}^{m} \theta_u I(w_u \leq t_i) \mathcal{W}_u, \tag{1.52}$$

where $n_u$ denotes the number of event times in $(w_u, w_{u+1}]$. We wish to estimate the regression coefficient vector $\boldsymbol{\beta}$ and the baseline hazard coefficient vector $\boldsymbol{\theta}$, subject to all $\theta_u \geq 0$ (recall this is denoted as $\boldsymbol{\theta} \geq 0$).

From the derivative

$$\frac{\partial l}{\partial \theta_u} = \frac{n_u}{\theta_u} - \mathcal{W}_u \sum_{i=1}^{n} e^{\mathbf{x}_i^\mathsf{T} \boldsymbol{\beta}} I(w_u \leq t_i),$$

the solution of $\theta_u$ from $\partial l/\partial\theta_u = 0$ can be easily obtained; this solution is given by

$$\theta_u = \frac{n_u}{\mathcal{W}_u \sum_{i=1}^{n} e^{\mathbf{x}_i^\mathsf{T}\boldsymbol{\beta}} I(w_u \le t_i)}, \tag{1.53}$$

and $\theta_u$ automatically satisfies the requirement $\theta_u \geq 0$. Next, substituting this expression of $\theta_u$ solution into the log-likelihood (1.52) we have the so-called *profile log-likelihood*, given by:

$$\begin{aligned}
l_p(\boldsymbol{\beta}) &= \sum_{i=1}^{n} \delta_i \mathbf{x}_i^\mathsf{T}\boldsymbol{\beta} + \sum_{u=1}^{m} n_u \log n_u - \sum_{u=1}^{m}\Big(n_u \log \sum_{i=1}^{n} e^{\mathbf{x}_i^\mathsf{T}\boldsymbol{\beta}} I(w_u \le t_i) + n_u \log \mathcal{W}_u\Big) \\
&\quad - \sum_{u=1}^{m} \frac{n_u}{\sum_{i=1}^{n} e^{\mathbf{x}_i^\mathsf{T}\boldsymbol{\beta}} I(w_u \le t_i)} \sum_{i=1}^{n} e^{\mathbf{x}_i^\mathsf{T}\boldsymbol{\beta}}\, I(w_u \le t_i) \\
&\propto \sum_{i=1}^{n} \delta_i\, \mathbf{x}_i^\mathsf{T}\boldsymbol{\beta} - \sum_{u=1}^{m} n_u \log \sum_{i=1}^{n} e^{\mathbf{x}_i^\mathsf{T}\boldsymbol{\beta}}\, I(t_i \ge w_u) \\
&\quad \sum_{i=1}^{n} \delta_i \mathbf{x}_i^\mathsf{T}\boldsymbol{\beta} - \sum_{u=1}^{m} n_u \log \sum_{i\in\mathcal{R}_u} e^{\mathbf{x}_i^\mathsf{T}\boldsymbol{\beta}}.
\end{aligned} \tag{1.54}$$

Here, the terms independent of $\boldsymbol{\beta}$ are ignored in the log-likelihood. When there are no tied event times, this profile likelihood becomes:

$$l_p(\boldsymbol{\beta}) = \sum_{i=1}^{n} \delta_i \Big(\mathbf{x}_i^\mathsf{T}\boldsymbol{\beta} - \log \sum_{i\in\mathcal{R}_i} e^{\mathbf{x}_i^\mathsf{T}\boldsymbol{\beta}}\Big). \tag{1.55}$$

This expression indicates the profile log-likelihood $l_p(\boldsymbol{\beta})$ is equivalent to the log-partial likelihood under certain conditions.

### 1.5.3 Limitations of partial likelihood

While partial likelihood has the advantage of enabling the widely successful Cox semi-parametric models used in clinical medicine, there are several considerations about this method:

- The partial likelihood method is not easily extended to interval-censored survival data.
- The Cox semi-parametric model does not itself provide estimates of its unspecified baseline hazard function; it only provides estimates of regression coefficients. To estimate the baseline hazard, the Breslow estimator, as given in (1.53), is commonly employed.
- The Breslow estimate of the baseline hazard function, however, is often extremely noisy, making it not very useful for interpretation. This noisy estimate is due to the fact that each partitioned sub-interval contains only a single event time.

A more general approach is to use the full likelihood. A full likelihood method can easily accommodate more general interval censoring. In Chapter 2, we will discuss a full likelihood-based method for fitting semi-parametric Cox models where the baseline hazard is approximated using basis functions. Furthermore, to ensure smoothness in the baseline hazard estimate, a penalty function is employed, creating a penalized log-likelihood function. In fact, the maximum penalized likelihood method is utilized throughout the entirety of this book.

## 1.6 The method of penalized likelihood: approaches

The penalized log-likelihood can be regarded as a log-posterior function, where the penalty function is analogous to the logarithm of a prior distribution density function. Consequently, the maximum penalized likelihood (MPL) method is also known as the maximum a posteriori (MAP) method, with the latter often categorized as a Bayesian approach.

From a frequentist perspective, penalties are seen as constraints, and the penalized log-likelihoods are considered Lagrange functions, as discussed in Silverman (1982).

Three key reasons for using a penalty function in the likelihood method for a semi-parametric survival regression model are:

- Smoothing (or regularization) of the baseline hazard estimate.
- Discouraging unnecessary knots.
- Simplifying the challenging task of selecting the optimal locations and number for the knot sequence.

Several MPL methods have been developed to fit Cox models, including those introduced by Joly et al. (1998), Cai & Betensky (2003), Ma et al. (2014) and Ma et al. (2021). All of these methods utilize basis functions to approximate $h_0(t)$.

In the next chapter, we will delve into the details of penalized likelihood methods for estimating the semi-parametric Cox model parameters, including the baseline hazard, when observed survival times are partly interval-censored (Kim 2003). This approach encompasses survival times which are subject to left or right censoring, as well as interval censored times.

## 1.7 R packages

The authors have provided packages `survivalMPL`, `survivalMPLdc` and `bnc`, as well as some other R functions that are available at Github, for the purposes of this book. Package `survivalMPL` provides MPL estimation methods described in later chapters. Package `bnc` provides probability imputations for the bivariate-Normal censored linear model introduced in Chapter 9.

## 1.8 Summary

This chapter serves the purpose of introducing basic concepts that are needed for the rest of this book. Particularly, we introduced different censoring types, summarized parametric and semi-parametric regression models for survival analysis and discussed the partial likelihood and full likelihood-based estimation methods. We also explained the reason for why we opt to use the maximum penalized likelihood method for parameters of a semi-parametric model.

In this book, we apply the MPL method to the Cox, extended Cox, Cox with truncation, Cox with dependent censoring and additive hazards model to estimate baseline hazards and regression coefficients simultaneously. We consider many of these models under a very general censoring scheme, namely the partly interval censoring as defined in Kim (2003).

# 2

# *Semi-parametric Cox Model with Interval Censoring*

In this chapter, we will explore how to fit semi-parametric Cox models when survival times are subject to more general censoring types. That is, apart from right censoring, they may also include left- and interval-censored data. In this context, the conventional partial likelihood (PL) method is difficult to implement directly. Instead, this chapter will discuss a full likelihood-based approach. However, since the baseline hazard of the semi-parametric Cox model is unspecified (non-parametric) and must be non-negative, this presents computational challenges in (i) handling the non-parametric baseline hazard function, and (ii) imposing the non-negative constraint.

## 2.1 Introduction

We explained in Chapter 1 that a key feature of survival analysis is that some event times are not fully observed. In such cases, these event times are censored. In this chapter, we extend right censoring considered in Chapter 1 to a more general censoring scheme so that, apart from event and right censoring times, we also allow for left and interval censoring in the observed survival times. Left censoring occurs when the event time is located earlier than a given time point, while interval censoring occurs when the event time falls between two known time points.

As before, the random variable $Y_i$ represents the event time of interest (also called the survival time) for individual $i$. In this chapter, our focus is on fitting the semi-parametric Cox model:

$$h_{Y_i}(t|\mathbf{x}_i) = h_0(t)e^{\mathbf{x}_i^\top \boldsymbol{\beta}}, \tag{2.1}$$

where $\mathbf{x}_i$ is a vector of values of $p$ time-fixed covariates associated with individual $i$, $h_{Y_i}(t|\mathbf{x}_i)$ is the hazard function for individual $i$, $h_0(t)$ represents the baseline hazard and its functional form is *unspecified*, and $\boldsymbol{\beta}$ is a $p$-vector of regression coefficients.

Throughout this chapter, we assume that the covariates are measured only at the baseline time, meaning that the covariate values are not time-varying. The case of time-varying covariates will be studied in Chapter 6.

Note that when there is no confusion, we will drop the subscript and write $h_{Y_i}(t|\mathbf{x}_i)$ as $h(t|\mathbf{x}_i)$ or $h_i(t)$. The latter two notations clearly indicate that the hazard

DOI: 10.1201/9781351109710-2

is related to individual $i$. Such a simplification in notation will also be applied to $S_{Y_i}(t|\mathbf{x}_i)$ and $H_{Y_i}(t|\mathbf{x}_i)$.

Since the function form of $h_0(t)$ is unspecified, model (2.1) a semi-parametric model. Clearly, we must require $h_0(t) \geq 0$ in order for $h_0(t)$, and hence also for $h_{Y_i}(t|\mathbf{x}_i)$, to be a valid hazard function (note that $\exp(\mathbf{x}_i^\top \boldsymbol{\beta})$ is always positive).

Before we discuss how to fit model (2.1), let us reemphasize that our analysis includes survival times with fully observed event times as well as interval, left and right censoring times. In chapter 3, we will also explore truncation, in addition to interval censoring.

Due to the presence of interval censoring, the traditional maximum partial likelihood method is difficult to adapt directly for estimating even the regression coefficients $\boldsymbol{\beta}$. In this context, one popular remedy in practice is to impute each left- or interval-censored survival time by a single value. For example, the middle point of the corresponding interval can be used. Then, the maximum partial likelihood method can still be applied to the modified data, where only right censoring presents. However, this approach should be used with caution as it can introduce biases in the $\boldsymbol{\beta}$ estimates, as demonstrated in the simulation results in Section 2.11. Multiple imputations can also be applied for interval-censored event times, as demonstrated in Pan (2000), and this approach can help to reduce biases but they are more computational demanding.

Full likelihood-based approaches, on the other hand, are more versatile and can be implemented easily to handle interval censoring. Here are some references on Cox hazard model estimation by likelihood methods: Wang et al. (2016), Finkelstein (1986), Sun (2006), Kim (2003), Pan (1999), Zhang et al. (2010), Huang (1996), Joly et al. (1998), Ma et al. (2014), Ma et al. (2021) and Cai & Betensky (2003).

Another important motivation for likelihood methods is that they are able to provide faster and more accurate predictive inferences on survival values, for example, survival probabilities, than the partial likelihood methods. This argument is supported by the fact that the baseline hazard and the regression coefficients are estimated simultaneously by full likelihood methods. As a result, the asymptotic covariance matrix of these estimates can be established accurately without resorting to computationally intensive re-sampling methods, such as bootstrapping. The estimation method we adopt throughout this chapter, and also for the rest of the book, is a more versatile penalized likelihood method. The penalty function is used to smooth the $h_0(t)$ estimate, and also to relax the impact of the number and location of knots that are used to define $h_0(t)$.

For the full likelihood methods, the baseline hazard $h_0(t)$, or a function of $h_0(t)$, such as the cumulative baseline hazard or log-cumulative baseline hazard, is typically approximated by a linear combination of *a finite number* of basis functions. In Section 2.3, we will explain the advantages and disadvantages of different options for approximating $h_0(t)$. If the number of basis functions is carefully controlled in accordance with the sample size, this approach becomes a special case of the method-of-sieves (e.g., Grenander (1981)).

In general, there are three reasons for having a penalty function in likelihood-based methods when the non-parametric function is approximated using basis functions. These reasons are summarized below:

1. to smooth (or regularize) the baseline hazard estimate,
2. to discourage unnecessary knots, and
3. to ease the difficult task of selecting the optimal location and number for the knots sequence.

In this chapter, we focus on the maximum penalized likelihood (MPL) method for estimation of the baseline hazard and regression coefficients. Several MPL methods have been developed to fit Cox models under interval censoring, such as Joly et al. (1998), Cai & Betensky (2003) and Ma et al. (2021), where all have used basis functions to approximate $h_0(t)$ or a function of $h_0(t)$.

If there is no penalty, the MPL method of this section is a special sieve maximum likelihood method. Furthermore, discussions in Section 2.6 explains that under certain conditions the MPL approach is also a sieve method. Descriptions of the sieve method can be found in, for example, Zhang et al. (2010) and Wang et al. (2016). Without a proper penalty function, however, sieve maximum likelihood can be sensitive to the locations and number of knots that are used to approximate $h_0(t)$, which may lead to numerical issues in practice. On the other hand, the estimated parameters of $h_0(t)$ can be located on the boundary of the parameter space, invalidating the conventional maximum likelihood asymptotic results.

In this chapter, we will present asymptotic normality results that allow some estimated parameters to be on the boundary of the parameter space. Comparisons of the penalized likelihood method with the partial likelihood method, where the partial likelihood method is built by replacing the interval-censored data with the mid-point of the censoring interval, will be made using simulations later in this chapter. This comparison study is important as it reveals the numerical and other benefits of the MPL methods. The R function for the MPL estimation from the "survivalMPL" R package is implemented throughout this chapter for simulation and data analysis.

## 2.2 Likelihood under interval censoring

We will begin with a survival dataset example featuring interval censoring. The dataset pertains to HIV/AIDS patients who were treated with the drug zidovudine. The primary event of interest is the time until resistance to zidovudine developed. This particular dataset was analyzed in Lindsay & Ryan (1998), where the Cox proportional hazards model was fitted as an example.

**Example 2.1** (Zidovudine dataset for interval censoring).
Table 2.1 below displays the first three records of this zidovudine dataset. The dataset consists of interval-, left- and right-censored observations, with no observed event

times. The wide intervals associated with interval censoring can make fitting any semi-parametric model a challenging task. Here, $t_i^L$ and $t_i^R$ denote, respectively, the left and right endpoints of censoring intervals, with $t_i^L = 0$ meaning a left censoring and $t_i^R = \text{inf}$ indicating a right censoring.

TABLE 2.1: Zidovudine example for a mixture of left, right and interval censoring.

| $t_i^L$ | $t_i^R$ | stage | dose | CD4: 100-399 | CD4: $\geq 400$ |
|---|---|---|---|---|---|
| 0 | 16 | 0 | 0 | 0 | 1 |
| 15 | inf | 0 | 0 | 0 | 1 |
| 12 | inf | 0 | 0 | 0 | 1 |

□

Recall that we use $Y_i$ to denote the time until the event-of-interest occurs for individual $i$. However, some or all of $Y_i$ may not be fully observed. We assume that they are interval-censored, where left and right censoring can be considered special cases.

Due to interval censoring, we adopt a random vector $\mathbf{T}_i = (T_i^L, T_i^R)^\mathsf{T}$ to denote a pair of independent random time points for individual $i$, where $T_i^L \leq T_i^R$ and satisfies $T_i^L \leq Y_i \leq T_i^R$. Assume $\mathbf{T}_i$ are independent for different $i$. Observed values for $T_i^L$ and $T_i^R$ are denoted by $t_i^L$ and $t_i^R$ respectively and we know that $t_i^L \leq Y_i \leq t_i^R$.

If $t_i^L = 0$ and $t_i^R$ is finite then $Y_i$ is left-censored at $t_i^R$, meaning that the event has occurred before this $t_i^R$; whilst if $t_i^L$ is finite (and non-zero) and $t_i^R = \infty$ then $Y_i$ is right-censored at $t_i^L$. If $t_i^L$ is non-zero and $t_i^R$ is finite then $Y_i$ is said to be interval-censored so that $Y_i \in [t_i^L, t_i^R]$. The case where $t_i^L = t_i^R$ indicates $Y_i$ is observed exactly so we have an event time in this case. Therefore, the observed survival times can have event times, as well as left, right and interval censoring times. These different censoring types will all be involved when fitting the Cox regression model. This type of survival data is called the *partly interval-censored* data (e.g. Kim (2003)).

In this chapter, we assume **independent interval censoring** given the covariates, meaning that $P(Y_i \leq t \,|\, T_i^L = t_i^L, T_i^R = t_i^R, T_i^L \leq Y_i < T_i^R) = P(Y_i \leq t \,|\, t_i^L \leq Y_i < t_i^R)$; see, for example, Chapter 1 of Sun (2006).

From the above notations, we can let $(t_i^L, t_i^R, \mathbf{x}_i)$ represent the observation associated with individual $i$, where $i = 1, \ldots, n$. To simplify mathematical expressions, we further adopt variables to indicate different censoring types. Let an indicator $\delta^A$ takes a value of 1 if event with censoring type $A$ occurs and 0 otherwise, and therefore $\delta_i^R$, $\delta_i^L$ and $\delta_i^I$ are the indicators for right-censoring, left-censoring, and interval-censoring times, respectively, all associated with individual $i$. Define $\delta_i = 1 - \delta_i^R - \delta_i^L - \delta_i^I$, which represents the indicator for an event time for $i$. Therefore, the entire partly interval-censored survival data can be amended to:

$$(t_i^L, t_i^R, \delta_i, \delta_i^L, \delta_i^R, \delta_i^I, \mathbf{x}_i),$$

for $i = 1, \ldots, n$.

Corresponding to the hazard function $h(t|\mathbf{x}_i)$, we can compute the cumulative hazard $H(t|\mathbf{x}_i) = \int_0^t h(w|\mathbf{x}_i)dw$ and the survival function, which is given by $S(t|\mathbf{x}_i) = \exp\{-H(t|\mathbf{x}_i)\}$. To simplify notation, we will let $h_i(t) = h(t|\mathbf{x}_i)$, $H_i(t) = H(t|\mathbf{x}_i)$ and $S_i(t) = S(t|\mathbf{x}_i)$ throughout this chapter.

Thus, using these indicators, the log-likelihood for individual $i$ can be expressed as:

$$\begin{aligned} l_i(\boldsymbol{\beta}, h_0(t)) =& \delta_i(\log h_0(t_i) + \mathbf{x}_i\boldsymbol{\beta} + \log S_i(t_i)) + \delta_i^R \log S_i(t_i^L) + \\ & \delta_i^L \log(1 - S_i(t_i^R)) + \delta_i^I \log(S_i(t_i^L) - S_i(t_i^R)). \end{aligned} \tag{2.2}$$

The log-likelihood from the entire dataset is then

$$l(\boldsymbol{\beta}, h_0(t)) = \sum_{i=1}^{n} l_i(\boldsymbol{\beta}, h_0(t)). \tag{2.3}$$

We comment that equation (2.2) will be repeatedly appeared throughout this book.

The maximum likelihood estimates of $\boldsymbol{\beta}$ and $h_0(t)$ of the Cox model (2.1) are given by the maximum of $l(\boldsymbol{\beta}, h_0(t))$, subject to the constraint $h_0(t) \geq 0$. Clearly, the requirement of $h_0(t) \geq 0$ will lead to the fact that $h(t|\mathbf{x}_i)$ (given by (2.1)) is valid hazard function. However, there are two obvious issues associated with this optimization problem:

**Issue 2.1** When fully unspecified, the baseline hazard $h_0(t)$ is an infinite dimensional parameter while the sample size $n$ is finite. Using a finite number of observations to estimate an infinite dimensional parameter is an **ill-posed** problem (Tikhonov & Arsenin 1977). Some restrictions must be imposed on $h_0(t)$ in order to facilitate estimation.

**Issue 2.2** The constraint $h_0(t) \geq 0$ can be difficult to impose.

Issue 2.1 is typically resolved by approximating $h_0(t)$ using a finite-dimensional functional space; this is discussed in Section 2.3 below. One example of such an approximation is the piecewise constant approximation to $h_0(t)$. Here, the domain for the survival times is divided into non-overlapping intervals, which are called bins, and the baseline hazard over each bin is approximated by a constant value; see Section 1.5.2 for an example. A special example of bin selection, when survival times are only subject to right censoring, is to assign each event time to a separate bin in the absence of ties. As a result, the cumulative baseline hazard $H_0(t)$ becomes a step function with jumps occurring at the event times. This approach is known as the nonparametric method for Cox model fitting in traditional survival analysis; see Huang (1996). In Section 1.5.2, we explained that, under this context, the maximum partial likelihood estimates are equivalent to the maximum profile likelihood estimates.

Care must be taken for Issue 2.2. The approach exhibited in Section 2.3 has certain computational advantages compared to a simple transformation, such as $h_0(t) = e^{a_0(t)}$, where $a_0(t)$ is unconstrained. This is because this transformation demands repeated evaluations of integrals (due to $H_0(t)$) during an optimization algorithm and thus it can be computationally intensive. Also, some transformations may create multiple local maxima for the log-likelihood $l(\boldsymbol{\beta}, h_0(t))$. These arguments will be further elaborated in Section 2.5.

In the next section, we discuss how to approximate $h_0(t)$ using a finite-dimension functional space.

## 2.3 Approximating the baseline hazard using basis functions

Since $h_0(t)$ is an infinite dimensional parameter, its estimation from a finite number of observations is an ill-posed problem. Since $h_0(t) \geq 0$, it is reasonable to adopt an approximating space, constructed from a finite-dimensional space of *non-negative* basis functions. Therefore, we approximate $h_0(t)$ by

$$h_0(t) = \sum_{u=1}^{m} \theta_u \psi_u(t), \tag{2.4}$$

where $\psi_u(t) \geq 0$ are non-negative basis functions and $m$ is the dimension of this approximating space. The basis functions are usually created with a "knot sequence" where the number of knots in this sequence is related to the sample size. Possible choices for basis functions include indicator functions, M-splines and Gaussian density functions: these examples will be discussed in more detail later. As a result, the requirement for $h_0(t) \geq 0$ can now be imposed more simply through $\boldsymbol{\theta} \geq 0$, where $\boldsymbol{\theta}$ is an $m$-vector for the $\theta_u$'s and $\boldsymbol{\theta} \geq 0$ is interpreted element-wise, that is, each element of $\boldsymbol{\theta}$ must be non-negative. This is a necessary condition for the constraint $h_0(t) \geq 0$.

From the baseline hazard approximation (2.4), the corresponding approximated cumulative baseline hazard is:

$$H_0(t) = \sum_{u=1}^{m} \theta_u \Psi_u(t), \tag{2.5}$$

where $\Psi_u(t) = \int_0^t \psi_u(w)dw$, the cumulative basis functions.

A distinctive advantage of approximating the baseline hazard $h_0(t)$ directly, rather than approximating a function of $h_0(t)$, such as $\log h_0(t)$ or $\log H_0(t)$, is that the pair $h_0(t)$ and $H_0(t)$ can be computed extremely easily.

For example, suppose we model $\alpha_0(t) = \log h_0(t)$ by a spline function so that there is no constraints on $\alpha_0(t)$ and $h_0(t)$ is automatically non-negative. Then, computations of $H_0(t) = \int_0^t e^{\alpha_0(w)}dw$ at different time points are no more simple unless a simple spline, such as a piecewise constant or linear spline, is adopted. Easy evaluation of both $h_0(t)$ and $H_0(t)$ is an important consideration for designing fast algorithms to fit the Cox model based on the likelihood function, this is because in each iteration, repeated evaluations of $h_0(t)$ and $H_0(t)$ are needed when updating the model parameters.

Substituting the approximation (2.4) into the Cox model (2.1) we then have the following approximated hazard function for individual $i$:

$$h(t|\mathbf{x}_i) = \left(\sum_{u=1}^{m} \theta_u \psi_u(t)\right) e^{\mathbf{x}_i^\top \boldsymbol{\beta}}. \tag{2.6}$$

The corresponding cumulative hazard and survival functions are computed according to

$$H(t|\mathbf{x}_i) = \left(\sum_{u=1}^{m} \theta_u \Psi_u(t)\right) e^{\mathbf{x}_i^\top \boldsymbol{\beta}}, \tag{2.7}$$

and

$$S(t|\mathbf{x}_i) = e^{-H(t|\mathbf{x}_i)}, \tag{2.8}$$

respectively.

Many researchers, including Zhang et al. (2010) for spline-based sieve maximum likelihood estimation, Cai & Betensky (2003) and Joly et al. (1998) for penalized linear spline and M-spline-based MPL estimation respectively, and Ma et al. (2014) and Ma et al. (2021) for constrained MPL estimation, have considered approximating $h_0(t)$, or $\log h_0(t)$, or $\log H_0(t)$ of the Cox model using basis functions.

The link between the likelihood-based method, where basis functions are used to approximate $h_0(t)$, and the sieve maximum likelihood method of Grenander (1981) has stimulated important theoretical developments in this approach, as seen in works such as Huang (1996), Huang & Rossini (1997) and Zhang et al. (2010). In particular, the sieve maximum likelihood theory has been adopted to derive strong consistent results for the sieve maximum likelihood estimate of $h_0(t)$ or $H_0(t)$ under certain regularity conditions. The rate of convergence for this estimate is also obtained, and it requires $m$ to grow very slowly with the sample size $n$, as otherwise, oscillations may appear in the hazard function estimation. The MPL method is able to dampen such unpleasant oscillations.

Next, we introduce several non-negative basis functions which are widely adopted in semi-parametric Cox models. They include indicator, M-spline and Gaussian basis functions as examples. Certainly, other basis functions are also possible.

### 2.3.1 Indicator basis functions

Indicator basis functions provide a piecewise constant approximation to $h_0(t)$. Recall that partly interval-censored survival data are given by $\{t_i^L, t_i^R\}$ for $i = 1, \ldots, n$. Let $a = \min\{t_i^L\}$ and $b = \max\{t_i^R : t_i^R < +\infty\}$. Using $m+1$ partition points $0 \le a = w_1 < \cdots < w_m < w_{m+1} = b < +\infty$, interval $\mathcal{T} = [a, b]$ is divided into $m$ mutually exclusive intervals, denoted as bins $(w_1, w_2], \ldots, (w_m, w_{m+1}]$. Let bin $\mathcal{B}u = (w_u, w_{u+1}]$. These bins are mutually exclusive and exhaustive, meaning that $\bigcup_u \mathcal{B}_u = \mathcal{T}$ and $\mathcal{B}_u \cap \mathcal{B}_v = \emptyset$ if $u \neq v$.

Following the same terminologies adopted for basis functions, we also called $w_1, \cdots, w_{m+1}$ as *knots* where $w_1$ and $w_{m+1}$ are the "boundary knots" and the rest are the "interior knots".

There are two common approaches to selecting bins: fixed bin width and fixed bin count; see Section 1.3.3. In the fixed bin width approach, each bin has a fixed width, while in the fixed bin count approach, the number of observations in each bin is fixed to a given quantity. The fixed bin count method is also called the "quantile method" for selecting bins.

One main issue with the fixed bin width approach is that some bins may contain zero counts, resulting in possible unstable algorithms for the parameter estimation. Our experience shows that the fixed bin count strategy usually works better.

To determine bin counts, we recommend using event, left and interval censoring times (right censoring is not used). An important question arises: how should we count the censored intervals? One common option is to count each left- or interval-censored observation as a single event. Another less common option is to consider the limits of censoring intervals, so that each interval censoring (excluding left and right censoring) contributes two counts while each left censoring contributes one count.

It is common to choose multiple counts per bin. The special case of one count per bin leads to the so-called non-parametric baseline hazard estimation method. Note that when $n$ is large, we may have a large number of $\theta_u$'s particularly when adopting one (or a small) count per bin. In this case, the multiplicative algorithm introduced in Section 2.5 can be extremely efficient.

Once the bins $\mathcal{B}_1, \ldots, \mathcal{B}_m$ are selected, indicator basis functions are given by

$$\psi_u(t) = \begin{cases} 1 & \text{if } t \in \mathcal{B}_u, \\ 0 & \text{otherwise,} \end{cases} \tag{2.9}$$

and the corresponding cumulative basis functions are

$$\Psi_u(t) = \begin{cases} 0 & \text{if } t \leq w_u, \\ t - w_u & \text{if } t \in \mathcal{B}_u, \\ w_{u+1} - w_u & \text{if } t > w_{u+1}, \end{cases} \tag{2.10}$$

for $u = 1, \ldots, m$.

When $t \in \mathcal{B}_u$, the baseline hazard and cumulative baseline hazard from these basis functions are, respectively,

$$h_0(t) = \theta_u$$

and

$$H_0(t) = \sum_{v=1}^{u-1} \theta_v (w_{v+1} - w_v) + \theta_u (t - w_u).$$

People often replace $t - w_u$ in the second term of the above $H_0(t)$ with $w_{u+1} - w_u$, so that when $t \in \mathcal{B}_u$

$$H_0(t) \approx \sum_{v=1}^{u} \theta_v (w_{v+1} - w_v).$$

**Example 2.2** (Piecewise constant approximation to a hazard).
In this example, we explain how to approximate the hazard $h_0(t) = 3t^2$ and its cumulative hazard $H_0(t) = t^3$ by a piecewise constant approximation, using a random sample generated from this distribution. The following steps are involved in this example.

(i) Generate a random sample of $n = 200$ from this distribution, and the formula for generating these random numbers is simple

$$Y_i = (-\log U_i)^{1/3},$$

where $U_i \sim unif(0, 1)$;
(ii) Select bins using equal bin counts, where $m = 10, 30, 100$.
(iii) Approximate $h_0(t)$ in each bin using the average $h_0(t)$ values for the sample points in this bin.
(iv) Plot the true $h_0(t)$ and the approximated $h_0(t)$ for different values of $m$.
(v) Also, plot the true $H_0(t)$ and the approximated $H_0(t)$.

```
# Generate a random sample of n=200
# from this distribution
neg.log.u <- -log(runif(200))
y <- (neg.log.u)^(1/3)
# Select bins using equal bin counts
# where m=10, 30, 100
bins_f <- function(m, y) {
    bins <- quantile(y, probs = seq(0, 1,
        length.out = (m + 1)))

    theta <- cum_theta <- rep(0, m)

    for (u in 1:m) {
        bin_u_obs <- y[which(bins[u] <= y &
            y < bins[u + 1])]
        # mean hazards in the bin
        theta[u] <- mean(3 * bin_u_obs^2)
        cum_theta[u] = mean(bin_u_obs^3)
    }

    out <- list(theta = theta, cum_theta = cum_theta,
        bins = bins)
    return(out)
}

h0t_m10 <- bins_f(10, y)
h0t_m30 <- bins_f(30, y)
h0t_m100 <- bins_f(100, y)

# Plot the true and approximated
# baseline hazard function for the
```

```
# different values of m
v <- seq(from = min(y), to = max(y), by = 0.01)
plot(3 * v^2 ~ v, type = "l", col = "black",
     main = "hazard and approxmations")
lines(c(h0t_m10$theta, h0t_m10$theta[10]) ~
      h0t_m10$bins, type = "s", lty = 2)
lines(c(h0t_m30$theta, h0t_m30$theta[30]) ~
      h0t_m30$bins, type = "s", lty = 3)
lines(c(h0t_m100$theta, h0t_m100$theta[100]) ~
      h0t_m100$bins, type = "s", lty = 4)
legend(0.3, 10, legend = c("true", "m=10",
      "m=30", "m=100"), lty = 1:4, cex = 0.9)

# Plot the true and approximated
# cumulative baseline hazard functions

plot(v^3 ~ v, type = "l", col = "black",
     main = "cumulative hazard and approxmations")
lines(c(h0t_m10$cum_theta, h0t_m10$cum_theta[10]) ~
      h0t_m10$bins, type = "s", lty = 2)
lines(c(h0t_m30$cum_theta, h0t_m30$cum_theta[30]) ~
      h0t_m30$bins, type = "s", lty = 3)
lines(c(h0t_m100$cum_theta, h0t_m100$cum_theta[100]) ~
      h0t_m100$bins, type = "s", lty = 4)
legend(0.3, 6, legend = c("true", "m=10",
      "m=30", "m=100"), lty = 1:4, cex = 0.9)
```

□

### 2.3.2 M-spline basis functions

M-splines provide another popular set of basis functions (Ramsay 1988). The M-splines can be viewed as scaled B-splines such that each M-spline integrates to 1. Thus, the M-splines of order $o(\geq 1)$ associated with a given knots sequence have the property that:

(i) Between any two consecutive knots, it is a polynomial of order $o$.

(ii) These polynomials are connected at the knots.

(iii) They have up to $(o-1)$-th order of continuous derivatives, and these derivatives at any interior knot have equal values for two neighbouring polynomials of this knot.

Some discussions on M-splines are available in Ramsay (1988). We summarize M-splines below. Let $a = w_1 < \cdots < w_{n_\alpha} < w_{n_\alpha+1} = b$ be a knot sequence where $a$ and $b$ have been defined previously and they are called the boundary knots. The boundary knots are repeated $o$ times at each end. Let $\mathbf{w}^\star$ be the vector for all the

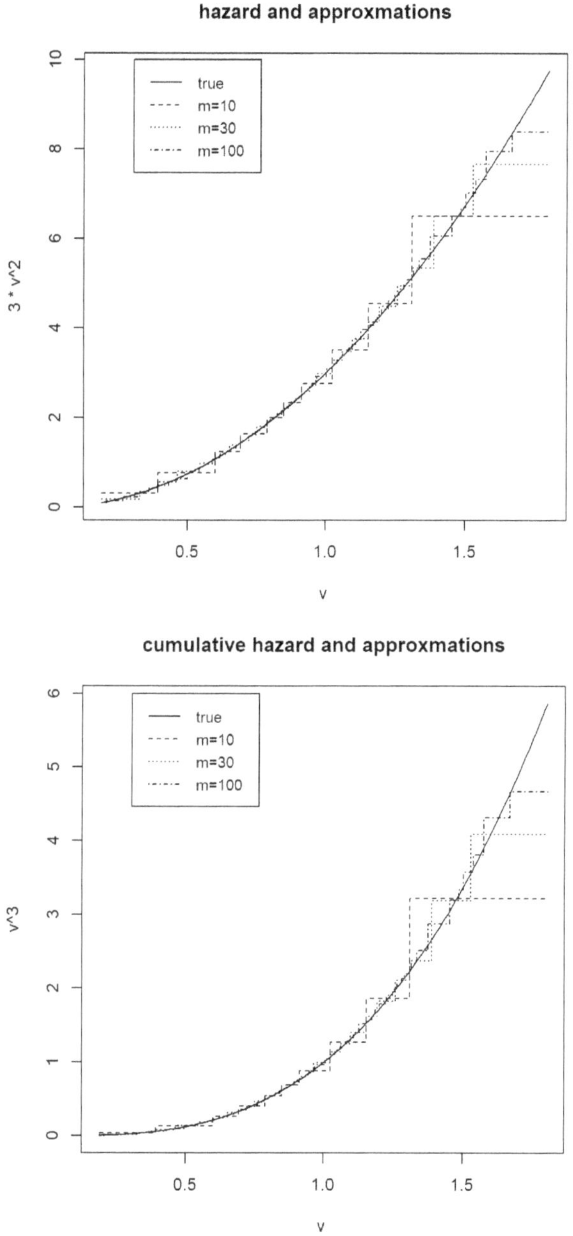


FIGURE 2.1: Plots of hazard and cumulative hazard, and their piecewise constant approximations using 10, 30 and 100 bins.

knots including boundary repeated knots. Vector $\mathbf{w}^\star$ has a length of $n_\alpha - 1 + 2o$. Let $I(A)$ denote the indicator function for condition $A$. Then, order $o$ M-splines $\psi_u^o(t)$, for $u = 1, \ldots, m$ with $m = n_\alpha - 1 + o$, are defined according to the following recursive formula:

$$\psi_u^o(t) = \begin{cases} \frac{I(w_u^\star \le t < w_{u+1}^\star)}{w_{u+1}^\star - w_u^\star} & \text{if } o = 1, \\ \frac{o}{o-1} \frac{I(w_u^\star \le t < w_{u+o}^\star)}{w_{u+o}^\star - w_u^\star} \left[ \left(t - w_u^\star\right) \psi_u^{o-1}(t) \right. \\ \left. \quad + \left(w_{u+o}^\star - t\right) \psi_{u+1}^{o-1}(t) \right] & \text{otherwise.} \end{cases} \tag{2.11}$$

The corresponding cumulative basis functions (also known as I-splines) are:

$$\Psi_u^o(t) = I(w_u^\star > t) \left[ \sum_{v=u+1}^{\min(u+o,m+1)} \frac{w_{v+o+1}^{\star\star} - w_v^{\star\star}}{o+1} \psi_v^{o+1}(t) \right]^{I(w_u^\star < t < w_{u+o}^\star)}, \tag{2.12}$$

where vector $\boldsymbol{w}^{\star\star} = (w_1, \boldsymbol{w}^{\star\mathsf{T}}, w_{n_\alpha+1})^\mathsf{T}$, an expansion of $\boldsymbol{w}^{\star\mathsf{T}}$ by the two boundary knots again. Note that $\Psi_u^o(t)$ is referred to as an I-spline in the literature. M-spline basis functions have the following properties: $\psi_u^o(t) = 0$ for $t \le w_u^\star$ or $t > w_{u+o}^\star$, and $\int_{w_u^\star}^{w_{u+o}^\star} \psi_u^o(v)dv = 1$.

The R package `spline2` can be used easily to compute M-spline and I-spline values. Here are some examples.

**Example 2.3** (M-spline calculation using R package "`spline2`"**).**
In this example we explain how to compute M-spline and I-spline values using "`spline2`".

(i) Generate a dataset $\{t_1, \ldots, t_n\}$ in the same as in Example 2.2, with $n = 200$.
(ii) We select 4 interior knots and cubic M-splines, then the number of splines $m = 4 - 1 + 3 = 6$. We may estimate the basis function coefficients $\theta_u$'s by minimizing

$$\sum_{i=1}^{n} \left( h_0(t_i) - \sum_{u=1}^{m} \theta_u \psi_u^o(t_i) \right)^2.$$

This can be solved using the least-squares approach but in the program below we use an `R` optimization function.
(iii) Plot the true $h_0(t)$ and M-spline approximated $h_0(t)$, and also the true and the approximated $H_0(t)$.

```
# Generate a data set of size n=200 as in the
# previous example
neg.log.u <- -log(runif(200))
y <- (neg.log.u)^(1/3)

# Select 4 interior knots, and estimate the
# theta's
library(splines2)
int.kn <- quantile(y, seq(0.1, 0.9, length.out = 4))
```

```
bound.kn <- c(
    0, max(y) +
       1e-04
)
psi <- mSpline(
    y, knots = int.kn, Boundary.knots = bound.kn,
    degree = 3
)
target_func <- function(theta, psi, t) {
    diff = sum((3 * t^2 - psi %*% (theta))^2)
    return(diff)
}

opts <- list(
    algorithm = "NLOPT_LN_COBYLA", xtol_rel = 1e-04,
    maxeval = 100
)

est_theta = nloptr(
    x0 = matrix(1, nrow = 7),
    eval_f = target_func, lb = (rep(0, 7)),
    ub = rep(Inf, 7),
    opts = opts, psi = psi, t = y
)$solution
est_theta

# Plot the true and estimated baseline hazard
# and cumulative baseline hazard functions
v <- seq(from = 0, to = 1.5, by = 0.01)
psi_v = mSpline(
    v, knots = int.kn, Boundary.knots = bound.kn,
    degree = 3
)
plot(3 * v^2 ~ v, type = "l")
est_h0t = psi_v %*% est_theta
lines(est_h0t ~ v, type = "l", col = "red")

Psi_v = mSpline(
    v, knots = int.kn, Boundary.knots = bound.kn,
    degree = 3, integral = TRUE
)
plot(v^3 ~ v, type = "l")
est_H0t = Psi_v %*% est_theta
lines(est_H0t ~ v, type = "l", col = "red")
```

□

**Example 2.4** (Comparing computational burden)**.**
In this example, we will compare the computational burden of evaluating the hazard

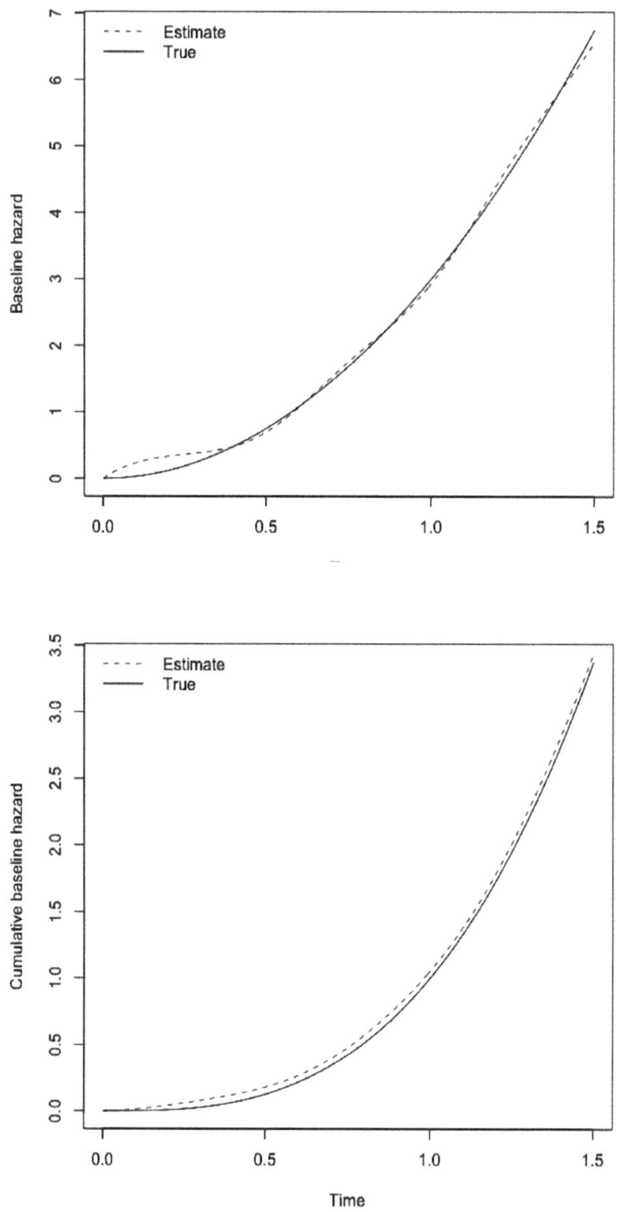


FIGURE 2.2: Plots of hazard and cumulative hazard, and their M-spline approximations.

and cumulative hazard functions using data from the previous example. We compare evaluations of $h_0(t)$ and $H_0(t)$ using M-spline and I-spline, respectively, against the transformation method where $h_0(t) = e^{\alpha_0(t)}$. Here, $\alpha_0(t)$ is approximated by B-splines, and then $H_0(t)$ is computed by integration.

(i) Compute the basis function matrix $\mathbf{A} = \left(\psi_u(t_i)\right)_{n\times m}$ and the cumulative basis function matrix $\mathbf{B} = \left(\Psi_u(t_i)\right)_{n\times m}$, where $\psi_u(t_i)$ and $\Psi_u(t_i)$ are M-spline and I-spline values obtained from the R package `splines2`.
(ii) The spline coefficients are then obtained by the least-squares method similar to the previous example.
(iii) Then, values of $h_0(t)$ and $H_0(t)$ at different $t_i$'s are simply given by $\mathbf{h}_0 = (h_0(t_1), \ldots, h_0(t_n))^\top = \mathbf{A}\boldsymbol{\theta}$ and $\mathbf{H}_0 = (H_0(t_1), \ldots, H_0(t_n))^\top = \mathbf{B}\boldsymbol{\theta}$, respectively. Record the total time used.
(iv) Now consider the transformation $h_0(t) = e^{\alpha_0(t)}$, where $\alpha_0(t) = \sum_{u=1}^{m} \xi_u \psi_u(t)$, where $\psi_u(t)$ are B-splines. We first estimate the coefficients $\xi_u$ in the same way as above.
(v) Then, calculate $\alpha_0(t_1), \ldots, \alpha_0(t_n)$ and then $h_0(t_1), \ldots, h_0(t_n)$. But for each $H_0(t_i)$, we need to use $H_0(t_i) = \int_0^{t_i} e^{\alpha_0(w)} dw$. Also, record the total time used. (vi) Compare the total time used for these two approaches and comment.

R codes for this example are given below.

```
# Generate data
neg.log.u <- -log(runif(200))
y <- (neg.log.u)^(1/3)

int.kn <- quantile(y, seq(0.1, 0.9, length.out = 4))
bound.kn <- c(
    0, max(y) +
        1e-04
)
psi <- mSpline(
    y, knots = int.kn, Boundary.knots = bound.kn,
    degree = 3
)

# Estimate theta to use for matrix
# multiplication

target_func <- function(theta, psi, t) {
    diff = sum((3 * t^2 - psi %*% (theta))^2)
    return(diff)
}

opts <- list(
    algorithm = "NLOPT_LN_COBYLA", xtol_rel = 1e-04,
    maxeval = 100
)
```

```
est_theta = nloptr(
    x0 = matrix(1, nrow = 7),
    eval_f = target_func, lb = (rep(0, 7)),
    ub = rep(Inf, 7),
    opts = opts, psi = psi, t = y
)$solution


# Create a grid of times to estimate the
# cumulative baseline hazard for
v <- seq(from = 0, to = 1.5, by = 0.01)
Psi_v = mSpline(
    v, knots = int.kn, Boundary.knots = bound.kn,
    degree = 3, integral = TRUE
)

# Record the time taken to estimate the
# cumulative baseline hazard
start.time <- Sys.time()
est_H0t = Psi_v %*% est_theta
end.time <- Sys.time()
time.taken <- round(end.time - start.time, 5)

# Estimate theta to use in exponential
# transformation

target_func_exp <- function(theta, psi, t) {
    alpha = psi %*% (theta)
    diff = sum((3 * t^2 - exp(alpha))^2)
    return(diff)
}

opts <- list(
    algorithm = "NLOPT_LN_COBYLA", xtol_rel = 1e-04,
    maxeval = 100
)

est_theta_alpha = nloptr(
    x0 = matrix(1, nrow = 7),
    eval_f = target_func, lb = (rep(-Inf, 7)),
    ub = rep(Inf, 7),
    opts = opts, psi = psi, t = y
)$solution


# Create functions which allow us to integrate
# the baseline hazard to get the cumulative
# baseline hazard
```

```
h0_t <- function(t, theta, int.kn, bound.kn) {
    psi <- bSpline(
        t, knots = int.kn, Boundary.knots = bound.kn,
        degree = 3
    )
    h0 <- exp(psi %*% est_theta_alpha)
}

H0_t <- function(upper_t, f, i, b) {
    integrate(f, 0, upper_t, int.kn = i, bound.kn = b)$value
}

# Record the time taken to estimate the
# cumulative baseline hazard
start.time <- Sys.time()
est_theta_H <- sapply(
    seq(from = 0, to = 1.5, by = 0.01),
    H0_t, f = h0_t, i = int.kn, b = bound.kn
)
est_H0t_exp = Psi_v %*% est_theta
end.time <- Sys.time()
time.taken <- round(end.time - start.time, 5)
```

TABLE 2.2: The time taken for each of the two methods compared in Example 2.4. Evidently, the direct evaluation is much faster (more than 50 times) than the transformation method.

| Method | Time taken (seconds) |
|---|---|
| Direct evaluation (part i of example) | 0.0005 |
| Transformation method (part ii of example) | 0.0322 |

Clearly, the computation time required by the transformation in this example is about sixty-four times that of the direct method. It is important to note that this only reflects their computing time difference for about one iteration. However, when a large number of iterations are needed to achieve convergence, this difference can accumulate quickly. Furthermore, other issues may arise from the transformation method, such as the presence of multiple local maxima. □

### 2.3.3 Gaussian basis functions

Another example of basis functions is given by truncated Gaussian distributions (i.e. truncated normal distributions) and we call them Gaussian basis function. For example, in Komárek et al. (2005), the Gaussian basis functions are used to approximate the baseline density function in the additive hazard model, and in Li & Ma (2020), the baseline hazard of the additive hazard model is approximated using Gaussian basis functions.

To be consistent with the terminologies we adopted for basis functions, we also define Gaussian basis functions using knots, where these knots are now basically the mean values of the corresponding Gaussian densities. Let $w_1, \ldots, w_m$ be knots that satisfy $a \leq w_1 < \ldots < w_m \leq b$. The first and last knots, i.e., $w_1$ and $w_m$, can be different from $a$ and $b$.

A Gaussian basis function $\psi_u(t)$ is a truncated Gaussian distribution density with location parameter $w_u$, scale parameter $\sigma_u$ ($> 0$) and range $[c, d]$. This leads to the following expressions of $\psi_u(t)$

$$\psi_u(t) = \frac{1}{\sigma_u \nabla_u} f_N\left(\frac{t - w_u}{\sigma_u}\right),$$

where $t \in [c, d]$, $f_N(\cdot)$ denotes the density of the standard Gaussian distribution and $\nabla_u = F_N((d - w_u)/\sigma_u) - F_N((c - w_u)/\sigma_u)$ with $F_N(\cdot)$ being the cumulative distribution function of the standard Gaussian distribution. This basis function clearly satisfies $\psi_u(t) \geq 0$ and $\int_c^d \psi_u(t)dt = 1$ for all $u$. The cumulative basis function $\Psi_u(t)$ is:

$$\Psi_u(t) = \int_c^t \psi_u(v)dv = \frac{1}{\nabla_u}\left[F_N\left(\frac{t - w_u}{\sigma_u}\right) - F_N\left(\frac{c - w_u}{\sigma_u}\right)\right].$$

In the next example we discuss how to use R to compute the Gaussian basis functions and cumulative basis functions.

**Example 2.5** (Gaussian basis and cumulative basis function).
In this example, we again adopt the simulated data from Example 2.3 to compute Gaussian basis and cumulative basis function values.

```
# Generate data as per
# the previous examples
neg.log.u <- -log(runif(200))
y <- (neg.log.u)^(1/3)

# Write functions to
# compute Gaussian basis
# functions
sigma <- 0.5
kn <- quantile(y, seq(0.01,
    0.99, length.out = 5))
phi.gauss <- NULL

phi_gauss_f <- function(y,
    kn, a, b) {
    phi.gauss <- NULL
    for (u in 1:length(kn)) {
        delta_u = pnorm((b -
```

```
            kn[u])/sigma) -
            pnorm((a - kn[u])/sigma)
        phi.gauss = cbind(phi.gauss,
            (1/(sigma * delta_u)) *
                dnorm((y -
                  kn[u])/sigma))
    }
    return(phi.gauss)
}

Phi_gauss_f <- function(y,
    kn, a, b) {
    Phi.gauss <- NULL
    for (u in 1:length(kn)) {
        delta_u = pnorm((b -
            kn[u])/sigma) -
            pnorm((a - kn[u])/sigma)
        Phi.gauss = cbind(Phi.gauss,
            (1/delta_u) *
                (pnorm((y -
                  kn[u])/sigma) -
                  pnorm((a -
                    kn[u])/sigma)))
    }
    return(Phi.gauss)
}

phi.gauss <- phi_gauss_f(y,
    kn, a = 0, b = max(y))
Phi.gauss <- Phi_gauss_f(y,
    kn, a = 0, b = max(y))

# Estimate theta

target_func_gauss <- function(theta,
    phi.gauss, t) {
    diff = sum((3 * t^2 -
        phi.gauss %*% (theta))^2)
    return(diff)
}

library(nloptr)
opts <- list(algorithm = "NLOPT_LN_COBYLA",
    xtol_rel = 1e-04, maxeval = 1000)

est_theta = nloptr(x0 = matrix(1,
    nrow = 5), eval_f = target_func_gauss,
    lb = (rep(0, 5)), ub = rep(Inf,
```

```
        5), opts = opts, phi.gauss = phi.gauss,
    t = y)$solution

# Plot baseline hazard
# function
v <- seq(from = 0, to = 1.5,
    by = 0.01)
plot(3 * v^2 ~ v, type = "l",
    main = "Hazard and Gaussian basis approx.")
psi_v <- phi_gauss_f(v, kn,
    a = 0, b = max(y))
est_h0t = psi_v %*% est_theta
lines(est_h0t ~ v, type = "l",
    lty = 2)
legend(0, 6.8, legend = c("true",
    "approx"), lty = 1:2,
    cex = 0.9)

# Plot cumulative
# baseline hazard
# function
Psi_v = Phi_gauss_f(v, kn,
    a = 0, b = max(y))
plot(v^3 ~ v, type = "l",
    main = "Cumulative hazard and Gaussian basis approx.")
est_H0t = Psi_v %*% est_theta
lines(est_H0t ~ v, type = "l",
    lty = 2)
legend(0, 3.4, legend = c("true",
    "approx"), lty = 1:2,
    cex = 0.9)
```

□

## 2.4 Constrained maximum penalized likelihood estimation

Since $h_0(t)$ is approximated using basis functions (as shown in equation (2.4)), we wish to estimate $\boldsymbol{\beta}$ and $\boldsymbol{\theta}$, subject to the constraint $\boldsymbol{\theta} \geq 0$ (where the inequality is interpreted element-wisely). Once we have an estimate of $\boldsymbol{\theta}$, we can obtain an estimate of $h_0(t)$. In this chapter (and also in other chapters), we use the constrained maximum penalized likelihood (MPL) method to estimate $\boldsymbol{\beta}$ and $\boldsymbol{\theta}$. We have briefly explained the motivation for incorporating a penalty function.

By substituting approximation (2.2) into (2.3), we obtain the corresponding approximation to the log-likelihood function, denoted by $l(\boldsymbol{\beta}, \boldsymbol{\theta})$ henceforth. We focus

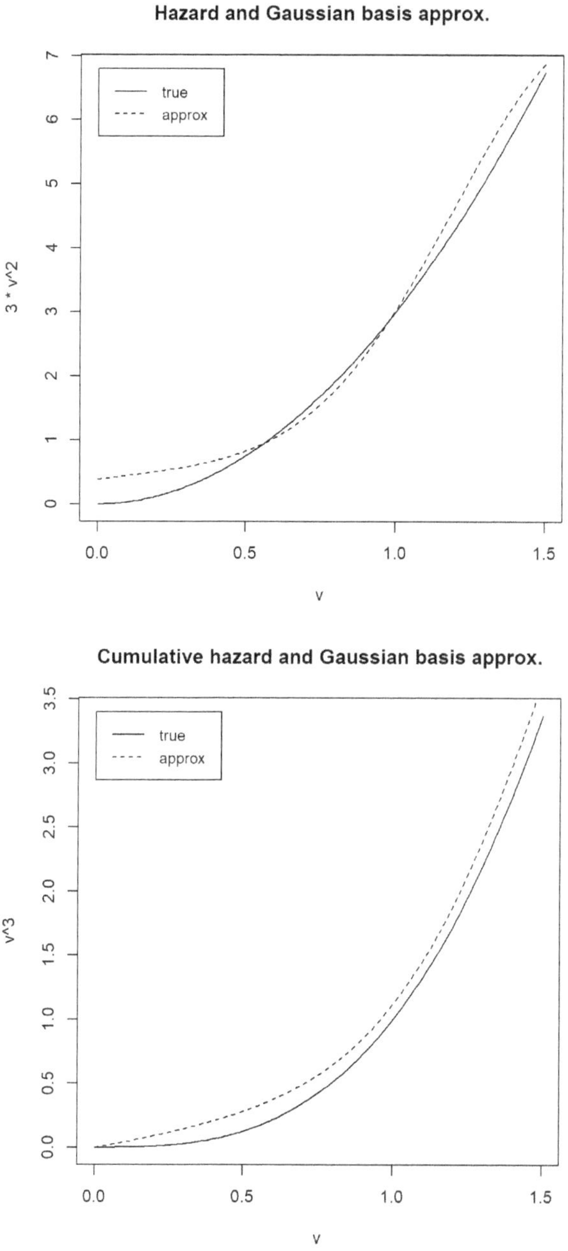


FIGURE 2.3: Plots of hazard and cumulative hazard, and their Gaussian basis approximations.

on the MPL estimate of $(\boldsymbol{\beta},\boldsymbol{\theta})$, given by

$$(\widehat{\boldsymbol{\beta}},\widehat{\boldsymbol{\theta}}) = \underset{\boldsymbol{\beta},\boldsymbol{\theta}}{\operatorname{argmax}}\{\Phi(\boldsymbol{\beta},\boldsymbol{\theta}) = l(\boldsymbol{\beta},\boldsymbol{\theta}) - \lambda J(\boldsymbol{\theta})\}, \tag{2.13}$$

subject to the constraint $\boldsymbol{\theta} \geq 0$. Here, $\lambda \geq 0$ is the smoothing (or regularization) parameter, and $J(\boldsymbol{\theta})$ is a penalty function. Note that the penalty $J(\boldsymbol{\theta})$ is a function of $\boldsymbol{\theta}$ alone, meaning only the $\boldsymbol{\theta}$ estimate is regularized. However, inclusion of $J(\boldsymbol{\theta})$ will also affect the estimation of $\boldsymbol{\beta}$ as the parameters $\boldsymbol{\beta}$ and $\boldsymbol{\theta}$ are generally non-orthogonal.

In general, inclusion of $J(\boldsymbol{\theta})$ will introduce biases to the estimate of $\boldsymbol{\beta}$, but these biases are negligible when $\lambda$ is small.

Although a penalty function can introduce bias, it also provides benefits that can exceed the costs when $\lambda$ is carefully selected. We will elaborate on the benefits of the penalty function below.

Adopting a penalty function on $\boldsymbol{\theta}$ when estimating $\boldsymbol{\beta}$ and $\boldsymbol{\theta}$ can have the following benefits. Some of them are key reasons for why the MPL method is capable of offering stable estimates for the semi-parametric Cox model when compared with the maximum likelihood method.

(1) A penalty function on $\boldsymbol{\theta}$ can be used to smooth (or regularize) the $h_0(t)$ estimate. A commonly used penalty to impose smoothness is the roughness penalty (e.g., Green & Silverman (1994)) which is

$$J(\boldsymbol{\theta}) = \int [h_0^{(\nu)}(t)]^2 dt, \tag{2.14}$$

where, for a positive integer $\nu$, $h_0^{(\nu)}(t)$ represents the $\nu$-th derivative of $h_0(t)$. This function gives a large value when $h_0(t)$ is a rough function and thus including this function in the penalized log-likelihood penalizes the roughness of the $h_0(t)$ estimate, and thus produces smoothed estimates. The penalty function (2.14) can be re-expressed in a quadratic form of $\boldsymbol{\theta}$. Actually, if we let $\Gamma_\nu(t)^\top = (\psi_1^{(\nu)}(t), \cdots, \psi_m^{(\nu)}(t))$ then

$$h_0^{(\nu)}(t) = \Gamma_\nu(t)^\top \boldsymbol{\theta}. \tag{2.15}$$

Plugging-in this expression into (2.14) we have

$$J(\boldsymbol{\theta}) = \boldsymbol{\theta}^\top \left( \int_a^b \Gamma_\nu(t)\Gamma_\nu(t)^\top dt \right) \boldsymbol{\theta} = \boldsymbol{\theta}^\top \mathbf{R} \boldsymbol{\theta}, \tag{2.16}$$

where matrix $\mathbf{R}$ has the dimension of $m \times m$ with the $(u,v)$-th element $r_{uv} = \int_a^b \psi_u^{(\nu)}(t)\psi_v^{(\nu)}(t)dt$. This roughness penalty demands that the approximated $h_0(t)$ is a continuous function and has up to $\nu\ (\geq 1)$ continuous derivatives. The continuity condition is not satisfied when the approximated $h_0(t)$ is a piecewise constant function (when indicator basis functions are assumed). In this case, since $\theta_u$ represents directly the value of $h_0(t)$ over bin $\mathcal{B}_u$, the smoothness constraint on $h_0(t)$ can be achieved by requiring the neighboring $\theta_u$ values are similar. For example, one possible approach is to use the first-order differences to impose smoothness on the $\theta_u$'s:

$$J(\boldsymbol{\theta}) = \sum_{u=2}^{m}(\theta_u - \theta_{u-1})^2 = \boldsymbol{\theta}^\mathsf{T}\mathbf{R}\boldsymbol{\theta}, \tag{2.17}$$

where $\mathbf{R} = \mathbf{C}^\mathsf{T}\mathbf{C}$ with matrix $\mathbf{C}$ being

$$\mathbf{C} = \begin{pmatrix} -1 & 1 & 0 & \cdots & 0 & 0 \\ 0 & -1 & 1 & \cdots & 0 & 0 \\ \vdots & \vdots & \vdots & \cdots & \vdots & \vdots \\ 0 & 0 & 0 & \cdots & -1 & 1 \end{pmatrix}_{(m-1)\times m} . \tag{2.18}$$

Another option is to use the second-order differences, which are also commonly used when $h_0(t)$ is piecewise constant. Thus,

$$J(\boldsymbol{\theta}) = \sum_{u=2}^{m-1}(\theta_{u+1} - 2\theta_u + \theta_{u-1})^2 = \boldsymbol{\theta}^\mathsf{T}\mathbf{R}\boldsymbol{\theta}, \tag{2.19}$$

where $\mathbf{R} = \mathbf{C}^\mathsf{T}\mathbf{C}$ but now with

$$\mathbf{C} = \begin{pmatrix} 1 & -2 & 1 & 0 & \cdots & 0 & 0 & 0 \\ 0 & 1 & -2 & 1 & \cdots & 0 & 0 & 0 \\ \vdots & \vdots & \vdots & \vdots & \cdots & \vdots & \vdots & \vdots \\ 0 & 0 & 0 & 0 & \cdots & 1 & -2 & 1 \end{pmatrix}_{(m-2)\times m} . \tag{2.20}$$

Since the above penalties all appear as quadratic forms in $\boldsymbol{\theta}$, they not only smooth $h_0(t)$ but also force $\boldsymbol{\theta}$ to zero. Thus, the $\theta_u$'s become diminished when $\lambda$ is getting larger.

If we wish to shrink all $\theta_u$ values towards 0 without smoothing $h_0(t)$, then the following penalty function can be adopted:

$$J(\boldsymbol{\theta}) = \boldsymbol{\theta}^\mathsf{T}\boldsymbol{\theta}. \tag{2.21}$$

If we wish to shrink only small (or non-important) $\theta_u$ values to zero, then the LASSO penalty can be used, given by (note that $\theta_u \geq 0$):

$$J(\boldsymbol{\theta}) = \sum_u \theta_u. \tag{2.22}$$

A penalty function that is capable of handling sudden changes in the baseline hazard function is the sparsity penalty, which is similar to the Total Variational (TV) penalty commonly used in image processing (Rudin et al. 1992). This penalty is built based on the fact that if $h_0(t)$ is smooth but has several sudden changes then the derivative of $h_0(t)$, denoted as $h_0'(t)$, is a sparse function. Therefore, we can use a $l_1$-norm penalty on $h_0'(t)$:

$$J(\boldsymbol{\theta}) = \int_a^b |h_0'(t)|\, dt. \tag{2.23}$$

(2) The second important reason for applying a penalty function on $\boldsymbol{\theta}$ is to reduce the influence of the number and position of knots on the estimation of $\boldsymbol{\beta}$ and $\boldsymbol{\theta}$. The penalty function can help to enhance the numerical stability of the estimation procedure. Our personal experience has shown that, when fitting the semi-parametric Cox model, the maximum likelihood estimation (without a penalty) can be highly sensitive to the number of knots used, and may easily fail if the number of knots exceeds the allowed quantity, which is often unknown. A penalty can stabilize numerical computations by forcing small $\theta_u$ values to zero, effectively eliminating unimportant knots from the computations. This serves as an automatic procedure for selecting the effective number and locations of knots.

(3) The maximum penalized likelihood method can also be viewed from a Bayesian perspective, specifically as an equivalent to the maximum a posteriori (MAP) method in the Bayesian approach. This connection becomes evident when the penalty $J(\boldsymbol{\theta})$ is considered as the logarithm of the prior density function $p(\boldsymbol{\theta})$:

$$J(\boldsymbol{\theta}) = \log(p(\boldsymbol{\theta})). \tag{2.24}$$

In the case of smoothness of $h_0(t)$, the roughness penalty given by (2.16) indicates that the prior distribution is a multivariate normal $N_m(\mathbf{0}_{m\times 1}, \sigma^2\mathbf{R}^{-1})$, where the variance component $\sigma^2$ is related with the smoothing parameter $\lambda$ through $\sigma^2 = 1/(2\lambda)$.

## 2.5 Computation of estimates

The model fitting task we have established so far can be formulated as a constrained optimization problem, where the penalized log-likelihood $\Phi(\boldsymbol{\beta}, \boldsymbol{\theta})$ defined in (2.13) is maximized subject to the constraint $\boldsymbol{\theta} \geq 0$. Here, the constraint notation should be interpreted element-wisely.

In the literature on Cox model fitting based on full likelihood, it is common to enforce the constraint $\boldsymbol{\theta} \geq 0$ through various transformations. However, these approaches can be *ad hoc* in nature. Existing transformation methods include, for example, those used by Joly et al. (1998) and Cai & Betensky (2003), and we call them the "square $\theta_u$" and "exponentiate $h_0(t)$" transformations respectively. We will now explain why a transformation approach to enforce $\boldsymbol{\theta} \geq 0$ or $h_0(t) \geq 0$ may not be ideal.

### 2.5.1 Transformations may not work

To simplify discussions, we assume a scalar $\theta$, and the function we wish to optimize is $\Phi(\theta)$. Let $\theta = g(\gamma)$ denote such a transformation where $g(\gamma) \geq 0$ but $\gamma$ is unconstrained. For example, $g(\gamma) = \gamma^2$ or $g(\gamma) = \exp\{\gamma\}$ are commonly used. Then, the problem of estimating $\theta$ with the constraint $\theta \geq 0$ becomes estimating $\gamma$ where $\gamma$ is unconstrained, thus transforming it into an unconstrained optimization problem. However, the solution obtained through $\gamma$ may not be the required optimal solution for $\theta$. Reasons for this are explained below.

Substituting this $\theta = g(\gamma)$ into the objective function gives $\Phi(g(\gamma))$, which becomes a function of $\gamma$ now and we denote this function by $\widetilde{\Phi}(\gamma) = \Phi(g(\gamma))$. The first two derivatives of $\widetilde{\Phi}$ are:

$$\frac{\partial \widetilde{\Phi}(\gamma)}{\partial \gamma} = \frac{\partial \Phi(\theta)}{\partial \theta} \frac{\partial g(\gamma)}{\partial \gamma}, \tag{2.25}$$

and

$$\frac{\partial^2 \widetilde{\Phi}(\gamma)}{\partial \gamma^2} = \frac{\partial^2 \Phi(\theta)}{\partial \theta^2} \left( \frac{\partial g(\gamma)}{\partial \gamma} \right)^2 + \frac{\partial \Phi(\theta)}{\partial \theta} \frac{\partial^2 g(\gamma)}{\partial \gamma^2}. \tag{2.26}$$

From these derivatives we can observe two potential problems with the transformation approach.

(1) If an optimal value of $\Phi(\theta)$ is located on the boundary of the domain $\mathcal{A} = \{\theta : \theta \geq 0\}$, then the transformation approach may not be able to provide a correct solution. Equation (2.25) indicates that $\partial \widetilde{\Phi}(\gamma)/\partial \gamma = 0$ when $\partial \Phi(\theta)/\partial \theta = 0$ or $\partial g(\gamma)/\partial \gamma = 0$. The condition $\partial \Phi(\theta)/\partial \theta = 0$ holds true only when the optimal $\theta$ is in the interior of $\mathcal{A}$. When the optimal $\theta$ is located on the boundary of $\mathcal{A}$ (thus an active constraint), it must satisfy $\partial \Phi(\theta)/\partial \theta < 0$ according to the Karush-Kuhn-Tucker (KKT) conditions. Therefore, the solution of $\partial \widetilde{\Phi}(\gamma)/\partial \gamma = 0$ must be obtained through $\partial g(\gamma)/\partial \gamma = 0$, and we denote this solution by $\widehat{\gamma}$. However, the corresponding $\widehat{\theta} = g(\widehat{\gamma})$ can not be guaranteed that $\partial \Phi(\widehat{\theta})/\partial \theta < 0$.

Another issue with using a transformation is the potential for creating multiple local maxima, as explained below.

(2) Assume $\Phi(\theta)$ is a concave function in $\theta$. We will explain that $\widetilde{\Phi}(\gamma)$ may not be concave in $\gamma$.

Since $\Phi(\theta)$ is concave in $\theta$, $\partial^2 \Phi(\theta)/\partial \theta^2 < 0$ for all $\theta$. Therefore, the first term of (2.26) must be non-positive, but its second term can be either positive or negative. This means that after a transformation, the Hessian $\partial^2 \widetilde{\Phi}(\gamma)/\partial \gamma^2$ may not be negative. As a result, $\widetilde{\Phi}(\gamma)$ may not be concave anymore, and consequently the objective function $\widetilde{\Phi}(\gamma)$ may possess multiple local maxima, leading to computational instability in the sense that the final estimate may depend on the initial value of the iterations.

### 2.5.2 Newton-MI-constrained optimization algorithm

The analysis above indicates that transformations may not be ideal for imposing the constraint $\boldsymbol{\theta} \geq 0$. A proper approach is to use a constrained optimization algorithm. In this section, We will explain a special constrained optimization algorithm to solve the optimization problem (2.13).

Certainly, other constrained optimization algorithms can also be used to solve this problem, including methods listed in Chapter 6 of Luenberger (1984), or other methods such as the projected Newton algorithm (Bertsekas 1982), or some of the convex optimization methods in Boyd & Vandenberghe (2004). However, one issue with some of these methods is their potential inefficiency when $m$ (the number of $\theta$ parameters) is large. In this section, we explain an efficient yet simple-to-compute algorithm, similar to the one developed in Ma et al. (2014) or Ma et al. (2021), that can estimate $\boldsymbol{\beta}$ and $\boldsymbol{\theta}$ under the constraint $\boldsymbol{\theta} \geq 0$, and can easily handle situations where $m$ is large.

This is basically an alternating iterative method. Each iteration involves two steps: the first step updates $\boldsymbol{\beta}$ using the Newton algorithm, and the second step computes $\boldsymbol{\theta}$ using the multiplicative iterative (MI) algorithm (e.g. Chan & Ma (2012)), where $\boldsymbol{\beta}$ is fixed at the value obtained from the first step. The aforementioned second step ensures that the estimates for $\boldsymbol{\theta}$ are non-negative. Since this algorithm involves both the Newton and the MI steps in each iteration, it is called the Newton-MI algorithm (Ma et al. 2014).

The Karush-Kuhn-Tucker (KKT) conditions are necessary conditions for a constrained optimization problem. For problem (2.13), the KKT conditions are

$$\frac{\partial \Phi}{\partial \beta_j} = 0, \tag{2.27}$$

$$\frac{\partial \Phi}{\partial \theta_u} = 0 \text{ if } \theta_u > 0, \text{ and } \frac{\partial \Phi}{\partial \theta_u} < 0 \text{ if } \theta_u = 0, \tag{2.28}$$

for $j = 1, \ldots, p$ and $u = 1, \ldots, m$. Clearly, these conditions specify that the gradient of $\theta_u$ must be strictly negative if $\theta_u = 0$ (an active constraint). Let $k$ denote the iteration number and $a^{(k)}$ represent the update of parameter $a$ after iteration $k$. We describe the Newton-MI algorithm next.

After iteration $k$, the updates $\boldsymbol{\beta}^{(k)}$ and $\boldsymbol{\theta}^{(k)}$ are obtained. Then at iteration $k+1$, we first estimate $\boldsymbol{\beta}$ by the Newton algorithm as follows:

$$\boldsymbol{\beta}^{(k+1)} = \boldsymbol{\beta}^{(k)} + \omega_1^{(k)} \left[ -\frac{\partial^2 \Phi(\boldsymbol{\beta}^{(k)}, \boldsymbol{\theta}^{(k)})}{\partial \boldsymbol{\beta} \partial \boldsymbol{\beta}^{\mathsf{T}}} \right]^{-1} \frac{\partial \Phi(\boldsymbol{\beta}^{(k)}, \boldsymbol{\theta}^{(k)})}{\partial \boldsymbol{\beta}}, \tag{2.29}$$

where $\omega_1^{(k)} \in (0, 1]$ is the line search step size such that the penalized log-likelihood increases from $(\boldsymbol{\beta}^{(k)}, \boldsymbol{\theta}^{(k)})$ to $(\boldsymbol{\beta}^{(k+1)}, \boldsymbol{\theta}^{(k)})$, i.e., $\Phi(\boldsymbol{\beta}^{(k+1)}, \boldsymbol{\theta}^{(k)}) \geq \Phi(\boldsymbol{\beta}^{(k)}, \boldsymbol{\theta}^{(k)})$. It is important to note that the requirement of the log-likelihood to increase is a crucial condition for the convergence of this algorithm.

In (2.29), expressions for the first and second derivatives can be easily derived. The first derivatives are given by:

$$\frac{\partial \Phi(\boldsymbol{\beta}, \boldsymbol{\theta})}{\partial \beta_j} = \sum_{i=1}^{n} x_{ij} \Bigg( \delta_i - \delta_i H_i(t_i) - \delta_i^R H_i(t_i^L) + \delta_i^L \frac{S_i(t_i^R) H_i(t_i^R)}{1 - S_i(t_i^R)} - \delta_i^I \frac{S_i(t_i^L) H_i(t_i^L) - S_i(t_i^R) H_i(t_i^R)}{S_i(t_i^L) - S_i(t_i^R)} \Bigg), \tag{2.30}$$

where $H_i(t)$ and $S_i(t)$ are the cumulative and survival functions of subject $i$ given by (2.7) and (2.8), respectively.

In (2.30), the time points associated with the right and left censoring are denoted by $t_i^L$ and $t_i^R$, respectively. This is because, under our interval censoring notation, a right-censored data is denoted by $[t_i^L, \infty)$ and a left-censored data by $(0, t_i^R]$. Similar expressions will be used throughout this book.

Elements of the Hessian matrix of $\Phi$ (the second derivative) are:

$$\frac{\partial^2 \Phi(\boldsymbol{\beta}, \boldsymbol{\theta})}{\partial \beta_j \partial \beta_t} = - \sum_{i=1}^{n} x_{ij} x_{it} \Bigg( \delta_i H_i(t_i) + \delta_i^R H_i(t_i^L) + \delta_i^L \frac{S_i(t_i^R) H_i(t_i^R)(H_i(t_i^R) + S_i(t_i^R) - 1)}{(1 - S_i(t_i^R))^2} + \delta_i^I \frac{S_i(t_i^L) S_i(t_i^R)(-H_i(t_i^L) + H_i(t_i^R))^2}{(S_i(t_i^L) - S_i(t_i^R))^2} + \delta_i^I \frac{-S_i(t_i^R) H_i(t_i^R) + S_i(t_i^L) H_i(t_i^L)}{S_i(t_i^L) - S_i(t_i^R)} \Bigg) \tag{2.31}$$

where $x_{ij}$ represents the $j$-th element of vector $\mathbf{x}_i$.

In (2.29), $\omega_1^{(k)}$ is the step size determined by a line search procedure. One simple and efficient search strategy is provided by Armijo's rule (e.g. Luenberger (1984)). The Armijo line search is a finite terminating algorithm. Briefly, it starts with $\omega_1 = 1$, and for each $\omega_1$ it checks if the following Armijo condition is satisfied:

$$\Phi(\boldsymbol{\beta}^{(k)} + \omega_1 \mathbf{b}_1^{(k)}, \boldsymbol{\theta}^{(k)}) \geq \Phi(\boldsymbol{\beta}^{(k)}, \boldsymbol{\theta}^{(k)}) + \xi_1 \omega_1 [\mathbf{b}_1^{(k)}]^\mathsf{T} \frac{\partial \Phi(\boldsymbol{\beta}^{(k)}, \boldsymbol{\theta}^{(k)})}{\partial \boldsymbol{\beta}}, \tag{2.32}$$

where

$$\mathbf{b}_1^{(k)} = \left[ - \frac{\partial^2 \Phi(\boldsymbol{\beta}^{(k)}, \boldsymbol{\theta}^{(k)})}{\partial \boldsymbol{\beta} \partial \boldsymbol{\beta}^\mathsf{T}} \right]^{-1} \frac{\partial \Phi(\boldsymbol{\beta}^{(k)}, \boldsymbol{\theta}^{(k)})}{\partial \boldsymbol{\beta}} \tag{2.33}$$

and $0 < \xi_1 < 1$ is a fixed parameter, such as $\xi_1 = 10^{-2}$. If (2.32) is true, then stop; otherwise, reset $\omega_1 = \rho_1 \omega_1$ (for example $\rho_1 = 0.8$) and reevaluate the Armijo condition (2.32). The first $\omega_1$ satisfying (2.32) will be the $\omega_1^{(k)}$ used in (2.29).

Next, we explain how to develop a multiplicative iterative (MI) (see Chan & Ma (2012)) scheme to update $\boldsymbol{\theta}$. Since the length of $\boldsymbol{\theta}$, denoted as $m$, can potentially be large (especially when $h_0(t)$ is approximated by a piecewise constant function), the MI scheme presented below is designed to be efficient for large $m$, while also respecting the non-negative constraint.

Firstly, the gradient with respect to $\boldsymbol{\theta}$ is:

$$\begin{aligned}\frac{\partial\Phi(\boldsymbol{\beta},\boldsymbol{\theta})}{\partial\theta_u} = &\sum_{i=1}^{n}\Bigg(\delta_i\frac{\psi_u(t_i)}{h_0(t_i)} - \delta_i\Psi_u(t_i)e^{\mathbf{x}_i\boldsymbol{\beta}} + \delta_i^L\frac{S_i(t_i^R)\Psi_u(t_i^R)}{1-S_i(t_i^R)}e^{\mathbf{x}_i\boldsymbol{\beta}} \\ &- \delta_i^R\Psi_u(t_i^L)e^{\mathbf{x}_i\boldsymbol{\beta}} - \delta_i^I\frac{S_i(t_i^L)\Psi_u(t_i^L)-S_i(t_i^R)\Psi_u(t_i^R)}{S_i(t_i^L)-S_i(t_i^R)}e^{\mathbf{x}_i\boldsymbol{\beta}}\Bigg) \\ &- \lambda\frac{\partial J(\boldsymbol{\theta})}{\partial\theta_u}. \end{aligned} \tag{2.34}$$

The MI algorithm demands to separate $\partial\Phi(\boldsymbol{\beta},\boldsymbol{\theta})/\partial\theta_u$ into positive and negative components.

We need the following notations in the discussions: for a number $c$, let $c^+ = \max\{0,c\}$ and $c^- = \min\{0,c\}$ so that $c = c^+ + c^-$. Let $\zeta_{1u}(\boldsymbol{\beta},\boldsymbol{\theta}) \geq 0$ and $-\zeta_{2u}(\boldsymbol{\beta},\boldsymbol{\theta}) \leq 0$ be the positive and negative parts of $\partial\Phi(\boldsymbol{\beta},\boldsymbol{\theta})/\partial\theta_u$ respectively, such that

$$\frac{\partial\Phi(\boldsymbol{\beta},\boldsymbol{\theta})}{\partial\theta_u} = \zeta_{1u}(\boldsymbol{\beta},\boldsymbol{\theta}) - \zeta_{2u}(\boldsymbol{\beta},\boldsymbol{\theta}), \tag{2.35}$$

where $\zeta_{1u}$ is given by

$$\begin{aligned}\zeta_{1u}(\boldsymbol{\beta},\boldsymbol{\theta}) = &\sum_{i=1}^{n}\left\{\delta_i\frac{\psi_u(t_i)}{h_0(t_i)} + \delta_i^L\frac{S_i(t_i^R)\Psi_u(t_i^R)}{1-S_i(t_i^R)}e^{\mathbf{x}_i\boldsymbol{\beta}} + \delta_i^I\frac{S_i(t_i^R)\Psi_u(t_i^R)}{S_i(t_i^L)-S_i(t_i^R)}e^{\mathbf{x}_i\boldsymbol{\beta}}\right\} \\ &- \lambda\left[\frac{\partial J(\boldsymbol{\theta})}{\partial\theta_u}\right]^- + \epsilon, \end{aligned} \tag{2.36}$$

and $\zeta_{2u}$ is given by

$$\begin{aligned}\zeta_{2u}(\boldsymbol{\beta},\boldsymbol{\theta}) = &\sum_{i=1}^{n}\left\{\delta_i\Psi_u(t_i)e^{\mathbf{x}_i\boldsymbol{\beta}} + \delta_i^R\Psi_u(t_i^L)e^{\mathbf{x}_i\boldsymbol{\beta}} + \delta_i^I\frac{S_i(t_i^L)\Psi_u(t_i^L)}{S_i(t_i^L)-S_i(t_i^R)}e^{\mathbf{x}_i\boldsymbol{\beta}}\right\} \\ &+ \lambda\left[\frac{\partial J(\boldsymbol{\theta})}{\partial\theta_u}\right]^+ + \epsilon. \end{aligned} \tag{2.37}$$

Here, $\epsilon$ is a small value (e.g., $10^{-3}$) used to avoid $\zeta_{2u}(\boldsymbol{\beta},\boldsymbol{\theta}) = 0$ in (2.38). Using a small $\epsilon$ does not significantly affect the convergence rate of the MI algorithm. It is important to note that the same $\epsilon$ should be applied to $\zeta_{1u}$ and $\zeta_{2u}$ as shown in equations (2.36) and (2.37).

Based on the gradient separation provided in (2.35), the MI algorithm updates $\boldsymbol{\theta}$ first using the following equation:

$$\theta_u^{(k+\frac{1}{2})} = \theta_u^{(k)}\frac{\zeta_{1u}(\boldsymbol{\beta}^{(k+1)},\boldsymbol{\theta}^{(k)})}{\zeta_{2u}(\boldsymbol{\beta}^{(k+1)},\boldsymbol{\theta}^{(k)})}. \tag{2.38}$$

for $u = 1, \ldots, m$. This is a temporary estimate as it may not necessarily increase the value of the $\Phi(\boldsymbol{\beta}, \boldsymbol{\theta})$ function from $\Phi(\boldsymbol{\beta}^{(k+1)}, \boldsymbol{\theta}^{(k)})$. This is why we denote the result of this update by $\boldsymbol{\theta}^{(k+\frac{1}{2})}$, basically to reflect that this is an intermediate step.

The updating formula (2.38) for $\boldsymbol{\theta}$ can also be expressed using the gradient of $\Phi(\boldsymbol{\beta}, \boldsymbol{\theta})$. In fact, it is easy to verify that (2.38) can be re-written as

$$\boldsymbol{\theta}^{(k+\frac{1}{2})} = \boldsymbol{\theta}^{(k)} + \mathbf{D}^{(k)} \frac{\partial \Phi(\boldsymbol{\beta}^{(k+1)}, \boldsymbol{\theta}^{(k)})}{\partial \boldsymbol{\theta}}, \tag{2.39}$$

where $\mathbf{D}^{(k)}$ is a diagonal matrix with diagonals $\theta_u^{(k)} / \zeta_{2u}(\boldsymbol{\beta}^{(k+1)}, \boldsymbol{\theta}^{(k)})$ for $u = 1, \ldots, m$, and they are non-negative. Usually, an initial value $\boldsymbol{\theta}^{(0)}$ for the MI scheme is selected to be significantly away from zero, such as $\boldsymbol{\theta}^{(0)} = \mathbf{1}_{m \times 1}$. Then, for a $\theta_u$, its value at iteration $k+1$ will be smaller than the value at iteration $k$ only when $\partial \Phi(\boldsymbol{\beta}^{(k+1)}, \boldsymbol{\theta}^{(k)}) / \partial \theta_u < 0$.

To achieve an increment in the penalized log-likelihood function $\Phi$, a line search can be used again. Let $\omega_2^{(k)} \in (0, 1]$ be the step size for the line search, then the final update for $\boldsymbol{\theta}$ is given by:

$$\boldsymbol{\theta}^{(k+1)} = \boldsymbol{\theta}^{(k)} + \omega_2^{(k)} \mathbf{D}^{(k)} \frac{\partial \Phi(\boldsymbol{\beta}^{(k+1)}, \boldsymbol{\theta}^{(k)})}{\partial \boldsymbol{\theta}}. \tag{2.40}$$

Here, the line search step size $\omega_2^{(k)}$ may also be calculated based on Armijo's rule. The Armijo line search is executed in a similar way as before, starting with $\omega_2 = 1$. For each trial of an $\omega_2$ value, we need to check if the following Armijo condition is satisfied:

$$\Phi(\boldsymbol{\beta}^{(k+1)}, \boldsymbol{\theta}^{(k)} + \omega_2 \mathbf{b}_2^{(k)}) \geq \Phi(\boldsymbol{\beta}^{(k+1)}, \boldsymbol{\theta}^{(k)}) + \xi_2 \omega_2 [\mathbf{b}_2^{(k)}]^{\mathsf{T}} \frac{\partial \Phi(\boldsymbol{\beta}^{(k+1)}, \boldsymbol{\theta}^{(k)})}{\partial \boldsymbol{\theta}}, \tag{2.41}$$

where

$$\mathbf{b}_2^{(k)} = \mathbf{D}^{(k)} \frac{\partial \Phi(\boldsymbol{\beta}^{(k)}, \boldsymbol{\theta}^{(k)})}{\partial \boldsymbol{\theta}}, \tag{2.42}$$

and $0 < \xi_2 < 1$ is a fixed parameter (e.g. $\xi_2 = 10^{-2}$). If (2.41) is true then stop and the corresponding $\omega_2^{(k)}$ is obtained; otherwise, reset $\omega_2 = \rho_2 \omega_2$ (e.g. $\rho_2 = 0.8$) and re-evaluate the Armijo condition (2.41).

Alternatively, we may re-write (2.40) as

$$\boldsymbol{\theta}^{(k+1)} = \boldsymbol{\theta}^{(k)} + \omega_2^{(k)} (\boldsymbol{\theta}^{(k+\frac{1}{2})} - \boldsymbol{\theta}^{(k)}). \tag{2.43}$$

The updating formula (2.38) guarantees $\boldsymbol{\theta}^{(k+\frac{1}{2})} \geq 0$ if $\boldsymbol{\theta}^{(k)} \geq 0$, whilst (2.43) shows if $\boldsymbol{\theta}^{(k+\frac{1}{2})} \geq 0$, $\boldsymbol{\theta}^{(k)} \geq 0$ and $\omega_2^{(k)} \leq 1$, then we will have $\boldsymbol{\theta}^{(k+1)} \geq 0$. Therefore, a line search over the MI updating result will not violate the non-negativity constraint.

After updating both $\boldsymbol{\beta}$ and $\boldsymbol{\theta}$, we have $\Phi(\boldsymbol{\beta}^{(k+1)}, \boldsymbol{\theta}^{(k+1)}) \geq \Phi(\boldsymbol{\beta}^{(k)}, \boldsymbol{\theta}^{(k)})$. This is a very important condition to ensure the convergence of this algorithm. Since the algorithm involves both the Newton and MI steps, we call it the **Newton-MI algorithm**. This algorithm is easy to implement, especially because the MI step only involves the first derivative, which greatly reduces the complexity in mathematical derivations. Additionally, the MI step does not involve any large matrix inversions, making it efficient when $m$ is large.

Following the same argument as in Chan & Ma (2012), we can show the convergence results of the Newton-MI algorithm stated in Theorems 2.1–2.3 . Firstly, we need to introduce two regularity conditions which are required for the convergence results.

C2.1 $\Phi(\boldsymbol{\beta}, \boldsymbol{\theta})$ is bounded, continuous and differentiable for $\boldsymbol{\beta} \in R^p$ and $\boldsymbol{\theta} \in R^m_+$, where $R^p$ denotes the $p$-dimensional real space and $R^m_+$ the non-negative orthant of the $m$-dimensional real space.

C2.2 $\partial\Phi(\boldsymbol{\beta}, \boldsymbol{\theta})/\partial\beta_j$ are bounded and continuous over $\boldsymbol{\beta} \in R^p$ for $j = 1, \ldots, p$, and $\partial\Phi(\boldsymbol{\beta}, \boldsymbol{\theta})/\partial\theta_u$ are bounded and continuous over $\boldsymbol{\theta} \in R^m_+$ for $u = 1, \ldots, m$.

For the estimation problem we discussed in this chapter, these conditions are clearly satisfied.

**Theorem 2.1.** Assume that Assumptions C2.1 and C2.2 hold for $\Phi(\boldsymbol{\beta}, \boldsymbol{\theta})$. Let $\mathcal{S}$ be the set of points of $\Phi(\boldsymbol{\beta}, \boldsymbol{\theta})$ satisfying the KKT conditions (called the stationary points), where $\boldsymbol{\beta} \in R^p$ and $\boldsymbol{\theta} \in R^m_+$. For sequences $\{\boldsymbol{\beta}^{(k)}, \boldsymbol{\theta}^{(k)}\}$ generated by the Newton-MI algorithm, all the limit points of $\{\boldsymbol{\beta}^{(k)}, \boldsymbol{\theta}^{(k)}\}$ are in $\mathcal{S}$. Moreover, there exist some $(\boldsymbol{\beta}^*, \boldsymbol{\theta}^*) \in \mathcal{S}$ such that $\Phi(\boldsymbol{\beta}^{(k)}, \boldsymbol{\theta}^{(k)})$ converges monotonically to $\Phi(\boldsymbol{\beta}^*, \boldsymbol{\theta}^*)$. □

Theorem 2.1 mainly shows that the limit of any convergent subsequence of $\{\boldsymbol{\beta}^{(k)}, \boldsymbol{\theta}^{(k)}\}$ is in the solution set $\mathcal{S}$, and that $\Phi(\boldsymbol{\beta}^{(k)}, \boldsymbol{\theta}^{(k)})$ converges to $\Phi(\boldsymbol{\beta}^*, \boldsymbol{\theta}^*)$ for some $(\boldsymbol{\beta}^*, \boldsymbol{\theta}^*) \in \mathcal{S}$. The next theorem proves that the sequence $\{\boldsymbol{\beta}^{(k)}, \boldsymbol{\theta}^{(k)}\}$ is convergent itself. This result is acquired through Theorems 2.2 and 2.3.

**Theorem 2.2.** Assume that Assumptions C2.1 and C2.2 hold for $\Phi(\boldsymbol{\beta}, \boldsymbol{\theta})$. Then the Newton-MI sequence $\{\boldsymbol{\beta}^{(k)}, \boldsymbol{\theta}^{(k)}\}$ satisfy $||\boldsymbol{\beta}^{(k+1)} - \boldsymbol{\beta}^{(k)}|| \to 0$ and $||\boldsymbol{\theta}^{(k+1)} - \boldsymbol{\theta}^{(k)}|| \to 0$ as $k \to \infty$. □

**Theorem 2.3.** Assume that Assumptions C2.1 and C2.2 hold for $\Phi(\boldsymbol{\beta}, \boldsymbol{\theta})$. According to Theorem 2.1, for the Newton-MI sequence $\{\boldsymbol{\beta}^{(k)}, \boldsymbol{\theta}^{(k)}\}$ there exist some stationary points $(\boldsymbol{\beta}^*, \boldsymbol{\theta}^*) \in \mathcal{S}$ such that $\Phi(\boldsymbol{\beta}^{(k)}, \boldsymbol{\theta}^{(k)}) \to \Phi(\boldsymbol{\beta}^*, \boldsymbol{\theta}^*)$ as $k \to \infty$. If the set for stationary points $\mathcal{S}$ is discrete then $\{\boldsymbol{\beta}^{(k)}, \boldsymbol{\theta}^{(k)}\}$ converges to a $(\boldsymbol{\beta}^*, \boldsymbol{\theta}^*) \in \mathcal{S}$. □

The above theorems explain that if there exists only one optimal solution (e.g., when $\Phi(\boldsymbol{\beta}, \boldsymbol{\theta})$ is concave), then the Newton-MI algorithm will generate a sequence $\{\boldsymbol{\beta}^{(k)}, \boldsymbol{\theta}^{(k)}\}$ converging to this solution. If there are multiple local maxima that are fully separated, then updates from the Newton-MI algorithm will converge to one of these local maxima, depending on the starting value for the sequence.

The results stated in Theorems 2.1 to 2.3 only require Assumptions C2.1 and C2.2, which are very general with respect to the objective function $\Phi(\boldsymbol{\beta}, \boldsymbol{\theta})$. Therefore, these results are general enough to be applied to the Newton-MI algorithms explained in other chapters of this book.

## 2.6 Association with the method of sieves

The non-parametric component in any semi-parametric model presents challenges when developing theoretical and computational results. Xue et al. (2004) summarized two common methods for addressing the non-parametric component. One approach is to use a non-parametric tool, such as the kernel estimator for the partly linear model (Speckman 1988) or the empirical cumulative baseline hazard approach for the Cox model (e.g., Huang (1996)), in the case of semi-parametric Cox models. Another approach is to approximate the non-parametric component using a flexible parametric function, where the complexity of this function increases with sample size. Our basis function approximation to $h_0(t)$ in this chapter falls into the latter approach. Our basis function approximation method is also known as the method of sieves (see Grenander (1981)), or simply the sieve method. For further discussions on the sieve method in survival analysis models, please see Huang & Rossini (1997). We will briefly introduce the sieve method in the context of the Cox model shortly, but first, we need to introduce some required concepts and notations.

For the partly interval-censored data considered in this chapter, Recall we have defined $a = \min_i\{t_i^L\}$ and $b = \max_i\{t_i^R\}$, where $t_i^R$ are finite. Let $C^r[a, b]$ denote the set of functions over $[a, b]$ that have $r$ $(\geq 1)$ continuous derivatives. We assume that $h_0(t)$ is bounded and $h_0(t) \in C^r[a, b]$. Let the space for $h_0(t)$ be denoted by

$$A = \{h_0(t) : h_0(t) \in C^r[a, b], 0 \leq h_0(t) \leq C_2 < \infty, \forall t \in [a, b]\}.$$

Assume all regression coefficients $\beta_j$ are bounded. Let the space for $\boldsymbol{\beta}$ be denoted by

$$B = \{\boldsymbol{\beta} : |\beta_j| \leq C_1 < \infty, \forall j \in \{1, \ldots, p\}\},$$

which is a compact subset of $R^p$. Then, the parameter space for $\boldsymbol{\tau} = (\boldsymbol{\beta}, h_0(t))$ is

$$\mathcal{B} = \{\boldsymbol{\tau} : \boldsymbol{\beta} \in B, h_0(t) \in A\} = B * A.$$

In order to emphasize its dependence on sample size $n$, we denote the approximating function to $h_0(t)$ by $h_n(t)$. For the approximation using basis functions, as discussed before, the approximating function is given by

$$h_n(t) = \sum_{u=1}^{m} \theta_u \psi_u(t), \tag{2.44}$$

where $m$ varies with $n$ and thus $m$ is the main reason why $h_n(t)$ depends on $n$. We assume that the coefficients $\theta_u$ are bounded and non-negative, and that $\psi_u(t)$ are bounded for $t \in [a, b]$. These assumptions are listed as Assumption A3.3 in Section 2.7.

Let $\boldsymbol{\theta} = (\theta_1, \ldots, \theta_m)^{\mathsf{T}}$. The space for $h_n(t)$ is

$$A_n = \{h_n(t) : 0 \leq h_n(t) \leq C_3 < \infty, \forall t \in [a, b]\}.$$

The infinite-dimensional parameter $\boldsymbol{\tau}$ is now approximated by a finite-dimensional parameter $\boldsymbol{\tau}_n = (\boldsymbol{\beta}, h_n(t))$. The parameter space for $\boldsymbol{\tau}_n$ is:

$$\mathcal{B}_n = \{\boldsymbol{\tau}_n : \boldsymbol{\beta} \in B, h_n \in A_n\} = B * A_n,$$

which approximates the parameter space $\mathcal{B}$. For a fixed $n$, the MPL estimator of $\boldsymbol{\tau}_n$ is denoted by $\widehat{\boldsymbol{\tau}}_n = (\widehat{\boldsymbol{\beta}}, \widehat{h}_n(t))$, where

$$\widehat{h}_n(t) = \sum_{u=1}^{m} \widehat{\theta}_u \psi_u(t). \tag{2.45}$$

For a sieve estimator of a non-parametric or semi-parametric parameter, it requires approximating the infinite-dimensional parameter space $\mathcal{B}$ with *a series* of finite-dimensional parameter spaces as the sample size $n$ goes to infinite. In the case of the Cox model, the basis approximation approach discussed in this chapter defines such a series of finite-dimensional spaces, namely:

$$\{\mathcal{B}_1, \ldots, \mathcal{B}_n, \ldots\}.$$

The sieve method requires that, under certain conditions, these finite-dimensional spaces converge to the true parameter space $\mathcal{B}$ as $n$ approaches infinity.

Thus, we need to explicate the concept of convergence first. "Convergence" is established under the concept of *distance* or *pseudo-distance*, which we will define below. Note that $\boldsymbol{\tau}$ contains both finite- (i.e., $\boldsymbol{\beta}$) and infinite- (i.e., $h_0(t)$) dimensional parameters. For a finite-dimensional parameter space, the Euclidean (or $L_2$) norm is commonly used to measure its distance, while for an infinite-dimensional (functional) space, one can adopt the $L_\infty$ norm to measure the distance. Combining these two norms, it is natural to define a distance between two points $\boldsymbol{\tau}_1, \boldsymbol{\tau}_2 \in \mathcal{B}$ as follows:

$$\rho(\boldsymbol{\tau}_1, \boldsymbol{\tau}_2) = \left\{ \|\boldsymbol{\beta}_1 - \boldsymbol{\beta}_2\|_2^2 + \sup_{t \in [a,b]} |h_{01}(t) - h_{02}(t)|^2 \right\}^{1/2}, \tag{2.46}$$

where $\boldsymbol{\tau}_1 = (\boldsymbol{\beta}_1, h_{01}(t)) \in \mathcal{B}$ and $\boldsymbol{\tau}_2 = (\boldsymbol{\beta}_2, h_{02}(t)) \in \mathcal{B}$. Clearly, this distance satisfies $\rho(\boldsymbol{\tau}_1, \boldsymbol{\tau}_2) \geq 0$ where the equality holds only when $\boldsymbol{\tau}_1 = \boldsymbol{\tau}_2$. Also, it is symmetric since $\rho(\boldsymbol{\tau}_1, \boldsymbol{\tau}_2) = \rho(\boldsymbol{\tau}_2, \boldsymbol{\tau}_1)$.

Let $\boldsymbol{\tau}_0 = (\boldsymbol{\beta}_0, h_0(t)) \in \mathcal{B}$ be the true parameter. The sieve estimator we will explore in this section can be conveniently defined using the expected log-likelihood operator and its empirical version. Let $\boldsymbol{W}_i, i = 1, \ldots, n$, be i.i.d. random vectors. Let $\boldsymbol{W}$ represent a general $\boldsymbol{W}_i$, whose density function and cumulative distribution function (CDF) are denoted by $f(\boldsymbol{w}; \boldsymbol{\tau})$ and $F(\boldsymbol{w}; \boldsymbol{\tau})$ respectively. Define $l(\boldsymbol{\tau}; \boldsymbol{W}) = \log f(\boldsymbol{W}; \boldsymbol{\tau})$ for $\boldsymbol{\tau} \in \mathcal{B}$ and $l(\boldsymbol{\tau}_n; \boldsymbol{W}) = \log f(\boldsymbol{W}; \boldsymbol{\tau}_n)$ for $\boldsymbol{\tau}_n \in \mathcal{B}_n$. For $\boldsymbol{\tau} \in \mathcal{B}$, let $P$ be the operator such that

$$Pl(\boldsymbol{\tau}; \boldsymbol{W}) = E_{\boldsymbol{\tau}_0}(l(\boldsymbol{\tau}; \boldsymbol{W})) = \int l(\boldsymbol{\tau}; \boldsymbol{W}) dF(\boldsymbol{W}; \boldsymbol{\tau}_0), \tag{2.47}$$

where $E_{\boldsymbol{\tau}_0}$ means the expectation is taken under the true parameter $\boldsymbol{\tau}_0$. The empirical version of $P$, denoted by $P_n$, is defined for a given sample $\widetilde{\mathbf{W}}_n = (\boldsymbol{W}_1, \ldots, \boldsymbol{W}_n)$ as follows:

$$P_n l(\boldsymbol{\tau}; \widetilde{\mathbf{W}}_n) = \frac{1}{n} \sum_{i=1}^{n} l(\boldsymbol{\tau}; \boldsymbol{W}_i). \tag{2.48}$$

For $\boldsymbol{\tau}_n \in \mathcal{B}_n$, $Pl(\boldsymbol{\tau}_n)$ and $P_n l(\boldsymbol{\tau}_n)$ can be similarly defined.

For the basis functions $\psi_u(t)$ we selected to approximate $h_0(t)$, if assuming that these functions are bounded and each has $r$ continuous derivatives, then the approximating spaces $\mathcal{B}_1, \ldots, \mathcal{B}_n$ are sub-spaces of $\mathcal{B}$. Now we are ready to define sieve estimates.

**Definition 2.1.** If for any $\boldsymbol{\tau} \in \mathcal{B}$ there exists a sequence of $\boldsymbol{\tau}_n \in \mathcal{B}_n \subset \mathcal{B}$ such that $\rho(\boldsymbol{\tau}_n, \boldsymbol{\tau}) \to 0$ when $n \to \infty$, then such a sequence of approximating spaces $\{\mathcal{B}_n : n \geq 1\}$ is called a sieve. Specifically, if we select a special estimator, denoted by $\widetilde{\boldsymbol{\tau}}_n$, which satisfies

$$P_n l(\widetilde{\boldsymbol{\tau}}_n; \widetilde{\mathbf{W}}_n) \geq P_n l(\widehat{\boldsymbol{\tau}}_n^*; \widetilde{\mathbf{W}}_n) - \varepsilon_n, \tag{2.49}$$

where

$$\widehat{\boldsymbol{\tau}}_n^* = \underset{\boldsymbol{\tau}_n \in \mathcal{B}_n}{\operatorname{argmax}}\, P_n l(\boldsymbol{\tau}_n; \widetilde{\mathbf{W}}_n) \tag{2.50}$$

and where $\varepsilon_n \geq 0$ satisfying $\varepsilon_n \to 0$ as $n \to \infty$, then $\widetilde{\boldsymbol{\tau}}_n$ is called the sieve MLE of $\boldsymbol{\tau}$. □

Similar definitions of the sieve estimator can also be found in Wong & Severini (1991)) or Grenander (1981).

Clearly, the MLE $\widehat{\boldsymbol{\tau}}_n^*$ is itself a sieve MLE of $\boldsymbol{\tau}$. For the MPL estimator, if its smoothing parameter $\lambda$ is controlled carefully as $n \to \infty$, then it becomes another example of sieve MLE. This fact is further elaborated next.

For the MPL estimator to be a particular sieve MLE it requires $\lambda/n \to 0$ when $n \to \infty$ (meaning that the smoothing parameter $\lambda$ becomes negligible when $n$ is large). To simplify notations we let $\mu_n = \lambda/n$. Then, the MPL estimate, denoted by $\widehat{\boldsymbol{\tau}}_n$, is given by

$$\widehat{\boldsymbol{\tau}}_n = \underset{\boldsymbol{\tau}_n \in \mathcal{B}_n}{\operatorname{argmax}} \{P_n l(\boldsymbol{\tau}_n; \widetilde{\mathbf{W}}_n) - \mu_n J(\boldsymbol{\tau}_n)\}. \tag{2.51}$$

Recall the MLE of $\boldsymbol{\tau}_n$ is $\widehat{\boldsymbol{\tau}}_n^*$ in (2.50), then estimators $\widehat{\boldsymbol{\tau}}_n$ and $\widehat{\boldsymbol{\tau}}_n^*$ satisfy

$$\begin{aligned} P_n l(\widehat{\boldsymbol{\tau}}_n; \widetilde{\mathbf{W}}_n) &\geq P_n l(\widehat{\boldsymbol{\tau}}_n^*; \widetilde{\mathbf{W}}_n) - \mu_n (J(\widehat{\boldsymbol{\tau}}_n^*) - J(\widehat{\boldsymbol{\tau}}_n)) \\ &\geq P_n l(\widehat{\boldsymbol{\tau}}_n^*; \widetilde{\mathbf{W}}_n) - \mu_n |J(\widehat{\boldsymbol{\tau}}_n^*) - J(\widehat{\boldsymbol{\tau}}_n)|. \end{aligned} \tag{2.52}$$

Let $\varepsilon_n = \mu_n |J(\widehat{\boldsymbol{\tau}}_n^*) - J(\widehat{\boldsymbol{\tau}}_n)|$. Since the penalty function $J(\cdot)$ is assumed bounded (see Assumption 3.2) and $\mu_n \to 0$, we have $\varepsilon_n \to 0$ as $n \to \infty$. This explains that the MPL estimate $\widehat{\boldsymbol{\tau}}_n$, when $\mu_n \to 0$, is also a sieve-MLE of $\boldsymbol{\tau}$.

Now we establish the fact that the MPL estimator is a special sieve ML estimator under very general conditions. Due to this fact, the theoretical results of the sieve method can be directly applied to the MPL estimator. There exist theoretical results for sieve estimators of general semi-parametric models, as documented in Shen & Wong (1994), Shen (1997), Wong & Severini (1991), Huang (1996) and Xue et al. (2004). These results can be readily adopted to obtain asymptotic results for the MPL estimate of the Cox model discussed in this chapter, such as asymptotic normality and rate of convergence.

However, implementing the asymptotic normality results can be challenging, as the formula for the asymptotic covariance matrix of the regression coefficient estimates is difficult to compute. Additionally, the asymptotic normality results in these papers do not account for the non-parametric component, making them impractical for some inferences such as inferences on hazard or survival functions.

In the next section, we first develop the strong consistency of the MPL estimators $\widehat{\beta}$ and $\widehat{h}_n(t)$ under certain regularity conditions. This result can also be derived from the general consistency result for sieve estimators of semi-parametric models given in Shen (1997). However, we provide a different proof for this consistency result following the approach of Xue et al. (2004), and this proof is given in Appendix A. We then establish the asymptotic normality for both MPL estimates $\widehat{\boldsymbol{\beta}}$ and $\widehat{\boldsymbol{\theta}}$, where the elements of $\widehat{\boldsymbol{\theta}}$ are all non-negative. Note that for the MPL estimate $\widehat{\boldsymbol{\theta}}$, it is common for some $\widehat{\boldsymbol{\theta}}$ elements to be exactly zero, which means that active constraints can exist. Active constraints must be carefully addressed when developing the asymptotic normality result; otherwise, improper outcomes such as negative asymptotic variances can be obtained.

## 2.7 Asymptotic results of the MPL estimates

Corresponding to the partly interval-censored survival data considered in this chapter, we define the random vector $\boldsymbol{W}_i = (T_i^L, T_i^R, \mathbf{x}_i^\top)^\top$ for $i = 1, \ldots, n$. We assume that these $\boldsymbol{W}_i$ are i.i.d., and the distribution functions of $\mathbf{x}_i$ are independent of the parameters of the Cox model.

According to the expression in (2.2), the density function of $\boldsymbol{W}_i$ is proportional to

$$f(\boldsymbol{w}_i; \boldsymbol{\tau}_n) = (h_i(t_i)S_i(t_i))^{\delta_i}(1 - S_i(t_i^R))^{\delta_i^L} S_i(t_i^L)^{\delta_i^R}(S_i(t_i^L) - S_i(t_i^R))^{\delta_i^I}\gamma(\mathbf{x}_i), \tag{2.53}$$

where $\delta_i^L, \delta_i^R, \delta_i^I$ and $\delta_i$ are indicators as defined previously, and $\gamma$ denotes the density function of $\mathbf{x}_i$, which is assumed independent of $\boldsymbol{\tau}_n = (\boldsymbol{\beta}, h_n(t))^\top$.

Traditional inferences for a Cox model mainly focus on conducting hypotheses testing and constructing confidence intervals on regression coefficients, i.e. elements of $\boldsymbol{\beta}$. However, nowadays, due to increased interests in risk assessment and survival probability evaluation, inferences made on hazard or survival functions have become important as well. For example, one may wish to construct piecewise- or simultaneous-confidence intervals on the hazard or survival functions of an individual. In different examples we will learn how to make these inferences using the MPL estimates $\widehat{\boldsymbol{\beta}}$ and $\widehat{\boldsymbol{\theta}}$, but in this section, we will only provide the asymptotic results for $\widehat{\boldsymbol{\beta}}$ and $\widehat{\boldsymbol{\theta}}$.

In this section, we focus on the asymptotic results for the MPL estimates. We first introduce the asymptotic consistency result of $(\widehat{\boldsymbol{\beta}}, \widehat{h}_n(t))$ and then the asymptotic normality result of $\widehat{\boldsymbol{\beta}}$ and $\widehat{\boldsymbol{\theta}}$.

The asymptotic consistency for the MPL estimates of $\widehat{\boldsymbol{\beta}}$ and $\widehat{h}_n(t)$ is obtained when the number of basis functions $m \to \infty$, but it goes to $\infty$ at a slower rate than $n$ goes to $\infty$ in the sense that $m/n \to 0$. For this result, we also need to demand the smoothing parameter $\lambda$ is negligible when $n \to \infty$, i.e. $\mu_n = \lambda/n \to 0$ when $n \to \infty$.

For the asymptotic normality result, the constraint $\boldsymbol{\theta} \geq 0$ must be considered carefully as there are good chances that some of such constraints are active, i.e. $\theta_u = 0$ for some $u$. Ignoring active constraints can cause unpleasant results such as negative variances.

### 2.7.1 Asymptotic consistency

Assume $m \to \infty$ when $n \to \infty$, but $m/n \to 0$. We further assume $\mu_n = \lambda/n \to 0$ when $n \to \infty$. Let $\widehat{\boldsymbol{\beta}}$ and $\widehat{\boldsymbol{\theta}}$ be respectively the MPL estimators of $\boldsymbol{\beta}$ and $\boldsymbol{\theta}$ and the corresponding baseline hazard estimator $\widehat{h}_n(t)$ is given by (2.45).

We will state the consistency results in Theorem 2.4 for estimators $\widehat{\boldsymbol{\beta}}$ and $\widehat{h}_n(t)$. These results require the following regularity conditions.

A2.1 Matrix $\mathbf{X}$ is bounded and $E(\mathbf{X}\mathbf{X}^\top)$ is non-singular.

A2.2 The penalty function $J(\boldsymbol{\theta})$ is bounded.

A2.3 For function $h_n(t)$, assume its coefficient vector $\boldsymbol{\theta}$ is in a compact subset of $R^m$, and moreover, assume its basis functions $\psi_u(t)$ are bounded for $t \in [a, b]$.

A2.4 The knots and basis functions are selected in a way such that for any $h_0(t) \in A$ there exists a $h_n(t) \in A_n$ such that $\max_t |h_n(t) - h_0(t)| \to 0$ as $n \to \infty$.

The Assumption A2.4 above can be guaranteed under certain regularity conditions, such as those stated in Proposition 2.8 in DeBoor & Daniel (1974).

**Theorem 2.4.** Assume Assumptions A2.1–A2.4 hold and $h_0(t)$ has up to $r \geq 1$ derivatives. Assume $m$ is selected according to $m = n^\upsilon$, where $0 < \upsilon < 1$ and $\mu_n = \lambda/n \to 0$ as $n \to \infty$. Let $\boldsymbol{\beta}_0$ be the vector of true regression coefficients and $h_0(t)$ the true baseline hazard. Then, when $n \to \infty$:

(1) $\|\widehat{\boldsymbol{\beta}} - \boldsymbol{\beta}_0\| \to 0$ almost surely.

(2) $\sup_{t\in[a,b]} |\widehat{h}_n(t) - h_0(t)| \to 0$ almost surely.

□

A prove to the results of this theorem is given in Appendix A.

Recall that the MPL estimator in this chapter is a special sieve MLE. Using the general sieve MLE results of, for example Huang (1996) and Zhang et al. (2010), we can extend the consistency results in Theorem 2.4 to further develop rates of convergence for $\widehat{\boldsymbol{\beta}}$ and $\widehat{h}_n(t)$, and then an asymptotic normality result can be obtained, due to the profile likelihood result of Murphy & van der Vaart (2000), but only for $\widehat{\boldsymbol{\beta}}$.

This asymptotic normality result for $\widehat{\boldsymbol{\beta}}$, although important theoretically, is less useful in practice for two reasons:

(i) the covariance matrix of $\widehat{\boldsymbol{\beta}}$ is difficult to compute due to its involvement of the *efficient score* function; and

(ii) the asymptotic distribution of $\widehat{\boldsymbol{\beta}}$ does not involve $\widehat{h}_n(t)$, making predictive inferences impractical.

In the following section, we will develop more practically useful asymptotic normality results for both $\widehat{\boldsymbol{\beta}}$ and $\widehat{\boldsymbol{\theta}}$ (which relates to $\widehat{h}_n(t)$), assuming that $m$ is fixed. The rationale behind this assumption is that the magnitude of changes in $m$ is much smaller than $n$, due to the slow convergence rate of $\widehat{h}_n(t)$. For example, in Huang (1996), the rate of convergence is $n^{1/3}$ when estimating the baseline cumulative hazard $H_0(t)$ using the non-parametric estimator of Groeneboom & Wellner (1992). In Zhang et al. (2010), the rate is $n^{r/(1+2r)}$ for estimating $\log H_0(t)$ using spline basis functions, where $r$ denotes the number of bounded derivatives of $\log H_0(t)$.

A fixed $m$ makes computation of the covariance matrix of $(\widehat{\boldsymbol{\beta}}, \widehat{\boldsymbol{\theta}})$ a feasible task, even when $n$ is large.

This strategy of keeping $m$ fixed works remarkably well as demonstrated from the simulation example in Section 2.11. We furnish in the next section the asymptotic results for constrained MPL estimates of $\boldsymbol{\beta}$ and $\boldsymbol{\theta}$ where $\mu_n = o(n^{-1/2})$ and $m$ is small relative to $n$.

### 2.7.2 Asymptotic normality

To simplify discussions we let $\boldsymbol{\eta} = (\boldsymbol{\theta}^\top, \boldsymbol{\beta}^\top)^\top$, where the length of vector $\boldsymbol{\eta}$ is $m+p$. We can rewrite the penalized likelihood in (2.13) as

$$\Phi(\boldsymbol{\eta}) = \sum_{i=1}^{n} \phi_i(\boldsymbol{\eta}),$$

where $\phi_i(\boldsymbol{\eta}) = l_i(\boldsymbol{\eta}) - \mu_n J(\boldsymbol{\eta})$ with $l_i(\boldsymbol{\eta}) = \log f(\mathbf{w}_i; \boldsymbol{\eta})$ and $J(\boldsymbol{\eta}) = J(\boldsymbol{\theta})$. The log-likelihood function is given by $l(\boldsymbol{\eta}) = \sum_{i=1}^{n} l_i(\boldsymbol{\eta})$.

The constrained MPL estimate of $\boldsymbol{\eta}$, denoted by $\widehat{\boldsymbol{\eta}}$, is obtained by maximizing $\Phi(\boldsymbol{\eta})$ subject to the constraint $\boldsymbol{\theta} \geq 0$. One special phenomena here is that some elements of $\boldsymbol{\theta} \geq 0$ can be active, particularly when the number, as well as the location, of knots were selected inappropriately. If a constraint $\theta_u \geq 0$ is active, then the estimate should be $\widehat{\theta}_u = 0$. Active constraints have to be taken into consideration when developing asymptotic results; otherwise, a non-positive definite information matrix can be obtained.

Let $\boldsymbol{\eta}_0$ represent the "true value" of parameter $\boldsymbol{\eta}$. We first state the following assumptions needed for the asymptotics. Since the scaled smoothing parameter $\mu_n \to 0$ as $n \to \infty$, most of these assumptions are related to the log-likelihood $l(\boldsymbol{\beta}, \boldsymbol{\theta})$.

B2.1 Assume random vector $\boldsymbol{W}_i = (T_i^L, T_i^R, \mathbf{x}_i^\mathsf{T})^\mathsf{T}, i = 1, \ldots, n$, are independently and identically distributed, and the distribution of $\mathbf{x}_i$ is independent of $\boldsymbol{\eta}$.

B2.2 Assume $E_{\boldsymbol{\eta}_0}[n^{-1}l(\boldsymbol{\eta})]$ exists and has a unique maximum at $\boldsymbol{\eta}_0 \in \Omega$, where $\Omega$ is the parameter set for $\boldsymbol{\tau}$. Assume $\Omega$ is a compact subspace in $R^{p+m}$.

B2.3 Assume $l(\boldsymbol{\eta})$ has a finite upper bound, $l(\boldsymbol{\eta})$ is twice continuously differentiable in a neighbourhood of $\boldsymbol{\eta}_0$ and the matrix

$$n^{-1}E_{\boldsymbol{\eta}_0}\left( - \frac{\partial^2 l(\boldsymbol{\eta})}{\partial \boldsymbol{\eta} \partial \boldsymbol{\eta}^\mathsf{T}} \right)$$

exists and is bounded.

B2.4 The penalty function $J(\boldsymbol{\theta})$ is bounded and twice continuously differentiable with respect to $\boldsymbol{\theta}$.

The literature on asymptotic results for constrained maximum likelihood estimates can be found in references such as Crowder (1984) and Moore et al. (2008). In the following discussions, we will closely follow the latter reference. To elucidate the discussions, we assume, without loss of generality, that the first $q$ constraints of $\boldsymbol{\theta} \geq 0$ are active in the MPL solution. Correspondingly, we define

$$\mathbf{U} = [0_{(m-q+p)\times q}, \mathbf{I}_{(m-q+p)\times(m-q+p)}]^\mathsf{T}, \tag{2.54}$$

which satisfies $\mathbf{U}^\mathsf{T}\mathbf{U} = \mathbf{I}_{(m-q+p)\times(m-q+p)}$. It is clear that the KKT conditions for MPL estimates of $\boldsymbol{\eta}$ can be written as:

$$\mathbf{U}^\mathsf{T} \frac{\partial \Phi(\boldsymbol{\eta})}{\partial \boldsymbol{\eta}} = \mathbf{0}. \tag{2.55}$$

Now we are ready to give asymptotic normality result for the constrained MPL estimates of $\boldsymbol{\eta}$.

**Theorem 2.5.** Assume Assumptions B2.1 – B2.4 hold. Assume the scaled smoothing value $\mu_n = o(n^{-1/2})$. Assume there are $q$ active constraints in the MPL estimate of $\boldsymbol{\theta}$ and the corresponding $\mathbf{U}$ matrix can be defined in a similar way as (2.54). Let $\mathbf{F}(\boldsymbol{\eta}) = n^{-1}E(-\partial^2 l(\boldsymbol{\eta})/\partial\boldsymbol{\eta}\partial\boldsymbol{\eta}^\mathsf{T})$ and let $\boldsymbol{\eta}_0$ be the true value of $\boldsymbol{\eta}$. Then, when $n \to \infty$:

(1) The constrained MPL estimate $\widehat{\boldsymbol{\eta}}$ is consistent for $\boldsymbol{\eta}_0$.

(2) $\sqrt{n}(\widehat{\boldsymbol{\eta}} - \boldsymbol{\eta}_0)$ converges in distribution to a multivariate normal distribution $N(\mathbf{0}, \widetilde{\mathbf{F}}(\boldsymbol{\eta}_0)^{-1}\mathbf{F}(\boldsymbol{\eta}_0)[\widetilde{\mathbf{F}}(\boldsymbol{\eta}_0)^{-1}]^{\mathsf{T}})$, where

$$\widetilde{\mathbf{F}}(\boldsymbol{\eta})^{-1} = \mathbf{U}(\mathbf{U}^{\mathsf{T}}\mathbf{F}(\boldsymbol{\eta})\mathbf{U})^{-1}\mathbf{U}^{\mathsf{T}}.$$

□

When there are no active constraints, $\mathbf{U}$ becomes an identity matrix. In this case, the results in Theorem 2.5 coincide with the traditional unconstrained MLE (note $\mu_n \to 0$ as $n \to \infty$) asymptotic normality result.

Implementing the results of Theorem 2.5 requires a deeper understanding of the asymptotic covariance matrix. We provide some comments below regarding this matrix.

**Remark 2.1** We comment that matrix $\widetilde{\mathbf{F}}(\boldsymbol{\eta})^{-1}$ is, in fact, straightforward to compute. Firstly, $\mathbf{U}^{\mathsf{T}}\mathbf{F}(\boldsymbol{\eta})\mathbf{U}$ is obtained simply by deleting the rows and columns of $\mathbf{F}(\boldsymbol{\eta})$ associated with the active constraints. The inverse of $\mathbf{U}^{\mathsf{T}}\mathbf{F}(\boldsymbol{\eta})\mathbf{U}$ is then calculated. Finally, $\widetilde{\mathbf{F}}(\boldsymbol{\eta})^{-1}$ is obtained by inserting the inverse of $\mathbf{U}^{\mathsf{T}}\mathbf{F}(\boldsymbol{\eta})\mathbf{U}$ with zeros in the deleted rows and columns.

**Remark 2.2** In practice, $\boldsymbol{\eta}_0$ is unknown and the expected information matrix $\mathbf{F}(\boldsymbol{\eta})$ can be difficult to compute. We can replace $\boldsymbol{\eta}_0$ by $\widehat{\boldsymbol{\eta}}$ and $\mathbf{F}(\boldsymbol{\eta})$ by the negative Hessian matrix.

**Remark 2.3** To implement the results in Theorem 2.5 (and also for Corollary 2.1), one must identify the active constraints. We suggest the following process for this task. At the end of running the Newton-MI algorithm, often some $\widehat{\theta}_u$ values are almost exactly zero with negative gradients, indicating that they are active constraints. It is also possible that some $\widehat{\theta}_u$ values may be close to, but not "exactly", zero. In such cases, we need to check the corresponding gradient values to see if they are negative and non-zero. In our R program, we define active constraints as "$|\widehat{\theta}_u| < 10^{-3}$ while the corresponding gradient is less than 0".

**Remark 2.4** Applications often require making inferences when the sample size $n$ is finite, and as a result, the smoothing value should be included in the covariance matrix formula. This implies that we need an approximate distribution for $\widehat{\boldsymbol{\eta}}$ when $n$ is large, and this result also includes $\lambda$ in the covariance matrix. The results are provided in Corollary 2.1 below.

**Corollary 2.1.** Assume Assumptions B2.1 – B2.4 hold. When $n$ is large, the distribution of $\widehat{\boldsymbol{\eta}} - \boldsymbol{\eta}_0$ can be approximated by a multivariate normal distribution with zero mean and an approximate covariance matrix

$$\widehat{\mathrm{Var}}(\widehat{\boldsymbol{\eta}}) = \mathbf{A}(\widehat{\boldsymbol{\eta}})^{-1}\Big(-\frac{\partial^2 l(\widehat{\boldsymbol{\eta}})}{\partial\boldsymbol{\eta}\partial\boldsymbol{\eta}^{\mathsf{T}}}\Big)\mathbf{A}(\widehat{\boldsymbol{\eta}})^{-1}, \tag{2.56}$$

where

$$\mathbf{A}(\widehat{\boldsymbol{\eta}})^{-1} = \mathbf{U}\Big(\mathbf{U}^{\mathsf{T}}\Big(-\frac{\partial^2 l(\widehat{\boldsymbol{\eta}})}{\partial\boldsymbol{\eta}\partial\boldsymbol{\eta}^{\mathsf{T}}} + \lambda\frac{\partial^2 J(\widehat{\boldsymbol{\theta}})}{\partial\boldsymbol{\theta}\partial\boldsymbol{\theta}^{\mathsf{T}}}\Big)\mathbf{U}\Big)^{-1}\mathbf{U}^{\mathsf{T}}.$$

□

Clearly, this result accommodates nonzero smoothing values and active constraints. These results allow us to make inferences with respect to various parameters, such as regression coefficients, baseline hazard, cumulative baseline hazard, and survival probabilities. Additionally, inferences on predictive values can be made. Simulation results reported in Ma et al. (2021) demonstrate that biases in the MPL estimates are usually negligible when smoothing values are small.

## 2.8 Hessian matrix elements

Matrix $\mathbf{F}$ involves Hessian of the log likelihood $l(\boldsymbol{\beta}, \boldsymbol{\theta})$. Equation (2.31) already gives $\partial^2 l/\partial\beta_j\partial\beta_t$. Other parts of $\mathbf{F}$ are provided below.

$$\begin{aligned}\frac{\partial^2 l(\boldsymbol{\beta},\boldsymbol{\theta})}{\partial\beta_j\partial\theta_u} =& -\sum_{i=1}^{n} x_{ij}e^{\mathbf{x}_i\boldsymbol{\beta}}\Bigg(\delta_i\Psi_u(t_i) + \delta_i^R\Psi_u(t_i) + \delta_i^L\frac{S_i(t_i)}{1-S_i(t_i)}\left[\frac{H_i(t_i)}{1-S_i(t_i)} - 1\right]\\ &+ \delta_i^I\frac{S_i(t_i^L)S_i(t_i^R)(-H_i(t_i^L)+H_i(t_i^R))(-\Psi_u(t_i^L)+\Psi_u(t_i^R))}{(S_i(t_i^L)-S_i(t_i^R))^2}\\ &+ \delta_i^I\frac{S_i(t_i^L)\Psi_u(t_i^L) - S_i(t_i^R)\Psi_u(t_i^R)}{S_i(t_i^L)-S_i(t_i^R)}\Bigg) \end{aligned} \tag{2.57}$$

$$\begin{aligned}\frac{\partial^2 l(\boldsymbol{\beta},\boldsymbol{\theta})}{\partial\theta_u\partial\theta_v} =& -\sum_{i=1}^{n}\Bigg(\delta_i\frac{\psi_u(t_i)\psi_v(t_i)}{h_0^2(t_i)} + e^{2\mathbf{x}_i\boldsymbol{\beta}}\Bigg[\delta_i^L\frac{S_i(t_i)}{(1-S_i(t_i))^2}\Psi_u(t_i)\Psi_v(t_i)\\ &+ \delta_i^I\frac{S_i(t_i^L)S_i(t_i^R)}{(S_i(t_i^L)-S_i(t_i^R))^2}(\Psi_u(t_i^R)-\Psi_u(t_i^L))(\Psi_v(t_i^R)-\Psi_v(t_i^L))\Bigg]\Bigg) \end{aligned} \tag{2.58}$$

## 2.9 Inference on baseline functions

The asymptotic variance matrix of $\widehat{\boldsymbol{\eta}}$ enables us to derive the asymptotic variance formulae for the following baseline estimates at a time $t$: baseline hazard $\widehat{h}_0(t)$ and baseline survival $\widehat{S}_0(t) = \exp\{-\widehat{H}_0(t)\}$, where $\widehat{H}_0(t) = \int_0^t \widehat{h}_0(s)ds$. These results are useful for constructing point-wise confidence intervals (CI).

Firstly, the estimated baseline hazard at $t$ can be written as

$$\widehat{h}_0(t) = \boldsymbol{\psi}(t)^{\mathsf{T}}\widehat{\boldsymbol{\theta}},$$

where $\boldsymbol{\psi}(t) = (\psi_1(t), \ldots, \psi_m(t))^\mathsf{T}$. Hence

$$\mathrm{Var}(\widehat{h}_0(t)) = \boldsymbol{\psi}(t)^\mathsf{T} \mathrm{var}(\widehat{\boldsymbol{\theta}}) \boldsymbol{\psi}(t). \tag{2.59}$$

For the estimated baseline survival function $\widehat{S}_0(t)$, its variance can be approximated by

$$\mathrm{Var}(\widehat{S}_0(t)) \approx \left( \frac{\partial \widehat{S}_0(t)}{\partial \widehat{\boldsymbol{\theta}}} \right)^\mathsf{T} \mathrm{var}(\widehat{\boldsymbol{\theta}}) \frac{\partial \widehat{S}_0(t)}{\partial \widehat{\boldsymbol{\theta}}}, \tag{2.60}$$

where $\partial \widehat{S}_0(t)/\partial \widehat{\theta}_u = -\widehat{S}_0(t)\Psi_u(t)$.

Confidence intervals (CIs) for $h_0(t)$ and $S_0(t)$ at a time point $t$ are better constructed using transformations to ensure these CIs are within the boundary of these functions. We adopt the logarithm transformation for the baseline hazard $h_0(t)$ and use the logit transformation for $S_0(t)$.

More specifically, let

$$\kappa(t) = \log \widehat{h}_0(t), \tag{2.61}$$

$$\xi(t) = \log \frac{\widehat{S}_0(t)}{1 - \widehat{S}_0(t)}. \tag{2.62}$$

Then, from the Delta method, approximate variances of $\kappa(t)$ and $\xi(t)$ are:

$$\mathrm{Var}(\kappa(t)) \approx \frac{1}{\widehat{h}_0(t)^2} \mathrm{Var}\left(\widehat{h}_0(t)\right), \tag{2.63}$$

$$\mathrm{Var}(\xi(t)) \approx \frac{1}{\widehat{S}_0^2(t)(1 - \widehat{S}_0(t))^2} \mathrm{Var}\left(\widehat{S}_0(t)\right). \tag{2.64}$$

Therefore, the $100(1-\alpha)\%$ CIs for $h_0(t)$ and $S_0(t)$ are, respectively,

$$\widehat{h}_0(t) e^{\pm z_{\alpha/2} \frac{1}{\widehat{h}_0(t)} \mathrm{std}(\widehat{h}_0(t))}, \tag{2.65}$$

$$\frac{\widehat{S}_0(t) e^{\pm z_{\alpha/2} \frac{1}{\widehat{S}_0(t)(1-\widehat{S}_0(t))} \mathrm{std}(\widehat{S}_0(t))}}{(1 - \widehat{S}_0(t)) + \widehat{S}_0(t) e^{\pm z_{\alpha/2} \frac{1}{\widehat{S}_0(t)(1-\widehat{S}_0(t))} \mathrm{std}(\widehat{S}_0(t))}}, \tag{2.66}$$

where "std" denotes the standard deviation.

## 2.10 Smoothing parameter estimation

Automatic smoothing parameter selection is crucial for the successful implementation of the penalized likelihood method in semi-parametric models, especially for users who may be less experienced in manually selecting a smoothing value.

There are several options for estimating the smoothing value automatically, such as Akaike's information criterion (AIC) (Hurvich et al. 1998), cross-validation (CV) (Eubank 1999) or generalized cross-validation (GCV) (Wahba 1985) method. In this section, however, we will focus on a marginal likelihood-based approach, as specified in, for example, Wood (2011) or Cai & Betensky (2003), to estimate the smoothing parameter $\lambda$ in the penalized log-likelihood $\Phi(\boldsymbol{\beta}, \boldsymbol{\theta})$. The benefits of this marginal likelihood approach are explained in Wood (2011).

Recall that we have adopted the roughness penalty to restrain the baseline hazard. This leads to a quadratic penalty function $J(\boldsymbol{\theta}) = \boldsymbol{\theta}^{\mathsf{T}}\mathbf{R}\boldsymbol{\theta}$. This quadratic expression of $\boldsymbol{\theta}$ can be related to a normal distribution for $\boldsymbol{\theta}$ (this is actually a prior distribution from Bayesian perspective), given by:

$$N(\mathbf{0}_{m\times 1}, \sigma^2\mathbf{R}^{-1}),$$

where the variance parameter $\sigma^2$ is related to the smoothing parameter $\lambda$ through $\sigma^2 = 1/(2\lambda)$.

Thus, after omitting the terms independent of $\boldsymbol{\beta}$, $\boldsymbol{\theta}$ and $\sigma^2$, the log-posterior is

$$l_{pos}(\boldsymbol{\beta}, \boldsymbol{\theta}, \sigma^2) = -\frac{m}{2}\log\sigma^2 + l(\boldsymbol{\beta}, \boldsymbol{\theta}) - \frac{1}{2\sigma^2}\boldsymbol{\theta}^{\mathsf{T}}\mathbf{R}\boldsymbol{\theta}. \tag{2.67}$$

After integrating out $\boldsymbol{\beta}$ and $\boldsymbol{\theta}$ from $L_{pos}(\boldsymbol{\beta}, \boldsymbol{\theta}, \sigma^2) = \exp\{l_{pos}(\boldsymbol{\beta}, \boldsymbol{\theta}, \sigma^2)\}$, the log-marginal likelihood for $\sigma^2$ is

$$l_m(\sigma^2) = -\frac{m}{2}\log\sigma^2 + \log\int\exp\left(l(\boldsymbol{\beta}, \boldsymbol{\theta}) - \frac{1}{2\sigma^2}\boldsymbol{\theta}^{\mathsf{T}}\mathbf{R}\boldsymbol{\theta}\right)d\boldsymbol{\beta}d\boldsymbol{\theta}. \tag{2.68}$$

The integration component in (2.68) generally does not have closed-form expressions; therefore, it must be approximated. Laplace's approximation is often used for this purpose, where estimates of $\boldsymbol{\beta}$ and $\boldsymbol{\theta}$ maximizing

$$\Phi(\boldsymbol{\beta}, \boldsymbol{\theta}) = l(\boldsymbol{\beta}, \boldsymbol{\theta}) - \frac{1}{2\sigma^2}\boldsymbol{\theta}^{\mathsf{T}}\mathbf{R}\boldsymbol{\theta},$$

namely the MPL estimates of $\boldsymbol{\beta}$ and $\boldsymbol{\theta}$, are required. After applying Laplace's approximation and plugging-in the MPL estimates for $\boldsymbol{\beta}$ and $\boldsymbol{\theta}$, denoted by $\widehat{\boldsymbol{\beta}}$ and $\widehat{\boldsymbol{\theta}}$ below, we have

$$l_m(\sigma^2) \approx -\frac{m}{2}\log\sigma^2 + l(\widehat{\boldsymbol{\beta}}, \widehat{\boldsymbol{\theta}}) - \frac{1}{2\sigma^2}\widehat{\boldsymbol{\theta}}^{\mathsf{T}}\mathbf{R}\widehat{\boldsymbol{\theta}} - \frac{1}{2}\log\left|-\widehat{\mathbf{H}} + \mathbf{Q}(\sigma^2)\right|, \tag{2.69}$$

where $\widehat{\mathbf{H}}$ is matrix $\mathbf{H}$ evaluated at the MPL estimates $\widehat{\boldsymbol{\beta}}$ and $\widehat{\boldsymbol{\theta}}$ with $\mathbf{H} = \partial^2 l(\boldsymbol{\eta})/\partial\boldsymbol{\eta}\partial\eta^{\mathsf{T}}$ being the Hessian matrix from the log-likelihood $l(\boldsymbol{\eta})$, and

$$\mathbf{Q}(\sigma^2) = \begin{pmatrix} \mathbf{0}_{p\times p} & \mathbf{0}_{p\times m} \\ \mathbf{0}_{m\times p} & \frac{1}{\sigma^2}\mathbf{R} \end{pmatrix}.$$

The solution of $\sigma^2$ maximizing (2.69) satisfies (note the following expression does not fully solve (2.69))

$$\widehat{\sigma}^2 = \frac{\widehat{\boldsymbol{\theta}}^{\mathsf{T}}\mathbf{R}\widehat{\boldsymbol{\theta}}}{m - \nu}, \tag{2.70}$$

where

$$\nu = \text{tr}\{(-\widehat{\mathbf{H}} + \mathbf{Q}(\widehat{\sigma}^2))^{-1}\mathbf{Q}(\widehat{\sigma}^2)\},$$

which can be conceived as the model degrees of freedom. Since $\boldsymbol{\beta}$ and $\boldsymbol{\theta}$ depend on $\sigma^2$, expression (2.70) naturally suggests an iterative procedure:

1. *Inner iterations:* with $\sigma^2$ being fixed at its current estimate, the corresponding MPL estimates of $\boldsymbol{\beta}$ and $\boldsymbol{\theta}$ are obtained, and we denote these estimates by $\widehat{\boldsymbol{\beta}}$ and $\widehat{\boldsymbol{\theta}}$;
2. *Outer iterations:* then $\sigma^2$ is updated by (2.70) with the newly obtained $\widehat{\boldsymbol{\beta}}$ and $\widehat{\boldsymbol{\theta}}$.

This process is continued until the degrees-of-freedom $\nu$ is stabilized. More specifically, stability in $\nu$ can be judged by observing that the changes in $\nu$ between two consecutive outer iterations are not greater than, say, 1 or 0.5. Initial values for the iterations may affect the speed of convergence of the algorithm. Typical initial values for $\boldsymbol{\beta}$, $\boldsymbol{\theta}$ and $\sigma^2$ are $\mathbf{0}_{p\times 1}$, $\mathbf{1}_{m\times 1}$ and $10^{-6}$, respectively. Our experience with this iterative algorithm is that it usually converges quickly.

Finally, we wish to give the following remarks about this algorithm.

**Remark 2.5** Active constraints of $\boldsymbol{\theta} \geq 0$ during an inner iteration loop may create troubles to the matrix inversion needed in $\nu$. In this case, a modification similar to the covariance matrix in Theorem 2.5 (i.e. using the $\mathbf{U}$ matrix) can be adopted. Accordingly, a modified degrees-of-freedom is

$$\nu = \text{tr}\{\mathbf{U}[\mathbf{U}^{\mathsf{T}}(-\widehat{\mathbf{H}} + \mathbf{Q}(\widehat{\sigma}^2))\mathbf{U}]^{-1}\mathbf{U}^{\mathsf{T}}\mathbf{Q}(\widehat{\sigma}^2)\}.$$

**Remark 2.6** Although it is very rare, it can occur that $\nu < m$ in formula (2.69). One possible reason for this is a convergence issue, which prevents finding a close solution to the MPL estimates after the inner iteration loop. To resolve this issue, reducing the number of knots used for approximating $h_0(t)$ can be attempted.

**Remark 2.7** Ma et al. (2021) have conducted an intensive simulation study to evaluate the Newton-MI algorithm when applied to fit the semi-parametric Cox model. Their findings are that this algorithm is numerically very stable when compared with some other existing methods. The obtained large sample standard deviations are generally accurate, and the coverage probabilities are generally close to their nominated values.

## 2.11 R package '`survivalMPL`' for Cox model fitting

The R package '`survivalMPL`' implements the Newton-MI algorithm discussed in this chapter. This package can be used to obtain MPL estimates for the Cox model regression coefficient vector $\boldsymbol{\beta}$ and for the baseline hazard $h_0(t)$. The penalty function smooths the $h_0(t)$ estimate and reduces its sensitivity to the number and location of knots. This package offers options for basis functions, particularly, the users can select M-spline, indicator (leading to a piecewise constant approximation to $h_0(t)$), or Gaussian basis functions.

There also exist other R packages for fitting semi-parametric Cox models with interval-censored survival data. However, these packages in general do not approximate the baseline hazard $h_0(t)$ directly; instead, they approximate either the cumulative baseline $H_0(t)$ or $\log H_0(t)$. Thus, these packages estimate $\boldsymbol{\beta}$ and $H_0(T)$ simultaneously.

In this section, we will run a simple simulation example to implement '`survivalMPL`' or fitting Cox models under partly interval censoring. Another purpose of this example is to demonstrate how to compute some commonly reported quantities for a statistical methodology or computational paper.

**Example 2.6** (R example: a simulation study).
In this example, we will run a simulation study to compare the MPL method discussed in this chapter against the partial likelihood (PL) method. The true model we generate data from the following Cox model:

$$h_i(t) = h_0(t)e^{0.75X_1 - 0.5X_2},$$

where $h_0(t)$ is a Weibull hazard given by $h_0(t) = 3t^2$, $X_1 \sim unif(0,1)$ and $X_2 \sim unif(0,5)$. Random numbers from this distribution can be generated simply using

$$y_i = S_0^{-1}\left(u_i^{1/\exp\{\mathbf{x}_i\boldsymbol{\beta}\}}\right),$$

where $S_0(t) = \exp\{-t^3\}$ being the baseline survival and $u_i \sim unif(0,1)$. Observed partly interval-censored survival times $(t_i^L, t_i^R)$ (including event and censoring times) are obtained through

$$\begin{aligned} t_i^L =& y_i^{\delta(u_i^E < \pi^E)} \left(\gamma_L u_i^L\right)^{\delta(\pi^E \le u_i^E, \gamma_L u_i^L \le y_i \le \gamma_R u_i^R)} \left(\gamma_R u_i^R\right)^{\delta(\pi^E \le u_i^E, \gamma_R u_i^R < y_i)} \\ & 0^{\delta(\pi^E \le u_i^E, y_i < \gamma_L u_i^L)}, \\ t_i^R =& y_i^{\delta(u_i^E < \pi^E)} \left(\gamma_L u_i^L\right)^{\delta(\pi^E \le u_i^E, y_i < \gamma_L u_i^L)} \left(\gamma_R u_i^R\right)^{\delta(\pi^E \le u_i^E, \gamma_L u_i^L \le y_i \le \gamma_R u_i^R)} \\ & \infty^{\delta(\pi^E \le u_i^E, \gamma_R u_i^R < y_i)}, \end{aligned}$$

where $y_i$ denotes the event time (which may not be observed exactly due to censoring) generated above, $\pi^E$ denotes the event proportion, $u_i^L$, $u_i^R$ and $u_i^E$ denote independent standard uniform variables, $\gamma_L$ and $\gamma_R$ are two scalars help to define interval censoring values, and $\delta(\cdot)$ represents the indicator function. Note that we have adopted the convention $0^0 = 1$ and $\infty^0 = 1$.

In Table 2.3, we list some important parameters for setting up this simulation, including: the regression coefficients, $\mathbf{X}$ matrix, baseline hazard function, sample size and event proportion used in the simulation. Using the values specified in this

TABLE 2.3: Parameters used in Example 2.6, where $\mathbf{u}_1$ and $\mathbf{u}_2$ are random vectors with their elements from the unif$(0, 1)$ distribution and $\gamma_L$ and $\gamma_R$ are constants, defined in Example 2.6, for generating interval censored survival times.

| | |
|---|---|
| **Parameters** | |
| $\boldsymbol{\beta}$ vector | $\boldsymbol{\beta} = [0.75, -0.50]^T$ |
| $\mathbf{X}$ matrix | $\mathbf{X} = [\mathbf{u}_1, 5\mathbf{u}_2]$ |
| $Y$ distribution | Weibull |
| Baseline hazard | $h_0(t) = 3t^2$ |
| $\gamma_L$ and $\gamma_R$ | $\gamma_L = 0.9, \gamma_R = 1.3$ |
| **Scenarios** | |
| Sample sizes | $n = 100, 1000$ |
| Percentages of events | $\pi^E = 0\%, 50\%$ |
| **Basis functions** | |
| MPL M-spline | $3^{rd}$ order M-splines |
| | $m = 4, 8$ for $n = 100, 1000$ |

table, the simulated data achieves the following proportions for the censoring types: left censoring 18%; interval censoring 44%; and right censoring 38%.

For each simulated dataset, we implement the following methods to fit the semi-parametric Cox model.

(1) The Partial Likelihood (PL) estimator uses the middle point to replace left- or interval-censored times. It is important to note that this method does not provide a direct estimate of the baseline hazard, but the Breslow method (Breslow 1972) can be used for this purpose.

(2) The MPL methods discussed in this chapter, with M-spline basis functions.

We start by generating 100 repeated simulated datasets. For each dataset, we obtain estimates of the Cox model regression coefficients $\widehat{\boldsymbol{\beta}}$ and baseline hazard parameters $\widehat{\boldsymbol{\theta}}$. From $\widehat{\boldsymbol{\theta}}$ we derive an estimate of $h_0(t)$ through

$$\widehat{h}_0(t) = \boldsymbol{\psi}(t)^\top \widehat{\boldsymbol{\theta}}, \qquad (2.71)$$

where $\boldsymbol{\psi}(t)^\top$ is a row vector for all the basis functions at time $t$, i.e., $\boldsymbol{\psi}(t)^\top = (\psi_1(t), \ldots, \psi_m(t))$. To plot the estimated function $\widehat{h}_0(t)$, we select 200 values of $t$ between the minimum and maximum simulated survival times and apply (2.71) to obtain 200 corresponding values of $\widehat{h}_0(t)$. For the MPL method, we approximate $h_0(t)$ using M-spline basis functions.

The "`survivalMPL`" package has a built-in automatic smoothing parameter selection as outlined in Section 2.10. The default knots placement option adopted by this package is the quantile-based knots selection (i.e. equal number of events between two consecutive knots) as this option performs well for the MPL approach. We select automatic smoothing and default quantile knots in this simulation. Note that this package also allows users to use their own knots and smoothing values.

Since we are adopting the MPL approach with M-spline basis functions, we will refer to our method as the MPL M-spline in the following discussions. Let $m$ be the number of basis functions. We select $m = 4, 8$ respectively for $n = 100, 1000$ for the MPL M-spline method. The readers may also wish to try a larger $m$, such as $m = 8$ for $n = 100$ and $m = 12$ for $n = 1000$.

In this simulation, there are four scenarios, corresponding to four combinations of sample size and event proportion. We will generate 100 samples for each scenario.

To assess the estimates of the regression coefficients $\boldsymbol{\beta}$, we will calculate and report the following: biases, means of the asymptotic standard errors, Monte Carlo standard errors obtained from the repeated samples (displayed in brackets), and coverage probabilities of the 95% confidence intervals (CIs). Biases are calculated as the difference between the average of $\widehat{\boldsymbol{\beta}}$ values and the true $\boldsymbol{\beta}$, while Monte Carlo standard errors are calculated as the standard deviation of $\widehat{\boldsymbol{\beta}}$ values obtained from the repeated samples.

To assess the estimates of the baseline hazard $h_0(t)$, we first select three-time values $t_1$, $t_2$ and $t_3$, respectively corresponding to the 25th, 50th and 75th percentile of $h_0(t)$. Then, we calculate and report, for each of $t_1$, $t_2$ and $t_3$, the biases, mean of the asymptotic standard errors, Monte Carlo standard errors (in bracket) and coverage probabilities from the 95% CIs.

The partial likelihood and the MPL M-spline method are respectively implemented by means of the `survival` and `survivalMPL` R packages.

The R code for this simple simulation is given below. It includes the generation of partly interval-censored survival times, implementation of `survivalMPL` to estimate $\boldsymbol{\beta}$ and $h_0(t)$, and the calculation of the reported quantities as specified above. For comparison, we also apply the conventional partial likelihood method to the simulated datasets, but with the middle points of the censoring intervals to replace the left- and interval-censored data.

```
# Write simulate() function to repeatedly generate a sample
simulate <- function(n, pi_E, g1, g2){

  # Simulate covariates
  x1 <- runif(n, 0, 1)
  x2 <- 5 * runif(n, 0, 1)
  X <- cbind(x1, x2)
  beta_true <- c(0.75, -0.5)
  eXtB <- exp(X %*% beta_true)

  # Simulate true event times
  neg.log.u <- - log(runif(n, 0, 1))
  y_i <- (neg.log.u/(eXtB))^(1/3)

  # Simulate censoring types & times
```

```
  u_E <- runif(n, 0, 1)
  u_L <- runif(n, 0, 1)
  u_R <- runif(n, u_L, 1)

  t_L <- (y_i^as.numeric(u_E < pi_E))
    * (g1 * u_L)^as.numeric(pi_E < u_E &
    g1*u_L <= y_i & y_i <= g2*u_R) *
    (g2 * u_R)^as.numeric(pi_E < u_E & g2*u_R < y_i) *
    0^as.numeric(pi_E < u_E & y_i < g1*u_L)

  t_R <- (y_i^as.numeric(u_E < pi_E)) *
    (g1 * u_L)^as.numeric(pi_E < u_E & y_i < g1*u_L) *
    (g2 * u_R)^as.numeric(pi_E < u_E
     & g1*u_L <= y_i & y_i <= g2*u_R) *
     Inf^(pi_E < u_E & g2*u_R < y_i)

  # Compute midpoint and censoring type for midpoint imputation

  midpoint <- t_L + (t_R - t_L)/2
  event <- rep(1, n)
  event[which(is.infinite(midpoint))] <- 0
  midpoint[which(is.infinite(midpoint))]
          <- t_L[which(is.infinite(midpoint))]

  df = data.frame(t_L, t_R, x1, x2, midpoint, event)
  return(df)
}

# Run a simulation study

ch2_ex26_save = matrix(0, nrow = 100, ncol = 8)
baseline_h0 = NULL

for(s in 1:100){

  # Generate data set
  dat <- simulate(n=500, pi_E = 0.5, g1 = 0.9, g2 = 1.3)

  # Fit MPL model
  fit.mpl <- coxph_mpl(Surv(t_L, t_R, type = "interval2") ~
                  x1 + x2, data = dat,
                     basis = "m", n.knots = c(5,0))


  # Save coefficient estimates & SE
  ch2_ex26_save[s,1] <- fit.mpl$coef$Beta[1]
  ch2_ex26_save[s,2] <- fit.mpl$coef$Beta[2]
```

```
  ch2_ex26_save[s,3] <- fit.mpl$se$Beta$M2HM2[1]
  ch2_ex26_save[s,4] <- fit.mpl$se$Beta$M2HM2[2]

  # Fit PH model
  fit.ph <- coxph(Surv(midpoint, event) ~ x1 + x2, data = dat)

  # Save coefficient estimates & SE
  ch2_ex26_save[s,5] <- fit.ph$coefficients[1]
  ch2_ex26_save[s,6] <- fit.ph$coefficients[2]

  ch2_ex26_save[s,7] <- sqrt(diag(fit.ph$var))[1]
  ch2_ex26_save[s,8] <- sqrt(diag(fit.ph$var))[2]

  # Save baseline hazard times and estimates
  psi <- mSpline(survfit(fit.ph)$time,
    knots = fit.mpl$knots$Alpha[2:6],
    Boundary.knots = c(fit.mpl$knots$Alpha[1],
    fit.mpl$knots$Alpha[7]))
  h0t_MPL = psi %*% fit.mpl$coef$Theta

  baseline_h0 <- cbind(baseline_h0,
    basehaz(fit.ph)$time, basehaz(fit.ph)$hazard, h0t_MPL)

}
```

Tables 2.4 and 2.5 report the simulation results, respectively, on $\boldsymbol{\beta}$ and $h_0(t)$. The method of partial likelihood with mid-point imputation displays large biases in all the simulations, and it sometimes also produces extremely poor coverage probabilities, such as for $\beta_2$. The MPL method generally has the smallest biases and its mean asymptotic standard errors agree closely with their Monte Carlo counterpart. Furthermore, MPL provides good 95% coverage probabilities. In general, the coverage probabilities of MPL confidence intervals tend to be close to the 95% nominal value.

Results on the $h_0(t)$ estimates are contained in Table 2.5. This table contains biases, mean of the asymptotic standard errors, and Monte Carlo standard errors (in brackets) for estimating the baseline hazard function for three time values $t_1$, $t_2$ and $t_3$, respectively, corresponding to the 25th, 50th and 75th percentile of $T$. We observe that the Breslow estimates provide large biases and standard errors, while the MPL M-spline estimates give reasonable biases and small standard errors in all cases of interest. The coverage probabilities of 95% confidence intervals for the baseline hazard estimates at the chosen percentiles of $T$ are also reported in this table. The MPL M-spline estimates have coverage probabilities close to 95 while the coverage probabilities from the Breslow method are generally poor. It seems the coverage levels of our MPL estimates are poor for $h_0(t_3)$. These poor coverage probabilities are caused mainly by small standard deviations of the MPL estimates of baseline hazards. These probabilities can be improved when using different smoothing parameters. □

TABLE 2.4: Simulation results for $\boldsymbol{\beta}$, where $\boldsymbol{\beta} = [\beta_1, \beta_2]^\mathsf{T} = [0.75, -0.50]^\mathsf{T}$.

| | | $\pi^E = 0\%$ | | $\pi^E = 50\%$ | |
|---|---|---|---|---|---|
| | | $n = 100$ | $n = 1000$ | $n = 100$ | $n = 1000$ |
| **Biases** | | | | | |
| $\beta_1$ | PL | -0.064 | -0.038 | 0.023 | -0.017 |
| | MPL-M | -0.011 | -0.002 | -0.035 | -0.014 |
| $\beta_2$ | PL | 0.013 | -0.016 | -0.006 | -0.003 |
| | MPL-M | -0.021 | -0.002 | 0.048 | 0.001 |
| **Standard errors** | | | | | |
| $\beta_1$ | PL | 0.591<br>(0.548) | 0.182<br>(0.173) | 0.448<br>(0.408) | 0.135<br>(0.121) |
| | MPL-M | 0.631<br>(0.573) | 0.189<br>(0.179) | 0.448<br>(0.351) | 0.136<br>(0.120) |
| $\beta_2$ | PL | 0.130<br>(0.129) | 0.042<br>(0.044) | 0.101<br>(0.105) | 0.031<br>(0.032) |
| | MPL-M | 0.143<br>(0.143) | 0.042<br>(0.046) | 0.096<br>(0.097) | 0.030<br>(0.031) |
| **95% CP** | | | | | |
| $\beta_1$ | PL | 0.96 | 0.95 | 0.98 | 0.95 |
| | MPL-M | 0.97 | 0.97 | 0.99 | 0.97 |
| $\beta_2$ | PL | 0.95 | 0.91 | 0.98 | 0.92 |
| | MPL-M | 0.94 | 0.94 | 0.97 | 0.94 |

**Example 2.7** (A pseudo melanoma study example).
In this example, we utilize the MPL method with M-spline bases to fit a Cox model to a pseudo melanoma dataset. This dataset comprises observations randomly generated from a model fitted on a real melanoma dataset, provided to the authors courtesy of the Melanoma Institute of Australia. The dataset is accessible on the book GitHub web page.

The dataset comprises the time of the first local melanoma recurrence for patients diagnosed with melanoma between 1998 and 2016 in Australia; for further information about a similar dataset, refer to Morton et al. (2014). Our objective here is to showcase the MPL method in real data applications without making comparisons with other methods. This dataset includes the date of melanoma diagnosis ($t_d$) and the date of the last follow-up ($t_f$), along with recurrence status for $n = 300$ patients. If a melanoma recurrence was observed, it also indicates when the first recurrence was diagnosed ($t_r$), as well as the date of the last negative check before recurrence ($t_n$), if available.

TABLE 2.5: Simulation results for $h_0(t)$, at 25% ($t_1$), 50% ($t_2$) and 75% ($t_3$) percentiles.

| | | $\pi^E = 0\%$ | | $\pi^E = 50\%$ | |
|---|---|---|---|---|---|
| | | $n = 100$ | $n = 1000$ | $n = 100$ | $n = 1000$ |
| **Biases** | | | | | |
| $t_1$ | PL | 0.872 | 0.947 | 1.306 | 1.302 |
| | MPL-M | -0.233 | -0.343 | -0.401 | -0.517 |
| $t_2$ | PL | 1.815 | 1.850 | 2.411 | 2.428 |
| | MPL-M | -0.183 | -0.287 | -1.037 | -1.246 |
| $t_3$ | PL | 2.768 | 2.784 | 3.867 | 3.876 |
| | MPL-M | 0.186 | 0.136 | -2.303 | -2.786 |
| **Standard errors** | | | | | |
| $t_1$ | PL | - | - | - | - |
| | | (0.026) | (0.009) | (0.024) | (0.008) |
| | MPL-M | 0.694 | 0.194 | 0.398 | 1.811 |
| | | (0.352) | (0.100) | (0.313) | (0.123) |
| $t_2$ | PL | - | - | - | – |
| | | (0.051) | (0.017) | (0.045) | (0.016) |
| | MPL-M | 0.617 | 0.168 | 0.388 | 1.306 |
| | | (0.807) | (0.227) | (0.513) | (0.243) |
| $t_3$ | PL | - | - | - | – |
| | | (0.099) | (0.167) | (0.102) | (0.035) |
| | MPL-M | 0.575 | 1.288 | 0.353 | 1.166 |
| | | (1.597) | (0.463) | (0.976) | (0.296) |
| **95% CP** | | | | | |
| $t_1$ | PL | 0.39 | 0.36 | 0.24 | 0.00 |
| | MPL-M | 0.99 | 0.95 | 0.98 | 1.00 |
| $t_2$ | PL | 0.02 | 0.00 | 0.00 | 0.00 |
| | MPL-M | 0.97 | 0.95 | 0.81 | 1.00 |
| $t_3$ | PL | 0.00 | 0.00 | 0.00 | 0.00 |
| | MPL-M | 0.95 | 0.96 | 0.41 | 0.01 |

Melanoma recurrence was observed in 37% of the patients. At the time of the last follow-up, 70.5% of the patients were alive and 29.5% were deceased. Among the living patients, 95% had no melanoma, 4% had melanoma and 1% had an unknown melanoma status. Among the deceased patients, 18% had no melanoma, 71% had melanoma and 11% had an unknown melanoma status. We set the melanoma diagnosis time as the time origin for each patient. Times of the first recurrence are typically interval-censored, occurring between patient visits to the doctor. For a patient

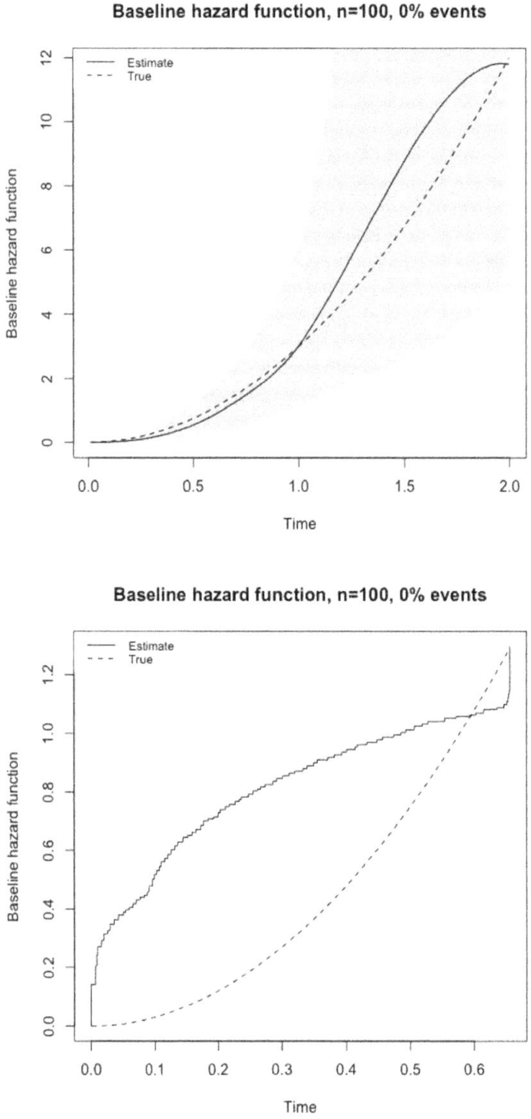


FIGURE 2.4: Plots of baseline hazard estimates and their 95% pointwise CIs. The top panel displays the average estimated (solid line) $h_0(t)$, where the estimates are from the MPL method with M-spline basis functions. The bottom panel plots the average $h_0(t)$ estimate (solid line) from the Breslow method based i=on the partial likelihood method. In both panel, the broken line curves represent the true $h_0(t)$.

with non-missing $t_n$ and $t_r$, the first melanoma recurrence is censored in the interval $[t_n - t_d, t_r - t_d]$. If a patient's $t_n$ is missing but $t_r$ is available, then melanoma recurrence is censored in the interval $[0, t_r - t_d]$. If $t_r$ is missing and the patient had melanoma at $t_f$, then the recurrence time is censored in the interval $[0, t_f - t_d]$. If $t_r$ is missing and the patient had no melanoma at $t_f$, the recurrence time is right-censored in the interval $[t_f - t_d, \infty)$. Cases with no observed recurrence and no known status at the time of the last follow-up were considered missing.

We considered the following covariates in our model:

(1) melanoma location at first diagnostic, a categorical variable with levels 'Head and neck' (19.1%), 'Arm' (14.4%), 'Leg' (28.7%), 'Trunk' (37.8%);
(2) melanoma stage at first diagnostic according to Breslow's thickness scale, an ordinal variable with levels '[0,1) mm.' (15.2%), '[1,2) mm.' (42.5%), '[2,4) mm.' (29.2%) and '4 mm. and more' (13.2%);
(3) gender, a categorical variable with levels 'Men' (58.1%), 'Women' (41.9%);
(4) (centered) age in years at first diagnostic, where the range of the non-centred ages is [5, 94] and the mean of non-centered ages equals 55.7 years.

The contrasts were chosen so that the baseline hazard corresponds to the instantaneous risk of experiencing a first melanoma recurrence on the head/neck for a 55.7-year-old male initially diagnosed with a melanoma of small size (<1mm). We modeled the baseline hazard function using 10 M-spline bases (with no specific effort made to optimize this number). Two of them were placed at the extremities of the time range of interest, while the others were positioned at equidistant interval mid-points.

(1) Download the data from Gihub or simulate data replicating the above-described study.
(2) Run the following R code to fit the Cox model using the R package 'survivalMPL'.

```
simulate_mel <- function(n, pi_E, a1, a2) {
    neglogU <- -log(runif(n))

    location <- sample(c(1, 2, 3, 4), n, replace = TRUE,
        prob = c(0.2, 0.15, 0.3, 0.35))
    Arm <- as.numeric(location == 2)
    Leg <- as.numeric(location == 3)
    Trunk <- as.numeric(location == 4)

    thickness <- sample(c(1, 2, 3, 4), n, replace = TRUE,
        prob = c(0.15, 0.4, 0.3, 0.15))
    mm1to2 <- as.numeric(thickness == 2)
    mm2to4 <- as.numeric(thickness == 3)
    mm4plus <- as.numeric(thickness == 4)

    Female <- rbinom(n, 1, 0.4)
```

```
    x8 <- rnorm(n, 56, 12)
    x8 <- x8 - mean(x8)
    Age_centred <- x8/10

    X <- cbind(Arm, Leg, Trunk, mm1to2, mm2to4,
        mm4plus, Female, Age_centred)
    beta_true = c(-0.56, 0.01, -0.22, 0.22, 0.87,
        1.13, -0.17, 0.14)
    eXtB = exp(X %*% beta_true)

    y <- as.numeric((neglogU/(2 * eXtB))^(2))

    # uniform variables
    U_E <- runif(n)
    U_L <- runif(n, 0, 1)
    U_R <- runif(n, U_L, 1)

    t_L <- y^((U_E < pi_E)) * (a1 * U_L)^((pi_E <=
        U_E & a1 * U_L <= y & y <= a2 * U_R)) *
        (a2 * U_R)^((pi_E <= U_E & a2 * U_R < y)) *
        (0)^((pi_E <= U_E & y < a1 * U_L))

    t_R <- y^((U_E < pi_E)) * (a1 * U_L)^((pi_E <=
        U_E & y < a1 * U_L)) * (a2 * U_R)^((pi_E <=
        U_E & a1 * U_L <= y & y <= a2 * U_R)) *
        Inf^((pi_E <= U_E & a2 * U_R < y))

    df = data.frame(t_L, t_R, X)
    return(df)

}

ch2_mel_dat <- simulate_mel(300, 0.37, 0.6, 1.2)
max(ch2_mel_dat$t_L)

mela_fit = coxph_mpl(formula = Surv(t_L, t_R, "interval2") ~
    Arm + Leg + Trunk + mm1to2 + mm2to4 + mm4plus +
        Female + Age_centred, data = ch2_mel_dat,
    basis = "m", tol = 1e-05, n.knots = c(7, 0),
    max.iter = c(1000, 5000, 1e+05))
summary(mela_fit)
plot(mela_fit)
```

The hazard ratio estimates are presented in Table 2.6. Compared to melanomas first diagnosed at the head and neck, melanomas on the arms or trunk show a significantly lower risk of recurrence. Additionally, initial melanoma thickness represents another strong risk factor for melanoma recurrence.

TABLE 2.6: Hazard ratio estimates ($\widehat{e^{\beta}}$), hazard ratio 95% confidence intervals and $p$-values of the significant tests.

| | | HR estimates | HR 95% CI | $p$-value |
|---|---|---|---|---|
| **Location** | Arm | 0.570 | [0.427; 0.761] | 0.0001 |
| | Leg | 1.008 | [0.811; 1.252] | 0.9446 |
| | Trunk | 0.802 | [0.655; 0.982] | 0.0327 |
| **Thickness** | 1 to 2 mm. | 1.245 | [0.939; 1.650] | 0.1278 |
| | 2 to 4 mm. | 2.390 | [1.807; 3.159] | <0.0001 |
| | 4 mm. and more | 3.108 | [2.305; 4.189] | <0.0001 |
| **Gender** | Female | 0.843 | [0.715; 0.993] | 0.0406 |
| **Centered Age** (10 years) | – | 1.148 | [1.090; 1.208] | <0.0001 |

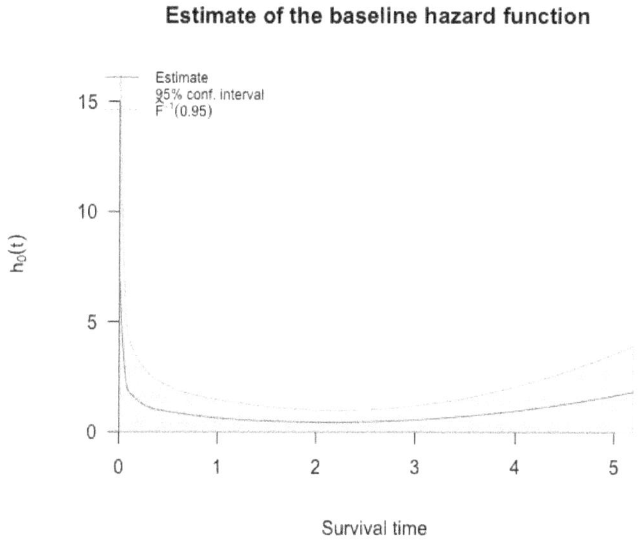


FIGURE 2.5: Plots of baseline hazard estimate and their 95% CI.

A 10-year increase in age corresponds to a significant (9% to 21%) increase in the risk of melanoma recurrence. Gender shows marginal significance, with women exhibiting a lower risk of melanoma recurrence compared to men.

The estimates of the baseline hazard function, along with its 95% pointwise confidence interval, are displayed in Figure 2.5. This plot indicates that when the covariates are set to their baseline values, the risk of melanoma recurrence decreases strongly and monotonically during the first 5 years. Subsequently, over the next decade, the risk continues to steadily decrease to a level close to 0. □

## 2.12 Summary

This chapter focuses on fitting semi-parametric Cox models when survival times are partly interval-censored, this means these survival times can include event times and left, right and interval censoring times.

This chapter explains a full likelihood-based approach in which the unknown baseline hazard is approximated using positive basis functions. Due to the constraint that the baseline hazard must be non-negative, we must demand that the coefficients of the basis functions are also non-negative. To solve the constrained optimization problem, a Newton-MI algorithm is explained in this chapter.

This chapter also justifies that MPL estimates are special cases of sieve ML estimates under the condition that the penalty term becomes negligible as the sample size approaches infinity. Hence, some asymptotic properties of the MPL estimates are readily available from the theoretical results of the method of sieves. To make inferences on, for example, the baseline hazard function, the cumulative hazard function, an individual survival function or a predictive survival function, we need a large sample normality result for the regression coefficients and basis function coefficients jointly. It is important to note that we must take into consideration the possibility of active constraints on the basis function coefficients when developing this large sample normality result.

The R package '`survivalMPL`' was developed based on the Newton-MI algorithm. This package can handle active constraints and performs automatic smoothing parameter selection as a default option. A simulation study is facilitated in this chapter to explain how to use this package. Also, this package is applied to a pseudo melanoma data to demonstrate its practice in a real data setting.

# 3

# *Extension to Include Truncation*

This chapter extends the last chapter by including truncation. While left truncation is common in survival data, right truncation is also frequently encountered. In this chapter, we will delve into the most comprehensive truncation setting, known as double truncation, where both left and right truncation are special cases. In this chapter, we will address interval censoring within the context of double truncation.

The simplest combination of truncation and censoring is left truncation and right censoring. In this scenario, if the Cox model baseline hazard is approximated by a piecewise constant function, we can devise a profile likelihood method to obtain maximum likelihood estimates. This approach is discussed in Section 3.2. More complex scenarios involve double truncation and interval censoring, and they are covered in Sections 3.3 through 3.8, with R examples provided in Section 3.9.

## 3.1 Introduction

Truncation is a phenomenon that is different from censoring, and truncation also frequently appears in survival data. For instance, individuals may never have the opportunity to participate in a clinical trial if they fail to meet the required condition for recruitment into this study.

It is essential to comprehend the difference between "censoring" and "truncation". Fundamentally, censoring means that a censored survival time is *partially* observed as it still contains information about the model parameters. In contrast, truncation means that an event time that could have been observed becomes entirely unobservable if this event time falls into the truncated time interval. We will refer to such an interval as the *truncation interval* from now on. For example, an event time $Y$ is left truncated at $t = 1$ (so that the interval $(0, 1]$ is the truncation interval for $Y$) implies $Y$ cannot be observed at all when it is not greater than 1. In comparison, a left censoring time at $t = 1$ means the event has already occurred before $t = 1$, so that $Y \leq 1$.

Truncation can manifest in various studies, including clinical trials, often stemming from specific conditions imposed during the collection of survival data. For instance, consider a clinical trial designed to investigate the effects of a particular drug. Participants might be enrolled only if they remain symptom-free after one month.

DOI: 10.1201/9781351109710-3

This design introduces left truncation, which is sometimes referred to as *delayed entry* (e.g. Jin et al. (2022)).

Neglecting truncation can introduce biases into the estimates of model parameters. In the case of delayed entry, if one reset the time origin (i.e. time 0) of each individual to the time point when the follow-up for this individual starts, and then analyzing the survival data based on these new time origins, then the left truncation (or delayed entry) is basically ignored. The bias arising from disregarding left truncation is categorized as a form of selection bias, as elaborated in references such as Schisterman et al. (2013). We will consider in this chapter a general truncation scheme which includes left, or right or both left and right truncation.

Depending on the study design, each individual may possess a distinct truncation interval, or otherwise a fixed truncation interval may be shared among all participants in the study. Left truncation, the most elementary and frequently encountered truncation type, pertains to scenarios in which the truncation interval for an event time $Y$ is defined as $(0, L]$, with $L$ being a known truncation time. Here, $L$ is termed the left truncation time. Correspondingly, right truncation (or truncation from above) involves situations where $Y$ remains entirely unobservable if it falls within the interval $[R, +\infty)$, where $R$ denotes the right truncation time.

It is possible that left and right truncation can co-exist in a study, and this is called double truncation. An example of double truncation in a clinical trial on dementia is given in Rennert & Xie (2019).

Consideration of truncation entails that the event time density function, as well as the hazard, cumulative hazard and survival functions, need to be adjusted. These adjustments are required to account for the absence of event times within the truncation intervals. Consequently, this complexity extends to the resulting likelihood function, rendering the task of obtaining likelihood-based estimates more challenging.

In this chapter, we will build upon the partly interval-censored survival data discussed in the last chapter by introducing truncation. This means that in addition to dealing with interval-censored survival times, we will also consider scenarios where truncation affects the event time of interest. Our focus will be on a general truncation scheme that encompasses both left and right truncation, often referred to as double truncation. It is important to note that either left truncation or right truncation can be seen as special cases within this general truncation framework.

Our emphasis in this chapter will center around devising a computational approach for obtaining the maximum penalized likelihood estimates of the Cox model parameters. These parameters encompass the regression coefficients and the nonparametric baseline hazard, all within the context of observed survival times that involve truncation and interval censoring. Here, truncation can be left, or right, or double truncation. We will also delve into practical R examples illustrating the fitting of Cox models using survival times that are interval-censored and subject to general truncation.

The discussions in this chapter are based on the assumption that truncation is non-informative about the event time. Thus, the likelihood function for parameter estimation is built without utilizing the marginal density of the truncation time. However, it is important to highlight that incorporating the marginal density of the truncation time holds the promise of yielding more efficient estimates (i.e. with smaller variance). For a deeper exploration of this approach, readers can consult references like Wang et al. (2021).

## 3.2 Profile likelihood for left truncation and right censoring

We start this chapter with the simplest context of left-truncated and right-censored survival times. In this context and with a piecewise constant approximation to the baseline hazard, a profile log-likelihood method can be developed, where the final model-fitting algorithm is very similar to the partial likelihood method that accommodates both left truncation and right censoring.

Left truncation is the most common truncation type, where the event time of an individual happens before a certain time point (often due to imposed condition(s) when collecting data) will not be collected. For example, an example given in Dörre & Emura (2019) states that "if an electric machine has been broken and discarded before the initiation of life testing, the machine and its lifetime is said to be left-truncated".

We denote the event time for individual $i$ by $Y_i$, where $Y_i$ is subject to left truncation and right censoring. We let $L_i$ denote the left truncation time for $i$ and $C_i$ its right censoring time. We assume both $L_i$ and $C_i$ are independent of $Y_i$ given the covariates of subject $i$. Let $Y_i^*$ be the left truncated $Y_i$, namely $Y_i^* \equiv \{Y_i \mid Y_i > L_i\}$. Then, the observable survival time is $T_i^* = \min(Y_i^*, C_i)$. Here, we have assumed implicitly that $C_i > L_i$. An observed $T_i^*$ is denoted by $t_i^*$.

We let $(L_i, t_i^*, \delta_i, \mathbf{x}_i)$ be the observations for individual $i$, where $\mathbf{x}_i$ represents the vector of values for $p$ covariates, and $L_i$ is the left truncation (or delayed entry) time for individual $i$. Although the observations we have are from $Y_i^*$, we wish to infer properties of the distribution of $Y_i$, rather than the distribution of $Y_i^*$. This means the Cox model we wish to fit is:

$$h_{Y_i}(t|\mathbf{x}_i) = h_0(t)e^{\mathbf{x}_i^\top \boldsymbol{\beta}}. \tag{3.1}$$

Before starting explanations on parameter estimation for the model (3.1) using truncated and censored data, let us first explore an example demonstrating how to simulate survival data with left truncation and right censoring.

**Example 3.1** (Simulate left-truncated and right-censored survival times).
In this example, we will explain how to generate Cox model survival times subject to left truncation and right censoring. The adopted model is a Cox model with a sample size of $n = 200$, where:
(i) the baseline hazard is from a log-logistic distribution (see Example 1.15 to understand how to generate random event times from a Cox model with a log-logistic baseline hazard);

(ii) covariates $X_1 \sim Binomial(1, 0.7)$ and $X_2 \sim unif(1, 2)$, with coefficients $\beta_1 = 1$ for $X_1$ and $\beta_2 = -0.5$ for $X_2$;
(iii) left truncation times $L_i \sim$ unif$(1, 3)$; and
(iv) the goal is to have a 60% right-censoring proportion.

```
trunc_sim_data3.1 <- function(nsample,
    beta, prob_left_0, right_cens_par1,
    right_cens_par2) {

    n <- 0  #set current sample size to 0
    dat <- NULL
    n_all <- 0  #set total number of simulated data sets to 0

    while (n < nsample) {
        # generate covariates
        X <- cbind(rbinom(1, 1,
            0.7), runif(1, 1, 2))
        n_all <- n_all + 1

        # generate event time
        neg.log.u <- -log(runif(1))
        mu_term <- exp(X %*% beta)
        y_i <- as.numeric((((4 *
            neg.log.u/(2 * mu_term)) +
            1)^(1/2) - 1)^(1/2))

        # generate left
        # truncation time
        L_i <- rbinom(1, 1, 1 -
            prob_left_0) * runif(1,
            1, 3)

        # generate right
        # truncation time
        R_i <- Inf

        # generate
        # independent right
        # censoring time
        cens_i <- runif(1, right_cens_par1,
            right_cens_par2)

        TL <- TR <- censor_type <- NA

        if (y_i > cens_i) {
            # right censored
            TL <- cens_i
            TR <- Inf
            censor_type <- 0
        } else {
            TL <- y_i
            TR <- y_i
            censor_type <- 1
        }

        if (censor_type == 1 &
```

```
            L_i < TL & TR <= R_i) {
            # if event time
            # is inside the
            # truncation
            # interval then
            # individual is
            # in the sample
            dat <- rbind(dat, c(y_i,
                cens_i, L_i, R_i,
                X, TL, TR))
            n <- n + 1  #increase current sample size by 1
        } else if (censor_type ==
            0 & L_i < TL & TR <=
            R_i) {
            # if right
            # censoring time
            # is inside the
            # truncation
            # interval then
            # individual is
            # in the sample
            dat <- rbind(dat, c(y_i,
                cens_i, L_i, R_i,
                X, TL, TR))
            n <- n + 1  #increase current sample size by 1
        }
    }

    trunc_data <- data.frame(dat)
    colnames(trunc_data) <- c("y_i",
        "cens_i", "Ltrunc", "Rtrunc",
        "x1", "X2", "Lcen", "Rcen")

    # censoring type
    trunc_data$censor_type <- NA
    trunc_data$censor_type[which(is.infinite(trunc_data$Rcen))] <- 0
    trunc_data$censor_type[which(trunc_data$Lcen ==
        trunc_data$Rcen)] <- 1

    # truncation type
    trunc_data$trunc_type = NA
    trunc_data$trunc_type[which(trunc_data$Ltrunc ==
        0 & trunc_data$Rtrunc ==
        Inf)] <- 1
    trunc_data$trunc_type[which(trunc_data$Ltrunc !=
        0 & trunc_data$Rtrunc ==
        Inf)] <- 2

    out <- list(trunc_data = trunc_data,
        n_all = n_all)
    return(out)
}
```

We then run this R program and the results are displayed below.

```
> trunc_data_event40  <- trunc_sim_data3.1(nsample = 200,
beta = c(1, -0.5), prob_left_0 = 0, right_cens_par1 = 0.2,
right_cens_par2 = 1.4)
>
> data <- trunc_data_event40$trunc_data
>
> format(head(data), digits = 4)
    y_i cens_i Ltrunc Rtrunc x1    X2  Lcen  Rcen cen_typ trun_typ
1 1.444  1.371  1.347    Inf  0 1.091 1.371   Inf       0        2
2 1.508  1.373  1.182    Inf  0 1.891 1.373   Inf       0        2
3 1.309  1.227  1.182    Inf  0 1.150 1.227   Inf       0        2
4 1.278  1.292  1.136    Inf  0 1.924 1.278 1.278       1        2
5 1.223  1.160  1.037    Inf  1 1.428 1.160   Inf       0        2
6 1.134  1.111  1.091    Inf  0 1.498 1.111   Inf       0        2
>
> #check censoring level
> table(data$cen_typ)

  0   1
109  91
>
> #check truncation types (2 = left truncated)
> table(data$trun_typ)

  2
200
>
> #check truncation proportion
> (trunc_data_event40$n_all - 200)/trunc_data_event40$n_all
[1] 0.9941382
```

□

The estimation of $\beta$ and $h_0(t)$ of the Cox model (3.1) can be achieved by maximum likelihood. Two types of likelihood functions are commonly discussed in the context of truncated survival times. The first type, called the conditional likelihood and as illustrated by, for example, Efron & Petrosian (1999) and Klein & Moeschberger (2003), is constructed under the assumption that the truncation time is independent of the event time and does not convey information about the model parameters of interest. In this context, the likelihood is constructed using the conditional distribution of $Y_i \mid Y_i > L_i$, which is represented as the distribution of $Y_i^*$. The second type, as illustrated for example in Shao et al. (2023), is constructed using the joint density of $(L_i, Y_i)$. A specific motivation behind using this type of likelihood is that conditional likelihood approaches may not always provide good efficiency, potentially leading to larger variances in the estimated parameters.

In this chapter, however, we will focus on conditional likelihood approaches, as our experience suggests that the gains in efficiency are not substantial, while the costs associated with optimizing the likelihood, which is derived from the joint density of $(L_i, Y_i)$, are significant.

To calculate the conditional likelihood, we first need to determine the density and survival functions of $Y_i^*$, a task that is not particularly challenging. In fact, for $t > L_i$ we have

$$f_{Y_i^*}(t) = f_{Y_i}(t|Y_i > L_i) = \frac{f_{Y_i}(t)}{S_{Y_i}(L_i)}, \text{ and}$$
$$S_{Y_i^*}(t) = \frac{S_{Y_i}(t)}{S_{Y_i}(L_i)}.$$

Then, the likelihood function from the given data $(L_i, t_i^*, \delta_i, \mathbf{x}_i)$, where $i = 1, \ldots, n$, is:

$$L^* = \prod_{i=1}^{n} \left(\frac{f_{Y_i}(t_i^*)}{S_{Y_i}(L_i)}\right)^{\delta_i} \left(\frac{S_{Y_i}(t_i^*)}{S_{Y_i}(L_i)}\right)^{1-\delta_i}. \tag{3.2}$$

Substituting the relationship $f_{Y_i}(t) = h_{Y_i}(t)S_{Y_i}(t)$ into (3.2) and taking logarithm, we obtain the log-likelihood, given by:

$$\begin{aligned} l^* &= \sum_{i=1}^{n} \Big(\delta_i \log h_{Y_i}(t_i^*) + \log S_{Y_i}(t_i^*) - \log S_{Y_i}(L_i)\Big) \\ &= \sum_{i=1}^{n} \delta_i(\log h_0(t_i^*) + \mathbf{x}_i^\top \boldsymbol{\beta}) - \sum_{i=1}^{n} e^{\mathbf{x}_i^\top \boldsymbol{\beta}} H_0(L_i, t_i^*). \end{aligned} \tag{3.3}$$

Here, $H_0(L_i, t_i^*) = H_0(t_i^*) - H_0(L_i) = \int_{L_i}^{t_i^*} h_0(s)ds$, where $H_0(t)$ is the cumulative baseline hazard.

Next, We will introduce a simple, but efficient, profile likelihood approach to estimate $\boldsymbol{\beta}$ when $h_0(t)$ is approximated by a piecewise constant function. We will borrow the notations used for piecewise constant approximation in Chapter 2 in the following discussions.

To define a piecewise constant approximation to $h_0(t)$ we divide the support of $h_0(t)$, denoted by $[a, b]$ where $a = \min_i\{t_i^*\}$ and $b = \max_i\{t_i^*\}$, into $m$ mutually exclusive and exhaustive bins $\mathcal{B}_1, \ldots, \mathcal{B}_m$ such that $[a, b] = \bigcup_{u=1}^{m} \mathcal{B}_u$. The knots we used to perform this partition are denoted by: $w_1, w_2, \ldots, w_m, w_{m+1}$, where the boundary knots are $w_1 = a$ and $w_{m+1} = b$. Hence, $\mathcal{B}_u = (w_u, w_{u+1}]$, where $u = 1, \ldots, m$.

Based on this partition, we can approximate $h_0(t)$ by

$$\begin{aligned} h_0(t) &\approx \sum_{u=1}^{m} \theta_u I\{t \in \mathcal{B}_u\} \\ &= \theta_u, \quad \text{if } t \in \mathcal{B}_u, \end{aligned}$$

where $u = 1, \ldots, m$. From such an approximated $h_0(t)$, the $H_0(L_i, t_i^*)$ function in (3.3) can be expressed as

$$H_0(L_i, t_i^*) = \sum_{u=1}^{m} \theta_u W_u I\{w_u \le t_i^*, w_{u+1} \ge L_i\}, \tag{3.4}$$

where $\mathcal{W}_u = w_{u+1} - w_u$, the bin-width of $\mathcal{B}_u$. From this, the log-likelihood in (3.3) becomes

$$l^* = \sum_{i=1}^{n} \delta_i \mathbf{x}_i^\mathsf{T} \boldsymbol{\beta} + \sum_{u=1}^{m} n_u \log \theta_u - \sum_{i=1}^{n} e^{\mathbf{x}_i^\mathsf{T} \boldsymbol{\beta}} \sum_{u=1}^{m} \theta_u \mathcal{W}_u I\{w_u \le t_i^*, w_{u+1} \ge L_i\},$$

where $n_u = \{\text{number of event times in bin } \mathcal{B}_u\}$. Now, since

$$\frac{\partial l^*}{\partial \theta_u} = \frac{n_u}{\theta_u} - \sum_{i=1}^{n} e^{\mathbf{x}_i^\mathsf{T} \boldsymbol{\beta}} \mathcal{W}_u I\{w_u \le t_i^*, w_{u+1} \ge L_i\},$$

the solution of $\theta_u$ from solving the equation $\partial l^* / \partial \theta_u = 0$ can be expressed in a closed form as

$$\theta_u = \frac{n_u}{\mathcal{W}_u \sum_{i=1}^{n} e^{\mathbf{x}_i^\mathsf{T} \boldsymbol{\beta}} I\{w_u \le t_i^*, w_{u+1} \ge L_i\}}. \tag{3.5}$$

We comment that:

(i) The solution for $\theta_u$ provided in (3.5) always satisfies $\theta_u \ge 0$, where equality is met only when $n_u = 0$.

(ii) The summation in the denominator in (3.5) clarifies that an individual $i$ contributes to the calculation of $\theta_u$ (associated with interval $\mathcal{B}_u$) only if (a) their observation is greater than $w_u$, the left limit of $\mathcal{B}_u$, and (b) their left truncation time is located before $w_{u+1}$, the right limit of $\mathcal{B}_u$. These conditions define a risk set for estimating $\boldsymbol{\beta}$ using the profile log-likelihood, with further discussions given below.

Therefore, by substituting the solutions for all $\theta_u$ back into the log-likelihood $l^*$, we obtain the so-called profile log-likelihood, which is a function of $\boldsymbol{\beta}$ only and can be written as

$$\begin{aligned} l_p^*(\boldsymbol{\beta}) &= \sum_{i=1}^{n} \delta_i \mathbf{x}_i^\mathsf{T} \boldsymbol{\beta} + \sum_{u=1}^{m} n_u \log \theta_u - \sum_{u=1}^{m} \theta_u \mathcal{W}_u \sum_{i=1}^{n} e^{\mathbf{x}_i^\mathsf{T} \boldsymbol{\beta}} I\{w_u \le t_i^*, w_{u+1} \ge L_i\} \\ &\propto \sum_{i=1}^{n} \delta_i \mathbf{x}_i^\mathsf{T} \boldsymbol{\beta} - \sum_{u=1}^{m} n_u \log \Big( \sum_{i=1}^{n} e^{\mathbf{x}_i^\mathsf{T} \boldsymbol{\beta}} I\{w_u \le t_i^*, w_{u+1} \ge L_i\} \Big), \end{aligned}$$

where the terms independent of $\boldsymbol{\beta}$ are ignored.

Using the profile log-likelihood, we can estimate $\boldsymbol{\beta}$ by

$$\widehat{\boldsymbol{\beta}} = \underset{\boldsymbol{\beta}}{\operatorname{argmax}}\, l_p^*(\boldsymbol{\beta}). \tag{3.6}$$

To achieve this, the Newton algorithm can be used to calculate $\widehat{\boldsymbol{\beta}}$ iteratively through:

$$\boldsymbol{\beta}^{(k+1)} = \boldsymbol{\beta}^{(k)} - \left[ \frac{\partial^2 l_p^*(\boldsymbol{\beta}^{(k)})}{\partial \boldsymbol{\beta} \partial \boldsymbol{\beta}^\mathsf{T}} \right]^{-1} \frac{\partial l_p^*(\boldsymbol{\beta}^{(k)})}{\partial \boldsymbol{\beta}}. \tag{3.7}$$

To implement this algorithm, we need to work out the gradient $\partial l_p^*(\boldsymbol{\beta}^{(k)})/\partial\boldsymbol{\beta}$ and the Hessian matrix $\partial^2 l_p^*(\boldsymbol{\beta}^{(k)})/\partial\boldsymbol{\beta}\partial\boldsymbol{\beta}^\mathsf{T}$.

Elements of the gradient vector are

$$\begin{aligned}\frac{\partial l_p^*}{\partial \beta_j} &= \sum_{i=1}^n \delta_i x_{ij} - \sum_{u=1}^m n_u \sum_{i=1}^n x_{ij} \frac{e^{\mathbf{x}_i^\mathsf{T}\boldsymbol{\beta}} I\{w_u \le t_i^*, w_{u+1} \ge L_i\}}{\sum_{i=1}^n e^{\mathbf{x}_i^\mathsf{T}\boldsymbol{\beta}} I\{w_u \le t_i^*, w_{u+1} \ge L_i\}} \\ &= \sum_{i=1}^n x_{ij}\Big(\delta_i - \sum_{u=1}^m n_u \widetilde{w}_{iu}\Big), \qquad (3.8)\end{aligned}$$

where $\widetilde{w}_{iu}$ is given by

$$\widetilde{w}_{iu} = \frac{e^{\mathbf{x}_i^\mathsf{T}\boldsymbol{\beta}} I\{w_u \le t_i^*, w_{u+1} \ge L_i\}}{\sum_{i=1}^n e^{\mathbf{x}_i^\mathsf{T}\boldsymbol{\beta}} I\{w_u \le t_i^*, w_{u+1} \ge L_i\}}.$$

As $\widetilde{w}_{iu} \ge 0$ and $\sum_i \widetilde{w}_{iu} = 1$, these $\widetilde{w}_{iu}$'s can be conceived as weights assigned from each bin $\mathcal{B}_u$ to different individuals in the sample. Now, the gradient vector can be expressed in a matrix form:

$$\frac{\partial l_p^*}{\partial \boldsymbol{\beta}} = \mathbf{X}^\mathsf{T}(\boldsymbol{\delta} - \widetilde{\mathbf{W}}\mathbf{n}_m), \qquad (3.9)$$

where $\mathbf{X}$ is the model matrix (also called the design matrix), $\boldsymbol{\delta} = (\delta_1, \ldots, \delta_n)^\mathsf{T}$, $\widetilde{\mathbf{W}} = (\widetilde{w}_{iu})_{n\times m}$ and $\mathbf{n}_m = (n_1, \ldots, n_m)^\mathsf{T}$.

Elements of the Hessian are:

$$\begin{aligned}\frac{\partial^2 l_p^*}{\partial \beta_j \partial \beta_t} &= -\sum_{u=1}^m n_u \Big(\sum_{i=1}^n x_{ij} x_{it} \widetilde{w}_{iu} - \overline{x}_j^u \overline{x}_t^u\Big) \qquad (3.10) \\ &= -\sum_{u=1}^m n_u \sum_{i=1}^n (x_{ij} - \overline{x}_j^u)\widetilde{w}_{iu}(x_{it} - \overline{x}_t^u), \qquad (3.11)\end{aligned}$$

where $\overline{x}_j^u = \sum_i \widetilde{w}_{iu} x_{ij}$, and $\overline{x}_t^u$ is similarly defined. Hence, the Hessian can be expressed using matrices:

$$\frac{\partial^2 l_p^*}{\partial \boldsymbol{\beta}\partial \boldsymbol{\beta}^\mathsf{T}} = -\mathbf{X}^\mathsf{T}\mathrm{diag}\Big(\sum_{u=1}^m n_u \widetilde{\mathbf{W}}_{[,u]}\Big)\mathbf{X} - \overline{\mathbf{X}}\mathrm{diag}(\mathbf{n}_m)\overline{\mathbf{X}}^\mathsf{T}, \qquad (3.12)$$

where $\widetilde{\mathbf{W}}_{[,u]}$ denotes the $u$-th column of matrix $\widetilde{\mathbf{W}}$ and $\overline{\mathbf{X}} = \mathbf{X}^\mathsf{T}\widetilde{\mathbf{W}}$. Note that the dimension of $\overline{\mathbf{X}}$ is $p \times m$.

The gradient and Hessian formulas given in (3.9) and (3.12) make the Newton algorithm extremely easy to implement. The example given below contains a complete R function for fitting Cox models with left truncation and right censoring survival data.

**Example 3.2** (R program for profile MLE ).
In this example, we provide a comprehensive R program for performing maximum profile likelihood estimation of the regression coefficients vector $\boldsymbol{\beta}$ in a Cox model under left truncation and right censoring.

```
mpleltrc <- function(Y, X, bin_num, beta0,
  maxiter=2000, tol=1e-5) {
  Y <- as.matrix(Y)
  X <- as.matrix(X)
  n <- nrow(X)
  p <- ncol(X)

  if(missing(beta0)) {beta0 <- matrix(1,p,1)}
  else {beta0 <- matrix(beta0,p,1)}

  if(missing(bin_num)) {bin_num <- floor(sqrt(sum(Y[,3])))}

  num_boundary <- bin_num+1
  boundary <- unname(quantile(unique(Y[Y[,3]==1,2]),
    probs = seq(0, 100, length.out=num_boundary)/100))
  boundary[1] <- min(Y[,2])
  boundary[num_boundary] <- max(Y[,2])
  upper_b <- boundary[2:num_boundary]
  lower_b <- boundary[1:(num_boundary-1)]
  binwid <- upper_b-lower_b

  h <- hist(Y[Y[,3]==1,2], breaks = boundary, right = F,
    plot = F)
  binCount <- matrix(h$counts,bin_num,1)
  bin_ind <- t(sapply(1:n, function(i){Y[i,1]
    <=upper_b & Y[i,2]>lower_b}))


  Xbeta <- X%*%beta0
  A_mat <- bin_ind*matrix(rep(exp(Xbeta),bin_num),n,bin_num)
  sumA_mat <- apply(A_mat,2,sum)
  sumA_mat[sumA_mat==0] <- 1e-10
  weights <- A_mat/matrix(rep(sumA_mat,n),n,bin_num,
    byrow = T)
  allpl <- NULL

  betaold <- beta0
  for(k in 1:maxiter) {
    proflik <- Y[,3]%*%Xbeta-log(sumA_mat)%*%binCount
    allpl <- c(allpl, proflik)
    gradient <- t(X)%*%(Y[,3]-weights%*%binCount)
    Xbar_mat <- t(X)%*%weights
    xtwx <- matrix(0,p,p)
    for(u in 1:bin_num) {
      xtmp <- t(X)%*%(matrix(rep(weights[,u],p),n,p)*X)
      xtwx <- xtwx+binCount[u,]*xtmp
    }
    profhess <- xtwx-Xbar_mat
      %*% diag(as.numeric(binCount))%*%t(Xbar_mat)

    betanew <- betaold + solve(profhess)%*%gradient
```

```
    Xbeta <- X%*%betanew
    A_mat <- bin_ind*matrix(rep(exp(Xbeta),bin_num),n,bin_num)
    sumA_mat <- apply(A_mat,2,sum)
    sumA_mat[sumA_mat==0] <- 1e-10
    weights <- A_mat/matrix(rep(sumA_mat,n),n,bin_num, byrow = T)
    if(all(abs(betaold-betanew)<tol)) {break}
    else{betaold <- betanew}

  }
  theta <- as.numeric(binCount)/binwid*sumA_mat
  retheta <- theta/binwid

  return(list(betanew, retheta, gradient, allpl,
    sqrt(diag(solve(profhess)))))
}
```

□

As per Murphy & van der Vaart (2000), the profile likelihood can be used as a complete likelihood, and it is common practice to employ the curvature of the profile likelihood function to estimate the variability of the estimated parameters. In fact, under some specific regularity conditions, the distribution of $\sqrt{n}(\widehat{\boldsymbol{\beta}} - \boldsymbol{\beta}_0)$, where $\boldsymbol{\beta}_0$ represents the true coefficient parameter and $\widehat{\boldsymbol{\beta}}$ is the maximum profile likelihood estimate, converges to a multivariate normal distribution with a mean vector $\mathbf{0}$ and a covariance matrix of

$$-\left[n^{-1}\frac{\partial^2 l_p^*(\widehat{\boldsymbol{\beta}})}{\partial\boldsymbol{\beta}\partial\boldsymbol{\beta}^\mathsf{T}}\right]^{-1}.$$

**Example 3.3** (An R simulation study).
In this example, we conduct a simple simulation study to compare the profile likelihood and partial likelihood approaches and to demonstrate the asymptotic properties of the profile likelihood estimate. In this example, we conduct the following tasks:

(1) The simulated survival data are left-truncated and right-censored, generated in the same way as Example 3.1, with 300 repetitions and sample sizes of $n = 100$ and $n = 1000$.
(2) The R programs used for model fitting in this example are as follows:
(i) for the profile likelihood method, we use the R program provided in Example 3.2;
(ii) for the partial likelihood method, we utilize the R `coxph` function in the `survival` package.

The results are displayed in Table 3.1 and Figure 3.1 below. The plots in Figure 3.1 clearly demonstrate that the distribution of each estimated $\beta$ is approximately normal.

```
## An R simulation study

# create empty matrix to save results
save_lefttrunc_sim <- matrix(0, nrow = 300, ncol = 8)

for (s in 1:300) {
    # simulate data
    trunc_data_event40 <- trunc_sim_data3.1(nsample = 1000,
        beta = c(1, -0.5), prob_left_0 = 0, right_cens_par1 = 0.55,
        right_cens_par2 = 1.5)

    # fit profile likelihood method
    trunc.y <- trunc_data_event40$trunc_data[, c(3,
        7, 9)]
    design <- trunc_data_event40$trunc_data[, 5:6]
    prof.trunc <- mpleltrc(trunc.y, design, 100)

    save_lefttrunc_sim[s, 1] <- prof.trunc[[1]][1]
    save_lefttrunc_sim[s, 2] <- prof.trunc[[1]][2]
    save_lefttrunc_sim[s, 3] <- prof.trunc[[5]][1]
    save_lefttrunc_sim[s, 4] <- prof.trunc[[5]][2]

    # fit partial likelihood method
    pl.trunc <- coxph(Surv(trunc_data_event40$trunc_data$Lcen,
        trunc_data_event40$trunc_data$censor_type) ~
        trunc_data_event40$trunc_data$x + trunc_data_event40$trunc_data$X2)

    save_lefttrunc_sim[s, 5] <- pl.trunc$coefficients[1]
    save_lefttrunc_sim[s, 6] <- pl.trunc$coefficients[2]
    save_lefttrunc_sim[s, 7] <- sqrt(pl.trunc$var[1,
        1])
    save_lefttrunc_sim[s, 8] <- sqrt(pl.trunc$var[2,
        2])

}
```

TABLE 3.1: Simulation results from Example 3.3 and comparison with the partial likelihood method.

| | | $n = 100$ | | | $n = 1000$ | | |
|---|---|---|---|---|---|---|---|
| | | Bias | SE | CP | Bias | SE | CP |
| Profile Lik. | $\beta_1$ | -0.028 | 0.326 (0.337) | 0.93 | -0.009 | 0.099 (0.092) | 0.97 |
| | $\beta_2$ | -0.088 | 0.516 (0.554) | 0.94 | 0.001 | 0.156 (0.167) | 0.93 |
| Partial Lik. | $\beta_1$ | 0.106 | 0.327 (0.314) | 0.94 | 0.108 | 0.099 (0.096) | 0.81 |
| | $\beta_2$ | -0.163 | 0.519 (0.572) | 0.92 | -0.066 | 0.156 (0.161) | 0.95 |

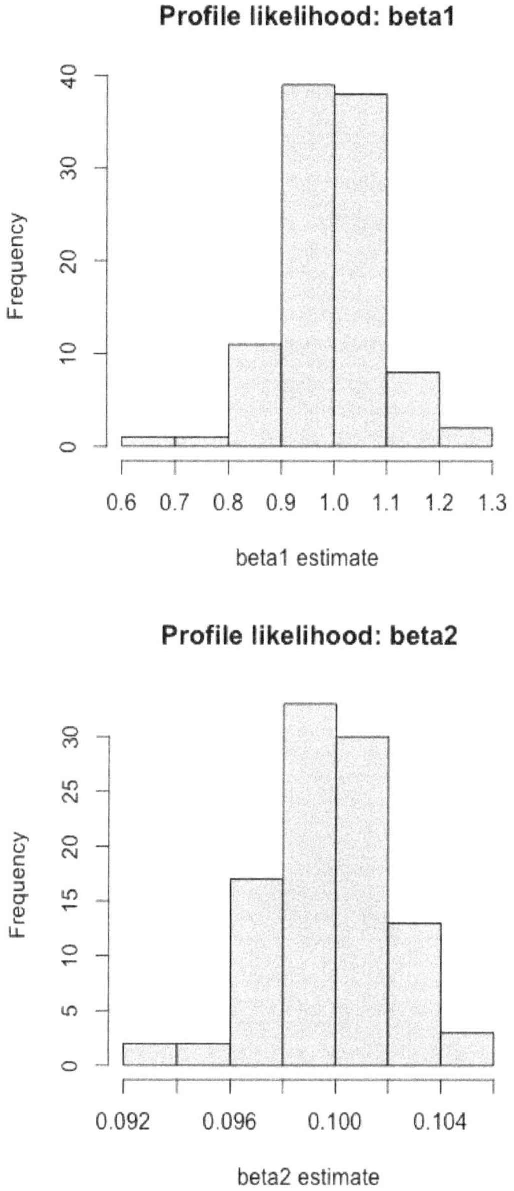


FIGURE 3.1: Simulation results from Example 3.3. Top panel: histogram of the $\beta_1$ estimates; buttom panel: histogram of the $\beta_2$ estimates.

□

## 3.3 Doubly truncated and interval-censored survival data

In this section, we extend the left truncation and right censoring discussed in the last section to more general double truncation and interval censoring. We start this section by introducing the notations necessary for our discussions.

For an individual $i$, as in Section 3.2, we let $Y_i$ denote the true event time. However, under truncation, $Y_i$ is not observed at all if it falls within the truncation interval. If $Y_i$ can be observed, we assume it is interval-censored, meaning $Y_i$ can be fully observed or interval-censored.

Under the assumption of double truncation, $Y_i$ will not be observed if either $Y_i \leq L_i$ (left truncation) or $Y_i \geq R_i$ (right truncation), where $L_i < R_i$. Thus, $Y_i$ can be observed only if $Y_i$ is in the non-truncation interval: i.e. $L_i < Y_i < R_i$, where $L_i$ and $R_i$ are assumed known. We introduce the notation $Y_i^* \equiv \{Y_i \,|\, L_i < Y_i < R_i\}$, so that $Y_i^*$ represents the doubly truncated $Y_i$.

It is clear that when $L_i = 0$ and $R_i = +\infty$, then there is no truncation for individual $i$ so that $Y_i^* = Y_i$ unconditionally. If $L_i$ is finite and non-zero, and $R_i = +\infty$, then we have a left truncation, while the situation of $L_i = 0$ and $R_i$ finite indicates a right truncation. When both $L_i \neq 0$ and $R_i \neq +\infty$, we have the double truncation situation. Therefore, left, right and double truncation can all be specified with special $L_i$ and $R_i$ values and they can co-exist in a dataset.

Recall $Y_i^*$ denotes the doubly truncated survival time for individual $i$, where $Y_i^* \in (L_i, R_i)$. For $Y_i^*$, we first need to workout its density, survival and hazard function. For simplicity, we assume $Y_i^*$ is a continuous random variable. The survival function of $Y_i^*$ is simply

$$S_{Y_i^*}(t) = \Pr(Y_i^* > t),$$

where $L_i < t < R_i$. This survival function can be computed from the survival function of $Y_i$ due to the relationship

$$S_{Y_i^*}(t) = \Pr(t < Y_i < R_i \,|\, L_i < Y_i < R_i). \tag{3.13}$$

The hazard function of $Y_i^*$ at time $t$ depicts the instantaneous chance of an event for $i$ just after time $t$ given it is event free up to $t$. Thus, the hazard of $Y_i^*$ is:

$$h_{Y_i^*}(t) = \lim_{\Delta t \to 0} \frac{\Pr(t < Y_i^* < t + \Delta t \,|\, Y_i^* > t)}{\Delta t}. \tag{3.14}$$

It is easy to show that the hazard function $h_{Y_i^*}(t)$, the survival function $S_{Y_i^*}(t)$ and the density function $f_{Y_i^*}(t)$ obey the usual relationship between them, namely

$$h_{Y_i^*}(t) = \frac{f_{Y_i^*}(t)}{S_{Y_i^*}(t)}, \tag{3.15}$$

$$\frac{d \log S_{Y_i^*}(t)}{dt} = -h_{Y_i^*}(t), \tag{3.16}$$

$$S_{Y_i^*}(t) = \exp\{-H_{Y_i^*}(t)\}, \tag{3.17}$$

where $H_{Y_i^*}(t)$ is the cumulative hazard function given by

$$H_{Y_i^*}(t) = \int_{L_i}^{t} h_{Y_i^*}(s)ds,$$

for $L_i < t < R_i$.

Observations are now collected on $Y_i^*$, but the Cox model one wishes to fit is still the model on $Y_i$ as given in (3.1). Hence, it necessitates to express the density, hazard and survival functions of $Y_i^*$ using the relevant functions of $Y_i$. The conditional log-likelihood function we will establish later is derived from the distribution of $Y_i^*$.

Firstly, according to (3.13), the survival function for $Y_i^*$ can be expressed as

$$S_{Y_i^*}(t) = \frac{S_{Y_i}(t) - S_{Y_i}(R_i)}{S_{Y_i}(L_i) - S_{Y_i}(R_i)}, \quad L_i < t < R_i, \tag{3.18}$$

where $S_{Y_i}(t)$ is the survival function of $Y_i$. Accordingly, the density function of $Y_i^*$ is expressed as a scaled density of $Y_i$:

$$f_{Y_i^*}(t) = \frac{f_{Y_i}(t)}{S_{Y_i}(L_i) - S_{Y_i}(R_i)}, \quad L_i < t < R_i, \tag{3.19}$$

where $f_{Y_i}(t)$ is the density function of $Y_i$. Combining (3.18) and (3.19), the hazard function for $Y_i^*$ is then

$$\begin{aligned} h_{Y_i^*}(t) &= \frac{f_{Y_i^*}(t)}{S_{Y_i^*}(t)} \\ &= \frac{f_{Y_i}(t)}{S_{Y_i}(t) - S_{Y_i}(R_i)} \\ &= \frac{h_{Y_i}(t)}{1 - S_{Y_i}(R_i)/S_{Y_i}(t)}, \end{aligned} \tag{3.20}$$

where $L_i < t < R_i$. This formula explains that the hazards for $Y_i^*$ and $Y_i$ at any time $t \in (L_i, R_i)$ are related nonlinearly with a scaling factor which is determined by a ratio of survival function values at $R_i$ and $t$.

It is interesting to note that the equation in (3.16) can also be verified using (3.18). In fact, from (3.18),

$$\begin{aligned} \frac{d \log S_{Y_i^*}(t)}{dt} &= -\frac{f_{Y_i^*}(t)}{S_{Y_i}(t) - S_{Y_i}(R_i)} \\ &= -h_{Y_i^*}(t). \end{aligned}$$

Next, we introduce the concept of interval censoring for $Y_i^*$ before deriving the conditional likelihood outlined in Section 3.4. This conditional likelihood is necessary for achieving the maximum conditional likelihood estimation. Similar to Section 2.2, for $Y_i^*$, we assume $Y_i^* \in [t_i^{*L}, t_i^{*R}]$, where $t_i^{*L}$ and $t_i^{*R}$ are a pair of values defining the censoring interval. Note that the censoring times $t_i^{*L}$ and $t_i^{*R}$

must be observable, so they should fall within the non-truncation interval, namely $L_i \leq t_i^{*L} \leq t_i^{*R} \leq R_i$.

For $Y_i^*$ (where $L_i \leq Y_i^* \leq R_i$), exactly observed $Y_i^*$ or interval-censored $Y_i^*$ can all be defined similarly to those of $Y_i$ in Section 2.2. Specifically, if $t_i^{*L} = t_i^{*R}$ $(= t_i^*)$, then we have that $Y_i^*$ has an exact event time at $t_i^*$. If $t_i^{*L} \neq t_i^{*R}$ and $L_i \neq 0$ and $R_i \neq \infty$, then $Y_i^*$ is interval-censored in the interval $[t_i^{*L}, t_i^{*R}]$.

Recall that in the context of double truncation, an event time $Y_i$ is unobservable if it occurs before $L_i$ or after $R_i$. If $t_i^{*L} = L_i$ and $t_i^{*R} \neq R_i$ then $Y_i^*$ is interpreted as being "left-censored" at $t_i^{*R}$ as $Y_i^* \in [L_i, t_i^{*R}]$. On the other hand, if $t_i^{*R} = R_i$ and $t_i^{*L} \neq L_i$ then $Y_i^* \in [t_i^{*L}, R_i]$ and we interpret this as $Y_i^*$ being "right-censored" at $t_i^{*L}$. When $L_i \neq 0$ and $R_i \neq \infty$, both of these scenarios can still be treated as interval censoring from a technical standpoint because the censoring intervals are finite. However, if $t_i^L = 0$, which is plausible only if $L_i = 0$, meaning $Y_i^*$ is now truly left-censored. Similarly, if $t_i^R = \infty$, then we must have $R_i = \infty$, and in this case $Y_i^*$ is truly right-censored.

## 3.4 Conditional log-likelihood

Although definitions for left, right and interval censoring have been established for truncated survival times, considering the involvement of truncation points $L_i$ and $R_i$ for each $i$, it becomes more convenient to formulate the log-likelihood function by treating left and right censoring as specific cases of interval censoring; see the discussions at the end of last section.

For individual $i$, where $i = 1, \dots, n$, truncated and partly interval-censored event times and covariates are denoted by $(L_i, t_i^{*L}, t_i^{*R}, R_i, \delta_i, \mathbf{x}_i)$, where $L_i \leq t_i^{*L} \leq t_i^{*R} \leq R_i$, $\delta_i$ is an indicator for an event, i.e. where $t_i^{*L} = t_i^{*R}$ and $\mathbf{x}_i$ is a vector containing the values of $p$ covariates for individual $i$. Let $\mathbf{X}$ be the matrix for covariate values of all $n$ individuals, with dimensions $n \times p$. Its $i$-th row is given by $\mathbf{x}_i^{\mathsf{T}}$.

Our objective is to estimate the parameters of the Cox model (3.1) using the provided truncated and interval-censored observations: $(L_i, t_i^{*L}, t_i^{*R}, R_i, \delta_i, \mathbf{x}_i)$, where $i = 1, \dots, n$.

It is important to acknowledge that the baseline hazard $h_0(t)$ remains unspecified (non-parametric) and is bound by the non-negative constraint, i.e. $h_0(t) \geq 0$. A non-parametric $h_0(t)$ implies that this parameter is of infinite dimensional, which presents a challenge when estimating it from a finite number of observations – an ill-posed problem.

To make the estimation problem well-posed, we approximate $h_0(t)$ using a finite set of non-negative basis functions as in Chapter 2, so that

$$h_0(t) = \sum_{u=1}^{m} \theta_u \psi_u(t), \tag{3.21}$$

where $\psi_u(t) \geq 0$ $(u = 1, \dots, m)$ are basis functions. These basis functions depend on a set of preselected knots. Now, the constraint $h_0(t) \geq 0$ can be replaced by a set of simpler constraints: $\theta_u \geq 0$ for all $u$.

From (3.21), the cumulative baseline hazard is

$$H_0(t) = \sum_{u=1}^{m} \theta_u \Psi_u(t), \tag{3.22}$$

where $\Psi_u(t) = \int_0^t \psi(s)ds$, denoting a cumulative basis function.

Let $\boldsymbol{\theta} = (\theta_1, \ldots, \theta_m)^\mathsf{T}$. We estimate $\boldsymbol{\beta}$ and $\boldsymbol{\theta}$ by maximizing a penalized log-conditional likelihood, which is constructed from the doubly-truncated and interval-censored survival data. The conditional likelihood is obtained from the relationships between $Y_i$ and $Y_i^*$ as explicated in (3.19)–(3.20).

In the next section, we will discuss how to obtain the penalized conditional likelihood estimate of $\boldsymbol{\beta}$ and $\boldsymbol{\theta}$. Here, we adopt a penalty function to achieve smoothness of the $h_0(t)$ estimate, following the approach in Chapter 2. The penalty function also serves to mitigate the influence arising from the number and placement of knots.

In the case of an event time, i.e. when $t_i^{*L} = t_i^{*R}$, we denote this common value by $t_i^*$. For individual $i$, the corresponding log-conditional likelihood, that is the log-density of $Y_i^* = \{Y_i | L_i \leq Y_i \leq R_i\}$, is given by

$$\begin{aligned} l_i^*(\boldsymbol{\beta}, \boldsymbol{\theta}) =& \delta_i \log f_{Y_i^*}(t_i^*) + (1 - \delta_i) \log(S_{Y_i^*}(t_i^{*L}) - S_{Y_i^*}(t_i^{*R})) \\ =& \delta_i \left\{\log h_0(t_i^*) + \mathbf{x}_i^\mathsf{T} \boldsymbol{\beta} - H_{Y_i}(t_i^*)\right\} \\ &+ (1 - \delta_i) \left\{\log(S_{Y_i}(t_i^{*L}) - S_{Y_i}(t_i^{*R}))\right\} - \log(S_{Y_i}(L_i) - S_{Y_i}(R_i)), \end{aligned} \tag{3.23}$$

where $H_{Y_i}(t)$ represents the cumulative hazard function of $Y_i$.

From the individual log-likelihood $l_i^*(\boldsymbol{\beta}, \boldsymbol{\theta})$, the log-likelihood from all the observations is simply

$$l^*(\boldsymbol{\beta}, \boldsymbol{\theta}) = \sum_{i=1}^{n} l_i^*(\boldsymbol{\beta}, \boldsymbol{\theta}). \tag{3.24}$$

In the following section, this log-likelihood function will be utilized to obtain the constrained penalized conditional likelihood estimates of $\boldsymbol{\beta}$ and $\boldsymbol{\theta}$, where $\boldsymbol{\theta} \geq 0$.

It is simple to verify that when the survival data only have left truncation and right censoring, that is when $R_i = \infty$ and $t_i^{*R} = \infty$, the log-likelihood given in (3.24) reduces to (3.3).

## 3.5 Cox model fitting under double truncation and interval censoring

The penalized log-conditional likelihood can be expressed as

$$\Phi^*(\boldsymbol{\beta}, \boldsymbol{\theta}) = l^*(\boldsymbol{\beta}, \boldsymbol{\theta}) - \lambda J(\boldsymbol{\theta}), \tag{3.25}$$

where $J(\boldsymbol{\theta})$ refers to a penalty function similar to the one described in Chapter 2 and $\lambda$ ($\geq 0$) represents the smoothing parameter. For choices of the function $J(\boldsymbol{\theta})$, readers are referred to Section 2.4.

The maximum penalized conditional likelihood (MPCL) estimates of $\boldsymbol{\beta}$ and $\boldsymbol{\theta}$ are then given by

$$(\widehat{\boldsymbol{\beta}}, \widehat{\boldsymbol{\theta}}) = \underset{\boldsymbol{\beta},\boldsymbol{\theta}}{\operatorname{argmax}}\{\Phi^*(\boldsymbol{\beta}, \boldsymbol{\theta})\}, \tag{3.26}$$

subject to $\boldsymbol{\theta} \geq 0$. This constrained optimization problem can be tackled using an algorithm akin to the constrained optimization algorithm Newton-MI, elucidated in Section 2.5.

To be specific, during iteration $k$ while keeping $\boldsymbol{\theta}$ fixed at its current estimate $\boldsymbol{\theta}^{(k)}$, the update for $\boldsymbol{\beta}$ follows the Newton methodNewton algorithm with line search. This update is expressed as follows:

$$\boldsymbol{\beta}^{(k+1)} = \boldsymbol{\beta}^{(k)} - \omega_1^{(k)} \left[\frac{\partial^2 \Phi^*(\boldsymbol{\beta}^{(k)}, \boldsymbol{\theta}^{(k)})}{\partial \boldsymbol{\beta} \partial \boldsymbol{\beta}^\mathsf{T}}\right]^{-1} \frac{\partial \Phi^*(\boldsymbol{\beta}^{(k)}, \boldsymbol{\theta}^{(k)})}{\partial \boldsymbol{\beta}}. \tag{3.27}$$

Further details about the gradient vector and the Hessian matrix with respect to $\boldsymbol{\beta}$ are provided in equations (3.32) and (3.34) below. In the equation (3.27), $\omega_1^{(k)}$ represents a step size that ensures an increase in the penalized log-likelihood $\Phi^*(\boldsymbol{\beta}, \boldsymbol{\theta}^{(k)})$ from $\boldsymbol{\beta}^{(k)}$ to $\boldsymbol{\beta}^{(k+1)}$. A value of $\omega_1^{(k)}$ can be determined using, for example, Armijo's rule.

In order to provide the details of the gradient and the Hessian, we need first to define the following expressions:

$$C_1(t_1, t_2) = \frac{-S_{Y_i}(t_1)H_{Y_i}(t_1) + S_{Y_i}(t_2)H_{Y_i}(t_2)}{S_{Y_i}(t_1) - S_{Y_i}(t_2)}, \tag{3.28}$$

$$C_2(t_1, t_2) = \frac{S_{Y_i}(t_1)(-H_{Y_i}(t_1) + H_{Y_i}(t_2))}{S_{Y_i}(t_1) - S_{Y_i}(t_2)}, \tag{3.29}$$

$$C_3(t_1, t_2) = \frac{S_{Y_i}(t_1)S_{Y_i}(t_2)(-H_{Y_i}(t_1) + H_{Y_i}(t_2))^2}{(S_{Y_i}(t_1) - S_{Y_i}(t_2))^2}. \tag{3.30}$$

Note that $C_2(t_1, t_2) \geq 0$ whenever $t_1 \leq t_2$, and $C_3(t_1, t_2) \geq 0$ for any $t_1$ and $t_2$.

Now we are ready to derive the gradient. It is important to note that for any $t_1$ and $t_2$,

$$\frac{\partial \log(S_{Y_i}(t_1) - S_{Y_i}(t_2))}{\partial \beta_j} = x_{ij} C_1(t_1, t_2). \tag{3.31}$$

Thus, the first derivative of $\Phi^*(\boldsymbol{\beta}, \boldsymbol{\theta})$ with respect to $\beta_j$ is given by

$$\frac{\partial \Phi^*(\boldsymbol{\beta}, \boldsymbol{\theta})}{\partial \beta_j} = \sum_{i=1}^n x_{ij} \left\{\delta_i(1 - H_{Y_i}(t_i^*)) + (1 - \delta_i) C_1(t_i^{*L}, t_i^{*R}) - C_1(L_i, R_i)\right\}. \tag{3.32}$$

Deriving the second derivatives of $\Phi^*$ with respect to $\beta_j$ and $\beta_t$ is a more difficult process. Since the gradient $\partial \Phi^* / \partial \beta_j$ involves $C_1(t_1, t_2)$, we need to first work out the derivative of $C_1$ with respect to $\beta_t$. One can differentiate $C_1(t_1,t_2)$ directly using the expression for $C_1$ provided in (3.28), but the following trick possibly yields an easier derivation for this derivative. Firstly, note that

$$C_1(t_1,t_2) = -H_{Y_i}(t_2) + C_2(t_1,t_2),$$

then the derivative of $C_1(t_1,t_2)$ with respect to $\beta_t$ can be expressed as

$$\frac{\partial C_1(t_1,t_2)}{\partial \beta_t} = x_{it}\left\{-H_{Y_i}(t_2) + C_2(t_1,t_2) - C_3(t_1,t_2)\right\}. \tag{3.33}$$

Based on this, the second derivative of $\Phi^*(\boldsymbol{\beta},\boldsymbol{\theta})$ with respect to $\beta_j$ and $\beta_t$ is:

$$\begin{aligned}\frac{\partial^2 \Phi^*(\boldsymbol{\beta},\boldsymbol{\theta})}{\partial \beta_j \partial \beta_t} = -\sum_{i=1}^{n} x_{ij}x_{it}\Big\{ &\delta_i H_{Y_i}(t_i^*) + (1-\delta_i)H_{Y_i}(t_i^{*R}) + H_{Y_i}(R_i) \\ &+ (1-\delta_i)[-C_2(t_i^{*L},t_i^{*R}) + C_3(t_i^{*L},t_i^{*R})] - C_2(L_i,R_i) + C_3(L_i,R_i)\Big\}.\end{aligned} \tag{3.34}$$

The expression given in (3.34) can be used to form a matrix expression for the Hessian matrix. This is given by

$$\frac{\partial^2 \Phi^*(\boldsymbol{\beta},\boldsymbol{\theta})}{\partial \boldsymbol{\beta} \partial \boldsymbol{\beta}^\mathsf{T}} = -\mathbf{X}^\mathsf{T}\mathbf{A}\mathbf{X}, \tag{3.35}$$

where $\mathbf{A} = \text{diag}(a_{ii})$, with dimension $n \times n$, and the diagonal elements $a_{ii}$ are given by:

$$\begin{aligned} a_{ii} = {} & \delta_i H_{Y_i}(t_i^*) + (1-\delta_i)H_{Y_i}(t_i^{*R}) + H_{Y_i}(R_i) \\ & + (1-\delta_i)[-C_2(t_i^{*L},t_i^{*R}) + C_3(t_i^{*L},t_i^{*R})] - C_2(L_i,R_i) + C_3(L_i,R_i).\end{aligned} \tag{3.36}$$

Assume the columns of the covariates matrix $\mathbf{X}$ are not linearly dependent. Then, if all $a_{ii}$ are positive, the above Hessian matrix is negative definite and can be directly employed in the Newton algorithm formula (3.27). However, in cases where some $a_{ii}$ are not positive, a quasi-Newton approach becomes necessitated. In this approach, the Hessian is substituted with a modified Hessian that ensures positive definiteness. We recommend to modify the Hessian by replacing its diagonal elements of the matrix $\mathbf{A}$ in (3.35) by the following $\widetilde{a}_{ii}$ that are ensured positive:

$$\begin{aligned} \widetilde{a}_{ii} = {} & \delta_i H_{Y_i}(t_i^*) + (1-\delta_i)H_{Y_i}(t_i^{*R}) + H_{Y_i}(R_i) \\ & + (1-\delta_i)C_3(t_i^{*L},t_i^{*R}) + C_3(L_i,R_i).\end{aligned} \tag{3.37}$$

Since this modified Hessian is negative definite, the corresponding iterative scheme in (3.27), where the Hessian is replaced with this modified Hessian, will produce a sequence of $\boldsymbol{\beta}$ updates with incremental profile penalized log-likelihood $\Phi^*(\boldsymbol{\beta},\boldsymbol{\theta}^{(k)})$ when moving from $\boldsymbol{\beta}^{(k)}$ to $\boldsymbol{\beta}^{(k+1)}$.

The step size employed in the line search of equation (3.27) plays a crucial role in ensuring that the penalized log-likelihood $\Phi^*(\boldsymbol{\beta}, \boldsymbol{\theta}^{(k)})$ increases from $\boldsymbol{\beta}^{(k)}$ to $\boldsymbol{\beta}^{(k+1)}$. Various methods are available for determining an appropriate value for this step size. One commonly used approach is the Armijo's rule, which is straightforward to apply and often yields favorable results. For details of the Armijo line search, the readers are referred to Section 2.4 of Chapter 2.

For the estimation of $\boldsymbol{\theta}$ ($\geq 0$), we can once again utilize the MI algorithm, which has been detailed in Chapter 2. During this phase of iteration $k+1$, we fix $\boldsymbol{\beta}$ at its latest estimate $\boldsymbol{\beta}^{(k+1)}$. Our objective is to determine an updated $\boldsymbol{\theta}$, denoted by $\boldsymbol{\theta}^{(k+1)}$, in a manner that increases the penalized log-likelihood $\Phi^*(\boldsymbol{\beta}^{(k+1)}, \boldsymbol{\theta})$ while ensuring that $\boldsymbol{\theta} \geq 0$.

Drawing from the MI algorithm, the updating formula for $\boldsymbol{\theta}$ closely resembles (2.40). This formula can be presented as:

$$\boldsymbol{\theta}^{(k+1)} = \boldsymbol{\theta}^{(k)} + \omega_2^{(k)} \mathbf{D}^{(k)} \frac{\partial \Phi^*(\boldsymbol{\beta}^{(k+1)}, \boldsymbol{\theta}^{(k)})}{\partial \boldsymbol{\theta}}, \tag{3.38}$$

where $\mathbf{D}^{(k)}$ is a diagonal matrix and its elements will be specified later, and $\omega_2^{(k)}$ is a line search step size.

The components of the gradient $\partial \Phi^*(\boldsymbol{\beta}^{(k+1)}, \boldsymbol{\theta}^{(k)})/\partial \boldsymbol{\theta}$ are derived as follows. Define

$$C_1(t_1, t_2; u) = \frac{-S_{Y_i}(t_1)\Psi_u(t_1) + S_{Y_i}(t_2)\Psi_u(t_2)}{S_{Y_i}(t_1) - S_{Y_i}(t_2)}.$$

Since

$$\frac{\partial \log(S_{Y_i}(t_1) - S_{Y_i}(t_2))}{\partial \theta_u} = e^{\mathbf{x}_i \boldsymbol{\beta}} C_1(t_1, t_2; u), \tag{3.39}$$

we have

$$\begin{aligned} \frac{\partial \Phi^*(\boldsymbol{\beta}, \boldsymbol{\theta})}{\partial \theta_u} &= \sum_{i=1}^{n} \left\{ \delta_i \left[ \frac{\psi_u(t_i^*)}{h_0(t_i^*)} - \Psi_u(t_i) e^{\mathbf{x}_i \boldsymbol{\beta}} \right] + (1 - \delta_i) C_1(t_i^{*L}, t_i^{*R}; u) e^{\mathbf{x}_i \boldsymbol{\beta}} \right. \\ &\quad \left. - C_1(L_i, R_i; u) e^{\mathbf{x}_i \boldsymbol{\beta}} \right\} - \lambda \frac{\partial J(\boldsymbol{\theta})}{\partial \theta_u}. \end{aligned} \tag{3.40}$$

In (3.38), the step size matrix $\mathbf{D}^{(k)}$ is diagonal. Its diagonal elements are given by $\theta_u^{(k)} / \zeta_{2u}^*(\boldsymbol{\beta}^{(k+1)}, \boldsymbol{\theta}^{(k)})$, where $u = 1, \ldots, m$. Here,

$$\begin{aligned} \zeta_{2u}^*(\boldsymbol{\beta}, \boldsymbol{\theta}) &= \sum_{i=1}^{n} \left\{ \delta_i \Psi_u(t_i^*) e^{\mathbf{x}_i \boldsymbol{\beta}} + (1 - \delta_i) \frac{S_{Y_i}(t_i^{*L}) \Psi_u(t_i^{*L})}{S_{Y_i}(t_i^{*L}) - S_{Y_i}(t_i^{*R})} e^{\mathbf{x}_i \boldsymbol{\beta}} \right. \\ &\quad \left. + \frac{S_{Y_i}(R_i) \Psi_u(R_i)}{S_{Y_i}(L_i) - S_{Y_i}(R_i)} e^{\mathbf{x}_i \boldsymbol{\beta}} \right\} + \lambda \left[ \frac{\partial J(\boldsymbol{\theta})}{\partial \theta_u} \right]^+ + \epsilon. \end{aligned} \tag{3.41}$$

Note that in (3.41), $\epsilon \geq 0$ is a small fixed quantity, such as $10^{-4}$, used to avoid $\zeta_{2u}^* = 0$. If all $\zeta_{2u}^*$ are non-zero then $\epsilon = 0$ can be chosen. It can be shown that the magnitude of this $\epsilon$ value will only influence the convergence rate of this algorithm, but will not alter the ultimate solution for $\boldsymbol{\theta}$.

The convergence of this algorithm to a stationary point can be illustrated in a manner analogous to Section 2.5.2. In fact, Theorems 2.1–2.3 can be suitably adapted to yield the global convergence outcome for the above algorithm.

## 3.6 Asymptotic results

Asymptotic outcomes for the MPCL estimates of Cox models with doubly truncated and interval-censored survival times can be formulated in a manner similar to Chapter 2. In fact, the asymptotic results presented in Section 2.7 are sufficiently comprehensive to be extended to the context addressed in this chapter. We consolidate certain asymptotic results within this section. Particularly, this section will present the asymptotic consistency and the asymptotic normality results.

The asymptotic consistency pertains to the estimates of $\boldsymbol{\beta}$ and $h_0(t)$, where $m$ (or the number of knots) grows as the sample size $n$ increases, but at a slower rate. On the other hand, the asymptotic normality results are focused on $\boldsymbol{\beta}$ and $\boldsymbol{\theta}$, with $m$ being held constant at a value smaller than $n$. However, the selection of $m$ remains adaptable.

The consistency results will be stated in Theorem 3.1 below for the maximum penalized conditional likelihood estimators $\widehat{\boldsymbol{\beta}}$ and $\widehat{h}(t)$. These outcomes necessitate certain conditions on $m$ and $\lambda$. Assume $m \to \infty$ when $n \to \infty$ but at a slow rate in the sense that $m = n^v$ with $0 < v < 1$ (this ensures $m/n \to 0$). Regarding the smoothing parameter $\lambda$, we assume it diminishes in significance as $n$ grows large, specifically, $\mu = \lambda/n \to 0$ when $n \to \infty$.

Let $\widehat{\boldsymbol{\beta}}$ and $\widehat{\boldsymbol{\theta}}$ be respectively the MPCL estimators of $\boldsymbol{\beta}$ and $\boldsymbol{\theta}$ and the corresponding baseline hazard estimator $\widehat{h}(t) = \sum_{u=1}^{m} \widehat{\theta}_u \psi_u(t)$. Where necessary, we will denote $\widehat{h}(t)$ by $\widehat{h}_n(t)$ to indicate its dependence on the sample size $n$. The regularity conditions A2.1 – A2.4 are still required for the asymptotic consistency results here.

**Theorem 3.1.** Consider the maximum penalized conditional likelihood estimation of $\boldsymbol{\beta}$ and $h_0(t)$ from doubly truncated and partly interval-censored survival data, where the estimation of $h_0(t)$ is achieved through approximation given by (3.21). We assume that Assumptions A2.1 – A2.4 hold where $l(\boldsymbol{\beta}, \boldsymbol{\theta})$ is replaced by $l^*(\boldsymbol{\beta}, \boldsymbol{\theta})$, and that $h_0(t)$ has up to $r$ ($\geq 1$) derivatives. Additionally, we assume that $m = n^v$, where $0 < v < 1$, and that the scaled smoothing value $\mu = \lambda/n \to 0$ as $n \to \infty$. Let $\boldsymbol{\beta}_0$ be the vector of true regression coefficients and $h_0(t)$ be the true baseline hazard. Then, when $n \to \infty$,

(1) $\|\widehat{\boldsymbol{\beta}} - \boldsymbol{\beta}_0\| \to 0$ almost surely, and

(2) $\sup_{t \in [a,b]} |\widehat{h}_n(t) - h_0(t)| \to 0$ almost surely, where $a = \min_i\{t_i^{*L}\}$ and $b = \max_i\{t_i^{*R}\}$. □

The proof of this theorem can be supplied by following the proof outlined in Appendix A for Theorem 2.4. As such, we will not present in this chapter the proof for this theorem.

Next, we will summarize the results regarding asymptotic normality. Similar to Chapter 2, our goal is to provide more practically applicable asymptotic outcomes below for both $\widehat{\boldsymbol{\beta}}$ and $\widehat{\boldsymbol{\theta}}$, assuming that $m$ is fixed and less than $n$. These results allow for the inclusion of $\boldsymbol{\theta}$ within the asymptotic conclusions and accommodate the active constraints of $\boldsymbol{\theta} \geq 0$. The rationale behind using a fixed value of $m$ has already been discussed in Section 2.7. This is mainly due to the slow rate at which $m$ approaches $\infty$ when $n \to \infty$, meaning that changes in $m$ are negligible when compared to changes in $n$.

In practice, people often choose to use a small value of $m$ due to the slow rate of convergence of the cumulative hazard estimator, as discussed in Huang (1996), Groeneboom & Wellner (1992) and Zhang et al. (2010). We can also adopt this approach for our estimator, especially when using spline basis functions. However, the penalty function will increase the flexibility of $m$, allowing for a relatively larger $m$. This occurs because the penalty function pushes unimportant $\theta_u$ values towards zero.

The asymptotic normality results outlined in Theorem 3.2 apply to the constrained MPCL estimates of $\boldsymbol{\beta}$ and $\boldsymbol{\theta}$, assuming that the scaled smoothing value $\mu_n = o(n^{-1/2})$ and that $m$ is relatively small compared to $n$. Typically, a guideline for selecting $m$ is: $m = \sqrt[3]{n}$. To simplify discussions we define the vector $\boldsymbol{\eta} = (\boldsymbol{\theta}^{\mathsf{T}}, \boldsymbol{\beta}^{\mathsf{T}})^{\mathsf{T}}$, whose length is $m + p$. Let $\boldsymbol{\eta}_0$ be the true parameter vector corresponding to $m$ and $p$. The MPCL estimate of $\boldsymbol{\eta}$, denoted by $\widehat{\boldsymbol{\eta}}$, is obtained by maximizing the penalized conditional log-likelihood $\Phi^*(\boldsymbol{\eta})$, as defined in (3.25), under the constraint $\boldsymbol{\theta} \geq 0$. Since $\mu_n \to 0$ when $n \to \infty$, influence of the penalty function diminishes when $n$ is becoming larger.

To handle the active $\boldsymbol{\theta} \geq 0$ constraints{active constraints, we need to define matrix $\mathbf{U}$. This matrix is identical to $\mathbf{U}$ defined in equation (2.54) if we assume, without loss of generality, that the first $q$ constraints of $\boldsymbol{\theta} \geq 0$ are active.

To establish the asymptotic normality as presented in Theorem 3.2, we continue to require Assumptions B2.1 – B2.4. However, in these assumptions, the log-likelihood $l(\boldsymbol{\eta})$ is substituted with $l^*(\boldsymbol{\eta})$ as defined in (3.24).

**Theorem 3.2.** Assume Assumptions B2.1 – B2.4 hold, where $l(\boldsymbol{\eta})$ is replaced by $l^*(\boldsymbol{\eta})$. Assume the scaled smoothing parameter $\mu_n = o(n^{-1/2})$. Assume there are $q$ active constraints in the MPCL estimate of $\boldsymbol{\theta}$ and let $\mathbf{U}$ be the matrix required for handling the active constraints and this matrix can be developed similar to (2.54). Let $\mathbf{F}^*(\boldsymbol{\eta}) = -n^{-1}E(\partial^2 l^*(\boldsymbol{\eta})/\partial\boldsymbol{\eta}\partial\boldsymbol{\eta}^{\mathsf{T}})$. Then, when $n \to \infty$,

(1) the constrained MPCL estimate $\widehat{\boldsymbol{\eta}}$ is consistent for $\boldsymbol{\eta}_0$, and

(2) $\sqrt{n}(\widehat{\boldsymbol{\eta}} - \boldsymbol{\eta}_0)$ converges in distribution to a multivariate normal distribution $N(\mathbf{0}, \widetilde{\mathbf{F}}^*(\boldsymbol{\eta}_0)^{-1}\mathbf{F}^*(\boldsymbol{\eta}_0)[\widetilde{\mathbf{F}}^*(\boldsymbol{\eta}_0)^{-1}]^{\mathsf{T}})$, where $\widetilde{\mathbf{F}}^*(\boldsymbol{\eta})^{-1} = \mathbf{U}(\mathbf{U}^{\mathsf{T}}\mathbf{F}^*(\boldsymbol{\eta})\mathbf{U})^{-1}\mathbf{U}^{\mathsf{T}}$. □

In this theorem, matrix $\mathbf{F}^*(\boldsymbol{\eta}_0)$ is the Fisher information matrix of the log-conditional likelihood $l^*(\boldsymbol{\eta})$, scaled by $n$ and evaluated at the true parameter $\boldsymbol{\eta}_0$. Since the expectation is difficult to calculate and $\boldsymbol{\eta}_0$ is usually unavailable, this information matrix can be replaced with the negative Hessian matrix and $\boldsymbol{\eta}_0$ can be replaced by its MPCL estimate $\widehat{\boldsymbol{\eta}}$. Using the matrix $\mathbf{F}^*$ and following the description in Section 2.7.2, we can readily compute the inverse matrix $\widetilde{\mathbf{F}}^*(\widehat{\boldsymbol{\eta}})^{-1}$.

The matrix $\mathbf{U}$ in the above $\widetilde{\mathbf{F}}^*(\boldsymbol{\eta})$ formula can be similarly constructed as (2.54). It demands, however, to firstly identify the active constraints of $\boldsymbol{\theta} \geq 0$. We again can adopt the process specified in Section 2.7.2 to identify active constraints. At the end of running the Newton-MI algorithm, often some $\widehat{\theta}_u$'s are almost zero with negative gradients, which means they are active. In our R "`survivalMPL`" package, we adopt the criterion that "$|\widehat{\theta}_u| < 10^{-3}$ while the corresponding gradient less than $-10^{-2}$" to define active constraints.

In practice, $n$ is never infinite, so when penalized likelihood estimates are preferred, the smoothing parameter $\lambda$, although small, is likely to be non-zero. Therefore, we need to make inferences when $n$ is finite where $\lambda$ is also included. The approximate distribution, provided in Corollary 3.1, can be used for making inferences without relying on computationally intensive re-sampling methods, such as bootstrapping.

**Corollary 3.1.** Assume the Assumptions in Theorem 3.2 hold. When $n$ is large, the distribution for the MPCL estimate $\widehat{\boldsymbol{\eta}} - \boldsymbol{\eta}_0$ can be approximated by a multivariate normal distribution with zero mean vector and the approximated covariance matrix

$$\widehat{\mathrm{Var}}(\widehat{\boldsymbol{\eta}}) = -\mathbf{A}^*(\widehat{\boldsymbol{\eta}})^{-1}\frac{\partial^2 l^*(\widehat{\boldsymbol{\eta}})}{\partial\boldsymbol{\eta}\partial\boldsymbol{\eta}^\top}\mathbf{A}^*(\widehat{\boldsymbol{\eta}})^{-1}, \tag{3.42}$$

where

$$\mathbf{A}^*(\widehat{\boldsymbol{\eta}})^{-1} = \mathbf{U}\Big(\mathbf{U}^\top\Big(-\frac{\partial^2 l^*(\widehat{\boldsymbol{\eta}})}{\partial\boldsymbol{\eta}\partial\boldsymbol{\eta}^\top} + \lambda\frac{\partial^2 J(\widehat{\boldsymbol{\eta}})}{\partial\boldsymbol{\eta}\partial\boldsymbol{\eta}^\top}\Big)\mathbf{U}\Big)^{-1}\mathbf{U}^\top.$$

□

The results in Corollary 3.1 are important as they circumvent computationally expensive bootstrapping methods when conducting inferences related to various aspects, such as regression coefficients, baseline hazard, cumulative baseline hazard, and survival probabilities. A notable feature here is its ability to accommodate active constraints. Proper management of active constraints helps prevent undesirable outcomes, such as negative variances.

## 3.7 Other Hessian matrix elements

The large sample covariance matrix derived above requires the full Hessian matrix. In equation (3.34), we have already obtained the Hessian element $\partial^2 l^*(\boldsymbol{\beta},\boldsymbol{\theta})/\partial\boldsymbol{\beta}\partial\boldsymbol{\beta}^\top$, which is the same as $\partial^2\Phi^*(\boldsymbol{\beta},\boldsymbol{\theta})/\partial\boldsymbol{\beta}\partial\boldsymbol{\beta}^\top$. Other second derivatives of $\Phi^*(\boldsymbol{\beta},\boldsymbol{\theta})$ are somewhat tedious but can still be derived.

We first define the following notations. Let

$$D_2(t_1,t_2;u) = \frac{S_{Y_i}(t_1)S_{Y_i}(t_2)(-\Psi_u(t_1)+\Psi_u(t_2))(-H_{Y_i}(t_1)+H_{Y_i}(t_2))}{(S_{Y_i}(t_1)-S_{Y_i}(t_2))^2},$$

$$D_3(t_1,t_2;u,v) = \frac{S_{Y_i}(t_1)S_{Y_i}(t_2)(-\Psi_v(t_1)+\Psi_v(t_2))(-\Psi_u(t_1)+\Psi_u(t_2))}{(S_{Y_i}(t_1)-S_{Y_i}(t_2))^2}.$$

Note that

$$\frac{\partial D_1(t_1,t_2;u)}{\partial \beta_j} = -x_{ij}D_2(t_1,t_2;u), \tag{3.43}$$

$$\frac{\partial D_1(t_1,t_2;u)}{\partial \theta_v} = -e^{\mathbf{x}_i\boldsymbol{\beta}}D_3(t_1,t_2;u,v), \tag{3.44}$$

then we have the other second derivatives in the Hessian matrix:

$$\frac{\partial^2 l^*(\boldsymbol{\beta},\boldsymbol{\theta})}{\partial\beta_j\partial\theta_u} = -\sum_{i=1}^{n} x_{ij}e^{\mathbf{x}_i\boldsymbol{\beta}}\Big\{\delta_i\Psi_u(t_i) + (1-\delta_i)[-D_1(t_i^L,t_i^R;u)+D_2(t_i^L,t_i^R;u)] + D_1(L_i,R_i;u) - D_2(L_i,R_i;u)\Big\}, \tag{3.45}$$

and

$$\frac{\partial^2 l^*(\boldsymbol{\beta},\boldsymbol{\theta})}{\partial\theta_u\partial\theta_v} = \sum_{i=1}^{n}\Big\{-\delta_i\frac{\psi_u(t_i)\psi_v(t_i)}{h_0^2(t_i)} - (1-\delta_i)D_3(t_i^L,t_i^R;u,v)e^{2\mathbf{x}_i\boldsymbol{\beta}} + D_3(L_i,R_i;u,v)e^{2\mathbf{x}_i\boldsymbol{\beta}}\Big\}. \tag{3.46}$$

## 3.8 Automatic smoothing parameter selection

An automatic smoothing parameter selection procedure can be developed exactly the same way as Section 2.10 of Chapter 2.

Towards this, we first rewrite the penalized log-conditional likelihood, which can be considered as the log-posterior after identifying the penalty as a prior distribution, as

$$\ell_p^*(\boldsymbol{\beta},\boldsymbol{\theta},\sigma^2) = -\frac{m}{2}\log\sigma^2 + l^*(\boldsymbol{\beta},\boldsymbol{\theta}) - \frac{1}{2\sigma^2}\boldsymbol{\theta}^{\mathsf{T}}\mathbf{R}\boldsymbol{\theta}. \tag{3.47}$$

where $\sigma_{\boldsymbol{\theta}}^2 = 1/(2\lambda)$. Then the log-marginal likelihood for $\sigma^2$ is

$$l_m^*(\sigma^2) = -\frac{m}{2}\log\sigma^2 + \log\int\exp\left(l^*(\boldsymbol{\beta},\boldsymbol{\theta}) - \frac{1}{2\sigma^2}\boldsymbol{\theta}^{\mathsf{T}}\mathbf{R}\boldsymbol{\theta}\right)d\boldsymbol{\beta}d\boldsymbol{\theta}. \tag{3.48}$$

After a Laplace's approximation to the integral in (3.48) and plugging in the MPCL estimates $\widehat{\boldsymbol{\beta}}$ and $\widehat{\boldsymbol{\theta}}$, we have an approximated $l_m(\sigma^2)$:

$$l_m^*(\sigma^2) \approx -\frac{m}{2}\log\sigma^2 + l^*(\widehat{\boldsymbol{\beta}},\widehat{\boldsymbol{\theta}}) - \frac{1}{2\sigma^2}\widehat{\boldsymbol{\theta}}^{\mathsf{T}}R\widehat{\boldsymbol{\theta}} - \frac{1}{2}\log\left|-\widehat{\mathbf{H}}^* + \mathbf{Q}^*(\sigma^2)\right|, \tag{3.49}$$

where $\mathbf{H}^*$ is the Hessian matrix from the log-conditional likelihood $l^*(\boldsymbol{\beta}, \boldsymbol{\theta})$, $\widehat{\mathbf{H}}^*$ is $\mathbf{H}^*$ evaluated at the MPCL estimates $\widehat{\boldsymbol{\beta}}$ and $\widehat{\boldsymbol{\theta}}$, and

$$\mathbf{Q}^*(\sigma^2) = \begin{pmatrix} 0 & 0 \\ 0 & \frac{1}{\sigma^2}\mathbf{R} \end{pmatrix}.$$

The solution of $\sigma^2$ maximizing (2.69) satisfies

$$\widehat{\sigma}^2 = \frac{\widehat{\boldsymbol{\theta}}^{\mathsf{T}} \mathbf{R} \widehat{\boldsymbol{\theta}}}{m - \nu^*}, \tag{3.50}$$

where

$$\nu^* = \mathrm{tr}\{(-\widehat{\mathbf{H}}^* + \mathbf{Q}^*(\widehat{\sigma}^2_{\boldsymbol{\theta}}))^{-1}\mathbf{Q}^*(\widehat{\sigma}^2_{\boldsymbol{\theta}})\},$$

which can be conceived as the model degrees of freedom. Since $\boldsymbol{\beta}$ and $\boldsymbol{\theta}$ depend on $\sigma^2$, expression (3.50) naturally suggests an iterative procedure:

1. *Inner iterations:* with $\sigma^2$ being fixed at its current estimate, the corresponding MPCL estimates of $\boldsymbol{\beta}$ and $\boldsymbol{\theta}$ are obtained;
2. *Outer iterations:* then $\sigma^2$ is updated by (3.50) with the new $\widehat{\boldsymbol{\beta}}$ and $\widehat{\boldsymbol{\theta}}$.

This process continues until the degrees of freedom $\nu$ stabilize, as determined by our adopted criterion (see Section 2.10), which specifies that the changes in $\nu$ between two consecutive outer iterations should not exceed one.

## 3.9 R examples

We have explained, in the previous sections, a semi-parametric Cox model estimation method for MPCL when survival observations are doubly truncated and interval censored. In this section, we will provide two R examples to demonstrate this method.

The first example is a simulation study to (i) illustrate how to compute the Cox model MPCL estimation described in this chapter and (ii) compare this method with two existing R packages: `SurvTrunc`, currently on `CRAN`, based on the method described in Rennert & Xie (2018); and the package presented in the paper by Mandel et al. (2018), both designed for handling doubly truncated survival data.

**Example 3.4** (A simulation example).
In this example, we discuss a simple simulation designed to assess the approach discussed in this chapter for fitting the Cox model under doubly truncated and interval-censored survival data. The following aspects are covered:

(i) Generate survival data with double truncation and interval censoring, with a sample size of $n = 200$. Here, $\beta_1 = 0.5$, $\beta_2 = -0.5$, $X_1 \sim Binomial(1, 0.4)$ and $X_2 \sim N(0, 1)$. There are 100 repetitions.
(ii) Fit each dataset using the method described in this chapter (called MPCL), as

well as the methods presented by Rennert & Xie (2018) and Mandel et al. (2018) (these two methods are named IPW1 and IPW2 respectively).
(iii) Report biases, standard errors (both asymptotic and Monte Carlo) and coverage probabilities.

```
# write a function to
# simulate double truncated,
# partly interval censored
# survival function
trunc_sim_data <- function(nsample,
    beta, prob_left_0, prob_right_inf,
    right_cens_par1, right_cens_par2,
    event_length) {

    n = 0  #set current sample size to 0
    dat = NULL

    while (n < nsample) {
        # generate covariates
        X <- cbind(rbinom(1, 1,
            0.4), rnorm(1, 0, 1))

        # generate event time
        neg.log.u <- -log(runif(1))
        mu_term <- exp(X %*% beta)
        y_i <- as.numeric((neg.log.u/(mu_term))^(1/3))

        # generate left
        # truncation time
        L_i <- rbinom(1, 1, 1 -
            prob_left_0) * runif(1,
            0.1, 0.5)

        # generate right
        # truncation time
        R_i <- (Inf^rbinom(1, 1,
            prob_right_inf)) +
            L_i + runif(1, 0.5,
            3)

        # generate event
        # observation times
        obs_n_i <- rpois(1, 40) +
            1
        int_obs_i <- runif(obs_n_i,
            0.2, 0.4)
        t_obs_i <- cumsum(int_obs_i)

        # generate
        # independent dropout
        # time
        cens_i <- runif(1, right_cens_par1,
            right_cens_par2)

        # truncate
        # observation times
        # at dropout
        t_obs_i_trunc <- c(t_obs_i[which(t_obs_i <
            cens_i)], cens_i)
        TL <- TR <- censor_type <- NA

        if (y_i > max(t_obs_i_trunc)) {
            # right censored
```

```
    TL <- max(t_obs_i_trunc)
    TR <- Inf
    censor_type <- 0
} else if (y_i < min(t_obs_i_trunc)) {
    # left censored
    TL <- 0
    TR <- min(t_obs_i_trunc)
    censor_type <- 2
} else {
    tiL_temp <- t_obs_i_trunc[max(which(t_obs_i_trunc <
        y_i))]
    tiR_temp <- t_obs_i_trunc[min(which(t_obs_i_trunc >
        y_i))]

    if ((tiR_temp - tiL_temp) <
        event_length) {
        # event time
        # (if
        # interval is
        # small
        # enough)
        TL <- y_i
        TR <- y_i
        censor_type <- 1
    } else {
        # interval
        # censored
        TL <- tiL_temp
        TR <- tiR_temp
        censor_type <- 3
    }
}

if (censor_type == 1 &
    L_i < TL & TR < R_i) {
    # if event time
    # is inside the
    # truncation
    # interval then
    # individual is
    # in the sample
    dat <- rbind(dat, c(y_i,
        L_i, R_i, X, TL,
        TR))
    n <- n + 1  #increase current sample size by 1
} else if (censor_type ==
    0 & L_i < TL & TR <
    R_i) {
    # if right
    # censoring time
    # is inside the
    # truncation
    # interval then
    # individual is
    # in the sample
    dat <- rbind(dat, c(y_i,
        L_i, R_i, X, TL,
        TR))
    n <- n + 1  #increase current sample size by 1
} else if (censor_type ==
    2 & L_i < TL & TR <
    R_i) {
    # if left
    # censoring time
    # is inside the
    # truncation
    # interval then
```

```
            # individual is
            # in the sample
            dat <- rbind(dat, c(y_i,
                L_i, R_i, X, TL,
                TR))
            n <- n + 1  #increase current sample size by 1
        } else if (censor_type ==
            3 & L_i < TL & TR <
            R_i) {
            # if censoring
            # interval is
            # inside the
            # truncation
            # interval then
            # individual is
            # in the sample
            dat <- rbind(dat, c(y_i,
                L_i, R_i, X, TL,
                TR))
            n <- n + 1  #increase current sample size by 1
        }
    }

    trunc_data <- data.frame(dat)
    colnames(trunc_data) <- c("y_i",
        "trunc_time_L", "trunc_time_R",
        "x", "X2", "censor_time_L",
        "censor_time_R")

    # censoring type
    trunc_data$censor_type <- NA
    trunc_data$censor_type[which(is.infinite(trunc_data$censor_time_R))] <- 0
    trunc_data$censor_type[which(trunc_data$censor_time_L ==
        0)] <- 2
    trunc_data$censor_type[which(trunc_data$censor_time_L ==
        trunc_data$censor_time_R)] <- 1
    trunc_data$censor_type[is.na(trunc_data$censor_type)] <- 3

    # truncation type
    trunc_data$trunc_type = NA
    trunc_data$trunc_type[which(trunc_data$trunc_time_L ==
        0 & trunc_data$trunc_time_R ==
        Inf)] <- 1
    trunc_data$trunc_type[which(trunc_data$trunc_time_L !=
        0 & trunc_data$trunc_time_R ==
        Inf)] <- 2
    trunc_data$trunc_type[which(trunc_data$trunc_time_L ==
        0 & trunc_data$trunc_time_R !=
        Inf)] <- 3
    trunc_data$trunc_type[which(trunc_data$trunc_time_L !=
        0 & trunc_data$trunc_time_R !=
        Inf)] <- 4

    return(trunc_data)
}

# generate double truncated
# data with no censoring
trunc_data_event100 <- trunc_sim_data(nsample = 200,
    beta = c(0.5, -0.5), prob_left_0 = 0,
    prob_right_inf = 0, right_cens_par1 = 10,
    right_cens_par2 = 100, event_length = 100)


# generate double truncated
# data with only interval
# censoring
```

```
trunc_data_interval = trunc_sim_data(nsample = 200,
    beta = c(0.5, -0.5), prob_left_0 = 0,
    prob_right_inf = 0, right_cens_par1 = 10,
    right_cens_par2 = 100, event_length = 0.2)
```

Using the function above, we can generate double truncated data with a variety of censoring types, including a dataset containing only directly observed event times and a dataset containing only interval-censored observations. Note that the methods of Rennert & Xie (2018) and Mandel et al. (2018) assume that there is no interval censoring and only account for cases where event times are exactly observed.

TABLE 3.2: Double truncation regression parameter simulation results, comparing MPCL and IPW methods, with (for scenarios with $\pi^E = 0.5$ and $\pi^E = 0.0$) and without (for scenarios with $\pi^E = 1.0$) midpoint imputation of interval censored event times when using IPW. IPW1 denotes the method of Rennert & Xie (2018) and IPW2 denotes the method of Mandel et al. (2018).

| | | $n = 200, \pi^E = 1.0$ | | | $n = 200, \pi^E = 0.5$ | | | $n = 200, \pi^E = 0.0$ | | |
|---|---|---|---|---|---|---|---|---|---|---|
| | | Bias | SE | CP | Bias | SE | CP | Bias | SE | CP |
| $\beta_1$ | MPCL | -0.023 | 0.062 (0.069) | 0.93 | 0.023 | 0.269 (0.289) | 0.94 | 0.001 | 0.318 (0.270) | 0.96 |
| | IPW1 | -0.004 | 0.055 (0.054) | 0.95 | 0.048 | 0.166 (0.170) | 0.93 | 0.063 | 0.163 (0.165) | 0.95 |
| | IPW2 | -0.005 | 0.048 (0.053) | 0.93 | 0.048 | 0.147 (0.167) | 0.90 | 0.065 | 0.146 (0.164) | 0.89 |
| $\beta_2$ | MPCL | 0.021 | 0.193 (0.202) | 0.93 | 0.003 | 0.132 (0.122) | 0.93 | 0.005 | 0.156 (0.165) | 0.96 |
| | IPW1 | 0.010 | 0.167 (0.168) | 0.94 | -0.051 | 0.089 (0.092) | 0.89 | -0.066 | 0.087 (0.098) | 0.84 |
| | IPW2 | 0.008 | 0.147 (0.165) | 0.92 | -0.051 | 0.077 (0.091) | 0.84 | -0.067 | 0.078 (0.096) | 0.76 |

In Table 3.2 we give the bias, asymptotic and Monte Carlo standard errors and the 95% coverage probabilities for $\beta_1$ and $\beta_2$ as given in the simulation description above. The first column shows a case where datasets are simulated with only exact event times, while the second and third columns show results from scenarios where datasets were simulated with, respectively, 50% event times and 50% interval censoring times, and 100% interval censoring times, with midpoint imputation used to fit the IPW methods of Rennert & Xie (2018) and Mandel et al. (2018) to the data.

Clearly, when there is censoring (i.e. the event proportion $\pi^E < 1$), the maximum penalized conditional likelihood performs the best among the three methods in both bias and coverage probability (CP). □

**Example 3.5** (A Real data application).
In this example, we apply the method of this chapter to a dataset described in Mandel et al. (2018). The dataset contains observations of age at (i.e. time to) onset of Parkinson's disease along with information on two single nucleotide polymorphisms (SNPs) that were hypothesized to be associated with the disease. More specifically, these variables were

(1) SNP10398 with alleles A and G forming a binary variable and

(2) PGC-1a with alleles A, G and AG forming a categorical variable with three levels.

Here we concern ourselves only with the data of the "early onset" group, which contains $n = 99$ observations. Double truncation is present in this dataset because:

(1) participants were required to have had their DNA sample taken within 8 years of the onset of Parkinson's disease to avoid survival-related biases and
(2) participants were required to have had their DNA sample taken after the onset of the disease.

In this example, we will perform:
(i) Fit a double truncated Cox model to this dataset assuming that the age of onset is an exactly observed event time, firstly performing univariable analysis for each of the SNP variables and then fitting a multivariable model including both SNP variables.
(ii) Repeat the same analysis, but now assuming that age of onset is a censoring interval (i.e., if the original recorded age of onset was $t_i$, then suppose that we actually have an interval-censored event with $t_i^L = t_i$ and $t_i^R = t_i + 0.999$).
(iii) Inspect the fitted survival functions and associated 95% point-wise confidence intervals for each sub-group combination of allele levels, based on the interval-censored model fitting in (ii).

In Table 3.3 we give the results of fitting the univariate and multivariate Cox regression models to the early onset Parkinson's disease data. In the first column we show the results when assuming that age at onset is an exact event time. In the second column, we show the results when assuming that the age at onset constitutes a censoring interval. In both cases, our results reflect the findings of Mandel et al. (2018) who found that there was no significant association between either of the SNPs and age at onset of Parkinson's disease.

In Figure 3.2, we plot the estimated survival functions for each of the SNP subgroups along with their point-wise 95% confidence intervals. Note that these are estimates of the survival function $S_{Y_i}(t)$ for the *unobserved*, untruncated survival times $Y_i$, not estimates of the survival function $S_{Y_i^*}(t)$ for the *observed*, truncated survival times $Y_i^*$. As such, although the earliest observed event in the dataset was between $[35, 35.99]$, the predicted survival functions shown here account for the possibility that there were unobserved events occurring before $t = 35$ which were truncated. □

TABLE 3.3: Regression parameter estimates from the univariate (Model 1 and Model 2) and multivariate (Model 3) models fitted to the early-onset Parkinson's disease data, assuming age at onset is an exact event time (left column) and that age at onset is a censoring interval (right column).

| | | Exact event times | | | Interval censored event times | | |
|---|---|---|---|---|---|---|---|
| Model | Variables | $\beta$ | SE | 95% CI | $\beta$ | SE | 95% CI |
| 1 | SNP10398: G | -0.297 | 0.483 | -1.244, 0.649 | -0.250 | 0.379 | -0.995, 0.494 |
| 2 | PGC-1a: AG | -0.058 | 0.417 | -0.876, 0.759 | -0.127 | 0.401 | -0.912, 0.659 |
| | PGC-1a: G | 0.192 | 0.412 | -0.616, 1.00 | -0.041 | 0.410 | -0.845, 0.763 |
| 3 | SNP10398: G | -0.464 | 0.542 | -1.525, 0.599 | -0.335 | 0.423 | -1.165, 0.493 |
| | PGC-1a: AG | -0.133 | 0.356 | -0.949, 0.682 | -0.172 | 0.406 | -0.967, 0.624 |
| | PGC-1a: G | 0.248 | 0.374 | -0.583, 1.078 | 0.025 | 0.422 | -0.802, 0.852 |

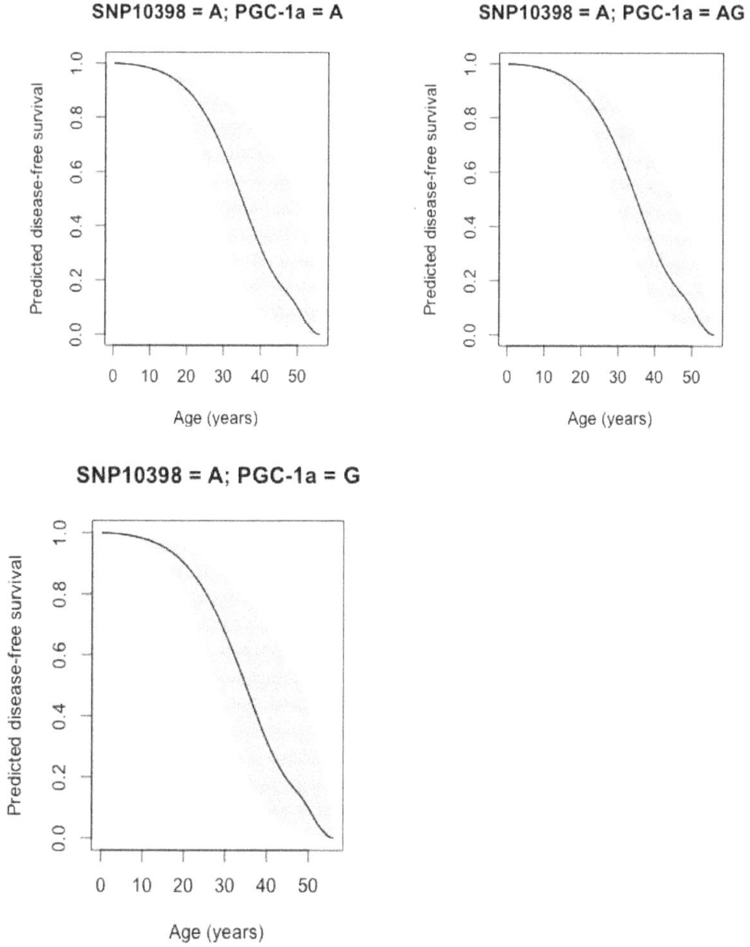


Predicted survival function for 'SNP10398=A'

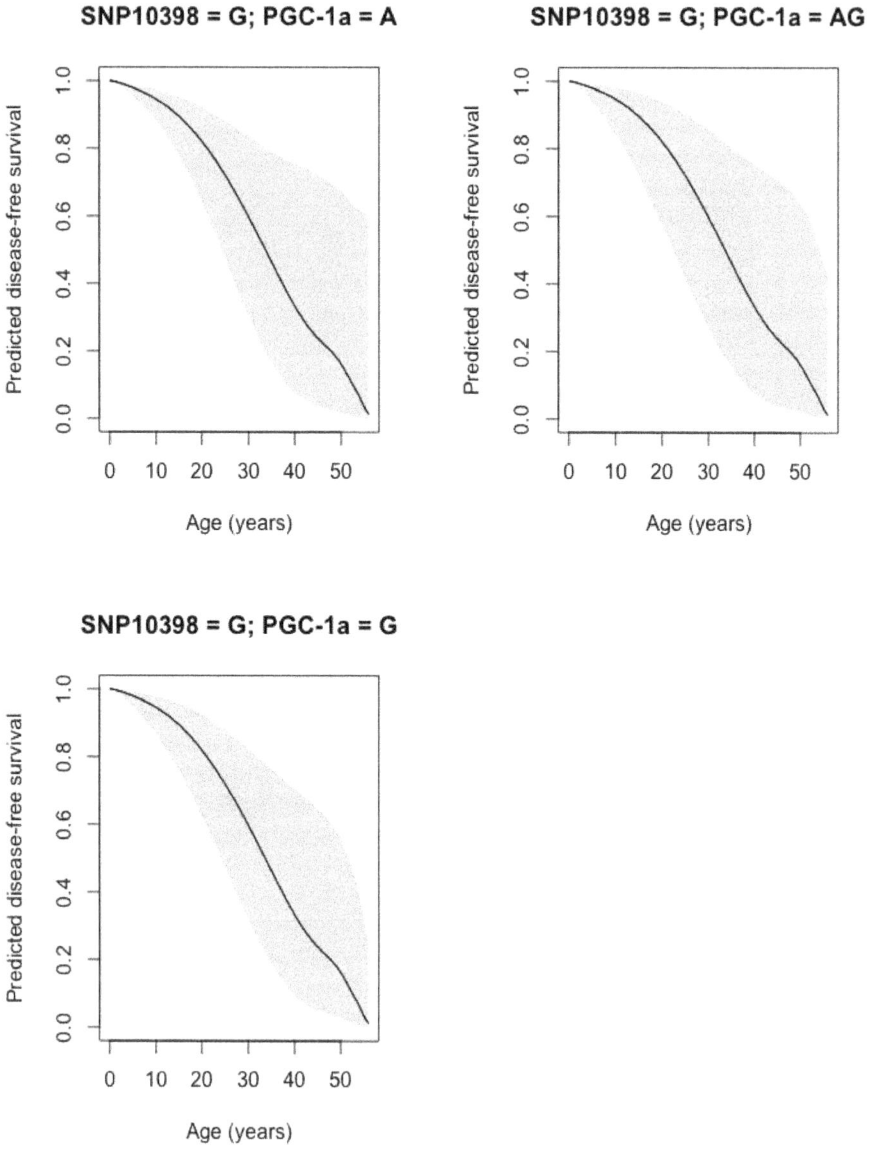


Predicted survival function for 'SNP10398=G'

FIGURE 3.2: Predicted survival functions and associated point-wise 95% confidence intervals for each allele sub-group, obtained from the multivariate model fitting assuming interval censoring.

## 3.10 Summary

This chapter supplements Chapter 2 by addressing an important case that was not covered in Chapter 2, namely truncation. Truncation is frequently encountered in practice, especially in data from clinical trials. Penalized conditional likelihood estimation of the semi-parametric Cox model under double truncation and interval censoring is explained in this chapter, which includes estimation of the baseline hazard. In the next chapter, we will address another complexity in practical survival data analysis – specifically, the presence of individuals in the cohort who may not experience the event of interest. These individuals constitute a subgroup known as the "cured fraction".

# 4

# *Extension to Include a Cured Fraction*

In this chapter, we will explore another practically important extension of the Cox model. This extension addresses situations where the population from which we collected survival data includes a cured fraction. This sub-population consists of individuals who will never experience the event of interest, regardless of the length of the waiting time (which is finite) for the event to occur. Ignoring such a fraction can often result in biased estimates in the regression model, leading to inaccurate predictions, for example, in survival probabilities.

## 4.1 Introduction

In the various Cox models discussed in the previous chapters, an implicit assumption made is that all members of the population under study are susceptible to the event of interest. That is, in the absence of right (or administrative) censoring, and with a finite follow-up time, all individuals would eventually experience the event of interest. In the context of, for instance, studies of chronic or recurrent diseases, it is clear that this assumption may not always hold, as some members of the population may never experience the event of interest. Survival datasets containing a so-called "cured" fractions are characterized by a high rate of right censoring even after a long follow-up time, and a long flat "tail" or plateau in the survival curve above zero at the end of the follow-up period.

For example, we may consider a study investigating cancer recurrence amongst patients diagnosed with thin primary cutaneous melanoma. For these patients, the rate of recurrence-free survival at 15 years post-diagnosis has been reported as upwards of 85%, indicating that a proportion of the patients are likely to never experience recurrence. In this context, ignoring the presence of the cured fraction can lead to survival probability estimates that are overestimated.

Two main approaches exist for performing survival analysis on datasets where some individuals may not ever experience the event. The first of these is the **mixture cure model** (Farewell 1982). This model assumes that the population of interest is made up of two sub-populations: the uncured population which is susceptible to the event of interest, and the cured population which is not. This model then models the probability of being uncured (the "incidence" model) separately to the time-to-event amongst the susceptible population (the "latency" model). Covariates can be

DOI: 10.1201/9781351109710-4

included in both the incidence and the latency models; the covariates used in these models may be identical, may have some overlap, or may be completely different. Importantly, the survival function obtained for the whole population (see Equation (4.3) below) will not be a "proper" survival function because it is not guaranteed to go to 0 as time goes to infinity, but this is exactly the feature we demand for a population containing a cured fraction. Additionally, if a proportional hazards model is used for the latency part, Cox's partial likelihood is not available as a method for estimating the regression parameters.

The other approach is the **promotion time cure model** (see Yakovlev et al. (1996), Tsodikov (1998), Tsodikov et al. (2003) among others). In contrast, this model assumes that the survival time (which will be $\infty$ for non-susceptible individuals) is a function of some latent process which may lead to the event occurring, and has particular application when considering events which are the result of biological processes. For example, when considering an event of interest related to a cancer recurrence, the latent process of interest may be the development of cancer cells remaining after a first treatment into active tumours. In this case, the number of cancer cells remaining after treatment for each individual may be considered to be Poisson distributed with some parameter $\theta$ which may be a function of some covariates, and the time taken for a cell to become an active tumour may have a distribution function $\lambda(\cdot)$. Then the survival function can be defined as $Pr(T > t) = \exp(-\theta\lambda(t))$. The key differences between the mixture cure model and the promotion time cure model are firstly its biological interpretation, and secondly that generally in the promotion time model the covariates effect both the cure probability and the survival through $\theta$ and cannot be separated, unlike in the mixture cure model.

In this chapter, we will focus on presenting the mixture cure model approach. We will first give an introduction to the model structure and discuss estimation methods for the model when only right censoring (and not left or interval censoring) is present in the sample. Following this, we focus on extending the penalized likelihood methods for fitting the proportional hazards model to partly interval-censored data to the mixture cure Cox model. We give details of regression parameter estimation and the approximation of the (conditional) baseline hazard function. We give an extended example of fitting the mixture cure Cox model to a simulated dataset based on data from a real study of partly interval-censored melanoma recurrence times. As part of this example, we give details of the computation of sub-population and population survival predictions. Finally, we give an overview of methods for assessing the predictive accuracy of these models.

## 4.2 The mixture cure Cox model

Here we present a model for time-to-event data with a cured fraction which uses a logistic regression model for the incidence and a semi-parametric Cox model for the latency. We first consider this model in the context of right censored data and later will extend this to a general partly interval censoring scheme.

We now introduce some notations needed for this model. Similar to the previous chapters, we let $Y_i$ be a random variable denoting the time to the event of interest for individual $i$, where $i = 1, ..., n$. Further, let $C_i$ be a random variable, independent of $Y_i$ for all $i$, denoting the right censoring time for individual $i$. We observe $T_i = \min(Y_i, C_i)$, and an observed value of $T_i$ is denoted by $t_i$. Additionally, let there be a partially observed random variable $U_i$ which indicates the cure status of individual $i$, such that $U_i = 1$ indicates that individual $i$ is susceptible to the event of interest and $U_i = 0$ indicates that individual $i$ is a member of the cured sub-population. The value of $U_i$ is partially observed because for any individual with $Y_i < C_i$, they are known to have $U_i = 1$, while an individual with $C_i < Y_i$ does not have an observed value of $U_i$. In other words, it is unknown whether individuals with a right-censored event time are susceptible to the event of interest or not. Similarly, an observed $U_i$ value is denoted by $u_i$.

The time-to-event among the susceptible sub-population can be modelled using a semi-parametric Cox model such that

$$h(t|\mathbf{x}_i, u_i = 1) = h_0(t)e^{\mathbf{x}_i^\top \boldsymbol{\beta}}, \tag{4.1}$$

where $h_0(t)$ is a non-parametric baseline hazard function, $\mathbf{x}_i$ is a vector of covariates, and $\boldsymbol{\beta}$ is a vector of regression coefficients. Additionally, the probability of an individual $i$ being uncured, namely $Pr(U_i = 1)$, denoted $\pi(\mathbf{z}_i)$, can be modelled using a logistic regression such that

$$\pi(\mathbf{z}_i) = \frac{\exp(\mathbf{z}_i^\top \gamma)}{1 + \exp(\mathbf{z}_i^\top \gamma)}, \tag{4.2}$$

where $\mathbf{z}_i$ is a vector of covariates and $\gamma$ is a vector of regression coefficients. Note that the covariate vectors $\mathbf{x}_i$ and $\mathbf{z}_i$ may be identical or have some shared elements, or may have no overlap.

Under the mixture cure model, the survival function can be expressed for the whole population (i.e. the combination of both the cured and uncured sub-populations) as

$$S_{pop}(t) = \pi(\mathbf{z})S(t|\mathbf{x}, u = 1) + 1 - \pi(\mathbf{z}). \tag{4.3}$$

where $S(t|\mathbf{x}, u = 1)$ is the survival function for the susceptible sub-population. If modeled using a Cox model, this survival function can be expressed as:

$$\begin{aligned} S(t|\mathbf{x}_i, u = 1) &= e^{-\exp(\mathbf{x}_i^\top \boldsymbol{\beta}) \int_0^t h_0(v)\, dv} \\ &= S_0(t)^{\exp(\mathbf{x}_i^\top \boldsymbol{\beta})}, \end{aligned}$$

where $S_0(t) = \exp\{-H_0(t)\}$ (with $H_0(t) = \int_0^t h_0(v)dv$) is the baseline survival function. Note that if there is no cured fraction, or in other words if $\pi(\mathbf{z}) = 1$, then Equation (4.3) reduces to the survival function for a normal Cox model.

## 4.3 Estimation methods under right censoring only

As discussed previously, the presence of left and/or interval censoring in survival data means that the partial likelihood estimation method for the Cox model cannot be applied, as the baseline hazard function cannot be left arbitrary. In the case of a mixture cure model, even if there is only right censoring, the partial likelihood method is also not available as the proportional hazards assumption for the overall hazard does not hold and so the baseline hazard function can no longer be left arbitrary without losing information about the parameters in the incidence model.

Instead of a partial likelihood, a common approach is to indirectly maximize the full likelihood function via an expectation-maximization (EM) algorithm, as in Sy & Taylor (2000). They denote the full observed likelihood for the cure model as

$$\begin{aligned} L(\boldsymbol{\beta},\boldsymbol{\gamma},h_0) = \prod_{i=1}^{n} &\{\pi_i(\mathbf{z}_i)h_0(t_i)e^{\mathbf{x}_i^{\mathsf{T}}\boldsymbol{\beta}}S(t|\mathbf{x}_i,u_i=1)\}^{\delta_i} \\ &\times \{(1-\pi_i(\mathbf{z}_i))+\pi_i(\mathbf{z}_i)S(t|\mathbf{x}_i,u_i=1)\}^{1-\delta_i}, \end{aligned} \tag{4.4}$$

and then the complete-date full likelihood, treating the partially unknown $u_i$ values as known, is

$$\begin{aligned} L_C(\boldsymbol{\beta},\boldsymbol{\gamma},h_0;u) &= \prod_{i=1}^{n} \pi_i(\mathbf{z}_i)^{u_i}(1-\pi_i(\mathbf{z}_i))^{1-u_i} \\ &\quad\times \left[\{h_0(t)e^{\mathbf{x}_i^{\mathsf{T}}\boldsymbol{\beta}}\}^{\delta_i}S(t|\mathbf{x}_i,u_i=1)\right]^{u_i} \\ &= L_1(\boldsymbol{\gamma};u)L_2(\boldsymbol{\beta},h_0(\cdot);u), \end{aligned} \tag{4.5}$$

which is then maximized with respect to the distribution of the partially unobserved $u_i$'s. In practice, this E-step of the EM algorithm requires replacing the $u_i$ values in Equation (4.5) with a conditional expectation, which is expressed as a weight $w_i^{(k)}$, which takes a value of 1 if $i$ is uncensored, and for censored $i$ is

$$\left.\frac{\boldsymbol{\pi}(\mathbf{z}_i;\boldsymbol{\gamma})S_0(t_i)^{\exp(\mathbf{x}_i^{\mathsf{T}}\boldsymbol{\beta})}}{1-\boldsymbol{\pi}(\mathbf{z}_i;\boldsymbol{\gamma})+\boldsymbol{\pi}(\mathbf{z}_i;\boldsymbol{\gamma})S_0(t_i)^{\exp(\mathbf{x}_i^{\mathsf{T}}\boldsymbol{\beta})}}\right|_{\boldsymbol{\eta}=\boldsymbol{\eta}^{(k)}},$$

where $\boldsymbol{\eta}^{(k)}$ denotes the values of the parameters $\boldsymbol{\beta}$, $\boldsymbol{\gamma}$, and $h_0(t)$ at the $k$-th iteration.

Following this, the M-step involves the maximization of the likelihood in (4.5) with respect to $\boldsymbol{\beta}$, $\boldsymbol{\gamma}$ and the function $h_0(t)$, given $w_i^{(k)}$. The method proposed by Sy & Taylor (2000) includes an additional maximization step to deal with the nuisance parameter $H_0(t)$, the cumulative function of $h_0(t)$, using a profile likelihood technique. These authors describe a Breslow-type estimator for $H_0(t)$ given $\boldsymbol{\beta}$, given by

$$\widetilde{H}_0^{(k)}(t) = \sum_{i:t_{(i)}\leq t} \frac{n_i}{\sum_{l\in\mathcal{R}_i} w_l^{(k)}e^{\mathbf{x}_l^{\mathsf{T}}\boldsymbol{\beta}}} \tag{4.6}$$

at iteration $k$, where $n_i$ is the number of events at time $t_i$, and $\mathcal{R}_i$ is the risk set at time $t_i^-$. Substituting $\widetilde{H}_0(t)$ into $L_2(\cdot)$ in Equation (4.5) leads to a partial likelihood for $\boldsymbol{\beta}$ given by

$$L_3(\boldsymbol{\beta}; w^{(k)}) = \prod_{i=1}^{n} \left( \frac{\exp(\mathbf{x}_i^\top \boldsymbol{\beta})}{\sum_{l \in \mathcal{R}_i} w_l^{(k)} \exp(\mathbf{x}_l^\top \boldsymbol{\beta})} \right)^{\delta_i}. \tag{4.7}$$

This is then maximized with respect to $\boldsymbol{\beta}$ given $w^{(k)}$, and the profile-likelihood estimator for the baseline survival function for the uncured fraction becomes $\exp\{-\widetilde{H}_0^{(k)}(t)\}$.

An important consideration in the estimation method outlined above is the identifiability of the right tail of the baseline survival function, as an accurate estimation of this right tail will enable the uncured proportion $\pi_i$ to be estimated accurately.

As noted above, a characteristic of survival data containing a cured fraction is a long, stable plateau in the right-hand tail of the baseline survival function. The identifiability conditions for the mixture cure model have been studied (for example, Amico & van Keilegom (2018)), and it has been pointed out that identifiability issues may arise unless an assumption is made that all uncured individuals have experienced their event before the beginning of the plateau. In practice, the method of Sy & Taylor (2000) imposes this assumption by forcing the baseline survival function to go to zero after the last observed event time. In the EM algorithm outlined above, this is incorporated by forcing $w_i^{(k)} = 0$ for any $i$ where $t_i$ is greater than the largest event time. As Peng (2003) has pointed out, this so-called "zero-tail constraint" can lead to an overestimation of the cured fraction in some cases, especially if the censoring times of the cured individuals are not well separated from those of the susceptible individuals.

**Example 4.1** (R Example: Right censored data with a cure fraction).
In this example, we will first explain, using an R program, how to generate a simulated survival dataset from a mixture cure Cox model. Then, we will fit this model using the R `smcure` package. More specifically, we will conduct the following activities:

1. Generate a sample of survival data of size $n = 500$, with a cured fraction, such that covariates $Z \sim unif(0, 1)$ for the logistic incidence model and $X \sim binomial(1, 0.5)$ for the non-cured fraction Cox model are distinct. The baseline hazard for the Cox model $h_0(t) = 3t^2$. In this simulation, we allow approximately $30\%$ of the sample to be cured, and additionally allow approximately $20\%$ of the uncured individuals to have their event times right censored. See, for example, Example 2.6 for how to generate random numbers from a Cox PH model.
2. Fit the mixture cure model to the data using the `smcure` package available in R and make comments on interpretation of the regression parameters.
3. Compute the values of $\pi(\cdot)$ for each $i$ and comment.
4. Obtain and plot the estimate of the survival function available in the `smcure` package.

```
# Write a function to generate
# survival data with a cured
# fraction
generate_cured_data <- function(n, b_true,
    g_true, censoringtimemax = 3) {

    # create X and Z
    X <- as.matrix(rbinom(n, 1, 0.5))
    Z <- as.matrix(runif(n, 0, 1))
    Z.mod <- cbind(1, Z)

    # generate results from
    # logistic regression
    pi_i <- exp(Z.mod %*% b_true)/(1 +
        exp(Z.mod %*% b_true))

    # generate cure indicator
    U_C <- runif(n)
    cure_i <- as.numeric(U_C < pi_i)  #1 is not cured

    # generate observation times
    TL_i <- TR_i <- Y_i <- censor <- rep(0,
        n)

    for (i in 1:n) {
        # generate right
        # censoring/'dropout' time
        max.followup <- runif(1, 0.5,
            censoringtimemax)
        if (cure_i[i] == 0) {
            # if they are cured
            Y_i[i] <- Inf  #they have no event time
            # they are right
            # censored their
            # dropout time
            TL_i[i] <- max.followup
            TR_i[i] <- Inf
        } else {
            # if they are not
            # cured generate an
            # event time
            U_Y <- runif(1)
            Y_i[i] <- (-log(U_Y)/exp(X[i,
                ] %*% g_true))^(1/3)
            if (Y_i[i] < max.followup) {
                # if their event
                # time is before
                # their
```

```
                # dropout/censoring
                # time
                TL_i[i] <- Y_i[i]
                TR_i[i] <- Y_i[i]
                # they have an
                # event time
            } else {
                # they are right
                # censored
                TL_i[i] <- max.followup
                TR_i[i] <- Inf
            }
        }
        censor[i] <- as.numeric(TR_i[i] !=
            Inf)
    }
    data <- data.frame(Y_i, TL_i, TR_i,
        censor, cure_i, Z, X)
    return(data)
}

dat <- generate_cured_data(n = 500,
    b_true = c(1, -0.5), g_true = 1,
    censoringtimemax = 1.5)

# Check simulated data
sum(dat$cure_i)/500  #should be ~70%
sum(dat$censor[dat$cure_i == 1])/sum(dat$cure_i ==
    1)
# should be ~80%

# Fit model using smcure package
library(smcure)
smc.fit <- smcure(Surv(TL_i, censor) ~
    X, cureform = ~Z, data = dat, model = "ph",
    link = "logit")

# Compute pi values for each i
pi_i <- exp(as.matrix(dat[, 6:7]) %*%
    as.matrix(smc.fit$b))/(1 + exp(as.matrix(dat[,
    6:7]) %*% as.matrix(smc.fit$b)))

# Compute and plot the baseline
# survival function
pred.smc <- predictsmcure(smc.fit, newX = c(0,
    0), newZ = c(0), model = "ph")
baseSurv.smc <- (pred.smc$prediction[,
    1])[order(pred.smc$prediction[,
```

```
    3]) ]
times.smc <- (pred.smc$prediction[,
    3])[order(pred.smc$prediction[,
    3])]
plot(baseSurv.smc ~ times.smc, type = "s",
    ylim = c(0, 1), main = "Predicted baseline survival",
    xlab = "Time", ylab = "Survival probability")
```

TABLE 4.1: Coefficient estimates for the intercept and effect of $Z$ on probability of being cured (logistic incidence model) and the effect of $X$ on event time (Cox regression latency model).

| Coefficient | Estimate | SE | p-value |
|---|---|---|---|
| $\gamma_0$ | 0.791 | 0.229 | $< 0.001$ |
| $\gamma_1$ | -0.276 | 0.416 | 0.505 |
| $\beta_1$ | 0.856 | 0.163 | $< 0.001$ |

Table 4.1 shows the coefficient estimates from the mixture cure Cox model fitted using the `smcure` package. The parameter $\gamma_0$ is the intercept in the logistic incidence model, and tells us an individual with $Z = 0$ has an estimated probability of $\exp(\gamma_0)/(1 + \exp(\gamma_0)) = \exp(0.791)/(1 + \exp(0.791)) = 0.688$ of being susceptible to the event of interest (recall we targeted a probability of $\sim 70\%$ in our simulation). The odds of an individual being susceptible to the event of interest when $Z = 1$ compared to $Z = 0$ decrease by a factor of $\exp(\gamma_1) = \exp(-0.276) = 0.759$. Then, *among those individuals who are susceptible to the event of interest*, the effect of $X = 1$ compared to $X = 0$ is to increase the risk of the event of interest at time $t$ by a factor of $\exp(\beta_1) = \exp(0.856) = 2.354$.

□

## 4.4 The penalized likelihood approach for partly interval-censored data

When a survival dataset contains a cured fraction and a combination of directly observed event times, right censoring times, left censoring times and interval censoring times, it is necessary to extend the likelihood function given in (4.4). For convenience and following the notations in the last two chapters, allow $\delta_i$, $\delta_i^R$, $\delta_i^L$ and $\delta_i^I$ to be indicator variables containing information about whether individual $i$ is non-censored, right-censored, left-censored or interval-censored respectively. Note that $\delta_i = 1 - \delta_i^R - \delta_i^L - \delta_i^I$.

We rewrite $S(t|\mathbf{x}_i, u_i = 1)$ by $S_i(t)$. Then, the likelihood function in Equation (4.4) can be extended to incorporate left- and interval-censored observations,

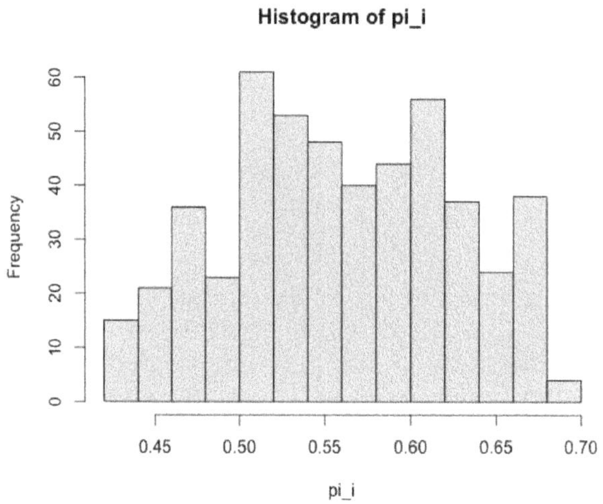


FIGURE 4.1: Histogram of estimated values of $\pi_i(\cdot)$, the probability of an individual being susceptible to the event of interest.

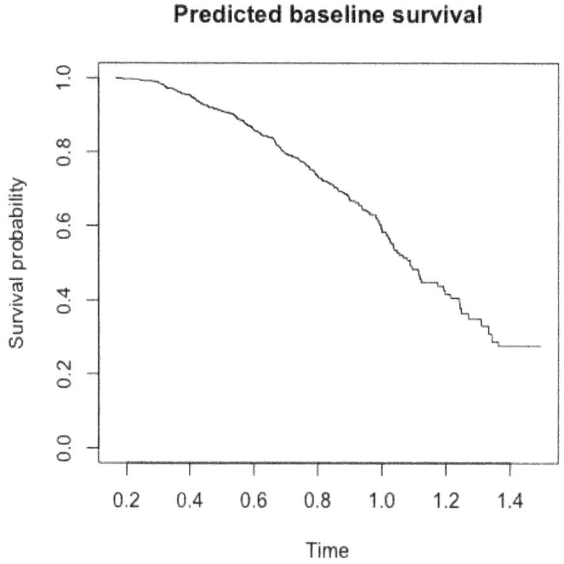


FIGURE 4.2: Survival function from R `smcure` package.

transforming the likelihood into

$$\begin{aligned} L(\boldsymbol{\beta}, \boldsymbol{\gamma}, h_0) = \prod_{i=1}^{n} & \{\pi_i(\mathbf{z}_i) h_0(t_i) e^{\mathbf{x}_i^\mathsf{T}\boldsymbol{\beta}} S_i(t_i)\}^{\delta_i} \\ & \times \{(1 - \pi_i(\mathbf{z}_i)) + \pi_i(\mathbf{z}_i) S_i(t_i^L)\}^{\delta_i^R} \{\pi_i(\mathbf{z}_i)(1 - S_i(t_i^R))\}^{\delta_i^L} \\ & \times \{\pi_i(\mathbf{z}_i)(S_i(t_i^L) - S_i(t_i^R))\}^{\delta_i^I}. \end{aligned}$$

Thus, the log-likelihood is:

$$\begin{aligned} l(\boldsymbol{\beta}, \boldsymbol{\gamma}, h_0) = \sum_{i=1}^{n} & \delta_i\{\log \pi_i(\mathbf{z}_i) + \log h_0(t_i) + \mathbf{x}_i^\mathsf{T}\boldsymbol{\beta} - H_i(t_i)\} \\ & + \delta_i^R \log\{(1 - \pi_i(\mathbf{z}_i)) + \pi_i(\mathbf{z}_i)S_i(t_i^L)\} \\ & + \delta_i^L\{\log \pi_i(\mathbf{z}_i) + \log(1 - S_i(t_i^R))\} \\ & + \delta_i^I\{\log \pi_i(\mathbf{z}_i) + \log(S_i(t_i^L) - S_i(t_i^R))\}. \end{aligned} \tag{4.8}$$

Since the baseline hazard $h_0(t)$ is unspecified in (4.8), maximization of (4.8) with respect to $h_0(t)$ becomes a difficult task. Similar to the previous chapters, this baseline hazard function is again approximated via non-negative basis functions. That is, we assume that the baseline hazard function for the uncured fraction can be approximated as

$$h_0(t) \approx \sum_{u=1}^{m} \theta_u \psi_u(t), \tag{4.9}$$

where $\theta_u \geq 0$ for all $u$ and the $\psi_u(t)$ are a set of non-negative basis functions, such as M-splines (see Chapter 2 for more in-depth discussions of the choice of basis functions).

We assume $h_0(t)$ is a smooth function. For the purpose of imposing smoothness on $h_0(t)$ and dampening the impacts of unnecessary knots, we include a penalty function into the logarithm of the likelihood function in (4.8). This penalized log-likelihood can be maximized to estimate the unknown parameters.

With the baseline hazard function approximation, the set of parameters to be estimated in the mixture cure Cox model becomes $\boldsymbol{\eta}^\mathsf{T} = (\boldsymbol{\theta}^\mathsf{T}, \boldsymbol{\beta}^\mathsf{T}, \boldsymbol{\gamma}^\mathsf{T})^\mathsf{T}$, where $\boldsymbol{\theta}$ denotes a vector for all $\theta_u$. The penalized log-likelihood is now given by

$$\Phi(\boldsymbol{\eta}) = l(\boldsymbol{\eta}) - \lambda J(\boldsymbol{\theta}), \tag{4.10}$$

where $J(\boldsymbol{\theta})$ is a roughness penalty function and $\lambda \geq 0$ is a smoothing parameter. The roughness penalty is given by $\int h_0''(v)^2 dv$. Since $h_0(t)$ is now given by (4.9), we can conveniently express this roughness penalty as

$$J(\boldsymbol{\theta}) = \boldsymbol{\theta}^\mathsf{T}\mathbf{R}\boldsymbol{\theta},$$

where $\mathbf{R}$ is an $m \times m$ matrix with the $(u, v)$-th element given by $r_{uv} = \int \psi_u''(t)\psi_v''(t)dt$.

Given the constraint that $\boldsymbol{\theta} \geq 0$, we have the following Karush-Kuhn-Tucker (KKT) conditions for a constrained optimal solution:

$$\frac{\partial\Phi(\boldsymbol{\eta})}{\partial\beta_t} = 0, \quad \frac{\partial\Phi(\boldsymbol{\eta})}{\partial\gamma_j} = 0;$$

$$\frac{\partial\Phi(\boldsymbol{\eta})}{\partial\theta_u} = 0 \text{ if } \theta_u > 0, \quad \frac{\partial\Phi(\boldsymbol{\eta})}{\partial\theta_u} < 0 \text{ if } \theta_u = 0.$$

These conditions are solved iteratively using an algorithm similar to the Newton-MI algorithm; see Ma et al. (2014) and Ma et al. (2021). This algorithm requires the score vector and the Hessian matrix when updating $\boldsymbol{\beta}$ and $\boldsymbol{\gamma}$, but for $\boldsymbol{\theta}$ it only demands its score vector. Details of score vector and the Hessian matrix can be found in Section 4.5.

Before describing this algorithm we first introduce some notations. Let $\boldsymbol{\beta}^{(k)}, \boldsymbol{\gamma}^{(k)}$ and $\boldsymbol{\theta}^{(k)}$ be, respectively, the estimates of $\boldsymbol{\beta}$, $\boldsymbol{\gamma}$ and $\boldsymbol{\theta}$ at iteration $k$. Also, for any function $a(x)$, we let $a(x)^+$ and $a(x)^-$ be respective the positive and negative components of $a(x)$, so that $a(x)^+ - a(x)^- = a(x)$. Iteration $k+1$ of this algorithm is obtained in a three-step process as follows.

Firstly, obtain $\boldsymbol{\beta}^{(k+1)}$ using a modified Newton algorithm:

$$\boldsymbol{\beta}^{(k+1)} = \boldsymbol{\beta}^{(k)} + \omega_1^{(k)} \left[ -\frac{\partial^2 \Phi(\boldsymbol{\beta}^{(k)}, \boldsymbol{\gamma}^{(k)}, \boldsymbol{\theta}^{(k)})}{\partial \boldsymbol{\beta} \partial \boldsymbol{\beta}^\mathsf{T}} \right]^{-1} \left[ \frac{\partial \Phi(\boldsymbol{\beta}^{(k)}, \boldsymbol{\gamma}^{(k)}, \boldsymbol{\theta}^{(k)})}{\partial \boldsymbol{\beta}} \right], \tag{4.11}$$

where $\omega_1 \in (0,1]$ is the line search step size, and it is used to ensure that $\Phi(\boldsymbol{\beta}^{(k+1)}, \boldsymbol{\gamma}^{(k)}, \boldsymbol{\theta}^{(k)}) \geq \Phi(\boldsymbol{\beta}^{(k)}, \boldsymbol{\gamma}^{(k)}, \boldsymbol{\theta}^{(k)})$. A value of the line search step size $\omega_1$ can be determined by using, for instance, Armijo's rule.

Secondly, we compute $\boldsymbol{\gamma}^{(k+1)}$ using again a modified Newton algorithm based on $\boldsymbol{\beta}^{(k+1)}, \boldsymbol{\gamma}^{(k)}$ and $\boldsymbol{\theta}^{(k)}$:

$$\boldsymbol{\gamma}^{(k+1)} = \boldsymbol{\gamma}^{(k)} + \omega_2^{(k)} \left[ -\frac{\partial^2 \Phi(\boldsymbol{\beta}^{(k+1)}, \boldsymbol{\gamma}^{(k)}, \boldsymbol{\theta}^{(k)})}{\partial \boldsymbol{\gamma} \partial \boldsymbol{\gamma}^\mathsf{T}} \right]^{-1} \left[ \frac{\partial \Phi(\boldsymbol{\beta}^{(k+1)}, \boldsymbol{\gamma}^{(k)}, \boldsymbol{\theta}^{(k)})}{\partial \boldsymbol{\gamma}} \right], \tag{4.12}$$

where $\omega_2$ is a line search step size defined similarly to $\omega_1$.

Finally, we get the update $\boldsymbol{\theta}^{(k+1)}$ from $\boldsymbol{\beta}^{(k+1)}, \boldsymbol{\gamma}^{(k+1)}$ and $\boldsymbol{\theta}^{(k)}$ using the multiplicative iterative algorithm:

$$\boldsymbol{\theta}^{(k+1)} = \boldsymbol{\theta}^{(k)} + \omega_3^{(k)} \mathbf{D}^{(k)} \frac{\partial \Phi(\boldsymbol{\beta}^{(k+1)}, \boldsymbol{\gamma}^{(k+1)}, \boldsymbol{\theta}^{(k)})}{\partial \boldsymbol{\theta}}, \tag{4.13}$$

where $\omega_3$ is defined similarly to $\omega_1$ and $\omega_2$ and $\mathbf{D}^{(k)}$ is a diagonal $m \times m$ matrix with elements $\theta_u^{(k)} / \zeta_{2u}^{(k)}$ for $u = 1, ..., m$, and where

$$\zeta_{2u} = \left[ \frac{\partial l(\boldsymbol{\beta}, \boldsymbol{\gamma}, \boldsymbol{\theta})}{\partial \theta_u} \right]^- + \lambda \left[ \frac{\partial J(\boldsymbol{\theta})}{\partial \theta_u} \right]^+ + \epsilon.$$

Referring to Section 4.5 for the score vector of $\boldsymbol{\theta}$, we can see that $\zeta_{2u}$ becomes:

$$\begin{aligned} \zeta_{2u} =& \delta_i \Psi_u(t_i) e^{\mathbf{x}_i^\mathsf{T} \boldsymbol{\gamma}} + \delta_i^R \frac{\pi(\mathbf{z}_i) S_i(t_i) \Psi_u(t_i)}{1 - \pi(\mathbf{z}_i) + \pi(\mathbf{z}_i) S_i(t_i)} e^{\mathbf{x}_i^\mathsf{T} \boldsymbol{\gamma}} \\ &+ \delta_i^I \frac{S_i(t_i^L) \Psi_u(t_i^L)}{S_i(t_i^L) - S_i(t_i^R)} e^{\mathbf{x}_i^\mathsf{T} \boldsymbol{\gamma}} + \lambda \left[ \frac{\partial J(\boldsymbol{\theta})}{\partial \theta_u} \right]^+ + \epsilon. \end{aligned}$$

Here, $\Psi_u(t)$ is the cumulative basis function taking the form $\Psi_u(t) = \int_0^t \psi_u(v) dv$. Note that $\epsilon \geq 0$ in the above expression is a small constant included simply to avoid the numerical issue of a zero denominator in the calculation of $\mathbf{D}^{(k)}$ and this value does not have any impact on the final solution for $\boldsymbol{\theta}$.

### 4.4.1 Smoothing parameter estimation

A marginal likelihood method that we have used in the last two chapters for the automatic selection of the smoothing parameter can be also implemented for the model of this chapter. In this method, a key feature is that the penalty function $J(\boldsymbol{\theta})$ is quadratic in $\boldsymbol{\theta}$:

$$J(\boldsymbol{\theta}) = \lambda \boldsymbol{\theta}^{\mathsf{T}} \mathbf{R} \boldsymbol{\theta},$$

so that $J(\boldsymbol{\theta})$ is related to a normal prior distribution for the vector $\boldsymbol{\theta}$ with $\boldsymbol{\theta} \sim N(0_{m\times 1}, \sigma^2 \mathbf{R}^{-1})$, where $\sigma^2 = 1/(2\lambda)$. We can then obtain the log-posterior:

$$l_p(\boldsymbol{\beta}, \boldsymbol{\gamma}, \boldsymbol{\theta}) = -\frac{m}{2}\log \sigma^2 + l(\boldsymbol{\beta}, \boldsymbol{\gamma}, \boldsymbol{\theta}) - \frac{1}{2\sigma^2}\boldsymbol{\theta}^{\mathsf{T}} \mathbf{R} \boldsymbol{\theta}. \tag{4.14}$$

The marginal likelihood for $\sigma^2$ may be difficult to obtain directly, and as such we can approximate it using Laplace's method. Applying the Laplace approximation and substituting in the MPL estimates of $\boldsymbol{\beta}$, $\boldsymbol{\gamma}$ and $\boldsymbol{\theta}$, we can obtain the approximated log-marginal likelihood for $\sigma^2$:

$$l_m(\sigma^2) \approx -\frac{m}{2}\log \sigma^2 + l(\widehat{\boldsymbol{\beta}}, \widehat{\boldsymbol{\gamma}}, \widehat{\boldsymbol{\theta}}) - \frac{1}{2\sigma^2}\widehat{\boldsymbol{\theta}}^{\mathsf{T}} \mathbf{R} \widehat{\boldsymbol{\theta}} - \frac{1}{2}\log \mid -\widehat{\mathbf{H}} + \mathbf{Q}(\sigma^2) \mid, \tag{4.15}$$

where $\widehat{\mathbf{H}}$ is the Hessian matrix from $l(\boldsymbol{\beta}, \boldsymbol{\gamma}, \boldsymbol{\theta})$ evaluated at the MPL estimates $\widehat{\boldsymbol{\beta}}$, $\widehat{\boldsymbol{\gamma}}$ and $\widehat{\boldsymbol{\theta}}$ with the current value of $\lambda$, and

$$\mathbf{Q}(\sigma^2) = \begin{pmatrix} \mathbf{0}_{p\times p} & \mathbf{0} & \mathbf{0} \\ \mathbf{0} & \mathbf{0}_{q\times q} & \mathbf{0} \\ \mathbf{0} & \mathbf{0} & \frac{1}{\sigma^2}\mathbf{R} \end{pmatrix}.$$

An approximate maximum marginal likelihood solution for $\sigma^2$ is:

$$\widehat{\sigma}^2 = \frac{\widehat{\boldsymbol{\theta}}^{\mathsf{T}} \mathbf{R} \widehat{\boldsymbol{\theta}}}{m - \nu}, \tag{4.16}$$

where $\nu = \mathrm{tr}\{(-\widehat{\mathbf{H}} + \mathbf{Q}(\widehat{\sigma}^2))^{-1}\mathbf{Q}(\widehat{\sigma}^2)\}$, which can be considered as equivalent to the model degrees of freedom. Given that the estimates of $\boldsymbol{\beta}$, $\boldsymbol{\gamma}$ and $\boldsymbol{\theta}$ depend on $\sigma^2$, this approximate solution for $\sigma^2$ allows for the development of an iterative procedure with two steps. Firstly, with a current $\sigma$, the corresponding MPL estimates for $\boldsymbol{\beta}$, $\boldsymbol{\gamma}$ and $\boldsymbol{\theta}$ are obtained. Then, $\sigma^2$ is updated using the current $\widehat{\sigma}^2$, and the just obtained $\widehat{\boldsymbol{\beta}}$, $\widehat{\boldsymbol{\gamma}}$ and $\widehat{\boldsymbol{\theta}}$ on the right-hand side of (4.16). These two steps are repeated until $\nu$ is stabilized, such as the difference between two consecutive $\nu$ values is less than 1.

We note that, in our experience, the choice of the smoothing parameter has implications for the identifiability of the model and the accuracy of the estimate for $\gamma_0$, the true intercept value in the logistic regression model for the incidence, in particular. In the paper by Corbière et al. (2009), who developed an alternative penalized likelihood approach to a mixture cure Cox model with right censored data, the authors comment that the zero-tail constraint discussed above is not needed if the smoothing parameter value is chosen well. To illustrate this for the model detailed in this chapter, we give the results of a small simulation study. The results in Table 4.2 show the bias in the estimate of $\gamma_0$ from simulated right censored data, comparing the fit of the smcure package (which imposes a zero-tail constraint as per Sy & Taylor (2000)), and the fit of the MPL model detailed in this chapter when the smoothing parameter $\lambda$ is set to 0, set to 0.1, or chosen automatically via the process detailed in this section. Note that differing values of $\gamma_0$ control the size of the cured fraction. Evidently, the results from smcure and the MPL method with automatic smoothing parameter selection are comparable, while the bias in the $\gamma_0$ estimate can be large when the smoothing parameter is inappropriate.

TABLE 4.2: Bias comparison of R smcure package and MPL in the estimate of the intercept in the logistic regression for incidence.

| | $n = 200$ | | | $n = 1000$ | | |
|---|---|---|---|---|---|---|
| | $\gamma_0 = 1.5$ | $\gamma_0 = 0$ | $\gamma_0 = -1.5$ | $\gamma_0 = 1.5$ | $\gamma_0 = 0$ | $\gamma_0 = -1.5$ |
| smcure | 0.027 | -0.041 | -0.032 | 0.011 | -0.023 | -0.019 |
| MPL, $\lambda = 0$ | -0.096 | -0.192 | -0.202 | -0.086 | -0.187 | -0.194 |
| MPL, $\lambda = 0.01$ | -0.054 | -0.587 | -0.697 | -0.043 | -0.612 | -0.654 |
| MPL, auto | 0.019 | 0.013 | -0.035 | 0.009 | 0.008 | -0.017 |

## 4.5 Score vector and Hessian matrix

The discussions in the last section involve the score vector and the Hessian matrix. In this section, we will provide the details about them.

The components of the score vector of the penalized likelihood are as follows. Let $z_{il}$ be element $l$ of vector $\mathbf{z}_i$ and let $x_{ij}$ be element $j$ of vector $\mathbf{x}_i$, for $l = 1, ..., q$ and $j = 1, ..., p$. Then the first derivative of $\Phi(\boldsymbol{\eta})$ with respect to $\gamma_l$ is

$$\frac{\partial \Phi(\boldsymbol{\eta})}{\partial \gamma_l} = \sum_{i=1}^{n} z_{il}\left((1-\delta_i^R)(1-\pi(\mathbf{z}_i)) + \delta_i^R \frac{(S_i(t_i^L)-1)\pi(\mathbf{z}_i)(1-\pi(\mathbf{z}_i))}{1-\pi(\mathbf{z}_i)+\pi(\mathbf{z}_i)S_i(t_i^L)}\right).$$

The first derivative of $\Phi(\boldsymbol{\eta})$ with respect to $\beta_j$ is

$$\frac{\partial \Phi(\boldsymbol{\eta})}{\partial \beta_j} = \sum_{i=1}^{n} x_{ij}\left(\delta_i(1-H_i(t_i)) - \delta_i^R \frac{\pi(\mathbf{z}_i)S_i(t_i^L)H_i(t_i^L)}{1-\pi(\mathbf{z}_i)+\pi(\mathbf{z}_i)S_i(t_i * L)}\right.$$
$$\left. + \delta_i^L \frac{S_i(t_i^R)H_i(t_i)}{1-S_i(t_i^R)} - \delta_i^I \frac{S_i(t_i^L)H_i(t_i^L) - S_i(t_i^R)H_i(t_i^R)}{S_i(t_i^L)-S_i(t_i^R)}\right).$$

The first derivative of $\Phi(\boldsymbol{\eta})$ with respect to $\theta_u$ is

$$\frac{\partial\Phi(\boldsymbol{\eta})}{\partial\theta_u} = \sum_{i=1}^{n}\left\{\delta_i\frac{\psi_u(t_i)}{h_0(t_i)} - e^{\mathbf{x}_i^\top\beta}\left(\delta_i\Psi_u(t_i) + \delta_i^R\frac{\pi(\mathbf{z}_i)S_i(t_i^L)\Psi_u(t_i^L)}{1-\pi(\mathbf{z}_i)+\pi(\mathbf{z}_i)S_i(t_i^L)}\right.\right.$$
$$\left.\left. - \delta_i^L\frac{S_i(t_i^R)\Psi_u(t_i^R)}{1-S_i(t_i^R)} + \delta_i^I\frac{S_i(t_i^L)\Psi_u(t_i^L) - S_i(t_i^R)\Psi_u(t_i^R)}{S_i(t_i^L)-S_i(t_i^R)}\right)\right\}.$$

The components on the Hessian matrix are as follows.

$$\frac{\partial^2\Phi(\boldsymbol{\eta})}{\partial\gamma_l\partial\gamma_r} = -\sum_{i=1}^{n} z_{il}z_{ir}\left((1-\delta_i^R)\pi(\mathbf{z}_i)(1-\pi(\mathbf{z}_i))\right.$$
$$\left. + \delta_i^R\frac{(S_i(t_i^L)-1)\pi(\mathbf{z}_i)(1-\pi(\mathbf{z}_i))^3}{(1-\pi(\mathbf{z}_i)+\pi(\mathbf{z}_i)S_i(t_i^L))^2}\right).$$

$$\frac{\partial^2\Phi(\boldsymbol{\eta})}{\partial\beta_j\partial\gamma_l} = -\sum_{i=1}^{n} x_{ij}z_{il}\delta_i^R\frac{S_i(t_i^L)H_i(t_i^L)\pi(\mathbf{z}_i)(1-\pi(\mathbf{z}_i))}{(1-\pi(\mathbf{z}_i)+\pi(\mathbf{z}_i)S_i(t_i^L))^2}.$$

$$\frac{\partial^2\Phi(\boldsymbol{\eta})}{\partial\beta_j\partial\theta_u} = -\sum_{i=1}^{n} x_{ij}e^{\mathbf{x}_i^\top\boldsymbol{\beta}}\delta_i^R\frac{S_i(t_i^L)\pi(\mathbf{z}_i)(1-\pi(\mathbf{z}_i))\Psi_u(t)}{(1-\pi(\mathbf{z}_i)+\pi(\mathbf{z}_i)S_i(t_i^L))^2}.$$

$$\frac{\partial^2\Phi(\boldsymbol{\eta})}{\partial\beta_j\partial\beta_s} = -\sum_{i=1}^{n} x_{ij}x_{is}\left(\delta_iH_i(t_i) + \delta_i^R\frac{\pi(\mathbf{z}_i)S_i(t_i^L)H_i(t_i^L)}{1-\pi(\mathbf{z}_i)+\pi(\mathbf{z}_i)S_i(t_i^L)}\right.$$
$$- \delta_i^R\frac{\pi(\mathbf{z}_i)(1-\pi(\mathbf{z}_i))S_i(t_i^R)H_i^2(t_i^L)}{(1-\pi(\mathbf{z}_i)+\pi(\mathbf{z}_i)S_i(t_i^L))^2} - \delta_i^L\frac{S_i(t_i^R)H_i(t_i^R)}{1-S_i(t_i^R)}$$
$$+ \delta_i^L\frac{S_i(t_i^R)H_i^2(t_i^R)}{(1-S_i(t_i^R))^2} + \delta_i^I\frac{S_i(t_i^L)H_i(t_i^L) - S_i(t_i^R)H_i(t_i^R)}{S_i(t_i^L)-S_i(t_i^R)}$$
$$\left. + \delta_i^I\frac{S_i(t_i^L)S_i(t_i^R)(H_i(t_i^L)-H_i(t_i^R))^2}{(S_i(t_i^L)-S_i(t_i^R))^2}\right).$$

$$\frac{\partial^2\Phi(\boldsymbol{\eta})}{\partial\beta_j\partial\theta_u} = -\sum_{i=1}^{n} x_{ij}e^{\mathbf{x}_i^\top\boldsymbol{\beta}}\left(\delta_i\Psi_u(t_i) + \delta_i^R\frac{\pi(\mathbf{z}_i)S_i(t_i^L)\Psi_u(t_i^L)}{1-\pi(\mathbf{z}_i)+\pi(\mathbf{z}_i)S_i(t_i^L)}\right.$$
$$- \delta_i^R\frac{\pi(\mathbf{z}_i)(1-\pi(\mathbf{z}_i))S_i(t_i^L)H(t)\Psi_u(t_i^L)}{(1-\pi(\mathbf{z}_i)+\pi(\mathbf{z}_i)S_i(t_i^L))^2} - \delta_i^L\frac{S_i(t_i^R)\Psi_u(t_i^R)}{1-S_i(t_i^R)}$$
$$+ \delta_i^L\frac{S_i(t_i^R)H_i(t_i^R)\Psi_u(t_i^R)}{(1-S_i(t_i^R))^2} + \delta_i^I\frac{S_i(t_i^L)\Psi_u(t_i^L) - S_i(t_i^R)\Psi_u(t_i^R)}{S_i(t_i^L)-S_i(t_i^R)}$$
$$\left. + \delta_i^I\frac{S_i(t_i^L)S_i(t_i^R)(H_i(t_i^L)-H_i(t_i^R))(\Psi_u(t_i^L)-\Psi_u(t_i^R))}{(S_i(t_i^L)-S_i(t_i^R))^2}\right).$$

$$\frac{\partial^2\Phi(\eta)}{\partial\theta_u\partial\theta_v} = -\sum_{i=1}^{n}\left\{\delta_i\frac{\psi_u(t_i)\psi_v(t_i)}{h_0^2(t_i)} - e^{\mathbf{x}_i^\top\beta}\left(\delta_i^R\frac{\pi(\mathbf{z}_i)(1-\pi(\mathbf{z}_i))S_i(t_i^L)\Psi_u(t_i^L)\Psi_v(t_i^L)}{(1-\pi(\mathbf{z}_i)+\pi(\mathbf{z}_i)S_i(t_i^L))^2}\right.\right.$$
$$-\delta_i^L\frac{S_i(t_i^L)\Psi_u(t_i^L)\Psi_v(t_i^L)}{(1-S_i(t_i^L))^2}$$
$$\left.\left.\delta_i^I\frac{S_i(t_i^L)S_i(t_i^R)(\Psi_u(t_i^L)-\Psi_u(t_i^R))(\Psi_v(t_i^L)-\Psi_v(t_i^R))}{(S_i(t_i^L-S(t_i^R))^2}\right)\right\}.$$

## 4.6 Asymptotic results and inference

Development of the asymptotic properties of the proposed model allows for large sample inference to be conducted without reliance on bootstrapping or other computationally intensive methods. Following from Ma et al. (2014) and Ma et al. (2021), it is possible to demonstrate asymptotic consistency for the MPL estimates of both sets of regression parameters, $\boldsymbol{\beta}$ and $\boldsymbol{\gamma}$, and the baseline hazard function $h_0(t)$. We adopt $\boldsymbol{\beta}_0$, $\boldsymbol{\gamma}_0$ and $h_0(t)$ to denote the true model parameters, i.e. the parameters that gave rise to the observed data. Theorem 4.1 states this asymptotic property; the proofs are similar to that of Theorem 2.4, and thus are omitted here.

The following theorem requires Assumptions 2.1–2.4 but with Assumption 2.1 replaced with the following assumption.

A4.1 Matrices **X** and **Z** are bounded and both $E(\mathbf{X}\mathbf{X}^\top)$ and $E(\mathbf{Z}\mathbf{Z}^\top)$ are non-singular.

With this adjustment to the assumptions, we are ready to state the asymptotic consistency results as given below.

**Theorem 4.1.** Assume that conditions A4.1 and A2.2–A2.4 hold, where the log-likelihood in A2.2–A2.4 is replaced with the log-likelihood in (4.8). Let $\boldsymbol{\beta}_0$, $\boldsymbol{\gamma}_0$ and $h_0(t)$ denote the true parameters. Let $a = \min_i\{t_i^L | t_i^L \neq 0 \text{ and } u_i = 1\}$ and $b = \max_i\{t_i^R | t_i^R \neq \infty \text{ and } u_i = 1\}$. Assume that $h_0(t)$ is bounded and has some number $r \geq 1$ derivatives over the interval $[a, b]$. Assume that $m = n^v$, where $0 < v < 1$. Then, when $n \to \infty$,

(1) $||\widehat{\boldsymbol{\beta}} - \boldsymbol{\beta}_0|| \to 0$ almost surely, and

(2) $||\widehat{\boldsymbol{\gamma}} - \boldsymbol{\gamma}_0|| \to 0$ almost surely, and

(3) $\sup_{t\in[a,b]} |\widehat{h}_0(t) - h_0(t)| \to 0$ almost surely.

□

Additionally, it is desirable to develop asymptotic normality results for all three parameters, $\boldsymbol{\beta}$, $\boldsymbol{\gamma}$ and $\boldsymbol{\theta}$ as this allows for inference to be made not only on regression parameters but also on other quantities, such as survival probabilities. In order to develop these results, however, it is necessary to restrict $m$ to be a finite number. Note that this fixed $m$ is not predetermined as it depends on the given sample size $n$. Usually, a practical guide for $m$ is $m = n_0^{1/3}$, where $n_0$ denotes the non-right censored sample size.

Another issue we face when developing asymptotic normality results is that we must take into account the possibility of encountering active constraints in the estimation of $\boldsymbol{\theta}$ ($\geq 0$). This is particularly likely to occur when the number of knots is larger than strictly necessary, or some knots are placed at non-important locations. The penalty function will push the corresponding $\theta_u$ to zero (Ma et al. 2021). Ignoring active constraints often leads to undesirable results, such as negative variances.

Recall that we have defined the parameter vector $\boldsymbol{\eta} = (\boldsymbol{\theta}^\top, \boldsymbol{\beta}^\top, \boldsymbol{\gamma}^\top)^\top$, which has a length of $p + q + m$, and that we can express the penalized likelihood function in terms of $\boldsymbol{\eta}$ such that

$$\Phi(\boldsymbol{\eta}) = l(\boldsymbol{\eta}) - \lambda J(\boldsymbol{\eta}),$$

where $J(\boldsymbol{\eta}) = J(\boldsymbol{\theta})$. Let $\widehat{\boldsymbol{\eta}}$ be the MPL estimate of $\boldsymbol{\eta}$, and let the true value of $\boldsymbol{\eta}$ be represented by $\boldsymbol{\eta}_0$. Without loss of generality, we assume that the estimates of first $r$ elements of $\boldsymbol{\theta}$ are zero, and so that they are actively constrained. Accordingly, define

$$\mathbf{U} = [\mathbf{0}_{(m-r+p+q)\times r}, \mathbf{I}_{(m-r+p+q)\times(m-r+p+q)}]^\top, \tag{4.17}$$

where $\mathbf{0}$ is a matrix of zeros, $\mathbf{I}$ is an identity matrix. Clearly, $\mathbf{U}^\top\mathbf{U} = \mathbf{I}_{(m-r+p+q)\times(m-r+p+q)}$ is satisfied. Theorem 4.2 below states the asymptotic normality results. We omit the proofs of this theorem as they are almost identical to the proofs of Theorem 2.5 which is available in Appendix A.

Let $T_i^L$ and $T_i^R$ be random variables representing the limits of the censoring interval for individual $i$. We require Assumptions B2.1–B2.4 for Theorem 4.2. However, Assumption B2.1 needs to be modified since we now have regression coefficients $\boldsymbol{\beta}$ and $\boldsymbol{\gamma}$.

B4.1 Assume random vector $W_i = (T_i^L, T_i^R, \mathbf{x}_i^\top, \mathbf{z}_i^\top)^\top, i = 1, \ldots, n$, are independently and identically distributed, and the distribution of both $\mathbf{x}_i$ and $\mathbf{z}_i$ are independent of $\boldsymbol{\eta}$.

**Theorem 4.2.** Assume the Assumption B4.1 and B2.2–B2.4 hold, where the log-likelihood $l$ in B2.2 – B2.4 is replaced with the log-likelihood $l(\boldsymbol{\eta})$ given in (4.8) with $h_0(t)$ replaced by (4.9). Let $\mu_n = \lambda/n$. Assume that $\mu_n = o(n^{-1/2})$. According to the positions of active constraints of $\boldsymbol{\theta}$, we can define matrix $\mathbf{U}$ similarly as (4.17). Let

$$\mathbf{F}(\boldsymbol{\eta}) = -E\left[n^{-1}\frac{\partial^2 l(\boldsymbol{\eta})}{\partial\boldsymbol{\eta}\partial\boldsymbol{\eta}^\top}\right].$$

Under these conditions, when $n \to \infty$, $\sqrt{n}(\widehat{\boldsymbol{\eta}} - \boldsymbol{\eta}_0)$ converges in distribution to a multivariate normal $N(\mathbf{0}, \widetilde{\mathbf{F}}(\boldsymbol{\eta}_0)^{-1}\mathbf{F}(\boldsymbol{\eta}_0)\widetilde{\mathbf{F}}(\boldsymbol{\eta}_0)^{-1})$, where $\widetilde{\mathbf{F}}(\boldsymbol{\eta}_0)^{-1} = \mathbf{U}(\mathbf{U}^\top\mathbf{F}(\boldsymbol{\eta})\mathbf{U})^{-1}\mathbf{U}^\top$. □

In order to implement the result of Theorem 4.2, it is necessary to define a method for identifying active constraints when they arise in the MPL estimation of $\boldsymbol{\theta}$. The method used here follows that proposed in Chapter 2. Active constraints can be identified by inspecting both the value of $\widehat{\theta}_u$ and the corresponding gradient for each $u$. After the Newton-MI algorithm has reached convergence, some $\widehat{\theta}_u$ may be almost exactly zero with negative gradients, and thus are clearly actively constrained. Furthermore, there may be some $\widehat{\theta}_u$ that are very close to, but not almost exactly, zero. For these $\widehat{\theta}_u$, a corresponding negative gradient value is indicative that they are also subject to an active constraint. In practice, active constraints are defined where, for a given $u$, $\widehat{\theta}_u < 10^{-3}$ and the corresponding gradient is less than $-\varepsilon$, where $\varepsilon$ is a positive threshold value such as $10^{-2}$. After the indices associated with active constraints are identified, obtaining the matrix $\widetilde{\mathbf{F}}(\boldsymbol{\eta}_0)^{-1}$ is a very straightforward computation. The matrix $\mathbf{U}^T\mathbf{F}(\boldsymbol{\eta})\mathbf{U}$ is obtained by removing the rows and columns of $\mathbf{F}(\boldsymbol{\eta})$ associated with the active constraints. The result is then inverted, and then padded with zeros in the deleted rows and columns to obtain $\widetilde{\mathbf{F}}(\boldsymbol{\eta}_0)^{-1}$.

To make use of these asymptotic results for inference on finite samples, it is necessary to approximate the distribution for $\widehat{\boldsymbol{\eta}}$ when $n$ is large. Doing so also incorporates non-zero values for the smoothing parameter $\lambda$ into the inference on the parameter estimates.

Assume that the smoothing parameter $\lambda \ll n$. Define

$$\mathbf{A}(\widehat{\boldsymbol{\eta}})^{-1} = \mathbf{U}\left(\mathbf{U}^\top\left(-\frac{\partial^2 l(\widehat{\boldsymbol{\eta}})}{\partial\boldsymbol{\eta}\partial\boldsymbol{\eta}^\top} + \lambda\frac{\partial^2 J(\widehat{\boldsymbol{\eta}})}{\partial\boldsymbol{\eta}\partial\boldsymbol{\eta}^\top}\right)\mathbf{U}\right)^{-1}\mathbf{U}^\top.$$

Then, when $n$ is large, the distribution for the MPL estimate $\widehat{\boldsymbol{\eta}} - \boldsymbol{\eta}_0$ can be approximated by a multivariate normal distribution having mean zero and covariance matrix

$$\widehat{\mathrm{Var}}(\widehat{\boldsymbol{\eta}}) = \mathbf{A}(\widehat{\boldsymbol{\eta}})^{-1}\left(-\frac{\partial^2 l(\widehat{\boldsymbol{\eta}})}{\partial\boldsymbol{\eta}\partial\boldsymbol{\eta}^\top}\right)\mathbf{A}(\widehat{\boldsymbol{\eta}})^{-1}.$$

These results allow for inferences to be made not only on both sets of regression parameters but also on quantities associated with the baseline hazard function.

## 4.7 A melanoma study example

In this section, we give an extended example of fitting the mixture cure proportional hazards model to interval-censored survival data with a cured fraction. The data used in this section is simulated, based on data from a real study of patients diagnosed with thin melanoma (see analysis of real dataset in Webb et al. (2022)). The simulated data used in this example is available on this book GitHub website: "https://github.com/MPL-book".

**Example 4.2** (A melanoma example with cured fraction).
In this analysis, the event of interest is any cancer recurrence after thin melanoma diagnosis. The prognosis of patients diagnosed with thin melanoma is generally excellent, with 10-year recurrence-free survival rates of upwards of 80%. Therefore,

any analysis of time to recurrence amongst thin melanoma patients must take account of this potentially very large cured fraction.

Surveillance of melanoma recurrence amongst these patients took place via follow-up appointment scheduled at various intervals and so there were no uncensored event times recorded. Instead, those who did experience a recurrence had their event time either interval-censored (i.e. the event occurred between two follow-up appointments) or left-censored (i.e. the time last recurrence-free follow-up appointment was unknown, and therefore the only information available was that the recurrence occurred some time after the diagnosis i.e. after $t = 0$). These interval or left censored individuals constituted less than 10% of the sample. All remaining individuals were right censored, as they had not had an observed recurrence by the end of their follow-up.

Covariates of interest in this study included the Breslow thickness of the initial melanoma tumour, whether the tumour was ulcerated, whether there was mitosis of the tumour, and the sex of the patient.

We have simulated a thin melanoma dataset which is very close to the real data, and this simulated data is available at the github web site for this book. A mixture cure Cox model was fitted to the simulated data. The results from this model are given in Table 4.3. For the purposes of comparison, we also give in this table the results of fitting a normal Cox regression model to this data using the `survivalMPL` package discussed in previous chapters.

TABLE 4.3: Model fitting outputs from a mixture cure Cox and a conventional Cox model.

| | Mixture cure Cox model | | | Typical Cox model | | |
|---|---|---|---|---|---|---|
| Cox regression | HR | 95% CI | p-value | HR | 95% CI | p-value |
| Breslow thickness > 0.8mm | 1.00 | 0.67, 1.49 | 0.9996 | 1.26 | 1.02, 1.56 | 0.0321 |
| Male sex | 0.36 | 0.16, 0.82 | 0.0147 | 0.71 | 0.57, 0.88 | 0.0018 |
| Tumor mitosis | 0.88 | 0.64, 1.21 | 0.4264 | 1.64 | 1.32, 2.03 | 0.0001 |
| Tumor ulceration | 1.07 | 0.72, 1.58 | 0.7483 | 1.30 | 1.05, 1.60 | 0.0170 |
| Logistic regression | OR | 95% CI | p-value | | | |
| Breslow thickness > 0.8mm | 1.29 | 1.01, 1.66 | 0.0460 | | | |
| Male sex | 0.85 | 0.60, 1.20 | 0.3659 | | | |
| Tumor mitosis | 1.80 | 1.40, 2.31 | 0.0001 | | | |
| Tumor ulceration | 1.34 | 1.05, 1.72 | 0.0196 | | | |

```
# Write a function to generate
# data similar to the melanoma
# study
generate_melanoma_example <- function(n,
    b_true, g_true, event = FALSE, event_adj = 0,
    censoringtimemax = 3) {

    # create X and Z
    BreslowThickness <- rbinom(n, 1,
        0.5)
```

```
SexMale <- rbinom(n, 1, 0.5)
Mitosis <- rbinom(n, 1, 0.5)
Z <- (cbind(BreslowThickness, SexMale,
    Mitosis))

Ulceration <- rbinom(n, 1, 0.5)
X <- (cbind(Ulceration, SexMale))

# generate results from
# logistic regression
pi_i <- rep(0, n)
for (i in 1:n) {
    pi_i[i] <- exp(b_true[1] + sum(b_true[-c(1)] *
        Z[i, ]))/(1 + exp(b_true[1] +
        sum(b_true[-c(1)] * Z[i,
            ])))
}

# generate cure indicator
U_C <- runif(n)
cure_i <- as.numeric(U_C < pi_i)  #1 is not cured
uncure.n <- length(cure_i[cure_i])
# number of uncured in sample
cure.n <- length(cure_i[!cure_i])
# number of cured in sample

# generate observation times
TL_i <- TR_i <- rep(0, n)
Y_i <- rep(0, n)

for (i in 1:n) {
    # generate censoring time
    max.followup <- as.numeric(cure_i[i] ==
        1) * runif(1, min = 1, max = censoringtimemax) +
        as.numeric(cure_i[i] ==
            0) * runif(1, min = 1,
            max = 2 * censoringtimemax)

    # simulate event
    # monitoring process
    exam.num <- as.numeric(cure_i[i] ==
        1) * rpois(1, 8) + 1 + as.numeric(cure_i[i] ==
        0) * rpois(1, 16) + 1
    exam.int <- runif(exam.num,
        min = 0.3, max = 0.7)
    examtimes <- cumsum(exam.int)

    if (cure_i[i] == 0) {
```

```
        # if they are cured
        Y_i[i] <- Inf  #they have no event time
        max.ind <- max(which(examtimes <
            max.followup))
        # they are right
        # censored at the last
        # exam time before
        # they exit the study
        TL_i[i] <- examtimes[max.ind]
        TR_i[i] <- Inf

    } else {
        # if they are not
        # cured generate an
        # event time
        U_Y <- runif(1)
        Y_i[i] = (-log(U_Y)/exp(X[i,
            ] %*% g_true))^(1/3)
        # F^-1 method to
        # sample from
        # weibull(1, 3)

        if (max.followup < min(examtimes)) {
            # if they exit the
            # study before the
            # first exam
            TL_i[i] <- max.followup
            # record their
            # censoring time,
            # they are right
            # censored
            TR_i[i] <- Inf

        } else if (Y_i[i] < max.followup &
            max.followup > min(examtimes)) {
            # if they have at
            # least one exam
            # AND have their
            # event before
            # exiting the
            # study
            last.exam.ind <- max(which(examtimes <
              max.followup))
            # they might be
            # left censored
            if (Y_i[i] < min(examtimes)) {
              TL_i[i] <- -Inf
              TR_i[i] <- min(examtimes)
```

```
            } else if (examtimes[last.exam.ind] <
              Y_i[i]) {
              # they might be
              # right censored
              TL_i[i] <- examtimes[last.exam.ind]
              TR_i[i] <- Inf

            } else {
              # they might be
              # interval
              # censored
              left <- max(which(examtimes <
                Y_i[i]))
              right <- min(which(examtimes >
                Y_i[i]))
              TL_i[i] <- examtimes[left]
              TR_i[i] <- examtimes[right]

              # in some cases
              # they might
              # have an event
              # time
              if (event == TRUE &
                (TR_i[i] - TL_i[i]) <
                  event_adj) {
                TL_i[i] = Y_i[i]
                TR_i[i] = Y_i[i]
              }
            }
        } else if (Y_i[i] > max.followup &
            max.followup > min(examtimes)) {
            # if they have at
            # least one exam
            # but do not have
            # their event
            # before exiting
            last.exam.ind <- max(which(examtimes <
              max.followup))
            TL_i[i] <- examtimes[last.exam.ind]
            TR_i[i] <- Inf
        }
    }
}

data <- data.frame(Y_i, TL_i, TR_i,
    cure_i, Z, X)
return(data)
```

```
}

# Generate data set
mel.dat <- generate_melanoma_example(n = 2000,
    b_true = c(-1.5, 0.5, 2, 1), g_true = c(0.5,
        -2), censoringtimemax = 1.5)

mel.surv <- Surv(time = mel.dat$TL,
    time2 = mel.dat$TR, type = "interval2")

# Fit mixture cure model using MPL
# method
m1 <- phmc_mpl(mel.surv ~ mel.dat$BreslowThickness +
    mel.dat$SexMale + mel.dat$Mitosis +
    mel.dat$Ulceration, pi.formula = ~mel.dat$BreslowThickness +
    mel.dat$SexMale + mel.dat$Mitosis +
    mel.dat$Ulceration, data = mel.dat,
    control = phmc_mpl.control(smooth = NULL,
        new_constraint = 1, new_criteria = 1,
        new_knots = 1, maxIter = c(10,
            4000, 10000), n.knots = 3,
        conv_limit = 1e-05))


# Compare with non mixture cure
# model
control.mel <- coxph_mpl.control(n.obs = NA,
    basis = "mspline", smooth = NULL,
    n.knots = c(3, 0), max.iter = c(100,
        2000, 4000))
m2 <- coxph_mpl(mel.surv ~ mel.dat$BreslowThickness +
    mel.dat$SexMale + mel.dat$Mitosis +
    mel.dat$Ulceration, data = mel.dat,
    control = coxph_mpl.control(n.obs = NA,
        basis = "mspline", smooth = NULL,
        n.knots = c(5, 0), max.iter = c(100,
            2000, 4000)))
```

According to the logistic regression model results for the incidence, a "Breslow tumor thickness of $>$ 0.8mm", "tumor mitosis" and "tumour ulceration" all significantly increase the probability of an individual being susceptible to melanoma recurrence. There was no significant difference found between males and females in terms of the probability of being susceptible to melanoma recurrence. Among the susceptible sub-population, "Breslow thickness", "tumor mitosis" and "tumor ulceration" are not significant predictors of time-to-recurrence. Sex is a significant predictor of time-to-recurrence, with males having a significantly lower risk of recurrence at a given time $t$ (i.e. females who are susceptible to recurrence are likely to have a recurrence earlier than males who are susceptible to recurrence). In contrast, the standard

Cox model fitted to this data identifies all four covariates investigated as significant predictors of time-to-recurrence. □

The mixture cure Cox model fitted using the MPL method detailed in this chapter can be used to make predictions and inferences about survival quantities. For example, we can predict the value of the survival function for a given time $t$ and a set of covariates, and also construct a 95% confidence interval around this predicted value.

For the mixture cure model, it may be of interest to make predictions in terms of either the conditional survival function $S(t|u = 1)$ (i.e., the survival function of an individual among the non-cured sub-population), or in terms of the population survival function $S_{pop}(t)$ given in Equation (4.3). Interest in one or the other may depend on how much confidence we have about the ability of the logistic regression incidence model to accurately discriminate between individuals who are and are not in the cured sub-population (see further discussion of this in the following section of this chapter). In the next example, we demonstrate the prediction of both of these types of survival functions and their associated point-wise confidence intervals.

**Example 4.3** (Predictive survival functions and plots).
In this example, we use the fitted mixture model of Example 4.2 to demonstrate how to plot some predictive survival curves.

Firstly, we present plots of the conditional baseline hazard function $h_0(t)$, conditional baseline cumulative hazard function $H_0(t)$ and conditional baseline survival function $S_0(t)$. These plots are shown in Figure 4.3. Evidently, the risk of melanoma recurrence among those who are susceptible (i.e. the non-cured fraction) increases over time. From the conditional baseline survival function, it appears that among those who are susceptible to thin melanoma recurrence, the probability of having not experienced this event reaches 0 between $t = 1$ and $t = 1.5$.

Secondly, we present plots of the predicted conditional survival function $S(t|u = 1)$ and its point-wise confidence intervals for specified values of covariates of interest. Specifically, in Figure 4.4, we present the conditional survival functions for a male compared to a female, when all other covariates are set at their mean value. The shaded area shows the point-wise confidence intervals for each time point. Here we give some detail on how these confidence intervals are calculated. Firstly, we comment that because the survival function is restricted to values between 0 and 1, it is useful to use a logit transformation to ensure that estimates of the confidence interval, once back-transformed, are also restricted to the range of 0 to 1. Therefore, we find the variance of the logit transformed conditional survival function at a given time $t$ by taking

$$\widehat{\mathrm{Var}}\big(\mathrm{logit}(S(t))\big) = \left(\frac{\partial \mathrm{logit}(S(t))}{\partial \boldsymbol{\varrho}}\right)^{\mathsf{T}} \widehat{\mathrm{Var}}(\widehat{\boldsymbol{\varrho}}) \frac{\partial \mathrm{logit}(S(t))}{\partial \boldsymbol{\varrho}},$$

where we take $\boldsymbol{\varrho} = [\boldsymbol{\beta}^{\mathsf{T}}, \boldsymbol{\theta}^{\mathsf{T}}]^{\mathsf{T}}$ Noting that the conditional survival function $S(t)$ is not a function $\gamma$, and $\widehat{\mathrm{Var}}(\widehat{\boldsymbol{\varrho}})$ is the estimated covariance matrix of $\boldsymbol{\beta}$ and $\boldsymbol{\theta}$ taken from the fitted MPL model. Then the point-wise 95% confidence interval for $S(t)$ at

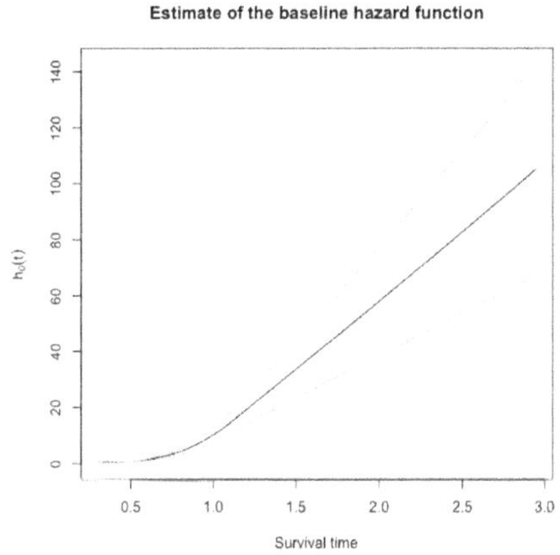


(a)

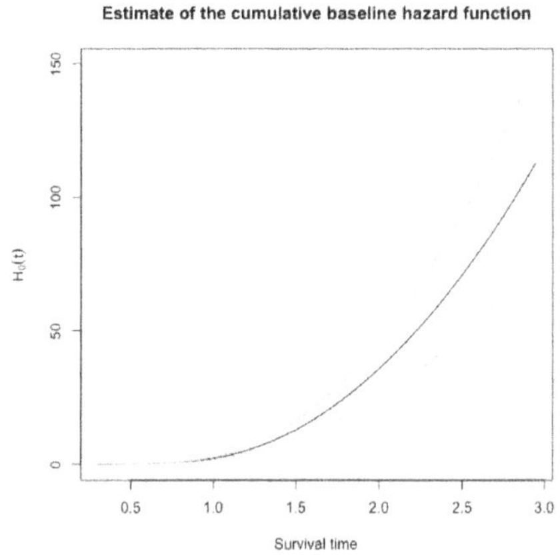


(b)

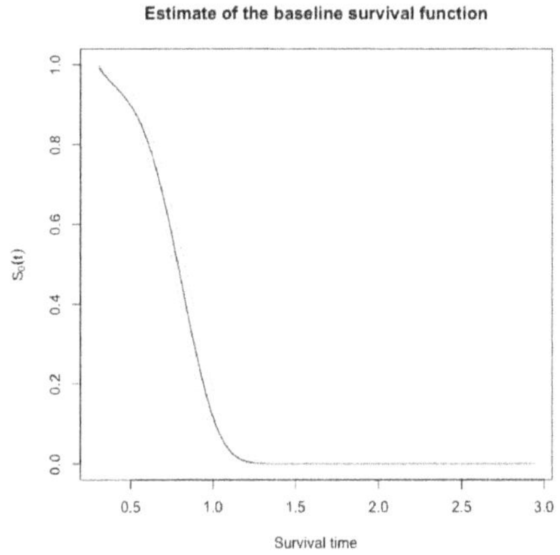


(c)

FIGURE 4.3: Estimates of baseline functions and 95% point-wise CIs.

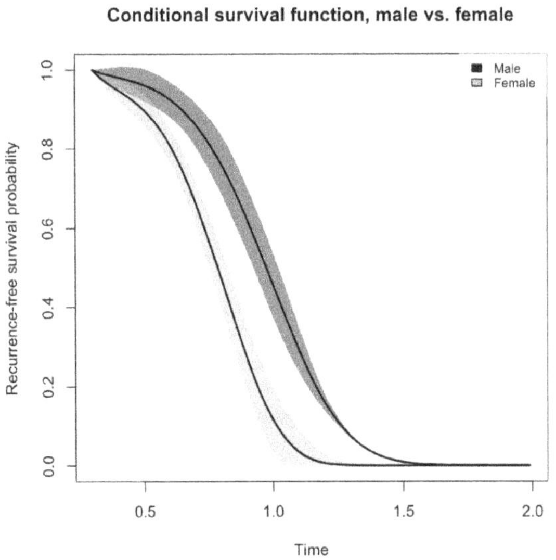


FIGURE 4.4: Predicted conditional survival function and its point-wise confidence intervals.

a given time $t$ is taken as

$$\text{logit}^{-1}\left(\text{logit}(S(t)) \pm 1.96 \times \sqrt{Var\big(\text{logit}(S(t))\big)}\right).$$

The above formulas give the conditional survival functions for males vs. females, and they are plotted in Figure 4.3 alongside their point-wise 95% confidence intervals. Clearly, from approximately $t = 0.6$ onwards, there is a significant difference between the survival probability of males compared to females within the susceptible sub-group, with females in the susceptible subgroup having a lower probability of survival across time.

Thirdly, we present plots of the predicted population survival function $S_{pop}(t)$ and its point-wise confidence intervals. The population survival function incorporates both the probability of an individual being unsusceptible to the event of interest, and the probability of the individual experiencing the event of interest at time $t$ given they are susceptible. We present the population survival functions for those with tumor mitosis compared to those without, with other covariates set at their mean values. The process for computing the point-wise confidence intervals is the same as detailed above, except that we take $\boldsymbol{\varrho} = [\boldsymbol{\gamma}^{\mathsf{T}}, \boldsymbol{\beta}^{\mathsf{T}}, \boldsymbol{\theta}^{\mathsf{T}}]^{\mathsf{T}}$ because $S_{pop}(t)$ is a function of all parameters. The estimated population survival functions and associated point-wise confidence intervals are given in Figure 4.5. We can see that although there was no significant difference in the latency Cox sub-model between tumor mitosis groups, in the overall population there may be a significant difference between the groups in

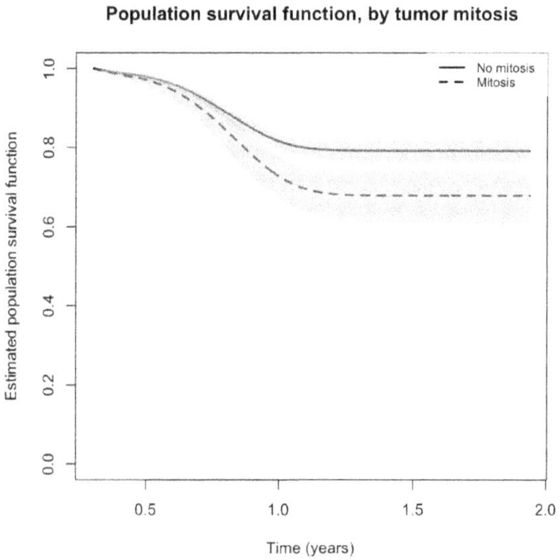


FIGURE 4.5: Predicted population survival function and its point-wise confidence intervals.

terms of risk of recurrence over time, due to the significant difference between the groups in the incidence model.

```
# Write 'predict' function and a
# function to compute the
# piecewise survival function and
# SEs

predict.phmc_mpl <- function(object,
    se = "M2QM2", type = "hazard", cov = NULL,
    i = NULL, time = NULL, prob = 0.95) {

    beta <- object$beta
    gamma <- object$gamma
    theta <- object$theta
    p <- object$dimensions$p
    q <- object$dimensions$q
    m <- object$dimensions$m
    covar <- object$covar[[se]]
    if (length(i) > 1) {
        warning("Only the first observation will be considered.\n",
            call. = FALSE)
    }
    if (is.null(time)) {
        n.x <- 1000
        V_x_X <- seq(object$knots$Alpha[1],
            max(object$knots$Alpha),
            length = n.x)
```

```
} else {
    n.x <- length(time)
    V_x_X <- time
}

if (is.null(i) & is.null(cov)) {
    xT <- object$data$X
} else if (!is.null(i) & is.null(cov)) {
    xT <- object$data$X[i[1], ]
} else {
    xT <- matrix(cov, nrow = 1)
}

Mu <- c(exp(xT %*% gamma))
out <- data.frame(time = V_x_X,
    mid = NA, se = NA, low = NA,
    high = NA)

if (type == "hazard") {
    psi = basis_phmc(V_x_X, knots = object$knots,
        order = object$control$order,
        which = 1)
    h0 <- psi %*% theta
    out$mid <- Mu * h0
    out$se <- sqrt(diag(psi %*%
        (covar[(p + q + 1):(p +
            q + m), (p + q + 1):(p +
            q + m)]) %*% t(psi)))
    out$low <- out$mid - 1.96 *
        out$se
    out$low[out$low < 0] <- 0
    out$high <- out$mid + 1.96 *
        out$se
} else {
    Psi <- basis_phmc(V_x_X, knots = object$knots,
        order = object$control$order,
        which = 2)
    H0 <- Psi %*% theta
    out$mid <- exp(-Mu * H0)
    out$se <- sqrt(diag((matrix(rep((exp(-H0)),
        m), ncol = m) * (-Psi)) %*%
        (covar[(p + q + 1):(p +
            q + m), (p + q + 1):(p +
            q + m)]) %*% t(matrix(rep((exp(-H0)),
        m), ncol = m) * (-Psi))))
    out$low <- out$mid - 1.96 *
        out$se
    out$low[out$low < 0] <- 0
    out$high <- out$mid + 1.96 *
        out$se
}

times <- c(object$data$time[, 1])
attributes(out)$inf = list(i = i[1],
    user.time = !is.null(time),
    prob = prob, upper.value = quantile(times,
```

```
            prob), max = max(times),
        m = m, risk = (type == "hazard"))
    colnames(out)[2] <- type
    class(out) = c("predict.phmc_mpl",
        "data.frame")
    out
}


phmc_piecewiseCI = function(obj, z_pos,
    x_pos, lvls = c(0, 1), CI = TRUE) {
    X_temp_save <- NULL

    if (is.null(x_pos)) {
        cov_obj <- apply(obj$data$X,
            2, mean)
        predict_obj <- predict(obj,
            type = "survival", cov = cov_obj)
        time <- predict_obj$time
        s <- rep(predict_obj$survival,
            length(lvls))
        s <- matrix(s, nrow = 1000,
            ncol = length(lvls), byrow = FALSE)
        X_temp_save <- matrix(rep(cov_obj,
            length(lvls)), nrow = length(lvls),
            byrow = TRUE)
    } else {
        s <- matrix(0, nrow = 1000,
            ncol = length(lvls))
        for (l in 1:length(lvls)) {
            cov_obj <- apply(obj$data$X,
                2, mean)
            cov_obj[x_pos] <- lvls[l]
            predict_obj <- predict(obj,
                type = "survival", cov = cov_obj)
            time <- predict_obj$time
            s[, l] <- predict_obj$survival
            X_temp_save <- rbind(X_temp_save,
                cov_obj)
        }
    }
    pop.survival <- s
    Z_temp_save <- NULL
    if (is.null(z_pos)) {
        Z_tmp <- matrix(apply(obj$data$Z,
            2, mean), nrow = 1)
        ZB <- Z_tmp %*% obj$beta
        pi <- as.numeric(exp(ZB)/(1 +
            exp(ZB)))
        pi_save <- rep(pi, length(lvls))
        pop.survival <- pi * s + (1 -
            pi)
        Z_temp_save <- matrix(rep(Z_tmp,
            length(lvls)), nrow = length(lvls),
            byrow = TRUE)
    } else {
```

```
        pi_save <- NULL
        for (l in 1:length(lvls)) {
            Z_tmp <- matrix(apply(obj$data$Z,
                2, mean), nrow = 1)
            Z_tmp[z_pos] <- lvls[l]
            ZB <- Z_tmp %*% obj$beta
            pi <- as.numeric(exp(ZB)/(1 +
                exp(ZB)))
            pi_save <- c(pi_save, pi)
            pop.survival[, l] <- pi *
                s[, l] + (1 - pi)
            Z_temp_save <- rbind(Z_temp_save,
                Z_tmp)
        }
    }

    CI_save <- matrix(0, nrow = 1000,
        ncol = length(lvls) * 2)

    for (l in 1:length(lvls)) {
        X_temp <- X_temp_save[l, ]
        Z_temp <- Z_temp_save[l, ]
        pi <- pi_save[l]
        mu_X <- exp(X_temp %*% obj$gamma)
        cov_eta <- obj$covar$M2QM2

        UL_population.survival.logit <- rep(0,
            length(time))
        LL_population.survival.logit <- rep(0,
            length(time))

        for (p in 2:length(time)) {
            t <- time[p]
            St <- s[p, l]
            logitSpop <- log(pop.survival[p,
                l]/(1 - pop.survival[p,
                l]))
            Psi_t <- basis_phmc(t, knots = obj$knots,
                order = obj$control$order,
                which = 2)

            dlogitSt_dbeta <- solve(pop.survival[p,
                l] * (1 - pop.survival[p,
                l])) %*% (St - 1) %*%
                pi %*% (1 - pi) %*%
                Z_temp
            dlogitSt_dgamma <- -solve(pop.survival[p,
                l] * (1 - pop.survival[p,
                l])) %*% (pi %*% St %*%
                (-log(St))) %*% X_temp
            dlogitSt_dtheta <- -mu_X %*%
                pi %*% St %*% solve(pop.survival[p,
                l] * (1 - pop.survival[p,
                l])) %*% Psi_t
            dlogitSt_deta <- matrix(c(dlogitSt_dbeta,
                dlogitSt_dgamma, dlogitSt_dtheta),
```

```
                  nrow = 1)

              var_logitSt <- dlogitSt_deta %*%
                  cov_eta %*% t(dlogitSt_deta)
              se_logitSt <- sqrt(var_logitSt)

              LL_population.survival.logit[p] <- logitSpop -
                  1.96 * se_logitSt
              UL_population.survival.logit[p] <- logitSpop +
                  1.96 * se_logitSt
          }
          LL <- exp(LL_population.survival.logit)/(1 +
              exp(LL_population.survival.logit))
          UL <- exp(UL_population.survival.logit)/(1 +
              exp(UL_population.survival.logit))

          CI_save[, 1] <- LL
          CI_save[, (1 + 2)] <- UL
    }
    out <- list(pop.survival = pop.survival,
        CI_save = CI_save, time = time)
    return(out)
}

# Calculate CONDITIONAL survival
# functions for males vs. females
# (All other covariates set to
# mean value)

covar <- apply(m1$data$X, 2, mean)
covar_male <- covar
covar_male[2] <- 1
covar_female <- covar
covar_female[2] <- 0

surv_conditional_male <- predict(m1,
    type = "survival", cov = covar_male)
surv_conditional_female <- predict(m1,
    type = "survival", cov = covar_female)

# Calculate POPULATION survival
# functions for males vs females
sex_ci <- phmc_piecewiseCI(m1, z_pos = 4,
    x_pos = 3, lvls = c(0, 1))
```

□

## 4.8 Predictive accuracy of the mixture cure model

When a survival regression model is used for prediction, it is important to assess its predictive accuracy. In this section, we will discuss predictive accuracy of the mixture cure model when it is used for prediction.

### 4.8.1 Discrimination of logistic regression model for incidence

A number of methods have been proposed that can be used to evaluate the ability of the logistic regression model used for the incidence to discriminate between individuals who are susceptible to the event and those who are not. These methods primarily consist of methods to estimate the sensitivity, specificity, receiver-operating characteristic (ROC) curve and the area under the ROC curve (AUC) in the presence of individuals whose true class membership is unknown. Though methods existing in the literature only specifically deals with right censoring times and directly observed event times, it is trivial to extend these methods to partly interval-censored data, as by definition left- and interval-censored event times can only be observed for individuals who are certainly not in the cured fraction, and so can be dealt with as if they were directly observed event times in all the methods presented here.

In the below examples, we assume that we are interested in calculating the sensitivity and specificity of the logistic regression model for predicting that an individual is susceptible to the event of interest, i.e., has a true value of $u_i = 1$. Some existing methods require the definition of a "clinical cure" time, $\tau$, which denotes a survival time at which the lack of an event would clinically indicate that an individual was not susceptible or was "cured". The inclusion of a "clinical cure" time can reduce the amount of uncertainty introduced by the fact that $u_i$ is unknown for all right-censored individuals; the idea is that if $t_i > \tau$ for a right-censored individual, then it is "known" that $u_i = 0$ i.e. the individual is cured.

Methods proposed by Asano et al. (2014) and Amico et al. (2021) require the definition of this "clinical cure" time. Once this time is defined, the individuals in the sample can be categorized into one of three groups, recalling that $\delta_i^R$ is an indicator variable for right censoring, and $u_i$ is a partially observed indicator variable where $u_i = 1$ indicates that individual $i$ is not cured:

- if $\delta_i^R = 0$ then individual $i$ is not cured and $u_i = 1$
- if $\delta_i^R = 1$ and $t_i^L > \tau$ (i.e. individual $i$ is right-censored at $t_i^L$ after the clinical cure time) then individual $i$ is cured and $u_i = 0$
- if $\delta_i^R = 1$ and $t_i^L \leq \tau$ then the cure status of individual $i$ is unknown and $u_i$ is missing

Then, based on some risk score derived from $\mathbf{z}_i^\mathsf{T}\gamma$ (often we may wish to use the susceptible probability $\hat{\pi}(\mathbf{z}_i)$), we can define estimators for the sensitivity and specificity. The methods of Asano et al. (2014) and Amico et al. (2021) differ quite substantially in their proposed estimators for the sensitivity and specificity incorporating uncertainty about the class of individuals where $u_i$ is unknown. In computing the sensitivity and specificity, Asano et al. (2014) use directly the estimated (unconditional) probability of individual $i$ being cured, $\pi_c(\mathbf{z}_i) = 1 - \pi(\mathbf{z}_i)$, whilst Amico et al. (2021) instead use a *conditional* probability of individual $i$ being cured.

Firstly, Asano et al. (2014) define a variable $v_i = 1$ if $u_i$ is known, and $v_i = 0$ if $u_i$ is unknown. Then, a proposed estimator for the sensitivity is

$$\widehat{Sen}(c) = \frac{\sum_i I(\widehat{\pi}(\mathbf{z}_i) \leq c)[v_i u_i + (1 - v_i)(1 - \widehat{\pi}_c(\mathbf{z}_i))]}{\sum_i [v_i u_i + (1 - v_i)(1 - \widehat{\pi}_c(\mathbf{z}_i))]},$$

and a proposed estimator for the specificity is

$$\widehat{Spe}(c) = 1 - \frac{\sum_i I(\widehat{\pi}(\mathbf{z}_i) \leq c)[v_i(1 - u_i) + (1 - v_i)\widehat{\pi}_c(\mathbf{z}_i)]}{\sum_i [v_i(1 - u_i) + (1 - v_i)\widehat{\pi}_c(\mathbf{z}_i)]},$$

for some cut-off $c$. In these estimators, individuals with a known cure status (i.e. with $u_i$ observed) have their value of $u_i$ used directly, and those with an unknown cure status have their unknown value of $u_i$ replaced with $\pi_c(\cdot)$, their estimated probability of being cured.

Alternatively, the method in Amico et al. (2021) uses a conditional probability of being cured. They propose the estimators

$$\widehat{Sen}(c) = \frac{1}{N_1} \sum_i w_{i1} I(l \leq c),$$

and

$$\widehat{Spe}(c) = 1 - \frac{1}{N_0} \sum_i w_{i0} I(l \leq c),$$

where $w_{i0} = I(u_i = 0)$, $w_{i1} = I(u_i = 1)$, $N_0 = \sum_i w_{i0}$ and $N_1 = n - N_0$. Because the values of $w_{i0}$ and $w_{i1}$ are not observed for some $i$, they can be estimated by $w_{i0} = (1 - \delta_i^R)\widehat{Pr}(u_i = 0)$ and $w_{i1} = 1 - w_{i0}$. This probability can be expressed as

$$\widehat{Pr}(u_i = 0) = \frac{Pr(t_i = \infty)}{Pr(t_i^L > \tau)} = \frac{1 - \pi(\mathbf{z}_i)}{1 - \pi(\mathbf{z}_i) + \pi(\mathbf{z}_i) S(t_i^L | u_i = 1)}.$$

With this, the values of $w_{i0}$ and $w_{i1}$ can be expressed as

$$\widehat{w}_{i0} = I(t_i^L > \tau) + \delta_i^R I(t_i^L \leq \tau) \frac{1 - \pi(\mathbf{z}_i)}{1 - \pi(\mathbf{z}_i) + \pi(\mathbf{z}_i) S(t_i^L | u_i = 1)},$$

and $\widehat{w}_{i1} = 1 - \widehat{w}_{i0}$.

The estimator proposed by Zhang et al. (2021) is somewhat similar to that proposed by Amico et al. (2021) in the sense that it uses a conditional probability of being cured. However, it differs in the sense that no "clinical cure" time is used. That is, they define the weight $w_i$ as

$$\widehat{w}_i(\eta) = (1 - \delta_i^R) + \delta_i^R \frac{\pi(\mathbf{z}_i^\top \widehat{\boldsymbol{\beta}}) \widehat{S}_0(t_i^{\cdot} \widehat{u}_i)^{\exp(\mathbf{x}_i^\top \widehat{\boldsymbol{\gamma}})}}{1 - \pi(\mathbf{z}_i^\top \widehat{\boldsymbol{\beta}}) + \pi(\mathbf{z}_i^\top \widehat{\boldsymbol{\beta}}) \widehat{S}_0(t, \widehat{u}_i)^{\exp(\mathbf{x}_i^\top \widehat{\boldsymbol{\gamma}})}}.$$

This is equivalent to $w_i = 1$ for any individuals without right-censored event times (who are known to be uncured), and for individuals with right-censored event times is equivalent to

$$\frac{Pr(y_i > t | y_i < \infty)}{Pr(y_i > t)},$$

i.e. the probability of an event occurring after the censoring time (such that the individual is not in the cured fraction) conditional on the probability of the event having not occurred before censoring. It therefore uses the information that the event did not occur before follow-up, unlike the conditional probability specified by the estimator proposed by Amico et al. (2021). Then the sensitivity and specificity estimators proposed by Zhang et al. (2021) are, respectively,

$$\widehat{Sen}(c, \widehat{\theta}) = \frac{\sum_i w_i(\widehat{\theta}) I(\widehat{l} \leq c)}{\sum_i w_i(\widehat{\theta})},$$

and

$$\widehat{Spe}(c, \widehat{\theta}) = \frac{\sum_i (1 - w_i(\widehat{\theta})) I(\widehat{l} > c)}{\sum_i (1 - w_i(\widehat{\theta}))},$$

The estimators of Amico et al. (2021) and Zhang et al. (2021) will be very similar as long as the "clinical cure" time is chosen well.

Whichever estimator is used for the sensitivity and specificity, the ROC curve can be estimated by

$$\widehat{ROC}(u) = \widehat{Sen}\big((1 - \widehat{Spe})^{-1}(u)\big),$$

for $0 < u < 1$. Then, the area under the ROC curve can be estimated by finding the (estimated) proportion of concordant pairs.

**Example 4.4** (Discrimination of the logistic model for the incidence).
In this example, we use the simulated melanoma sample analyzed in Examples 4.2 and 4.3 to explain how to calculate sensitivity and specificity after the mixture cure model parameters are estimated by the MPL method. The results of a fitted mixture cure Cox model are displayed in Table 4.3. We will conduct the following exercises:

1. Compute the estimated sensitivity and specificity using each of the three methods detailed in this section. For methods requiring a "clinical cure" time, investigate the difference when using $\tau = 1.5$ and $\tau = 2$, and for all methods use a cut-off value of $c = 0.25$.
2. Compare the estimates obtained and comment on whether the logistic regression model for incidence used above appears to discriminate between cured and uncured individuals.

```
# Calculate individual pi and 1 -
# pi from the fitted model
prob_uncured <- m1$pi_estimate
prob_cured <- 1 - m1$pi_estimate

# Calculate the survival function
```

```
# value at t
times <- mel.dat$TL_i
X <- m1$data$X
Z <- m1$data$Z

Psi <- mSpline(times, knots = c(m1$knots$Alpha[2:4]),
    Boundary.knots = c(m1$knots$Alpha[1],
        m1$knots$Alpha[5]), integral = TRUE)
H0_t <- Psi %*% m1$theta
H_t <- H0_t * exp(X %*% m1$gamma)
S_t <- exp(-H_t)
S_t[which(is.nan(S_t))] <- 0

# ASANO ET AL. METHOD clinical
# cure tau = 1.5 whether cure
# status is known exactly
v_i = as.numeric(m1$data$censoring >
    0)
v_i[which(m1$data$censoring == 0 & mel.dat$TL_i >
    1.5)] = 1

# cure status value
u_i = as.numeric(m1$data$censoring ==
    0)
u_i[which(m1$data$censoring == 0 & mel.dat$TL_i <
    1.5)] = prob_cured[which(m1$data$censoring ==
    0 & mel.dat$TL_i < 1.5)]

# sensitivity
sum(as.numeric(prob_uncured < 0.25) *
    (v_i * u_i + (1 - v_i) * (1 - prob_cured)))/sum((v_i *
    u_i + (1 - v_i) * (1 - prob_cured)))

# specificity
1 - sum(as.numeric(prob_uncured < 0.25) *
    (v_i * (1 - u_i) + (1 - v_i) * prob_cured))/sum(v_i *
    (1 - u_i) + (1 - v_i) * prob_cured)

# clinical cure tau = 2 whether
# cure status is known exactly
v_i = as.numeric(m1$data$censoring >
    0)
v_i[which(m1$data$censoring == 0 & mel.dat$TL_i >
    2)] = 1

## cure status value
u_i = as.numeric(m1$data$censoring ==
    0)
```

```
u_i[which(m1$data$censoring == 0 & mel.dat$TL_i <
    2)] = prob_cured[which(m1$data$censoring ==
    0 & mel.dat$TL_i < 2)]

# sensitivity
sum(as.numeric(prob_uncured < 0.25) *
    (v_i * u_i + (1 - v_i) * (1 - prob_cured)))/sum((v_i *
    u_i + (1 - v_i) * (1 - prob_cured)))

# specificity
1 - sum(as.numeric(prob_cured < 0.75) *
    (v_i * (1 - u_i) + (1 - v_i) * prob_cured))/sum(v_i *
    (1 - u_i) + (1 - v_i) * prob_cured)

## AMICO ET AL. METHOD clinical
## cure tau = 1.5
w_i0 = as.numeric(mel.dat$TL_i > 1.5) +
    (m1$data$censoring == 0) * (as.numeric(mel.dat$TL_i <
        1.5) * prob_cured/(prob_cured +
        prob_uncured * S_t))
w_i1 = 1 - w_i0

# sensitivity
1/((2000 - sum(w_i0))) * sum(w_i1 *
    as.numeric(prob_uncured < 0.25))

# specificity
1 - (1/sum(w_i0)) * sum(w_i0 * as.numeric(prob_uncured <
    0.25))

# clinical cure tau = 2
w_i0 = as.numeric(mel.dat$TL_i > 2) +
    (m1$data$censoring == 0) * (as.numeric(mel.dat$TL_i <
        2) * prob_cured/(prob_cured +
        prob_uncured * S_t))
w_i1 = 1 - w_i0

# sensitivity
1 - (1/sum(w_i0)) * sum(w_i0 * as.numeric(prob_cured <
    0.75))

# specificity
1/((2000 - sum(w_i0))) * sum(w_i1 *
    as.numeric(prob_cured < 0.75))

## ZHANG METHOD no clinical cure
w_i = as.numeric(m1$data$censoring !=
    0) + (m1$data$censoring == 0) *
```

```
    (prob_uncured * S_t)/(prob_cured +
    prob_uncured * S_t)

# sensitivity
sum(w_i * as.numeric(prob_uncured <
    0.25))/sum(w_i)

# specificity
sum((1 - w_i) * as.numeric(prob_uncured >
    0.25))/sum(1 - w_i)
```

TABLE 4.4: Comparison of different methods for computing sensitivity and specificity where model parameters were computed by MPL.

| Estimate | Clinical cure time $\tau$ | Sensitivity | Specificity |
|---|---|---|---|
| Asano et al. | 1.5 | 0.6792 | 0.2891 |
| Asano et al. | 2.0 | 0.6623 | 0.7143 |
| Amico et al. | 1.5 | 0.5993 | 0.2693 |
| Amico et al. | 2.0 | 0.7307 | 0.4007 |
| Zhang et al. | NA | 0.5993 | 0.2693 |

The values of the estimated sensitivity and specificity for each estimation method and each clinical cure time $\tau$ are given in Table 4.4. As the data used in this example was simulated we are able to compute the "true" sensitivity and specificity of this model (i.e. the proportion of non-cured individuals in the sample with $l < 0.25$ and the proportion of cured individuals in the sample with $l > 0.25$); these values are $Sen(0.25) = 0.6816$ and $Spec(0.25) = 0.2811$. □

## 4.9 Summary

Standard survival analysis methods need to be adjusted when the event of interest may not be experienced by all individuals in the population, giving rise to a so-called "cured fraction". The mixture cure Cox model, which used a logistic regression sub-model for the incidence (probability of being susceptible) and a conditional Cox regression sub-model for the latency (time-to-event among the susceptible) is a commonly used approach. The inclusion of the incidence sub-model means that a partial likelihood which leaves the non-parametric baseline hazard function arbitrary is not available. This chapter discussed a full likelihood-based approach. Asymptotic normality results are presented so that inferences on survival quantities can be made.

# 5

# *Stratified Cox models under interval censoring*

Stratified Cox models are commonly employed when the proportional hazards assumption of a Cox model is not met. In cases where only right censoring is present in the survival data, a modified partial likelihood can still be applied to estimate the regression coefficients. However, if the survival dataset contains other types of censoring, such as left or interval censoring, the partial likelihood method cannot be applied directly, and instead, one can utilize full likelihood-based methods.

In this chapter, we explore a maximum penalized likelihood approach for estimating the parameters of a stratified Cox model, which encompasses the regression coefficients and baseline hazards. The survival data we consider are again partly-interval censored data.

## 5.1 Introduction

The last three chapters studied Cox models where the *proportional hazards* (PH) assumption is assumed when the model covariates are not time-dependent.

In Section 1.4.2, we defined the proportional hazards assumption, which requires that the hazard ratio, for example for two levels of a categorical covariate variable, remains constant over time $t$. When applying a Cox model to fit survival data, it is crucial to thoroughly validate this PH assumption. Methods for checking this assumption can be found in, for example, Chapter 6 of Hosmer et al. (2008).

Clearly, this assumption can be unrealistic in some data applications, necessitating amendments to Cox models to accommodate non-proportional hazards situations.

When the proportional hazards (PH) assumption is violated in a Cox model, there are two common amendments:

(1) One approach involves stratifying covariates responsible for non-proportional hazards. This approach gives rise to the so-called "*stratified Cox models*" (e.g., see Andersen et al. (1993), Martinussen & Scheike (2006) and Kleinbaum & Klein (2005)). Stratified Cox models essentially consist of multiple Cox models, one for each stratum. In these Cox models, the baseline hazards can differ between strata, but their regression coefficients remain identical. It is important to note that the identical regression coefficients indicate the absence of interaction terms between the strata and other covariates; see Chapter 5 of Kleinbaum & Klein (2005).

DOI: 10.1201/9781351109710-5

(2) Another common approach for addressing the issue of non-proportional hazards is to incorporate time-dependent covariates, i.e., covariates that vary with time $t$, into the Cox model. For example, if a categorical covariate $x_1$ does not satisfy the proportional hazard assumption, we may include $x_1 t$ into the Cox model. Essentially, this approach leads to an extended Cox model that includes time-varying (or time-dependent) covariates (see, for example, Cox & Oakes (1984) and Kleinbaum & Klein (2005)). As a result, the right-hand side of equation (1.46) varies with time $t$, effectively avoiding the proportional hazards assumption. Cox models with time-varying covariates will be discussed further in Chapter 6.

In this chapter, we will review and study stratified Cox models under partly interval censoring. Specifically, we will explain how to obtain the maximum penalized likelihood (MPL) estimates for the baseline hazards and regression coefficients of stratified Cox models. Here, penalty functions are applied to the baseline hazards to smooth the hazard estimates and also to enhance numerical stability in computations. Once again, the computational challenge of the MPL estimates comes from a constrained optimization requirement similar to those encountered in Chapters 2–4.

In this chapter, we cover the following sections: Section 5.2 defines stratified Cox models, Section 5.3 summarizes the conventional partial likelihood approach, and Section 5.4 establishes the likelihood function for the parameters of the stratified Cox model (see (5.3)). In Section 5.5, we explain how to compute the constrained MPL estimates of the regression coefficients and baseline hazards, utilizing penalty functions to smooth the baseline hazard estimates. This method, similar to the MPL approach in Chapter 2, allows for the joint estimation of regression coefficients and baseline hazards. It is suitable for handling partly interval censoring, where event times and left, right and interval censoring times are all considered. Section 5.7 provides asymptotic results for the MPL estimates to facilitate inferences and Section 5.8 discusses an automatic smoothing parameter estimation process. In Section 5.9, two R examples will be provided. In the first example, a simple simulation study, designed to compare the penalized likelihood estimation method discussed in this chapter and the partial likelihood method where left and interval-censored data are replaced by "mid-point" imputation. In the second example, we will discuss a melanoma recurrence example and an application of the MPL method to this dataset. Finally, Section 5.10 offers summary remarks. R examples are presented throughout this chapter.

## 5.2 Stratified Cox models

The conventional Cox model assumes the PH assumption when covariates are not time-dependent. Clearly, requiring all covariates to satisfy the PH assumption can be unrealistic in some applications. When the PH assumption is not met, one option is to consider a stratified Cox model. Before introducing this model, let us briefly summarize how to test the PH assumption. More detailed descriptions of these tests can be found in Kleinbaum & Klein (2005).

### 5.2.1 Testing the proportional hazards assumption

There are three common approaches to test the PH assumption and they are illustrated below.

1. **Graphic Approaches**
   (a) **$-\log(-\log)$ survival curves**
   This approach works well when covariates are categorical and each has a moderate number of levels. We explain the rationale behind this approach using the following simple example. Suppose there is only one covariate which is categorical with two levels: $x_i = 0$ (the reference group) or $x_i = 1$. If the PH assumption is satisfied, then, as indicated by (1.46), the survival functions of individual $i$ corresponding to these two levels must satisfy

$$S_{Y_i}(t|x_i = 1) = [S_{Y_i}(t|x_i = 0)]^{\exp\{\beta\}}, \tag{5.1}$$

   so that

$$-\log(-\log S_{Y_i}(t|x_i = 1)) = -\log(-\log S_{Y_i}(t|x_i = 0)) + b, \tag{5.2}$$

   where $b = -\beta$, a quantity independent of $t$. Under the PH assumption, the two "$-\log(-\log)$" survival functions: $-\log(-\log S_{Y_i}(t|x_i = 1))$ and $-\log(-\log S_{Y_i}(t|x_i = 0))$ (both of which are monotonically decreasing) should be parallel. This observation can be utilized to assess the PH assumption, as explained next. To do so, we divide the data into two groups (as for this example the covariate only has two levels) according to "$x_i = 1$" and "$x_i = 0$". If the PH assumption holds true, the plot of $-\log(-\log)$ of the survival functions for these two groups against time $t$ should exhibit a parallel pattern. These survival functions can be conveniently obtained using the Kaplan-Meier method. However, this graphical method may not work effectively for categorical covariates with many levels, as the plot can become cluttered. For continuous covariates, they need to be discretized before applying this graphical method to assess the PH assumption.
   (b) **Observed v.s. expected survival curves**
   This graphical method is again well-suited for categorical covariates. Let us again consider a single covariate with two levels as an example. First, fit a Cox model that includes only this covariate. Then, plug in the two values of this covariate to obtain two survival curves, which are referred to as the "expected survival curves". Keep in mind that the two levels of this covariate divide the observations

into two groups, as mentioned earlier. You can then obtain the "observed survival curve" by applying the Kaplan-Meier method to these two groups. Finally, display both the Kaplan-Meier (observed) and Cox model (expected) survival curves on the same graph. If these observed and expected plots overlap well, it indicates that the proportional hazards assumption is satisfied for this covariate.

2. **Including Time-Dependent Covariates**
The idea of this approach is to introduce a "created" time-varying covariate into the Cox model and then assess the significance of its coefficient. To do this, a time-varying covariate is created by taking, for example, the product of a time-fixed covariate (the one being tested) and a function of time $t$. This created time-varying covariate is then incorporated into the Cox model under consideration. If the coefficient of this created time-varying covariate proves to be significant, it indicates that the corresponding time-fixed covariate does not satisfy the PH assumption. For instance, when assessing the PH assumption for the variable "AgeGroup", we can expand the Cox model to include the variable "AgeGroup $*\, t$" alongside "AgeGroup". If the coefficient of this product term is found to be significant, it suggests a violation of the PH assumption for "AgeGroup".

3. **Tests using Schoenfeld Residuals**
Residuals, such as the Schoenfeld residuals (Schoenfeld 1982), can be employed to assess the goodness-of-fit of a Cox model and, consequently, the PH assumption. Schoenfeld residuals are computed for each covariate to determine whether each covariate independently adheres to the assumptions of the Cox model. As demonstrated by Grambsch & Therneau (1994), if the regression coefficients of the Cox PH model remain constant over time, the mean of the scaled Schoenfeld residuals is zero. Therefore, one can conduct a test for a zero slope in a non-linear regression of the scaled Schoenfeld residuals against time $t$. Rejection of this hypothesis suggests that the PH assumption is likely violated. In addition to performing these tests, it is beneficial to visually inspect the non-linear regression graph. This visual examination can identify situations where a non-zero slope exists but is not detected by the test.

Next, we provide R examples to demonstrate the PH testing ideas explained above.

**Example 5.1** (Simulating data from a stratified Cox model).
In this example, we demonstrate the following activities related to stratified Cox models:

1. Simulate a dataset of size $n = 500$ from a stratified Cox model with 2 strata. In one stratum let the baseline hazard function be $h_0(t) = 3t^2$ and in the other let it be $h_0(t) = 1/3$. Let the probability of being in one of the given strata be equal to 0.5. For both strata, let $X_1 \sim Bernoulli(0.5)$ and $X_2 \sim N(1, 1)$, and let the true regression coefficient vector $\boldsymbol{\beta} = [1, -0.5]^{\mathsf{T}}$. In this example we only consider right

censoring and the censoring time distribution is $C_i \sim exp(rate = 1.5)$ (so the mean of this exponential distribution is $1/1.5 = 0.667$). This censoring distribution will give approximately 70% of right-censored survival times.
2. Fit a standard Cox model to this data using partial likelihood, ignoring the strata for now.
3. Check the proportional hazards assumption using each of the three methods outlined above.

```
gen_strat_data = function(nsample, nstrata,
    prob_strata, beta) {

    # beta = c(1, -0.5)

    istrata <- sample(1:nstrata, size = nsample,
        prob = prob_strata, replace = T)

    X <- cbind(rbinom(nsample, 1, 0.5),
        rnorm(nsample, 1, 1))

    y_i <- rep(0, nsample)

    for (i in 1:nsample) {
        neg.log.u = -log(runif(1))
        mu_term = exp(X[i, ] %*% beta)

        if (istrata[i] == 1) {
            y_i[i] = (neg.log.u/(mu_term))^(1/3)
        } else {
            y_i[i] = (3 * (neg.log.u)/(mu_term))
        }

    }

    t_i = delta_i = rep(0, nsample)

    for (i in 1:nsample) {
        c_i = rexp(1, 1.5)
        delta_i[i] = as.numeric(y_i[i] <
            c_i)
        t_i[i] = min(c_i, y_i[i])
    }

    data = data.frame(c(1:nsample),
        t_i, delta_i, X, istrata)
    return(data)

}
```

Then we apply this data generation R function, and fit the simulated data using R `coxph` function. Some R outputs are displayed below.

```
> dat.strat = gen_strat_data(500, 2, c(0.5, 0.5), c(1, -0.5))
> table(dat.strat$delta_i)

  0   1
369 131

> library(survival)
> unstrat.fit = coxph(Surv(t_i, delta_i) ~ X1 + X2
  + factor(istrata), data = dat.strat)
> summary(unstrat.fit)
Call:
coxph(formula = Surv(t_i, delta_i) ~ X1 + X2
  + factor(istrata), data = dat.strat)

  n= 500, number of events= 131

                     coef exp(coef) se(coef)      z Pr(>|z|)
X1                 0.9821    2.6701   0.1900  5.170 2.34e-07 ***
X2                -0.4742    0.6224   0.0954 -4.973 6.58e-07 ***
factor(istrata)2 -0.5187    0.5953   0.1888 -2.747  0.00601 **
---
Signif. codes: 0 '***' 0.001 '**' 0.01 '*' 0.05 '.' 0.1 ' ' 1

                 exp(coef) exp(-coef) lower .95 upper .95
X1                  2.6701     0.3745    1.8401    3.8746
X2                  0.6224     1.6068    0.5163    0.7503
factor(istrata)2    0.5953     1.6798    0.4112    0.8619

Concordance= 0.648  (se = 0.029 )
Likelihood ratio test= 53.44  on 3 df,   p=1e-11
Wald test            = 50.23  on 3 df,   p=7e-11
Score (logrank) test = 50.91  on 3 df,   p=5e-11

> km_strat <- survfit(Surv(t_i, delta_i) ~ istrata,
  data = dat.strat)
> plot(km_strat, fun = "cloglog", xlab = "Time using log",
  + ylab = "log-log survival",
  main = "log-log curves by istrat")
> unstrat.fit.tvc = coxph(Surv(t_i, delta_i) ~ X1 + X2
  + factor(istrata) + tt(istrata), data = dat.strat,
  + tt = function(x, t,...) t*(x-1))
> summary(unstrat.fit.tvc)
Call:
coxph(formula = Surv(t_i, delta_i) ~ X1 + X2 + factor(istrata)
 + tt(istrata), data = dat.strat, tt = function(x, t, ...) t
 * (x - 1))
```

```
  n= 500, number of events= 131

                      coef exp(coef) se(coef)      z Pr(>|z|)
X1                  1.1015   3.0086   0.1952   5.644 1.66e-08 ***
X2                 -0.4867   0.6147   0.0954  -5.097 3.45e-07 ***
factor(istrata)2    2.7946  16.3563   0.5302   5.271 1.36e-07 ***
tt(istrata)        -5.3783   0.0046   0.8338  -6.451 1.11e-10 ***
---
Signif. codes:  0 '***' 0.001 '**' 0.01 '*' 0.05 '.' 0.1 ' ' 1

                 exp(coef) exp(-coef) lower.95  upper.95
X1                  3.0086     0.3322    2.0523    4.4105
X2                  0.6147     1.6269    0.5098    0.7412
factor(istrata)2   16.3563     0.0611    5.7857   46.2394
tt(istrata)         0.0046   216.6601    0.0009    0.0237

Concordance= 0.751  (se = 0.024 )
Likelihood ratio test= 125  on 4 df,   p=<2e-16
Wald test            = 87.44  on 4 df,   p=<2e-16
Score (logrank) test = 119.5  on 4 df,   p=<2e-16

> cox.zph(unstrat.fit)
                 chisq df       p
X1               2.935  1   0.087
X2               0.509  1   0.476
factor(istrata) 66.191  1 4.1e-16
GLOBAL          66.876  3 2.0e-14
```

□

### 5.2.2 Setting up stratified Cox models

After assessing the PH assumption of a Cox model, one can identify the covariates that satisfy the PH assumption and those that do not. In the following discussions, we refer to the covariates that violate the PH assumption as *non-PH covariates* and the ones that do not as *PH covariates*.

To simplify notation, in this chapter, we use $X_j$'s to denote the PH covariates and $Z_j$'s for non-PH covariate.

The construction of stratified Cox models initially requires a "stratification process" that divides the non-PH covariates into strata, where different strata must be non-overlapping. This process will be described in the next paragraph. In R, this task can be easily accomplished using the "`strata`" function.

Now, we discuss how to stratify the non-PH covariates. Denote these covariates as $Z_1, \ldots, Z_q$, which can be either categorical or continuous. If all $Z_1, \ldots, Z_q$ are categorical, the stratification process will create a single categorical variable $Z^*$ that

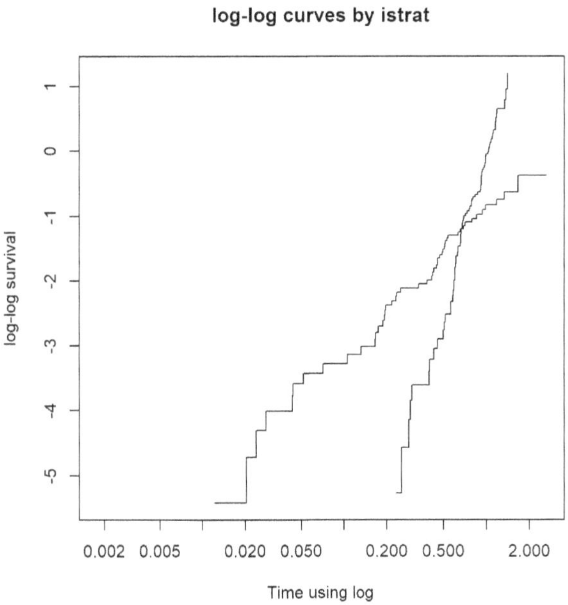


FIGURE 5.1: Plot of log-log survival functions to check the PH assumption.

collectively represents all of them. The total number of categories for $Z^*$ is determined by the product of the numbers of categories of all the $Z_v$'s.

For example, if $Z_1$ represents gender (Male and Female) and $Z_2$ represents age group ($< 50$ and $\geq 50$), then $Z^*$ becomes categorical with four categories: Male with age $< 50$; Male with age $\geq 50$; Female with age $< 50$; and Female with age $\geq 50$. However, if any of $Z_1, \ldots, Z_q$ is continuous, it needs to be first converted into a categorical variable, also called the categorization or discretization process.

After stratification, let us assume that $Z^*$ contains $s$ categories. Consequently, the dataset can now be divided into $s$ mutually exclusive and exhaustive subsets, which are referred to as "strata". We will denote $\mathcal{O}_r$ as the index set for observations belonging to stratum $r$, where $r = 1, \ldots, s$. Furthermore, let $X_1, \ldots, X_p$ represent the PH covariates (i.e. they do not violate the PH assumption). These covariates will be used to construct the linear predictor for the stratified Cox model as specified below.

In this context, the stratified Cox models consist of $s$ separate Cox models, each corresponding to one of these $s$ strata. The Cox model corresponding to stratum $r$ is given by:

$$h_r(t|\mathbf{x}_i) = h_{0r}(t)e^{\mathbf{x}_i^\top \boldsymbol{\beta}}, \tag{5.3}$$

where $i \in \mathcal{O}_r$, $\mathbf{x}_i$ is a $p$-vector for the observed values from the PH covariates $X_1, \ldots, X_p$ associated with individual $i$, $h_{0r}(t)$ is an unspecified baseline hazard (making it non-parametric) specific to each $r$, and $\beta$ is a $p$-vector of regression coefficients common to all strata.

Clearly, stratified Cox models (5.3), where $r = 1, \ldots, s$, exhibit the following characteristics:

(a) These Cox models, associated with different strata, share identical regression coefficients.

(b) They feature distinct baseline hazards, with one for each stratum.

As the regression coefficient vector $\beta$ is unchanged across all strata, the stratified Cox models implicitly assume the absence of interactions between the strata variable $Z^*$ and the PH-covariates $X_1, \ldots, X_p$; see Kleinbaum & Klein (2005).

From the stratum $r$ hazard function $h_r(t|\mathbf{x}_i)$, we can derive its cumulative hazard and survival functions, denoted as $H_r(t|\mathbf{x}_i)$ and $S_r(t|\mathbf{x}_i)$ respectively. The expressions of these functions can be found in (5.7) and (5.8) below.

For the stratified Cox models, when the observed survival data involve only *right censoring* and the primary focus is on estimating $\beta$, the conventional maximum partial likelihood method can still be applied; see, for example, Kleinbaum & Klein (2005) (Chapter 5) for details. Additionally, Section 5.3 below provides a summary of the partial likelihood approach for stratified Cox models. In this context, if an inference task also necessitates the estimation of the baseline hazards $h_{0r}(t)$ ($r = 1, \ldots, s$), such as when calculating the survival probability of an individual, one can combine the maximum partial likelihood method with the Breslow method (Breslow 1972). Specifically, for each stratum $r$, the Breslow method can be employed to estimate the cumulative baseline hazards $H_{0r}(t)$ by utilizing the data from this stratum. Subsequently, an estimate of $h_{0r}(t)$ can be derived from the estimated $H_{0r}(t)$.

There are, however, two drawbacks of this combined Breslow and partial likelihood method:

(i) It often results in a noisy estimate of $h_{0r}(t)$, necessitating an additional smoothing step to derive smoothed baseline hazard estimates. This smoothing step is necessary for obtaining a sensible interpretation from the $h_{0r}(t)$ estimate.

(ii) Obtaining the asymptotic covariance matrix of the estimates for both $\beta$ and $h_{0r}(t)$ can be challenging since $\beta$ and $h_{0r}(t)$ are estimated separately. To address this issue, computer-intensive methods like bootstrapping are typically employed to compute the covariance matrix. However, such methods can be time-consuming, making them less practical when dealing with large sample sizes.

When observed survival times involve more than just right censoring, the direct application of the partial likelihood method to fit stratified Cox models becomes challenging. In such scenarios, methods involving an imputation to replace the left- or interval-censored data are often employed, resulting in a modified dataset where only right censoring is present. Another commonly used approach is to *redefine* the event of interest, such as "time to the *diagnosis* of the event of interest" instead of the original "time to the event of interest". However, it is important to realize that these approaches either introduce bias into the coefficient estimates or regression coefficients do not link to the original event of interest.

In contrast, the full likelihood method is more versatile as it can handle arbitrary censoring types. Moreover, it allows for the simultaneous estimation of regression coefficients and baseline hazards, which simplifies the computation of the asymptotic covariance matrix. Likelihood-based methods not only enable the maintenance of the "semi-parametric" feature of the model (although an approximation to the non-parametric function is required), but also facilitate the computation of the asymptotic covariance matrix for both the $\boldsymbol{\beta}$ and $h_{0r}(t)$ estimates. Calculating the asymptotic covariance matrix is typically a fast process.

In the next section, we summarize the partial likelihood approach for stratified Cox models when only right censoring presents.

## 5.3 The partial likelihood method for right censored data

When survival data contain only right censoring, one can still conduct inference on the regression coefficients of stratified Cox models using the partial likelihood function. However, unlike the Cox model discussed in Chapter 1, the partial likelihood for stratified Cox models must include the partial likelihood contributions from different strata.

For an individual $i$ from stratum $r$, it can be denoted as $i \in \mathcal{O}_r$. Let $n_r$ represent the sample size of stratum $r$, then $i$ can be one of: $1, \ldots, n_r$. Assuming that the survival dataset contains only right censoring, we denote the survival time of this individual as $(t_i, \delta_i)$, where $t_i$ can be either an event time or a right censoring time, and $\delta_i$ is the event indicator for $t_i$, such that $\delta_i = 1$ if $t_i$ is an event time.

Corresponding to stratum $r$ (so that $i \in \mathcal{O}_r$ in the partial likelihood below), the partial likelihood is:

$$\widetilde{L}_r = \prod_{i=1}^{n_r} \left[ \frac{\exp\{\mathbf{x}_i^\top \boldsymbol{\beta}\}}{\sum_{a \in \mathcal{R}(t_i)} \exp\{\mathbf{x}_a^\top \boldsymbol{\beta}\}} \right]^{\delta_i}, \tag{5.4}$$

where $\mathcal{R}(t_i)$ denotes the risk set at time $t_i^-$, where $i \in \mathcal{O}_r$.

Then, the partial likelihood from all the observations (including all the strata) is:

$$\widetilde{L} = \prod_{r=1}^{s} \widetilde{L}_r. \tag{5.5}$$

Estimates of the regression coefficients can be obtained by maximizing the log-partial likelihood $\log \widetilde{L}$. They are called the maximum partial likelihood estimates.

Under certain conditions, the maximum of $\log \widetilde{L}$ represents a stationary point. In this case, the partial likelihood estimate of $\boldsymbol{\beta}$ is obtained by solving the system:

$$\frac{\partial \log \widetilde{L}}{\partial \boldsymbol{\beta}} = 0. \tag{5.6}$$

These equations are generally non-linear, and they can be solved iteratively using methods such as the Newton algorithm. The computational procedure for obtaining maximum partial likelihood estimates in stratified Cox models is very similar to that of the conventional Cox model (unstratified), as discussed in Chapter 1 where the Newton algorithm for maximum partial likelihood estimates is summarized.

Noticeable limitations of the partial likelihood method are that it **only** applies to survival data with right censoring and that the baseline hazards cannot be estimated directly. The latter issue can be addressed by a separate baseline hazards estimation procedure by, for example, the Breslow method.

The asymptotic covariance matrix for the $\beta$ estimate is given by inverse of the negative Hessian matrix, namely

$$\left[-\frac{\partial^2 \log \widetilde{L}}{\partial \beta \partial \beta^{\mathsf{T}}}\right]^{-1}.$$

The R package `survival` and R function `strata()` can be used to obtain the maximum partial likelihood estimate of $\beta$. The example given below explains how to use this package.

**Example 5.2** (Fitting stratified Cox using partial likelihood).
In this example, we demonstrate the partial likelihood-based approach to fit stratified Cox models using the dataset simulated in the previous example. In particular, we explain the stratified Cox model using the "`survival`" R package and function `strata()`.

```
strat.fit = coxph(Surv(t_i, delta_i) ~ X1 + X2 +
    strata(istrata), data = dat.strat)
summary(strat.fit)
cox.zph(strat.fit)
```

```
> strat.fit = coxph(Surv(t_i, delta_i) ~ X1 + X2 +
  + strata(istrata), data = dat.strat)
> summary(strat.fit)
Call:
coxph(formula = Surv(t_i, delta_i) ~ X1 + X2 + strata(istrata)
  , data = dat.strat)

  n= 500, number of events= 131

        coef exp(coef) se(coef)      z Pr(>|z|)
X1   1.08920   2.97189  0.19537  5.575 2.47e-08 ***
X2  -0.47920   0.61928  0.09527 -5.030 4.91e-07 ***
---
Signif. codes:  0 '***' 0.001 '**' 0.01 '*' 0.05 '.' 0.1 ' ' 1

   exp(coef) exp(-coef) lower .95 upper .95
X1    2.9719     0.3365    2.0265    4.3584
```

```
X2     0.6193      1.6148     0.5138     0.7464

Concordance= 0.684  (se = 0.027 )
Likelihood ratio test= 48.51  on 2 df,   p=3e-11
Wald test            = 45.71  on 2 df,   p=1e-10
Score (logrank) test = 47.31  on 2 df,   p=5e-11

> cox.zph(strat.fit)
         chisq df    p
X1      0.7407  1 0.39
X2      0.0261  1 0.87
GLOBAL 0.7480  2 0.69
```

□

## 5.4 Full likelihood under partly interval censoring

Given that this chapter also addresses partly interval censoring, we can reuse the interval censoring-related notations defined in Chapter 2.

For each individual $i$, where $i = 1, \ldots, n$, their event time of interest is denoted by $Y_i$, but it may not be observed fully. We accommodate the possibility of interval-censored observations for $Y_i$, where both left and right censoring are considered as special cases. We use $[t_i^L, t_i^R]$ to represent two observed time points such that $Y_i \in [t_i^L, t_i^R]$. This framework encompasses event time, left censoring, right censoring and interval censoring, with each of these being a special case of this interval. When $t_i^L = t_i^R = t_i$, $t_i$ is an observed time (event time) for $Y_i$. If $t_i^L = 0$, then the corresponding $t_i^R$ is a left censoring time for $Y_i$, and if $t_i^R = \infty$, its corresponding $t_i^L$ is a right censoring time for $Y_i$. The situation of $t_i^L \neq t_i^R$, and both are finite and non-zero, represents an interval-censored observation for $Y_i$.

The same as in Chapter 2, we assume independent censoring for the above-defined event time and censoring interval, namely $Y_i$ and its corresponding censoring interval are independent given the covariates; see Chapter 2.

The observations for individual $i$ can now be written as $(t_i^L, t_i^R, \mathbf{x}_i, z_i^*)$, where $z_i^*$ is an indicator for a stratum, taking a value from the stratum index set $1, \ldots, s$. It is important to note that covariate $z_i^*$ does not appear explicitly on the right-hand side of (5.3). Let $\delta_i$, $\delta_i^L$, $\delta_i^R$ and $\delta_i^I$ represent indicators for event, left censoring, right censoring and interval censoring, respectively. Let $\mathcal{O}_r$ denote the index set for observations belonging to stratum $r$. These notations will be used in the log-likelihood function provided in (5.9).

In correspondence with the stratified Cox models defined in (5.3), the cumulative hazard and survival functions for stratum $r$ are given by, respectively,

$$H_r(t|\mathbf{x}_i) = H_{0r}(t)e^{\mathbf{x}_i^\mathsf{T}\boldsymbol{\beta}} \tag{5.7}$$

and

$$S_r(t|\mathbf{x}_i) = [S_{0r}(t)]^{\exp\{\mathbf{x}_i^\top \boldsymbol{\beta}\}}. \tag{5.8}$$

Here, $H_{0r}(t)$ denotes the cumulative baseline hazard of stratum $r$ and $S_{0r}(t) = \exp\{-H_{0r}(t)\}$ represents the baseline survival function of stratum $r$.

To simplify notation, we will denote the hazard, cumulative hazard, and survival function for an individual $i$ in stratum $r$ as $h_{ir}(t)$, $H_{ir}(t)$ and $S_{ir}(t)$, respectively, wherever necessary.

From observations $(t_i^L, t_i^R, \delta_i, \delta_i^R, \delta_i^L, \delta_i^I, \mathbf{x}_i, z_i^*)$, where $i = 1, \ldots, n$, the log-likelihood is given by:

$$l = \sum_{r=1}^{s} \sum_{i \in \mathcal{O}_r} \{\delta_i(\log h_{0r}(t_i) + \mathbf{x}_i^\top \boldsymbol{\beta}) - \delta_i H_{ir}(t_i) - \delta_i^R H_{ir}(t_i^L)$$
$$+ \delta_i^L \log(1 - S_{ir}(t_i^R)) + \delta_i^I \log(S_{ir}(t_i^L) - S_{ir}(t_i^R))\}. \tag{5.9}$$

The maximum likelihood estimates of $\boldsymbol{\beta}$ and all $h_{0r}(t)$ can be obtained by maximizing this log-likelihood with respect to these unknown parameters. However, since each $h_{0r}(t)$ is non-parametric and thus is infinite dimensional, direct estimation of $h_{0r}(t)$ using (5.9), which is constructed based on a finite number of observations, poses an ill-posed problem. One common approach to address this issue, as adopted in Chapters 2 – 4, is to approximate each $h_{0r}(t)$ using a finite number of basis functions, where the number of basis functions is allowed to grow with the sample size but at a much slower rate. In this context, this approach is similar to the sieve method; see Grenander (1981).

In particular, to facilitate the imposition of constraints such as $h_{0r}(t) \geq 0$, non-negative basis functions will be employed. Commonly used basis functions include M-splines (Ramsay 1988), indicator functions, and Gaussian density functions; see Chapter 2 for the details of these basis functions.

For stratum $r$, assuming we have selected non-negative basis functions $\psi_{1r}(t), \ldots, \psi_{m_r r}(t)$, where the number of basis functions for stratum $r$ (i.e., $m_r$) or equivalently, the number of knots used to construct these basis functions, depends on the sample size $n_r$ of stratum $r$. Using these basis functions, we approximate $h_{0r}(t)$ by:

$$h_{0r}(t) = \sum_{u=1}^{m_r} \theta_{ur} \psi_{ur}(t). \tag{5.10}$$

Note that we still denote the approximated $h_{0r}(t)$ by $h_{0r}(t)$ for simplicity of notation.

The challenging functional constraints $h_{0r}(t) \geq 0$ for all $r$ are now replaced by much simpler constraints $\theta_{ur} \geq 0$ for all $u$ and $r$.

Corresponding to (5.10), the cumulative baseline hazard of stratum $r$ is

$$H_{0r}(t) = \sum_{u=1}^{m_r} \theta_{ur} \Psi_{ur}(t),$$

where $\Psi_{ur}(t) = \int_0^t \psi_{ur}(w)dw$, which is called the cumulative basis function. From

$H_{0r}(t)$, the baseline survival functions are simply

$$S_{0r}(t) = e^{-H_{0r}(t)}.$$

In Chapter 2, we explained that a major computational benefit of approximation (5.10) is that its cumulative hazard $H_{0r}(t)$ can be obtained remarkably easily.

Let $\boldsymbol{\theta}_r$ be the vector for coefficients $\theta_{ur}$ for a given stratum $r$, namely $\boldsymbol{\theta}_r = (\theta_{1r}, \ldots, \theta_{m_r r})^\mathsf{T}$. Let $\boldsymbol{\theta} = (\boldsymbol{\theta}_1^\mathsf{T}, \ldots, \boldsymbol{\theta}_s^\mathsf{T})^\mathsf{T}$. After replacing the non-parametric functions $h_{0r}(t)$ in (5.9) by their approximating functions obtained from the basis functions, the resulting log-likelihood function depends on $\boldsymbol{\beta}$ and $\boldsymbol{\theta}$. We denote this log-likelihood by $l(\boldsymbol{\beta}, \boldsymbol{\theta})$ henceforth. We wish to estimate $\boldsymbol{\beta}$ and $\boldsymbol{\theta}$.

We aim to develop the MPL estimates of $\boldsymbol{\beta}$ and $\boldsymbol{\theta}_r$ for all $r$. There will be $s$ penalty functions, one for each stratum, included in the objective function that we want to maximize. The penalty function for stratum $r$, denoted as $J_r(\boldsymbol{\theta}_r)$, serves primarily for the following two reasons:

(i) smoothing the estimate of $h_{0r}(t)$; and

(ii) reducing the sensitivity to the number of knots and knot locations, thus enhancing the numerical stability of the estimation process.

The MPL estimates of $\boldsymbol{\beta}$ and $\boldsymbol{\theta}$ are obtained by the following optimization problem:

$$(\widehat{\boldsymbol{\beta}}, \widehat{\boldsymbol{\theta}}) = \underset{\boldsymbol{\beta}, \boldsymbol{\theta}}{\operatorname{argmax}} \left\{ \Phi(\boldsymbol{\beta}, \boldsymbol{\theta}) = l(\boldsymbol{\beta}, \boldsymbol{\theta}) - \sum_{r=1}^{s} \lambda_r J_r(\boldsymbol{\theta}_r) \right\}, \tag{5.11}$$

subject to $\boldsymbol{\theta} \geq 0$ (here inequality is interpreted element-wisely), where $J_r(\boldsymbol{\theta}_r)$ are penalty functions and $\lambda_r \geq 0$ are smoothing parameters.

This presents a constrained optimization problem, and there are several existing methods suitable for solving it. Methods such as the "projected Newton method", the "primal-dual interior point method" and "active set methods" are all applicable. Detailed information on these methods can be found in references like Luenberger & Ye (2008) or Nocedal & Wright (2000). However, these methods may not be efficient when dealing with large values of $m_r$. For instance, if one chooses a piecewise constant approximation for $h_{0r}(t)$, and the pieces are determined such that each piece corresponds to either one observed event time or one observed interval (excluding left and right censoring), the number of pieces, denoted as $m_r$, can potentially be large, especially for large sample sizes.

In the next section, we will discuss the Newton-MI algorithm, similar to the method in Chapter 2, for computing the constrained MPL estimates of $\boldsymbol{\beta}$ and $\boldsymbol{\theta}$.

## 5.5 Computation of the MPL estimates

The Karush-Kuhn-Tucker (KKT) conditions for the constrained optimization problem (5.11) are specified as follows. For $j = 1, \ldots, p$, $u = 1, \ldots, m_r$, and $r =$

$1, \ldots, s$

$$\frac{\partial \Phi(\boldsymbol{\beta}, \boldsymbol{\theta})}{\partial \beta_j} = 0, \tag{5.12}$$

$$\frac{\partial \Phi(\boldsymbol{\beta}, \boldsymbol{\theta})}{\partial \theta_{ur}} = 0 \text{ if } \theta_{ur} \neq 0, \text{ and } \frac{\partial \Phi(\boldsymbol{\beta}, \boldsymbol{\theta})}{\partial \theta_{ur}} < 0 \text{ if } \theta_{ur} = 0. \tag{5.13}$$

We solve these equations iteratively using the Newton-MI algorithm similar to Chapter 2; also see Ma et al. (2014) and Ma et al. (2021).

This is an alternating algorithm where $\boldsymbol{\beta}$ and $\boldsymbol{\theta}_1, \ldots, \boldsymbol{\theta}_s$ are updated sequentially in each iteration. First, the regression coefficient vector $\boldsymbol{\beta}$ (common to all strata) is updated. Then, each stratum-specific baseline hazard spline parameter vector $\boldsymbol{\theta}_r$, for $r = 1, \ldots, s$, is updated based on the most currently available values of $\boldsymbol{\theta}$ and $\boldsymbol{\beta}$.

More specifically, $\boldsymbol{\beta}$ is updated using the Newton algorithm, which includes a line search step to ensure an increase in likelihood with the updated $\boldsymbol{\beta}$. Afterward, the $\boldsymbol{\theta}_r$'s are updated sequentially using the multiplicative-iterative (MI) algorithm (e.g., Chan & Ma (2012)). In this process, line search steps are adopted to achieve increments in the penalized log-likelihood. This MI algorithm is easy to implement and particularly efficient for enforcing the constraint $\boldsymbol{\theta}_r \geq 0$. The details of this Newton-MI algorithm are provided below.

Before implementing the Newton-MI algorithm to fit the stratified Cox models, we need to calculate some derivatives of $\Phi(\boldsymbol{\beta}, \boldsymbol{\theta})$. Specifically: For the Newton algorithm (with respect to $\boldsymbol{\beta}$):

1. First derivative: $\partial \Phi(\boldsymbol{\beta}, \boldsymbol{\theta}) / \partial \boldsymbol{\beta}$.
2. Second derivative: $\partial^2 \Phi(\boldsymbol{\beta}, \boldsymbol{\theta}) / \partial \boldsymbol{\beta} \partial \boldsymbol{\beta}^{\mathsf{T}}$.

For the MI algorithm with respect to $\boldsymbol{\theta}_r$, for $r = 1, \ldots, s$, we only need the first derivative: $\partial \Phi(\boldsymbol{\beta}, \boldsymbol{\theta}) / \partial \boldsymbol{\theta}_r$.

These derivatives are given below:

$$\begin{aligned}\frac{\partial \Phi}{\partial \beta_j} =& \sum_{r=1}^{s} \sum_{i \in \mathcal{O}_r} x_{ij} \left\{ \delta_i - \delta_i H_{ir}(t_i) + \delta_i^L \frac{S_{ir}(t_i^R) H_{ir}(t_i^R)}{1 - S_{ir}(t_i^R)} \right. \\ & \left. - \delta_i^R H_{ir}(t_i^L) + \delta_i^I \frac{-S_{ir}(t_i^L) H_{ir}(t_i^L) + S_{ir}(t_i^R) H_{ir}(t_i^R)}{S_{ir}(t_i^L) - S_{ir}(t_i^R)} \right\}; \end{aligned} \tag{5.14}$$

$$\begin{aligned}\frac{\partial^2 \Phi}{\partial \beta_j \partial \beta_t} =& - \sum_{r=1}^{s} \sum_{i \in \mathcal{O}_r} x_{ij} x_{it} \left\{ \delta_i H_{ir}(t_i) + \delta_i^L \frac{S_{ir}(t_i^R) H_{ir}(t_i^R) (H_{ir}(t_i^R) + S_{ir}(t_i^R) - 1)}{(1 - S_{ir}(t_i^R))^2} \right. \\ & + \delta_i^R H_{ir}(t_i^L) + \delta_i^I \frac{S_{ir}(t_i^L) S_{ir}(t_i^R) (-H_{ir}(t_i^L) + H_{ir}(t_i^R))^2}{(S_{ir}(t_i^L) - S_{ir}(t_i^R))^2} \\ & \left. + \delta_i^I \frac{-S_{ir}(t_i^R) H_{ir}(t_i^R) + S_{ir}(t_i^L) H_{ir}(t_i^L)}{S_{ir}(t_i^L) - S_{ir}(t_i^R)} \right\}; \end{aligned} \tag{5.15}$$

and

$$\frac{\partial \Phi}{\partial \theta_{ur}} = \sum_{i \in \mathcal{O}_r} \left\{ \delta_i \frac{\psi_{ur}(t_i)}{h_{0r}(t_i)} - \delta_i \Psi_{ur}(t_i) e^{\mathbf{x}_i^\top \boldsymbol{\beta}} + \delta_i^L \frac{S_{ir}(t_i^R)\Psi_{ur}(t_i^R)}{1 - S_{ir}(t_i^R)} e^{\mathbf{x}_i^\top \boldsymbol{\beta}} \right.$$
$$\left. -\delta_i^R \Psi_{ur}(t_i^L) e^{\mathbf{x}_i^\top \boldsymbol{\beta}} - \delta_i^I \frac{S_{ir}(t_i^L)\Psi_{ur}(t_i^L) - S_{ir}(t_i^R)\Psi_{ur}(t_i^R)}{S_{ir}(t_i^L) - S_{ir}(t_i^R)} e^{\mathbf{x}_i^\top \boldsymbol{\beta}} \right\} - \lambda_r \frac{\partial J_r(\boldsymbol{\theta}_r)}{\partial \theta_{ur}}. \tag{5.16}$$

To facilitate the implementation of the Newton algorithm, we can express the derivatives in (5.14) and (5.15) in matrix form. There are two ways to represent the gradient vector and the Hessian matrix in matrix form: the first method uses the model matrix of each stratum, while the second method combines the model matrices from all strata.

Recall that $n_r$ represents the number of observations in stratum $r$, and therefore, the total number of observations, namely the sample size, is given by $n = \sum_{r=1}^{s} n_r$. To segregate the observations based on the strata index $z_i^*$, we define $\mathbf{X}_r$ as the model matrix containing covariates values associated with stratum $r$. Matrix $\mathbf{X}_r$ has the dimension of $n_r \times p$. Let $\boldsymbol{\delta}_r$ be a vector of length $n_r$, where each element corresponds to $\delta_i$ for $i \in \mathcal{O}_r$. In other words, the elements of $\boldsymbol{\delta}_r$ indicate whether the observed survival times in stratum $r$ are event times (denoted as 1) or censored times (denoted as 0).

When using all the $\mathbf{X}_r$'s, the gradient vector (score function) for $\boldsymbol{\beta}$, as derived from (5.14), can be expressed as follows:

$$\frac{\partial \Phi(\boldsymbol{\beta}, \boldsymbol{\theta})}{\partial \boldsymbol{\beta}} = \sum_{r=1}^{s} \mathbf{X}_r^\top (\boldsymbol{\delta}_r - \mathbf{b}_r), \tag{5.17}$$

where $\mathbf{b}_r$ is an $n_r$-vector, and its elements, which are denoted by $b_{ir}$ $(i = 1, \dots, n_r)$, are specified according to the censoring type. Specifically:

- For event times $b_{ir} = H_{ir}(t_i)$.
- For left censoring times $b_{ir} = -S_{ir}(t_i^R)H_{ir}(t_i^R)/(1 - S_{ir}(t_i^R))$.
- For right censoring times $b_{ir} = H_{ir}(t_i^L)$.
- For interval censoring times $b_{ir} = (S_{ir}(t_i^L)H_{ir}(t_i^L) - S_{ir}(t_i^R)H_{ir}(t_i^R))/(S_{ir}(t_i^L) - S_{ir}(t_i^R))$.

From expression (5.15), the Hessian matrix for $\boldsymbol{\beta}$, where all $\mathbf{X}_r$'s are involved, is

$$\frac{\partial^2 \Phi(\boldsymbol{\beta}, \boldsymbol{\theta})}{\partial \boldsymbol{\beta} \partial \boldsymbol{\beta}^T} = -\sum_{r=1}^{s} \mathbf{X}_r^\top \mathbf{A}_r \mathbf{X}_r \tag{5.18}$$

where $\mathbf{A}_r$ are diagonal matrices, given by $\mathbf{A}_r = \text{diag}(a_{ir})$ for $i = 1, \dots, n_r$. Elements $a_{ir}$ again depend on the censoring type.

- For event times $a_{ir} = H_{ir}(t_i)$.
- For left censoring times $a_{ir} = S_{ir}(t_i^R)H_{ir}(t_i^R)(H_{ir}(t_i^R) + S_{ir}(t_i^R) - 1)/(1 - S_{ir}(t_i^R))^2$.
- For right censoring times $a_{ir} = H_{ir}(t_i^L)$.

• For interval censoring times $a_{ir} = S_{ir}(t_i^L)S_{ir}(t_i^R)(-H_{ir}(t_i^L)+H_{ir}(t_i^R))^2/(S_{ir}(t_i^L) - S_{ir}(t_i^R))^2 + (-S_{ir}(t_i^R)H_{ir}(t_i^R) + S_{ir}(t_i^L)H_{ir}(t_i^L))/(S_{ir}(t_i^L) - S_{ir}(t_i^R))$.

Alternatively, one can use a single model matrix for all strata to compute the gradient and Hessian. In fact, if letting $\mathbf{X} = (\mathbf{X}_1^\top, \ldots, \mathbf{X}_s^\top)^\top$, $\mathbf{b} = (\mathbf{b}_1^\top, \ldots, \mathbf{b}_s^\top)^\top$, $\boldsymbol{\delta} = (\boldsymbol{\delta}_1^\top, \ldots, \boldsymbol{\delta}_s^\top)^\top$ and $\mathbf{A} = \text{diag}(\mathbf{A}_1, \ldots, \mathbf{A}_s)$, the gradient and Hessian of $\Phi$ for $\boldsymbol{\beta}$ can be written as

$$\frac{\partial \Phi(\boldsymbol{\beta}, \boldsymbol{\theta})}{\partial \boldsymbol{\beta}} = \mathbf{X}^\top (\boldsymbol{\delta} - \mathbf{b}) \tag{5.19}$$

and

$$\frac{\partial^2 \Phi(\boldsymbol{\beta}, \boldsymbol{\theta})}{\partial \boldsymbol{\beta} \partial \boldsymbol{\beta}^\top} = -\mathbf{X}^\top \mathbf{A} \mathbf{X}. \tag{5.20}$$

From the above gradient and Hessian formulae (that is (5.17) and (5.18) or (5.19) and (5.20)), we first update $\boldsymbol{\beta}$ using the Newton algorithm with line search, namely we update $\boldsymbol{\beta}$ according to

$$\boldsymbol{\beta}^{(k+1)} = \boldsymbol{\beta}^{(k)} + \omega_1^{(k)} \left[ -\frac{\partial^2 \Phi(\boldsymbol{\beta}^{(k)}, \boldsymbol{\theta}^{(k)})}{\partial \boldsymbol{\beta} \partial \boldsymbol{\beta}^\top} \right]^{-1} \frac{\partial \Phi(\boldsymbol{\beta}^{(k)}, \boldsymbol{\theta}^{(k)})}{\partial \boldsymbol{\beta}}, \tag{5.21}$$

where $\omega_1$ is a line search step size selected to ensure $\Phi(\boldsymbol{\beta}^{(k+1)}, \boldsymbol{\theta}^{(k)}) \geq \Phi(\boldsymbol{\beta}^{(k)}, \boldsymbol{\theta}^{(k)})$. We comment that this $\boldsymbol{\beta}$ updating formula is identical to the $\boldsymbol{\beta}$ formula in Chapter 2.

After obtaining the updated value $\boldsymbol{\beta}^{(k+1)}$, we proceed to update $\boldsymbol{\theta}$. Each stratum $r$ has its own baseline hazard parameter $\boldsymbol{\theta}_r$ in the stratified Cox models. We have two strategies for updating the $\boldsymbol{\theta}_r$'s, both involve updating the $\boldsymbol{\theta}_r$'s sequentially.

The first strategy is to update each $\boldsymbol{\theta}_r$ independently using only the parameter estimates from the last iteration, without considering the most recently available $\boldsymbol{\theta}$ estimates from other strata.

The second strategy involves updating each $\boldsymbol{\theta}_r$ while taking into account the most current $\boldsymbol{\theta}$ estimates from other strata.

These two strategies are similar in spirit to respectively the Jacobi over-relaxation (JOR) and successive over-relaxation (SOR) methods for solving a set of linear systems. Based on our experience, the second method tends to be more stable and converge more quickly. Consequently, we will concentrate on this approach in the subsequent discussions.

To explain the second strategy, we first introduce a notation needed in the following discussions. For $r = 1, \ldots, s$, we divide $\boldsymbol{\theta}$ into two sub-vectors: $\boldsymbol{\theta}_{\leq r-1} = (\boldsymbol{\theta}_1^\top, \ldots, \boldsymbol{\theta}_{r-1}^\top)^\top$ and $\boldsymbol{\theta}_{\geq r} = (\boldsymbol{\theta}_r^\top, \ldots, \boldsymbol{\theta}_s^\top)^\top$. Clearly, when $r = 1$, the vector $\boldsymbol{\theta}_{\leq r-1}$ is empty and the vector $\boldsymbol{\theta}_{\geq r} = \boldsymbol{\theta}$.

The sequential computation of $\boldsymbol{\theta}^{(k+1)}$ is achieved by updating each $\boldsymbol{\theta}_r$ one after another, employing a MI scheme similar to the $\boldsymbol{\theta}$ updating formula discussed in Chapter 2:

$$\boldsymbol{\theta}_r^{(k+1)} = \boldsymbol{\theta}_r^{(k)} + \omega_{2r}^{(k)} \mathbf{D}_r(\boldsymbol{\beta}^{(k+1)}, \boldsymbol{\theta}_{\leq r-1}^{(k+1)}, \boldsymbol{\theta}_{\geq r}^{(k)}) \frac{\partial \Phi(\boldsymbol{\beta}^{(k+1)}, \boldsymbol{\theta}_{\leq r-1}^{(k+1)}, \boldsymbol{\theta}_{\geq r}^{(k)})}{\partial \boldsymbol{\theta}_r}, \tag{5.22}$$

for $r = 1, \dots, s$. Here, $\omega_{2r}^k$ is a line search step size for achieving

$$\Phi(\boldsymbol{\beta}^{(k+1)}, \boldsymbol{\theta}_{\leq r}^{(k+1)}, \boldsymbol{\theta}_{\geq r+1}^{(k)}) \geq \Phi(\boldsymbol{\beta}^{(k+1)}, \boldsymbol{\theta}_{\leq r-1}^{(k+1)}, \boldsymbol{\theta}_{\geq r}^{(k)})$$

when updating $\boldsymbol{\theta}_r$. In (5.22), the gradient for $\boldsymbol{\theta}_r$ is readily available from (5.16). In fact, its vector form can be written as:

$$\frac{\partial \Phi(\boldsymbol{\beta}, \boldsymbol{\theta})}{\partial \boldsymbol{\theta}_r} = \boldsymbol{\delta}_r * \mathbf{C}_r + \boldsymbol{\delta}_r^R * \mathbf{E}_r^R + \boldsymbol{\delta}_r^L * \mathbf{E}_r^L + \boldsymbol{\delta}_r^I * \mathbf{E}_r^I - \lambda_r \frac{\partial J_r(\boldsymbol{\theta}_r)}{\partial \boldsymbol{\theta}_r}, \tag{5.23}$$

where '$*$' denotes element-wise multiplication between two vectors, $\boldsymbol{\delta}_r$ is an indicator vector for event times of stratum $r$, and $\boldsymbol{\delta}_r^R$, $\boldsymbol{\delta}_r^L$ and $\boldsymbol{\delta}_r^I$ are indicator vectors for, respectively, right, left and interval censoring of stratum $r$. In (5.23),

- $\mathbf{C}_r$ is a vector with elements $\psi_{ur}(t_i)/h_{0r}(t_i) - \Psi_{ur}(t_i)\exp(\mathbf{x}_i^\top \boldsymbol{\beta})$;
- $\mathbf{E}_r^R$ is a vector with elements $\Psi_{ur}(t_i^L)\exp\{\mathbf{x}_i^\top \boldsymbol{\beta}\}$;
- $\mathbf{E}_r^L$ is a vector with elements $S_{ir}(t_i^R)\Psi_{ur}(t_i^R)/(1 - S_{ir}(t_i^R))\exp\{\mathbf{x}_i^\top \boldsymbol{\beta}\}$;
- $\mathbf{E}_r^I$ is a vector with elements $(-S_r(t_i^L)\Psi_{ur}(t_i^L) + S_r(t_i^R)\Psi_{ur}(t_i^R))/(S_r(t_i^L) - S_r(t_i^R))\exp(\mathbf{x}_i^\top \boldsymbol{\beta})$;

and all are associated with stratum $r$. The matrix $\mathbf{D}_r$ in (5.22) is a diagonal matrix with diagonal elements given by $\theta_{ur}/\zeta_{ur}$, where

$$\zeta_{ur} = \sum_{i \in \mathcal{O}_r} \left\{ \delta_i \Psi_{ur}(t_i) + \delta_i^R \Psi_{ur}(t_i^L) + \delta_i^I \frac{S_r(t_i^L)\Psi_{ur}(t_i^L)}{S_r(t_i^L) - S_r(t_i^R)} \right\} e^{\mathbf{x}_i^\top \boldsymbol{\beta}} \\ + \lambda_r \left[ \frac{\partial J_r(\boldsymbol{\theta}_r)}{\partial \boldsymbol{\theta}_r} \right]^+ + \epsilon.$$

Here, The notation $[a]^+$ represents the positive part of a number $a$, defined as $\max\{0, a\}$. The parameter $\epsilon$ is a small threshold, typically set to a small value like $10^{-3}$, used to prevent division by zero in the computation of $\zeta_{ur}$. As explained in Chapter 2, the choice of $\epsilon$ does not affect the final solution but can impact the convergence speed of the algorithm.

The algorithm described above is straightforward to implement and typically converges quickly. It efficiently enforces the constraint $\boldsymbol{\theta}_r \geq 0$ for all $r$, even when $m = \sum_r m_r$ is large. However, upon convergence, it may still be necessary to identify active constraints, which are those with $\theta_{ur} = 0$. This is because it would be computationally expensive to precisely drive all active $\theta_{ur}$ to exactly zero. To identify active constraints, the following rule is employed: for any $\theta_{ur}$, if its final estimate is less than $10^{-3}$ and the corresponding gradient $\partial \Phi / \partial \theta_{ur} < -10^{-2}$, then this $\theta_{ru}$ is considered actively constrained.

Active constraints, if not accounted for, can lead to issues like negative variances in some parameter estimates. In Section 5.7, we will develop an asymptotic result that can accommodate active constraints.

## 5.6 Other second derivatives in the Hessian matrix

In the last section, we provided the second derivative of $\Phi$ with respect to $\boldsymbol{\beta}$. The asymptotic normality results stated in the next section demand the full Hessian matrix, which involves the second derivatives with all the parameters of the model we study in this chapter. These second derivatives are provided below.

Clearly, if $r \neq q$, we have

$$\frac{\partial\Phi(\boldsymbol{\beta},\theta)}{\partial\theta_{ur}\partial\theta_{vq}} = 0. \tag{5.24}$$

Otherwise, if $r = q$, we have

$$\begin{aligned}\frac{\partial\Phi(\boldsymbol{\beta},\theta)}{\partial\theta_{ur}\partial\theta_{vr}} &= \sum_{i\in\mathcal{O}_r}\Bigg\{-\delta_i\frac{\psi_{ur}(t_i)\psi_{vr}(t_i)}{h_{0r}^2(t_i)} - \delta_i^L\frac{S_{ir}(t_i^R)\Psi_{vr}(t_i^R)\Psi_{ur}(t_i^R)}{(1-S_{ir}(t_i^R))^2}e^{2\mathbf{x}_i^\top\boldsymbol{\beta}} \\ &\quad - \delta_i^I\frac{S_{ir}(t_i^L)S_{ir}(t_i^R)(\Psi_{ur}(t_i^R)-\Psi_{ur}(t_i^L))(\Psi_{vr}(t_i^R)-\Psi_{vr}(t_i^L))}{(S_{ir}(t_i^L)-S_{ir}(t_i^R))^2}e^{2\mathbf{x}_i^\top\boldsymbol{\beta}}\Bigg\} \\ &\quad - \lambda_r\frac{\partial J_r(\boldsymbol{\theta}_r)}{\partial\theta_{ur}\partial\theta_{vr}}.\end{aligned} \tag{5.25}$$

Finally,

$$\begin{aligned}\frac{\partial\Phi(\boldsymbol{\beta},\theta)}{\partial\theta_{ur}\partial\beta_j} &= \sum_{i\in\mathcal{O}_r}\Bigg\{-\delta_i\Psi_{ur}(t_i) - \delta_i^R\Psi_{ur}(t_i^L) \\ &\quad - \delta_i^L\left(\frac{S_{ir}(t_i^R)H_{ir}(t_i^R)}{(1-S_{ir}(t_i^R))^2} - \frac{S_{ir}(t_i^R)}{(1-S_{ir}(t_i^R))}\right)\Psi_{ur}(t_i^R) \\ &\quad - \delta_i^I\Bigg(\frac{S_{ir}(t_i^L)S_{ir}(t_i^R)(H_{ir}(t_i^R)-H_{ir}(t_i^L))(\Psi_{ur}(t_i^R)-\Psi_{ur}(t_i^L))}{(S_{ir}(t_i^L)-S_{ir}(t_i^R))^2} \\ &\quad + \frac{S_{ir}(t_i^L)\Psi_{ur}(t_i^L)-S_{ir}(t_i^R)\Psi_{ur}(t_i^R)}{S_{ir}(t_i^L)-S_{ir}(t_i^R)}\Bigg)\Bigg\}e^{\mathbf{x}_i^\top\boldsymbol{\beta}}x_{ij}.\end{aligned} \tag{5.26}$$

## 5.7 Asymptotic results

In this section, we will first establish the asymptotic consistency of the estimates for $\boldsymbol{\beta}$ and $\boldsymbol{\theta}$ as the number of basis functions for each approximate $h_{0r}(t)$ approaches infinity when $n$ tends to infinity. While this result maintains the non-parametric nature of $h_{0r}(t)$, it is not practically useful.

The second asymptotic result will provide a large sample approximate normality for the estimates of $\boldsymbol{\beta}$ and $\boldsymbol{\theta}$ with a fixed number of basis functions for each $r$. In this result, the effects of active constraints will be removed from the asymptotic variance formula. This asymptotic normality result is particularly useful for inferences such as predicting survival probabilities.

Let $\boldsymbol{\tau} = (\boldsymbol{\beta}, h_{01}(t), \ldots, h_{0s}(t))$, representing the set of parameters that we aim to estimate. Denote the true $\boldsymbol{\tau}$ by $\boldsymbol{\tau}_0 = (\boldsymbol{\beta}_0, h_{01}(t), \ldots, h_{0s}(t))$. Let $B = \{\boldsymbol{\beta} : |\beta_j| < C_1 < \infty, \forall j\}$ and $A_r = \{h_{0r}(t) : 0 \leq h_{0r}(t) \leq C_2 < \infty, t \in [a, b]\}$, where $r = 1, \ldots, s$, $a = \min_i\{t_i^L\}$ and $b = \max_i\{t_i^R\}$ with $t_i^R \neq \infty$. The space for $\boldsymbol{\tau}$ is given by $B \otimes A_1 \otimes \cdots \otimes A_s$.

The approximation to $h_{0r}(t)$ we have adopted is denoted by $h_{nr}(t)$, namely $h_{nr}(t) = \sum_{u=1}^{m_r} \theta_{ur}\psi_{ur}(t)$, where the basis functions are employed. We let the basis function vector $\boldsymbol{\psi}_r(t) = (\psi_{1r}(t), \ldots, \psi_{m_r r}(t))^{\mathsf{T}}$ and the corresponding coefficient vector $\boldsymbol{\theta}_r = (\theta_{1r}, \ldots, \theta_{m_r r})^{\mathsf{T}}$. Let $\boldsymbol{\tau}_n = (\boldsymbol{\beta}, \boldsymbol{\psi}_1(t)^{\mathsf{T}}\boldsymbol{\theta}_1, \ldots, \boldsymbol{\psi}_s(t)^{\mathsf{T}}\boldsymbol{\theta}_s)$, and where $\boldsymbol{\beta}, \boldsymbol{\theta}_1, \ldots, \boldsymbol{\theta}_s$ are the parameters that we will estimate. Define $A_{nr} = \{\boldsymbol{\psi}_r(t)^{\mathsf{T}}\boldsymbol{\theta}_r : 0 \leq \theta_{ur} \leq C_3 < \infty, \forall u\}$. The space for $\boldsymbol{\omega}_n$ is then $\mathcal{B} \otimes A_{n1} \otimes \cdots \otimes A_{ns}$. Denote the MPL estimate of $\theta_{ur}$ by $\widehat{\theta}_{ur}$. The MPL estimate of $h_r(t)$ is then $\widehat{h}_{nr}(t) = \sum_{u=1}^{m_r} \widehat{\theta}_{ur}\psi_{ur}(t)$.

Let $n_r$ be the number of events for stratum $r$. Assume that when $n \to \infty$ then all $n_r \to \infty$. The results in Theorem 5.1 below state that, under certain regularity conditions, the MPL estimates of $\boldsymbol{\beta}$ and $h_{0r}(t)$ converge to their true values when $n \to \infty$. Here, the regularity conditions are similar to Assumptions A2.1–A2.4 of Theorem 2.4 in Chapter 2. In fact, we can retain the assumption A2.1 but need to replace A2.2–2.4 by the following assumptions.

- A5.2 The penalty functions $J_r(\boldsymbol{\theta}_r)$ are all bounded $\forall r$.
- A5.3 For each stratum $r$ and the associated approximation function $h_{nr}(t)$, assume its coefficient vector $\boldsymbol{\theta}_r$ is in a compact subset of $R^{m_r}$, and moreover, assume its basis functions $\psi_{ur}(t)$ are bounded for $t \in [a, b]$.
- A5.4 For each $r$, the knots and basis functions are selected in a way such that for any $h_{0r}(t) \in A_r$ there exists a $h_{nr}(t) \in A_{nr}$ such that $\max_{t\in[a,b]} |h_{nr}(t) - h_{0r}(t)| \to 0$ as $n \to \infty$.

Now, we can present the asymptotic consistency results below.

**Theorem 5.1.** Assume Assumptions A2.1 and A5.2–A5.4 hold. Assume each $h_{0r}(t)$ has up to $c$ ($\geq 1$) derivatives for $t \in [a, b]$ and $r = 1, \ldots, s$. Assume each $n_r \to \infty$ when $n$ goes to $\infty$. Suppose $m_r$ is selected according to $m_r = n_r^{\upsilon_r}$, where $0 < \upsilon_r < 1$, and assume $\mu_{rn} = \lambda_r/n_r \to 0$ when $n \to \infty$. Then, as $n \to \infty$, we have

(1) $\|\widehat{\boldsymbol{\beta}} - \boldsymbol{\beta}_0\|_2 \to 0$ (a.s.), and

(2) $\sup_{t\in[a,b]} |\widehat{h}_{nr}(t) - h_{0r}(t)| \to 0$ (a.s.) $\forall r$.

□

A proof of this theorem can be constructed similarly to the proof of Theorem 2.4 in Appendix A and is therefore omitted here.

The fact that $m_r$ increases extremely slowly compared to $n$ provides a rationale for the next asymptotic normality result for $\widehat{\boldsymbol{\beta}}$ and $\widehat{\boldsymbol{\theta}}$ where all $m_r$ are assumed to be fixed. This means that, when compared to changes in $n$, the changes in $m_r$ are almost negligible. As commented by Yu & Ruppert (2002), the asymptotic results obtained by fixed $m_r$ lie somewhere between parametric and non-parametric modeling. This is because the values for $m_r$ and knot locations have certain but limited freedom to vary. However, this freedom brings another challenge, namely, some of the $\theta_{ru} \geq 0$ constraints may become active in the MPL estimates (i.e., some $\theta_{ru} = 0$). This fact must be integrated into the asymptotic results; otherwise, issues may arise in the asymptotic covariance matrix, particularly leading to negative variances.

Now, denote the parameter vector by $\boldsymbol{\eta} = (\boldsymbol{\theta}^\mathsf{T}, \boldsymbol{\beta}^\mathsf{T})^\mathsf{T}$, which has the dimension of $(\sum_r m_r + p) \times 1$. Let $\boldsymbol{\eta}_0 = (\boldsymbol{\theta}_0^\mathsf{T}, \boldsymbol{\beta}_0^\mathsf{T})^\mathsf{T}$ denote the true value of $\boldsymbol{\eta}$.

The same as in the previous chapters, the results in Moore et al. (2008) are used to derive the asymptotic normality result for the constrained MPL estimates developed in this chapter. It requires a matrix $\mathbf{U}$ to handle the active constraints. To simplify the discussion, we assume, without loss of generality, that the first $q$ elements of $\boldsymbol{\theta}$ satisfy $\theta_{ru} = 0$ and therefore are active.

In this context, we define

$$\mathbf{U} = \left[\mathbf{0}_{(p+m-q)\times q}, \mathbf{I}_{(p+m-q)\times(p+m-q)}\right]^\mathsf{T}, \tag{5.27}$$

where $m = \sum_r m_r$. This matrix $\mathbf{U}$ satisfies $\mathbf{U}^\mathsf{T}\mathbf{U} = \mathbf{I}_{(p+m-q)\times(p+m-q)}$. For other cases of active constraints, $\mathbf{U}$ can be defined similarly.

It is important to note that, with this choice of $\mathbf{U}$ given in equation (5.27), the matrix $\mathbf{U}^\mathsf{T}\mathbf{F}(\boldsymbol{\eta})\mathbf{U}$ that will be mentioned in Theorem 5.2 is equivalent to an operation that deletes the first $q$ rows and first $q$ columns of the matrix $\mathbf{F}(\boldsymbol{\eta})$.

We will next define Assumptions B5.1 to B5.4, which are needed for Theorem 5.2. These assumptions are identical to B2.1 to B2.4, but with $l(\boldsymbol{\eta})$ replaced by the log-likelihood defined in (5.9) and $\lambda J(\boldsymbol{\theta})$ by $\sum_r \lambda_r J_r(\boldsymbol{\theta}_r)$.

- B5.1 Assume random vectors $\boldsymbol{W}_i = (T_i^L, T_i^R, \mathbf{x}_i^\mathsf{T}, z_i^*)^\mathsf{T}$, $i = 1, \ldots, n$, are independently and identically distributed, and the distribution of $\mathbf{x}_i$ is independent of the parameter vector $\boldsymbol{\eta}$.
- B5.2 For the log-likelihood defined in (5.9), assume $E[n^{-1}l(\boldsymbol{\eta})]$ exists and has a unique maximum at the true parameter value $\boldsymbol{\eta}_0$, where $\boldsymbol{\eta}_0 \in \Omega$, with $\Omega$ being the parameter set for $\boldsymbol{\tau}$. Assume $\Omega$ is a compact subspace in $R^{p+m}$, where $m = \sum_r m_r$.
- B5.3 Assume $l(\boldsymbol{\eta})$ has a finite upper bound, $l(\boldsymbol{\eta})$ is twice continuously differentiable in a neighbourhood of $\boldsymbol{\eta}_0$ and the matrix

$$n^{-1}E\left(-\frac{\partial^2 l(\boldsymbol{\eta})}{\partial\boldsymbol{\eta}\partial\boldsymbol{\eta}^\mathsf{T}}\right)$$

exists and is bounded.
- B5.4 The penalty functions $J_r(\boldsymbol{\theta})$, where $r = 1, \ldots, s$, are bounded and twice continuously differentiable with respect to $\boldsymbol{\theta}$.

**Theorem 5.2.** Assume Assumptions B5.1 to B5.4 hold. Assume $n_r \to \infty$ as $n \to \infty$. Let $\mu_{rn} = \lambda_r/n_r$, which represents the scaled smoothing parameter. Assume $\mu_{rn} = o(n_r^{-1/2})$ for all $r$. For the $\boldsymbol{\theta} \geq 0$ constrains, suppose there exist $q$ $(< m)$ active constraints. Let matrix $\mathbf{U}$ be defined similarly as in (5.27). Let $\mathbf{F}(\boldsymbol{\eta}) = -E(n^{-1}\partial^2 l/\partial\boldsymbol{\eta}\partial\boldsymbol{\eta}^\mathsf{T})$. Then, when $n \to \infty$,

(1) The constrained MPL estimate $\widehat{\boldsymbol{\eta}}$ is consistent for $\boldsymbol{\eta}_0$, and

(2) $\sqrt{n}(\widehat{\boldsymbol{\eta}} - \boldsymbol{\eta}_0)$ converges in distribution to a multivariate normal distribution with mean $\mathbf{0}_{(p+m)\times 1}$ and covariance matrix $\widetilde{\mathbf{F}}(\boldsymbol{\eta}_0)^{-1}\mathbf{F}(\boldsymbol{\eta}_0)\widetilde{\mathbf{F}}(\boldsymbol{\eta}_0)^{-1}$, where matrix $\widetilde{\mathbf{F}}(\boldsymbol{\eta})^{-1}$ is given by: $\widetilde{\mathbf{F}}(\boldsymbol{\eta})^{-1} = \mathbf{U}(\mathbf{U}^\mathsf{T}\mathbf{F}(\boldsymbol{\eta})\mathbf{U})^{-1}\mathbf{U}^\mathsf{T}$.

□

A proof to this theorem can be obtained in a similar way to the proof of Theorem 2.5 in Appendix A, and is therefore omitted here.

Note that $\boldsymbol{\eta}_0$ is generally unavailable, and it can be replaced by $\widehat{\boldsymbol{\eta}}$. On the other hand, computation of the expected information matrix $\mathbf{F}(\boldsymbol{\eta})$ can be challenging, and it can be simply replaced by the negative Hessian matrix.

In practice, since $n$ is always finite, a large sample approximate distribution is more appropriate, where the penalty terms should remain in the covariance matrix. We modify the above asymptotic covariance result Theorem 5.2 assuming $n$ is finite. In this context, the distribution for $\widehat{\boldsymbol{\eta}}$ is approximately multivariate normal but its covariance matrix becomes

$$\widehat{\mathrm{Var}}(\widehat{\boldsymbol{\eta}}) = -\mathbf{A}(\widehat{\boldsymbol{\eta}})^{-1}\frac{\partial^2 l(\widehat{\boldsymbol{\eta}})}{\partial\boldsymbol{\eta}\boldsymbol{\eta}^\mathsf{T}}\mathbf{A}(\widehat{\boldsymbol{\eta}})^{-1}, \tag{5.28}$$

where

$$\mathbf{B}(\widehat{\boldsymbol{\eta}})^{-1} = \mathbf{U}\left(\mathbf{U}^\mathsf{T}\left(-\frac{\partial^2 l(\widehat{\boldsymbol{\eta}})}{\partial\boldsymbol{\eta}\partial\boldsymbol{\eta}^\mathsf{T}} + \sum_r \lambda_r \frac{\partial^2 J_r(\widehat{\boldsymbol{\theta}}_r)}{\partial\boldsymbol{\eta}\partial\boldsymbol{\eta}^\mathsf{T}}\right)\mathbf{U}\right)^{-1}\mathbf{U}^\mathsf{T}.$$

Here, $\partial^2 J_r(\boldsymbol{\theta}_r)/\partial\boldsymbol{\eta}\partial\boldsymbol{\eta}^\mathsf{T}$ is a matrix of dimension $(p+m)\times(p+m)$, and this matrix is zero everywhere except in the rows and columns corresponding to $\boldsymbol{\theta}_r$, where this component is given by $\partial^2 J_r(\boldsymbol{\theta}_r)/\partial\boldsymbol{\theta}_r\partial\boldsymbol{\theta}_r^\mathsf{T}$.

## 5.8 Estimation of smoothing parameters

For smoothing parameters $\lambda_r$, where $r = 1, \ldots, s$, their estimation can be achieved using marginal likelihood functions, similar to the smoothing parameter estimation described in Chapter 2, Section 2.10.

Providing an accurate estimation procedure for the smoothing parameters, which allows the $\lambda_r$'s to be selected automatically, is crucial for the success of the MPL approach in estimating the regression vector $\boldsymbol{\beta}$ and baseline hazards $h_{01}(t), \ldots, h_{0s}(t)$ for the stratified Cox model.

When selecting roughness penalties, then each penalty function becomes quadratic: $J_r(\boldsymbol{\theta}_r) = \boldsymbol{\theta}_r^\mathsf{T} \mathbf{R}_r \boldsymbol{\theta}_r$; see Section 2.10. Thus, the penalty $\lambda_r J_r(\boldsymbol{\theta}_r)$ can be viewed as a log-normal prior, where the normal distribution is $N(\mathbf{0}_{m_r \times 1}, \sigma_r^2 \mathbf{R}_r^{-1})$ with $\sigma_r^2 = 1/(2\lambda_r)$. The corresponding log-posterior density is then

$$l_p(\boldsymbol{\beta}, \boldsymbol{\theta}, \sigma_1^2, \ldots, \sigma_s^2) = -\frac{1}{2} \sum_{r=1}^{s} m_r \log \sigma_r^2 + l(\boldsymbol{\beta}, \boldsymbol{\theta}) - \sum_{r=1}^{s} \frac{1}{2\sigma_r^2} \boldsymbol{\theta}_r^\mathsf{T} \mathbf{R}_r \boldsymbol{\theta}_r. \quad (5.29)$$

Let $\Phi(\boldsymbol{\beta}, \boldsymbol{\theta}, \sigma_1^2, \ldots, \sigma_s^2) = l(\boldsymbol{\beta}, \boldsymbol{\theta}) - \sum_{r=1}^{s} \boldsymbol{\theta}_r^\mathsf{T} \mathbf{R}_r \boldsymbol{\theta}_r / (2\sigma_r^2)$. The log marginal likelihood for $\sigma_1^2, \ldots, \sigma_s^2$, after integrating out $\boldsymbol{\beta}$ and $\boldsymbol{\theta}$ from $l_p(\boldsymbol{\beta}, \boldsymbol{\theta}, \sigma_1^2, \ldots, \sigma_s^2)$, is:

$$l_m(\sigma_1^2, \ldots, \sigma_s^2) = -\frac{1}{2} \sum_{r=1}^{s} m_r \log \sigma_r^2 + \log \int e^{\Phi(\boldsymbol{\beta}, \boldsymbol{\theta}, \sigma_1^2, \ldots, \sigma_s^2)} d\boldsymbol{\beta} d\boldsymbol{\theta}. \quad (5.30)$$

After employing Laplace's method to approximate the multiple integral in (5.30), we can derive an approximate marginal log-likelihood $l_m$. Let $\widehat{\boldsymbol{\beta}}$ and $\widehat{\boldsymbol{\theta}}$ denote respectively the $\boldsymbol{\beta}$ and $\boldsymbol{\theta}$ values maximizing $\Phi(\boldsymbol{\beta}, \boldsymbol{\theta})$ with fixed $\sigma_1^2, \ldots, \sigma_s^2$ values and let $\boldsymbol{\eta} = (\boldsymbol{\beta}^\mathsf{T}, \boldsymbol{\theta}^\mathsf{T})^\mathsf{T}$. Then, using Laplace's approximation and omitting the terms independent of all $\sigma_r^2$, we have

$$\begin{aligned} l_m(\sigma_1^2, \ldots, \sigma_s^2) \propto & -\frac{1}{2} \sum_{r=1}^{s} m_r \log \sigma_r^2 + \Phi(\widehat{\boldsymbol{\eta}}, \sigma_1^2, \ldots, \sigma_s^2) \\ & -\frac{1}{2} \log \left| -\frac{\partial^2 \Phi(\widehat{\boldsymbol{\eta}}, \sigma_1^2, \ldots, \sigma_s^2)}{\partial \boldsymbol{\eta} \partial \boldsymbol{\eta}^\mathsf{T}} \right|. \end{aligned} \quad (5.31)$$

Maximization of $l_m(\sigma_1^2, \ldots, \sigma_s^2)$ in (5.31) can be attained by solving, for $r = 1, \ldots, s$,

$$\begin{aligned} \frac{\partial l_m}{\partial \sigma_r^2} = & -\frac{m_r}{2\sigma_r^2} + \frac{1}{2\sigma_r^4} \boldsymbol{\theta}_r^\mathsf{T} \mathbf{R}_r \boldsymbol{\theta}_r \\ & + \frac{1}{2\sigma_r^2} \mathrm{tr} \left\{ \left( -\frac{\partial^2 \Phi(\widehat{\boldsymbol{\eta}}, \sigma_1^2, \ldots, \sigma_s^2)}{\partial \boldsymbol{\eta} \partial \boldsymbol{\eta}^\mathsf{T}} \right)^{-1} \mathbf{Q_r}(\sigma_r^2) \right\} = 0, \end{aligned}$$

where

$$\mathbf{Q_r}(\sigma_r^2) = \begin{pmatrix} 0 & 0 & \cdots & 0 & \cdots & 0 \\ 0 & 0 & \cdots & 0 & \cdots & 0 \\ \vdots & \vdots & \ddots & \vdots & \vdots & \vdots \\ 0 & 0 & \cdots & \frac{1}{\sigma_r^2} \mathbf{R}_r & \cdots & 0 \\ \vdots & \vdots & \vdots & \vdots & \ddots & \vdots \\ 0 & 0 & \cdots & 0 & \cdots & 0 \end{pmatrix}.$$

Clearly, the solutions of the $\sigma_r^2$'s satisfy

$$\sigma_r^2 = \frac{\widehat{\boldsymbol{\theta}}_r^{\mathsf{T}} \mathbf{R}_r \widehat{\boldsymbol{\theta}}_r}{m_r - \widehat{\nu}_r}, \tag{5.32}$$

where $\widehat{\nu}_r$ is given by

$$\widehat{\nu}_r = \operatorname{tr}\left\{ - \left( \frac{\partial^2 \Phi(\widehat{\boldsymbol{\eta}}, \widehat{\sigma}_1^2, \ldots, \widehat{\sigma}_s^2)}{\partial \boldsymbol{\eta} \partial \boldsymbol{\eta}^{\mathsf{T}}} \right)^{-1} \mathbf{Q}_{\mathrm{r}}(\widehat{\sigma}_r^2) \right\}.$$

The $\widehat{\sigma}_r^2$, $r = 1, \ldots, s$ on the right-hand side of (5.32) denote the estimates of $\sigma_r^2$'s from the last iteration loop; see below.

Note that both $\boldsymbol{\beta}$ and $\boldsymbol{\theta}$ depend on all $\sigma_r^2$, which means that their estimation requires different iterative procedures. Specifically, our algorithm involves inner and outer iterations.

**Inner iterations:** In each round of inner-outer iterations, $\boldsymbol{\beta}$ and $\boldsymbol{\theta}$ are first updated in inner iterations using the Newton-MI algorithm described above with $\sigma_r^2$ (or equivalently $\lambda_r$) fixed at its current value.

**Outer iterations:** Then, with $\boldsymbol{\beta}$ and $\boldsymbol{\theta}$ fixed at their current estimates, all $\sigma_r^2$ are updated using (5.32) during this outer iteration process.

One round of inner and outer iterations forms "one iteration loop". This process continues until the degrees-of-freedom $\widehat{\nu}_r$ are stabilized, which means that the differences between their values in consecutive iterations are less than 1 or 0.5. The simulation study reported in the next section reveals that this procedure converges quickly, and we adopted the M-splines basis functions in this study.

## 5.9 More R examples

In this section, we will provide two more R examples for stratified Cox models. The first example presents a simulation study designed to evaluate the full likelihood-based estimation method discussed in this chapter. The second example considers a simulated melanoma dataset, which is a simulated version of a real melanoma dataset.

**Example 5.3** (A simple simulation example).
In this example, we design a simple simulation study to compare the estimates obtained from stratified Cox models when interval censoring is present, using both the MPL method and the partial likelihood method. In the latter method, left- and interval-censored data are replaced by the mid-point of the censoring interval. The R code for this example includes the following steps:

(i) Data are generated from stratified Cox models, where the baseline hazards are:

$$h_{01}(t) = 3t^2$$

and

$$h_{02}(t) = t^{-2/3}/3,$$

representing the hazards of two Weibull distributions.
(ii) The covariates for the model are: $X_1 \sim N(1,1)$ and $X_2 \sim \text{Binomial}(1, 0.5)$.
(iii) Based on 200 repetitions, we calculate the following statistics: average bias, average asymptotic standard errors (computed using formula (5.28)), Monte Carlo standard errors and coverage probabilities determined using the Monte Carlo standard errors. These results are reported in Table 5.1.

(iv) Figure 5.2 displays plots of the average baseline hazard estimates for the two strata, alongside their true baseline hazards and 95% pointwise confidence intervals.

```
# Write a function to generate
# interval censored survival data
# from two strata
gen_strat_interval_censoring <- function(nsample,
    nstrata, pi_E, prob_strata, a1,
    a2) {

    # Set true coefficients
    beta <- c(1, -0.5)

    # Generate strata and
    # covariates
    istrata <- sample(1:nstrata, size = nsample,
        prob = prob_strata, replace = T)
    X <- cbind(rbinom(nsample, 1, 0.5),
        rnorm(nsample, 1, 1))

    # Generate true event times
    y_i <- rep(0, nsample)
    for (i in 1:nsample) {
        neg.log.u <- -log(runif(1))
        mu_term <- exp(X[i, ] %*% beta)
        if (istrata[i] == 1) {
            y_i[i] <- (neg.log.u/(mu_term))^(1/3)
        } else {
            y_i[i] <- (2 * (neg.log.u)/(mu_term))^(1/2)
        }
    }

    # Generate interval censoring
    # times
```

TABLE 5.1: Stratified Cox model fitted using MPL compared to PL under 30% and 70% censoring; both scenarios have $n = 200$ and an equal 0.5 probability of any individual belonging to each of the two strata.

| | | $\pi^E = 0.7$ | | | $\pi^E = 0.3$ | | |
|---|---|---|---|---|---|---|---|
| | | Bias | SE | CP | Bias | SE | CP |
| $\beta_1$ | MPL | 0.019 | 0.176 (0.152) | 0.97 | 0.007 | 0.204 (0.168) | 0.98 |
| | PL | 0.004 | 0.173 (0.162) | 0.97 | -0.046 | 0.193 (0.174) | 0.96 |
| $\beta_2$ | MPL | 0.012 | 0.082 (0.082) | 0.95 | 0.030 | 0.089 (0.084) | 0.94 |
| | PL | 0.004 | 0.087 (0.087) | 0.96 | 0.036 | 0.097 (0.086) | 0.96 |

```
TL <- TR <- rep(0, nsample)

for (i in 1:nsample) {
    U_E <- runif(1)
    U_L <- runif(1, 0, 1)
    U_R <- runif(1, U_L, 1)

    strata_i <- istrata[i]
    time <- y_i[i]

    event <- as.numeric(U_E < pi_E[strata_i])
    interval <- as.numeric(a1[strata_i] *
        U_L <= time & time <= a2[strata_i] *
        U_R & U_E >= pi_E[strata_i])
    right <- as.numeric(a2[strata_i] *
        U_R < time & U_E >= pi_E[strata_i])
    left <- as.numeric(time < a1[strata_i] *
        U_L & U_E >= pi_E[strata_i])

    TL[i] <- (time) * event + (a1[strata_i] *
        U_L) * interval + (a2[strata_i] *
        U_R) * right + (0) * left
    TR[i] <- (time) * event + (a2[strata_i] *
        U_R) * interval + (0) *
        right + (a1[strata_i] *
        U_L) * left

    if (TR[i] == 0) {
        TR[i] <- Inf
    }
    if (TL[i] == 0) {
```

```
            TL[i] <- -Inf
        }
    }

    data <- data.frame(c(1:nsample),
        TL, TR, X, istrata)
    return(data)
}

# Run simulations
save <- matrix(0, nrow = 200, ncol = 82)
save[, 1] <- c(1:200)

for (s in 1:200) {

    # Generate data
    data <- gen_strat_interval_censoring(200,
        2, pi_E = c(0.3, 0.3), c(0.5,
            0.5), c(0.4, 0.9), c(1.1,
            2))
    data.surv <- Surv(time = data$TL,
        time2 = data$TR, type = "interval2")

    # Save details of data
    save[s, 2] <- sum(as.numeric(data.surv[data$istrata ==
        1, 3] == 1))
    save[s, 3] <- sum(as.numeric(data.surv[data$istrata ==
        1, 3] == 0))
    save[s, 4] <- sum(as.numeric(data.surv[data$istrata ==
        1, 3] == 2))
    save[s, 5] <- sum(as.numeric(data.surv[data$istrata ==
        1, 3] == 3))

    save[s, 6] <- sum(as.numeric(data.surv[data$istrata ==
        2, 3] == 1))
    save[s, 7] <- sum(as.numeric(data.surv[data$istrata ==
        2, 3] == 0))
    save[s, 8] <- sum(as.numeric(data.surv[data$istrata ==
        2, 3] == 2))
    save[s, 9] <- sum(as.numeric(data.surv[data$istrata ==
        2, 3] == 3))

    # Fit MPL model
    try1 <- coxph_mpl(data.surv ~ data$X1 +
        data$X2, istrata = 6, data = data,
        control = coxph_mpl.control(basis = "msplines"))

    save[s, 10] <- mpl.fit$coef$Beta[1]
```

```
    save[s, 11] <- mpl.fit$coef$Beta[2]

    save[s, 12] <- mpl.fit$se$Beta$se.Eta_M2HM2[1]
    save[s, 13] <- mpl.fit$se$Beta$se.Eta_M2HM2[2]

    save[s, 14:18] <- mpl.fit$s_obj[[1]]$knots_strat$Alpha

    save[s, 26:31] <- mpl.fit$s_obj[[1]]$M_theta_m1
    save[s, 37:42] <- mpl.fit$se$Theta$se.Eta_M2HM2[1:6]

    save[s, 48:53] <- mpl.fit$s_obj[[2]]$M_theta_m1
    save[s, 59:64] <- mpl.fit$se$Theta$se.Eta_M2HM2[7:12]

    ## Fit PL model with midpoint
    ## imputation
    data$midpoint_time <- 0
    data$midpoint_time[which(data.surv[,
        3] == 0)] <- data$TL[which(data.surv[,
        3] == 0)]
    data$midpoint_time[which(data.surv[,
        3] == 1)] <- data$TL[which(data.surv[,
        3] == 1)]
    data$midpoint_time[which(data.surv[,
        3] == 3)] <- data$TL[which(data.surv[,
        3] == 3)] + (data$TR[which(data.surv[,
        3] == 3)] - data$TL[which(data.surv[,
        3] == 3)])/2

    data$midpoint_event <- 0
    data$midpoint_event[which(data.surv[,
        3] != 0)] <- 1

    data.surv.midpoint <- Surv(time = data$midpoint_time,
        event = data$midpoint_event)
    ph.fit <- coxph(data.surv.midpoint ~
        data$X1 + data$X2 + strata(data$istrata),
        data = data)

    save[s, 70] <- ph.fit$coefficients[1]
    save[s, 71] <- ph.fit$coefficients[2]

    save[s, 72] <- sqrt(diag(ph.fit$var))[1]
    save[s, 73] <- sqrt(diag(ph.fit$var))[2]

}
```

This example demonstrates both the MPL and mid-point imputation with the partial likelihood approaches for fitting stratified Cox models. It highlights that the mid-point method may perform reasonably well, especially when the censoring intervals are not wide. Readers are encouraged to conduct another simulation by modifying the provided R code in this example so that the simulated censoring intervals are wider. By doing so, one can observe that the mid-point method can be poor, potentially resulting in biased estimates when the proportion of censoring increases. □

In the next example, we will study a dataset which is simulated from a real melanoma data.

**Example 5.4** (A pseudo melanoma recurrence example).
Here, we discuss the application of a stratified Cox model to a set of simulated melanoma data concerning the time to melanoma recurrence. The original data were provided by the Melanoma Institute Australia from their prospectively maintained research database. This simulated dataset contains information on follow-up dates, recurrence status and baseline and pathological characteristics of 5330 patients diagnosed with melanoma between 2004 and 2014 who had a sentinel node biopsy performed. The simulated dataset is available on GitHub for this book.

The primary outcome of interest was the time to the first recurrence from the date of the initial melanoma diagnosis, and this recurrence time is generally interval censored. The dataset included information on demographic and clinical characteristics, such as

"Sex" (male; female),
"Age" at diagnosis (under 50 years; 50 years or over),
"Tumor ulceration" (yes; no),
"Tumor body site" (arm; head or neck; leg; trunk),
"Sentinel node status" (SNStatus: positive; negative), and
"Breslow tumor thickness" (T1; T2; T3; T4).

All of these variables have been identified as significant predictors of recurrence risk in previous research.

The proportional hazards assumption was initially tested for the covariates using mid-point imputation for recurrence times in a Cox regression model fitted with partial likelihood. The function `cox.zph()` identified variables where there was a significant relationship between the scaled Schoenfeld residuals and time. Both SNStatus and Breslow thickness failed this test, with p-values less than $10^{-3}$. The graphical checks of Schoenfeld residuals for these two variables are provided in Figure 5.3, indicating that the proportional hazards assumption is inappropriate.

Based on these results, strata were created for the interaction between SNStatus and Breslow thickness, resulting in four strata:

Breslow thickness T1 or T2 and SN negative (sample size $n_1 = 2421$);
Breslow thickness T1 or T2 and SN positive (sample size $n_2 = 315$);
Breslow thickness T3 or T4 and SN negative (sample size $n_3 = 1733$); and
Breslow thickness T3 or T4 and SN positive (sample size $n_4 = 650$).

These four strata were used to fit stratified Cox models for the melanoma recurrence data using the proposed method. Predictors in the stratified model included "Sex", "Age at diagnosis", "Tumor ulceration", and "Body site", as described above. This

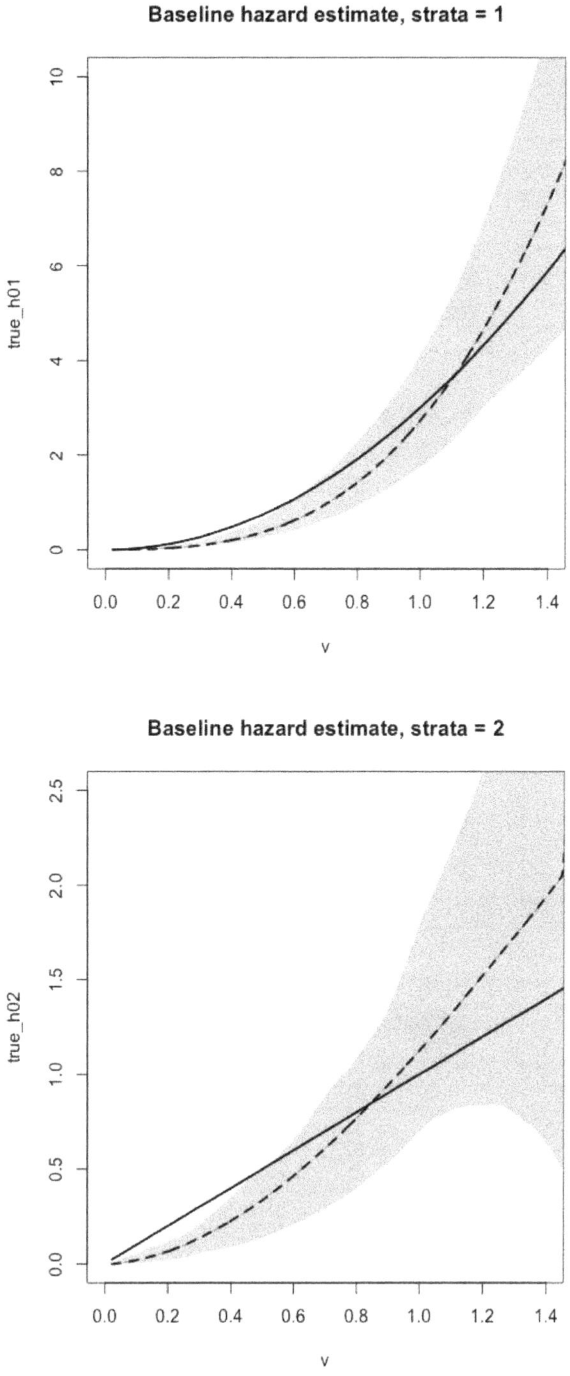


FIGURE 5.2: Estimated and true baseline hazards for two strata and their 95% pointwise confidence intervals.

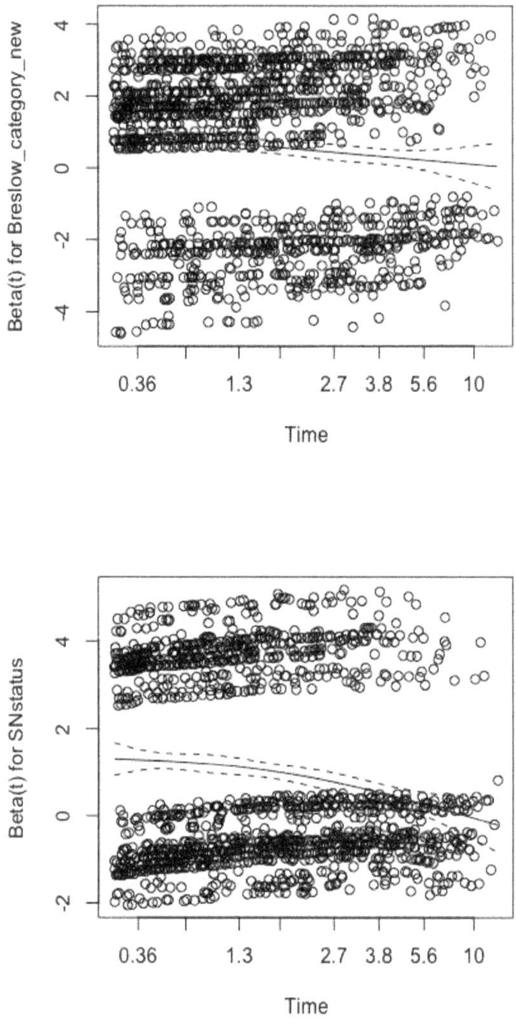


FIGURE 5.3: Schoenfeld residual plots for variables Breslow thickness and SNStatus from a Cox model for right censored data, fitted using midpoint imputation.

model was also fitted using mid-point imputation, allowing estimation via Cox's partial likelihood. The results are presented in Table 5.2. Additionally, the spline-approximated baseline survival functions for each stratum, estimated by the MPL model, are shown in Figure 5.4. The Breslow estimates of the stratified baseline survival functions obtained from the `coxph` fit are also included in this figure.

```
# Write a function to
# generate data similar to
# the melanoma data
gen_strat_IC_melanoma <- function(nsample,
    nstrata, pi_E, prob_strata,
    a1, a2) {

    # True parameter values
    beta <- c(0.25, 0.45, 0.55,
        0, 0.45, 0.45, 0.25)

    # Strata membership
    istrata <- sample(1:nstrata,
        size = nsample, prob = prob_strata,
        replace = T)

    # Shared covariates
    Body_Site = sample(1:4, size = nsample,
        prob = c(0.25, 0.25, 0.25,
            0.25), replace = T)
    X <- cbind(rbinom(nsample,
        1, 0.5), rbinom(nsample,
        1, 0.5), rbinom(nsample,
        1, 0.5), as.numeric(Body_Site ==
        1), as.numeric(Body_Site ==
        2), as.numeric(Body_Site ==
        3), as.numeric(Body_Site ==
        4))

    # Generate event times
    # from four different
    # distributions
    y_i <- rep(0, nsample)

    for (i in 1:nsample) {
        neg.log.u <- -log(runif(1))
        mu_term <- exp(X[i, ] %*%
            beta)

        if (istrata[i] == 1) {
            y_i[i] = 2 * ((neg.log.u/mu_term)^(1/1.5))
        } else if (istrata[i] ==
            2) {
            y_i[i] = ((neg.log.u/(0.5 *
                mu_term)))
        } else if (istrata[i] ==
            3) {
            y_i[i] = 1.8 * ((neg.log.u/mu_term))
        } else if (istrata[i] ==
            4) {
```

```
            y_i[i] = 2 * ((neg.log.u/mu_term)^(1/1.1))
        }
    }

    # Generate interval
    # censoring times
    TL <- TR <- rep(0, nsample)

    for (i in 1:nsample) {
        # uniform variables
        U_E <- runif(1)
        U_L <- runif(1, 0, 1)
        U_R <- runif(1, U_L, 1)

        strata_i <- istrata[i]
        time <- y_i[i]

        event <- as.numeric(U_E <
            pi_E[strata_i])
        interval <- as.numeric(a1[strata_i] *
            U_L <= time & time <=
            a2[strata_i] * U_R &
            U_E >= pi_E[strata_i])
        right <- as.numeric(a2[strata_i] *
            U_R < time & U_E >=
            pi_E[strata_i])
        left <- as.numeric(time <
            a1[strata_i] * U_L &
            U_E >= pi_E[strata_i])

        TL[i] <- (time) * event +
            (a1[strata_i] * U_L) *
                interval + (a2[strata_i] *
            U_R) * right + (0) *
            left
        TR[i] <- (time) * event +
            (a2[strata_i] * U_R) *
                interval + (0) *
            right + (a1[strata_i] *
            U_L) * left

        if (TR[i] == 0) {
            TR[i] <- Inf
        }
        if (TL[i] == 0) {
            TL[i] <- -Inf
        }
    }
    data <- data.frame(c(1:nsample),
        TL, TR, X, istrata)
    return(data)
}

# Generate a sample dataset
mel.data.sim <- gen_strat_IC_melanoma(5000,
    4, c(0, 0, 0, 0), c(0.5, 0.1,
```

```
        0.3, 0.1), c(0.8, 0.8,
        0.7, 0.7), c(1.2, 1.3,
        1.4, 1.5))

# calculate midpoint
mel.data.sim.mp <- mel.data.sim
id1 <- which(is.infinite(mel.data.sim.mp$TL))
id2 <- which(!is.infinite(mel.data.sim.mp$TR))
id3 <- which(!is.infinite(mel.data.sim.mp$TR))
id4 <- which(!is.infinite(mel.data.sim.mp$TR))
id5 <- which(!is.infinite(mel.data.sim.mp$TR))
mel.data.sim.mp$TL[id1] <- 0
mel.data.sim.mp$midpoint <- mel.data.sim.mp$TL
mel.data.sim.mp$midpoint[id2] <- mel.data.sim.mp$midpoint[id3]
+(mel.data.sim.mp$TR[id4] - mel.data.sim.mp$TL[id5])/2

# calculate event status
mel.data.sim.mp$event <- as.numeric(!is.infinite(mel.data.sim.mp$TR))

# create right censored Surv
# object
mp.mel.surv <- Surv(mel.data.sim.mp$midpoint,
    mel.data.sim.mp$event)

# use cox.zph() to check for
# violation of PH assumption
# by istrata
(cox.zph(coxph(mp.mel.surv ~ X1 +
    X2 + X3 + X5 + X6 + X7 + istrata,
    data = mel.data.sim.mp)))

# refit a stratified Cox
# model
ph.fit <- coxph(mp.mel.surv ~ X1 +
    X2 + X3 + X5 + X6 + X7 + strata(istrata),
    data = mel.data.sim.mp)
summary(ph.fit)

# fit using interval
# censoring MPL model

# create an interval censored
# Surv object
IC.data.surv <- Surv(mel.data.sim$TL,
    mel.data.sim$TR, type = "interval2")

# fit stratified model using
# MPL method
mpl.ic.fit <- coxph_mpl(IC.data.surv ~
    X1 + X2 + X3 + X5 + X6 + X7,
    istrata = 11, data = mel.data.sim,
    control = coxph_mpl.control(basis = "msplines",
        n.knots = c(8, 2)))
summary(mpl.ic.fit)
```

TABLE 5.2: Regression coefficients of the stratified Cox models for the melanoma data. Two sets of results are presented: the proposed MPL estimates and the PL estimates obtained using the middle-point imputation.

| | Partly interval-censored, MPL | | | Mid-point imputation, PL | | |
|---|---|---|---|---|---|---|
| Predictor | HR | 95% CI | $p$-value | HR | 95% CI | $p$-value |
| Sex: Male | 1.318 | 1.149, 1.515 | $< 0.001$ | 1.289 | 1.151, 1.444 | $< 0.001$ |
| Age at Diagnosis: Over 50 | 1.577 | 1.370, 1.817 | $< 0.001$ | 1.536 | 1.365, 1.728 | $< 0.001$ |
| Tumour Ulceration: Yes | 1.712 | 1.492, 1.964 | $< 0.001$ | 1.693 | 1.516, 1.890 | $< 0.001$ |
| Body Site: Head or Neck | 1.627 | 1.371, 2.040 | $< 0.001$ | 1.573 | 1.304, 1.897 | $< 0.001$ |
| Body Site: Leg | 1.596 | 1.441, 1.769 | $< 0.001$ | 1.489 | 1.250, 1.773 | $< 0.001$ |
| Body Site: Trunk | 1.331 | 1.174, 1.508 | $< 0.001$ | 1.229 | 1.036, 1.457 | 0.0177 |

In the MPL fitted model, all included covariates were statistically significant at the 5% level. Being male, over 50 years old at diagnosis, having tumour ulceration, and having a tumour on the head or neck, the leg or the trunk instead of the arm all significantly increased the risk of melanoma recurrence at any time $t$. The results were similar between the MPL and PL models, but notably the effect sizes of all covariates were smaller in the PL model than the MPL model. The estimated baseline survival functions indicate that negative SNStatus reduces the risk of recurrence, as does a Breslow thickness T1 or T2 compared to Breslow thickness T3 or T4.

The plot of stratified baseline survival functions from the MPL method in Figure 5.4 includes asymptotic 95% point-wise confidence intervals as shaded areas. In this plot, there is little overlap between these confidence intervals, suggesting that the four strata are significantly different in terms of baseline survival probability. Similar asymptotic confidence intervals are not directly available for the PL estimates. Comparing the plots, it appears that the survival probability estimates obtained from the PL decrease much more sharply in the first five years than those obtained from the MPL model, especially for the strata containing individuals with a positive SNStatus, although by 15 years the survival probability estimates are similar.

Figure 5.5 shows the predicted survival probabilities across all four strata for males and females obtained from the MPL model, along with their asymptotic 95% point-wise confidence intervals. Across all four strata, males have higher risks for melanoma recurrence across the follow-up period than females. For patients with negative SNStatus and Breslow thickness T1 or T2, the probability of being recurrence free at 10 years is 0.73 (95% CI: 0.70, 0.75) for males, and 0.78 (95% CI: 0.76, 0.81) for females. For patients with negative SNStatus but Breslow thickness T3 or T4, males have probability of 0.56 (95% 0.52, 0.59) for being recurrence free at 10 years compared to female's 0.64 (95% CI: 0.60, 0.68). A male patient with a positive SNStatus and a Breslow thickness of T1 or T2 has a probability of no recurrence at 10 years is 0.44 (95% CI: 0.37, 0.50), while a female patient in this strata has a probability of 0.53 (95% CI: 0.47, 0.59). Finally, if a patient has a positive SNStatus and a Breslow thickness of T3 or T4 and is male, his 10-year recurrence-free probability is 0.27 (95% CI: 0.22, 0.32), while for a female this probability is 0.37 (95% CI: 0.32, 0.42).

□

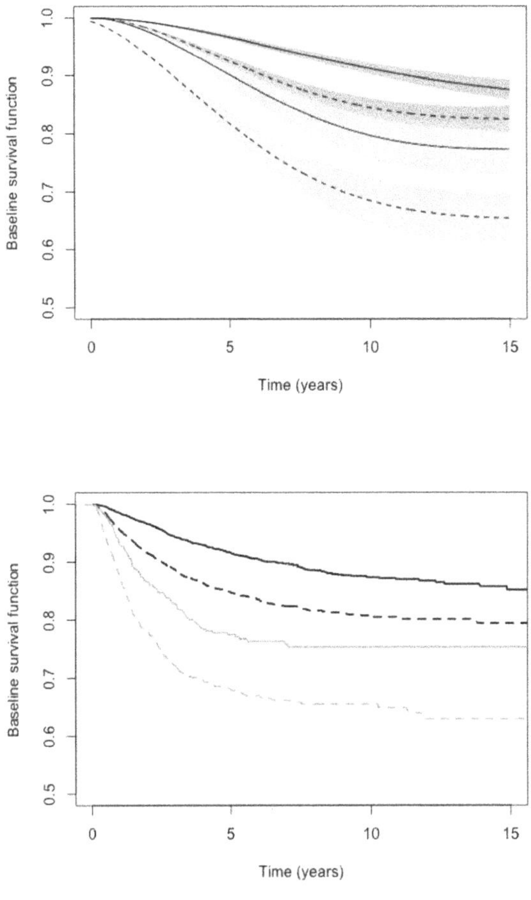


FIGURE 5.4: Baseline survival function estimates for the four strata, from the MPL (top) and PL (bottom) models. Dark grey colouring indicates a negative SNStatus while light grey indicates positive; a solid line indicates a Breslow thickness T1 or T2 while a dashed line indicates Breslow thickness T3 or T4.

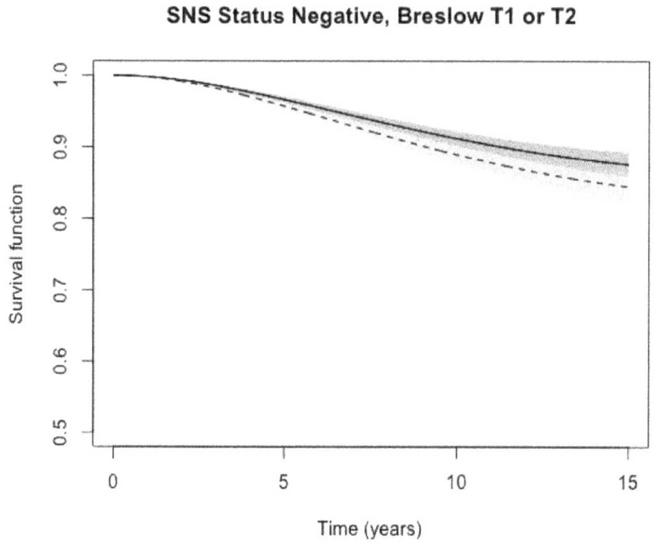
SNS Status Negative, Breslow T1 or T2
Survival function
0.5
0.6
0.7
0.8
0.9
1.0
0
5
10
15
Time (years)

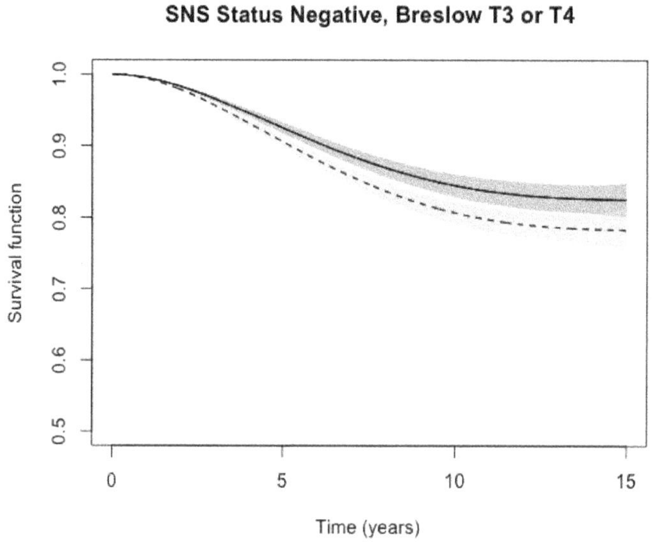
SNS Status Negative, Breslow T3 or T4
Survival function
0.5
0.6
0.7
0.8
0.9
1.0
0
5
10
15
Time (years)

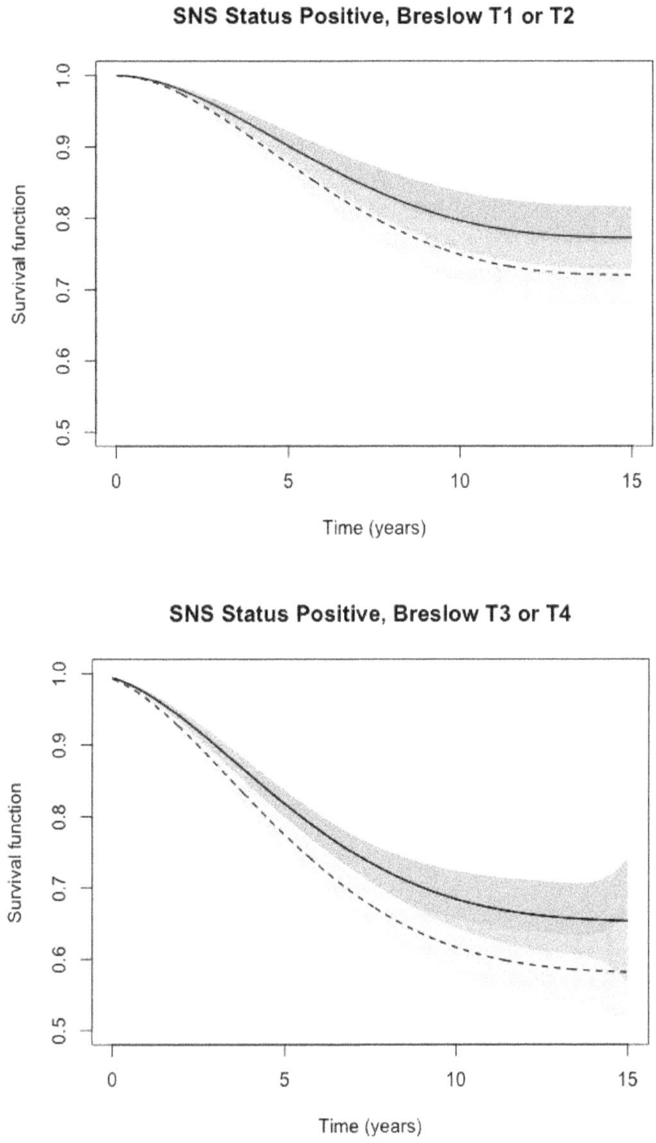


FIGURE 5.5: Baseline survival function estimates for the four strata, for females (dark grey, solid line) and males (light grey, dashed line).

## 5.10 Summary

In survival data analysis, the stratified Cox model becomes an option when the proportional hazards assumption of the Cox model does not hold for certain covariates. For the stratified Cox model, the maximum partial likelihood method is currently the most accepted method for estimation of regression coefficients when only right censoring times are included in the observed event times. If there exist interval-censored event times, the partial likelihood method is problematic if implemented directly. Moreover, the partial likelihood approach does not provide direct estimates to the baseline hazards, making prediction inferences more difficult. In this paper, we consider general interval censoring and discuss a penalized likelihood method for fitting the stratified Cox model where regression coefficients and baseline hazard are estimated simultaneously. The penalty functions are used to smooth the baseline hazard estimates. We allow partly interval-censored survival data which include event times and left, right and interval censoring times. We also provide asymptotic results on the estimates so that parameters of interest can be inferred. Effectiveness of our method is demonstrated through a simulation study.

# 6

# *Cox models with time-varying covariates under right censoring*

Incorporating time-varying covariates into a Cox model represents an important extension of the traditional Cox model. This extension is particularly valuable in fields like medical research, where patient data often include longitudinal covariates that varies with time, such as treatment history (e.g., dose changes). Including time-varying covariates in a Cox model can significantly enhance its predictive accuracy and yield more informative insights. Another important application of time-varying covariates is their frequent use in addressing the non-proportional hazards issue.

When a Cox model contains time-varying covariates, the process of likelihood-based estimation becomes more complex. This chapter will primarily focus on the computational aspects of MPL estimation for Cox models with time-varying covariates.

## 6.1 Introduction

In the previous chapters, we mainly considered Cox regression models for survival analysis. These models are used when dealing with survival data obtained from subjects who are followed for a specific period, with the anticipation that an event of interest will occur. The random variable under consideration is the time elapsed from the beginning of the follow-up period until the event of interest occurs. However, this follow-up time may not be observed exactly due to censoring. In Chapter 2, we discussed various types of censoring in addition to event times, including right censoring, left censoring and interval censoring. This combination of even times and different types of censoring times is referred to as partly interval-censored survival times.

Chapter 2 explains a full likelihood-based approach for fitting Cox models under partly interval censoring when covariates are time-fixed. Typically, time-fixed covariate values are collected at the baseline time and remain unchanged throughout the follow-up period. Chapters 3 and 5 extend the penalized likelihood approach presented in Chapter 2 to truncated and stratified Cox models, respectively, both with time-fixed covariates.

DOI: 10.1201/9781351109710-6

In practice, it often occurs that there are time-varying (or time-dependent) covariates whose values change over time. A more detailed description of time-varying covariates will be provided soon in this chapter. Due to technical complications, the discussions in this chapter will focus solely on right-censored survival data. For the estimation of the Cox model with time-varying covariates under more general partly interval censoring, the readers are referred to Webb & Ma (2023).

Time-fixed covariates typically have values at the baseline (e.g., study entry time 0) when the follow-up starts. In contrast, values of a time-varying covariate change over time. A time-varying covariate can be either a continuous function of time $t$, such as $\log(t)$, or a piecewise constant function of $t$. In the latter case, a vector of covariate values, collected at different time points (usually intermittent inspection time points), represents the time-varying covariate.

For example, when collecting data from patients who have cardiovascular disease, their cholesterol levels might be measured when visiting the doctor. Clearly, this "cholesterol level" covariate is time-varying as its value changes with time and is piecewise constant.

Next, we explain a data example of time-varying covariates. This example was studied in Crowley & Hu (1977).

**Example 6.1** (Stanford Heart Transplant data Crowley & Hu (1977)).
In this example, we briefly introduce the Stanford Heart Transplant study. The aim of this study is to investigate the effects of a number of covariates, including the heart transplant indicator variable which is time dependent, to see if heart transplant prolongs survival for patients with heart disease.

For patient $i$, let $z_i(t)$ be an indicator if this patient had received a heart transplant before time $t$ (including $t$). Let $b_i$ be the heart transplant time for patient $i$, then

$$z_i(t) = \begin{cases} 0, & t < b_i; \\ 1, & t \geq b_i. \end{cases} \tag{6.1}$$

This $z_i(t)$ is a step-function (therefore piecewise constant) with a change point at $b_i$. If patient $i$ did not receive a transplant before the follow-up time $y_i$, we had $b_i > y_i$ in this case, or even $b_i = \infty$. □

When there are time-varying covariates, it can be advantageous to incorporate them into the Cox regression model being investigated. The paper by Crowley & Hu (1977) is perhaps one of the first to introduce time-varying covariates into a Cox model. A Cox model that includes time-varying covariates is commonly known as an "extended Cox model", for obvious reasons.

Including time-varying covariates in a Cox model generally does not pose computational challenges when using the maximum partial likelihood method. However, for estimation approaches based on the full likelihood, the presence of time-varying covariates can lead to computational difficulties. In this chapter, we will explain these challenges and develop a computationally feasible algorithm to handle them.

In the context of the Cox model with time-varying covariates, a limitation of the partial likelihood method is again that it can only handle survival data with right censoring. However, for the likelihood-based method discussed in this chapter, it can be readily extended to handle partly interval censoring as well; see Webb & Ma (2023) for more details.

In this chapter, to simplify our discussions, we assume that there are no measurement errors in the time-varying covariates. However, it is worth noting that some time-varying covariates may indeed contain measurement errors. Handling measurement errors in covariates is possible through joint modeling of survival data and longitudinal time-varying covariates data. For more details on this approach, please refer to, for example, Rizopoulos (2012).

Next, we set up a general extended Cox model which includes both time-fixed and time-varying covariates.

For individual $i$, where $i = 1, \dots, n$, let $\mathbf{x}_i$ be a vector of observations of time-fixed covariates as before. To incorporate time-varying covariates into a Cox model, we introduce $\mathbf{z}_i(t) = (z_{i1}(t), \dots, z_{iq}(t))^\top$, a vector of $q$ time-varying covariates associated with time point $t$. We define the time-varying covariate process $\widetilde{\mathbf{z}}_i(t)$ for individual $i$ as follows:

$$\widetilde{\mathbf{z}}_i(t) = \{\mathbf{z}_i(s) : 0 \leq s \leq t\}. \tag{6.2}$$

Clearly, the process $\widetilde{\mathbf{z}}_i(t)$ represents the history of $\mathbf{z}_i(s)$ up to time point $t$. The extended Cox model considered in this chapter is defined as follows:

$$h(t|\mathbf{x}_i, \widetilde{\mathbf{z}}_i(t)) = h_0(t)e^{\mathbf{x}_i^\top \boldsymbol{\beta} + \mathbf{z}_i(t)^\top \boldsymbol{\gamma}}, \tag{6.3}$$

where $\boldsymbol{\beta}$ and $\boldsymbol{\gamma}$ are vectors representing regression coefficients (both are time-fixed) that require estimation, and $h_0(t) \geq 0$ is the non-parametric baseline hazard function that will also be estimated alongside $\boldsymbol{\beta}$ and $\boldsymbol{\gamma}$. When there is no confusion, we will denote $h(t|\mathbf{x}_i, \widetilde{\mathbf{z}}_i(t))$ by $h_i(t)$ in the following discussions.

If the conventional partial likelihood approach is adopted (only for right-censored survival data) for the *extended Cox model* (6.3), $h_0(t)$ is not estimated directly with $\boldsymbol{\beta}$ and $\boldsymbol{\gamma}$. While partial likelihood methods enjoy advantages such as faster convergence speed and problem-free asymptotic covariance matrices, they also have some limitations, including:

1. Potentially inaccurate estimates for regression coefficients of time-varying covariates in small samples with heavy censoring (Heinze & Dunkler 2008).
2. Inability to directly estimate the baseline hazard, requiring a separate estimation via the Breslow method.
3. Difficulty in extending this method to handle more general partly interval censoring scenarios.

Estimating only regression coefficients is not an issue when the primary goal of the analysis is to draw inferences about covariate effects or calculate hazard ratios. However, if the analysis also involves inferences about other quantities, such as survival probabilities, an estimate of the baseline hazard becomes necessary. The Breslow method offers a secondary estimation of the baseline hazard, requiring the estimated regression coefficients from the partial likelihood as inputs. The resulting baseline hazard estimates are often highly volatile (e.g., Hosmer et al. (2008), Thackham & Ma (2020)), and they frequently require additional smoothing.

The full likelihood-based MPL method largely avoids these limitations of partial likelihood methods, but it typically suffers from relatively slow convergence and the active constraints issue caused by estimating the baseline hazard, where the latter issue can possibly lead to unpleasant negative asymptotic variances. For the full likelihood method we will discuss in this chapter, an approximate $h_0(t)$ is estimated together with $\beta$ and $\gamma$, where an approximation to $h_0(t)$ is obtained using basis functions similar to Chapters 2–5.

Clearly, for model (6.3), the proportional hazards assumption for a Cox model is no longer needed as the hazard ratio corresponding to two sets of covariates is now a function of time $t$. For this reason, including a covariate involving time $t$ is a widely adopted strategy for testing the proportional hazards assumption; see Cox & Oakes (1984). Also, Chapter 6 of Kleinbaum & Klein (2005) provides an example of this kind of test, where a Cox model contains a single covariate "`Sex`". By including another covariate, "`Sex * t`", one can test the proportional hazards assumption by performing a significance test on the regression coefficient of "`Sex * t`".

In this chapter, we consider time-varying covariates as *discrete* functions (or piecewise constant) of time $t$. This is because, in practice, the values of time-varying covariates are often obtained at a finite set of intermittent inspection time points. For example, when a patient visits a doctor, measurements such as blood pressure and cholesterol levels are taken. These values remain unchanged until the next visit. Furthermore, we assume that there are *no measurement errors* in the time-varying covariates, as previously explained. While we will mainly focus on discrete time-varying covariates in the following discussions, the methods discussed in Sections 6.3 and 6.4 can be readily extended to continuous time-varying covariates.

There are two different types of time-varying covariates (e.g., Kalbfleisch & Prentice (2002)): *internal* time-varying covariates and *external* time-varying covariates. Internal time-varying covariates have values that depend on the event time of interest. For example, if the event time of interest is an individual's time of death, variables like blood pressure, cholesterol level, or general biomarkers become unavailable after the individual's time of death. Therefore, these variables are considered *internal* time-varying covariates. In contrast, external time-varying covariates are independent of the event time. An example of external time-varying covariates is daily pollution recordings made during the follow-up period leading to an event.

Internal time-varying covariates are not only possibly dependent on the event of interest, but also some of them may be influenced by other time-fixed or time-dependent covariates. For example, when assessing the effect of a treatment (for example a surgery), where a treatment assignment is time-varying, and if an internal time-varying covariate (such as a medicine dose level) has values after the treatment, then it is possible that the treatment may impact the event endpoint through this internal covariate. Such relationships, when they exist, need to be carefully modeled, and this is usually achieved through joint modeling of the event time and the internal time-varying covariates (Rizopoulos 2012).

For internal time-varying covariates, it is possible that their values become unavailable at, and after, the event (such as death) time. In this case, we can impute the missing time-varying covariates values using, for example, an imputation based on the "last value carry forward" strategy. Details of this assumption will be explained in Section 6.3.

In this chapter, we discuss how to obtain full likelihood-based estimates of the regression coefficients and the baseline hazard when time-varying covariates are present. Similar to previous chapters, we adopt a penalty function to regularize the baseline hazard estimate, and the estimation method used is maximum penalized likelihood. This method effectively addresses the shortcomings of the partial likelihood method.

Despite the long history of semi-parametric Cox models with time-varying covariates in survival analysis, the literature on full likelihood estimation for these models is relatively new. Some existing methods include: Wong et al. (2017), which developed a maximum likelihood approach for a piecewise constant proportional hazard model with time-varying covariates. However, their method is limited to special time-varying covariates where all covariates change from 0 to 1 at a preselected time point, and this time point is identical for all individuals in the dataset. On the other hand, the method by Thackham & Ma (2020) is more general, as it can handle both continuous and discrete time-dependent covariates, and the baseline hazard can be either continuous or discrete depending on the basis functions selected to approximate it. Moreover, the method of Webb & Ma (2023) deals with more general partly interval censoring when time-varying covariates are included in the extended Cox model.

In this chapter, we first summarize the partial likelihood method and provide a few R examples in Section 6.2. Then detailed explanations of the penalized likelihood estimation are given in Sections 6.3 and 6.4. Asymptotic results and other inferences are contained in Section 6.5. A smoothing parameter estimation process is discussed in Section 6.7. For the discussed MPL method in this chapter, a simulation R example and R examples on real data applications are exhibited in Section 6.8. A summary of this chapter is given in Section 6.11.

## 6.2 Partial likelihood approach under right censoring

Similar to the Cox model with time-fixed covariates, when observed survival times include only right censoring, the estimation of regression coefficients for time-varying covariates is commonly achieved using the partial likelihood method.

### 6.2.1 Partial likelihood

The partial likelihood method, when time-varying covariates are present, is almost identical to the partial likelihood method with only time-fixed covariates. More specifically, both methods involve a product of conditional probabilities at the observed event times, with each of these probabilities determined from the risk set at the corresponding event time. However, for the partial likelihood of the Cox model with time-varying covariates, the conditional probability at any event time is computed involving the values of the time-varying covariates for the individuals who experience the event at that time or for the individuals who are in the risk set at that time. This implies that the same individual who appears in different risk sets will contribute varying covariate values, a distinction from the scenario with time-fixed covariates.

To begin, we introduce some notations necessary for this chapter. Similar to Chapters 2 and 5, we let $Y_i$ represent the event time for subject $i$. In this chapter, we exclusively consider right censoring, and we use $C_i$ to denote the corresponding non-informative right censoring time. The observable survival time for individual $i$ is then defined as $T_i = \min(Y_i, C_i)$ for $i = 1, \ldots, n$. Observed survival times are denoted as $t_i$, which include both event and censoring times.

It is customary to represent $t_i$ as $(t_i, \delta_i)$, where $\delta_i = 1$ when an event of interest is observed for subject $i$, and $\delta_i = 0$ when the event time is right-censored. Consequently, observations from different subjects can be expressed as

$$(t_i, \delta_i, \mathbf{x}_i, \widetilde{\mathbf{z}}_i(t_i)),$$

where $i = 1, \ldots, n$.

Figure 6.1 illustrates an example with a total of 5 individuals, where 3 experienced the event (death) (marked by dots), while 2 were right-censored (marked by crosses) with one at $t_2$ and the other at $t_5$, which is the end-of-study time. The risk set is different at different event times. For example, at $t_1^-$, the risk set is: $R_1 = \{t_1, t_2, t_3, t_4, t_5\}$, and at $t_4^-$ the risk set is: $R_4 = \{t_4, t_5\}$. At time $t_1$, the conditional likelihood used in the product formula to compute the partial likelihood involves the time-varying covariates at $t_1$ for all the individuals in $R_1$, i.e. all the time-varying covariates are computed using time $t_1$. At time $t_4$, the time-varying covariates of individuals in $R_4$ are all set to their values at $t_4$.

The partial likelihood function that can be used to estimate the regression coefficient vectors $\boldsymbol{\beta}$ and $\boldsymbol{\gamma}$, the regression coefficient vectors for time-fixed and time-varying covariates respectively, is

$$\widetilde{L}(\boldsymbol{\beta}, \boldsymbol{\gamma}) = \prod_{i=1}^{n} \left\{ \frac{\exp(\mathbf{x}_i \boldsymbol{\beta} + \mathbf{z}_i(t_i)^\top \boldsymbol{\gamma})}{\sum_{r \in \mathcal{R}_i} \exp(\mathbf{x}_r \boldsymbol{\beta} + \mathbf{z}_r(t_i)^\top \boldsymbol{\gamma})} \right\}^{\delta_i}. \tag{6.4}$$

Here, $\mathcal{R}_i$ denotes the risk set at time $t_i^-$. The logarithm of this partial likelihood, namely $\widetilde{l}(\boldsymbol{\beta}, \boldsymbol{\gamma}) = \log \widetilde{L}(\boldsymbol{\beta}, \boldsymbol{\gamma})$, can be maximized using, for example, the Newton algorithm as usual to obtain the maximum partial likelihood estimates of $\boldsymbol{\beta}$ and $\boldsymbol{\gamma}$. The asymptotic normality, similar to the one in Chapter 1 Section 1.5, can be used

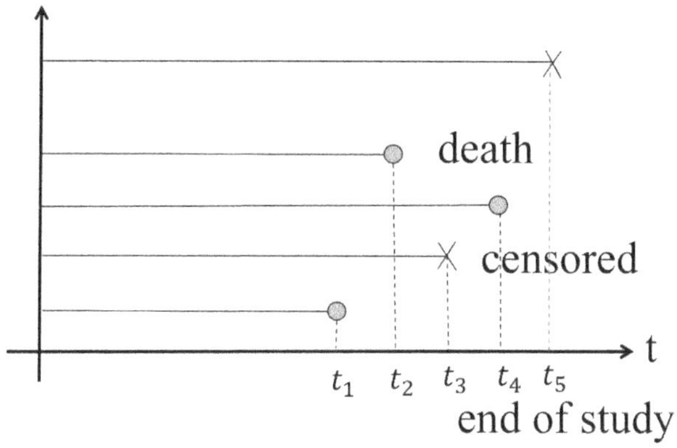


FIGURE 6.1: An example of how to determine risk set and determine the associated time-varying covariates values.

to perform hypothesis testing tasks involving $\beta$ and $\gamma$. Specifically, the likelihood ratio, Wald and Score tests can all be applied to test the coefficients of extended Cox models with time-varying covariates.

A direct estimation of the baseline hazard $h_0(t)$ is not available from the partial likelihood method. However, the Breslow method can be used again to estimate the cumulative baseline $H_0(t)$ in the context of time-varying covariates where the estimates of $\beta$ and $\gamma$ are used as inputs; see Lin (2007). Then, $h_0(t)$ can be estimated from the cumulative baseline $H_0(t)$.

### 6.2.2 Estimation using R "`survival`" package

In this example, we analysis the Heart Transplant data described in Crowley & Hu (1977) using R `survival` package.

**Example 6.2** (Stanford Heart Transplant).
In this example, we apply the partial likelihood method to fit an extended Cox model using the Stanford Heart Transplant data. This dataset has been studied several times, notably by Crowley & Hu (1977) who study the effects of a number of covariates as to whether heart transplant prolongs survival for patients with heart disease. The data follows 103 patients admitted to the Stanford Heart Transplant Program during which time a search for a suitable donor heart is undertaken. This search takes from between a few days to up to a year. Seventy-five patients undergo a transplant, however some patients die prior to a suitable donor heart is found. The R codes for analyzing this dataset are provided below. The output is displayed in Table 6.1

```
library(survival)
data("heart")
```

```
surv.obj <- Surv(time = heart$start, time2 = heart$stop,
    type = "counting", event = heart$event)
pl.fit <- coxph(surv.obj ~ heart$age + heart$year +
    heart$surgery + heart$transplant, data = heart,
    id = heart$id)
summary(pl.fit)
```

TABLE 6.1: Stanford Heart Transplant example.

| | coeff. | std. err. | z | Pr(Z>z) | 95% CI |
|---|---|---|---|---|---|
| Age | 0.027 | 0.014 | 1.981 | 0.024 | 0.001, 0.054 |
| Year | -0.146 | 0.070 | -2.074 | 0.019 | -0.284, -0.008 |
| Surgery | -0.636 | 0.367 | -1.731 | 0.042 | -1.356, -0.082 |
| Transplant | -0.015 | 0.313 | -0.046 | 0.482 | -0.625, 0.605 |

□

## 6.3 Penalized likelihood approach for time-varying covariates

Now we return to the main focus of this chapter, namely the full likelihood-based estimation of $\boldsymbol{\beta}$, $\gamma$ and $h_0(t)$. In particular, we will discuss a maximum penalized likelihood approach to fit the extended Cox model (6.3).

Note that in order for (6.3) to be a valid hazard function, it needs to satisfy the constraint $h_0(t) \geq 0$. This constraint has been imposed in all the MPL estimations of the various Cox models we have discussed so far. As usual, this requirement adds computational challenges to the full likelihood-based estimates.

Recall that in model (6.3), the covariates for subject $i$ were separated into time-fixed and time-varying components, denoted by $\mathbf{x}_i$ and $\mathbf{z}_i(t)$, respectively. We can re-express (6.3) as

$$h_i(t) = h_{0i}^*(t)e^{\mathbf{x}_i^\mathsf{T}\boldsymbol{\beta}}, \tag{6.5}$$

where $h_{0i}^*(t) = h_0(t)\exp\{\mathbf{z}_i(t)^\mathsf{T}\gamma\}$. These $h_{0i}^*(t)$'s can be loosely conceived as individual-specific "baseline" hazard functions, as they correspond to $\mathbf{x}_i = 0$.

Assume that $(T_i, \delta_i, \mathbf{x}_i^\mathsf{T}, \widetilde{\mathbf{z}}_i(T_i)^\mathsf{T})$ are independent for $i = 1, \ldots, n$. Let

$$H_{0i}^*(t) = \int_0^t h_{0i}^*(s)ds.$$

Corresponding to observations $(t_i, \delta_i, \mathbf{x}_i, \widetilde{\mathbf{z}}_i(t_i))$ for $i = 1, \ldots, n$, the log-likelihood function is

$$l(\boldsymbol{\beta}, \boldsymbol{\gamma}, h_0(t)) = \sum_{i=1}^{n} \delta_i (\log h_0(t_i) + \mathbf{x}_i^\top \boldsymbol{\beta} + \mathbf{z}_i(t_i)^\top \boldsymbol{\gamma}) - \sum_{i=1}^{n} H_{0i}^*(t_i) e^{\mathbf{x}_i^\top \boldsymbol{\beta}}. \quad (6.6)$$

As explained previously, direct maximization of such a likelihood with respect to $h_0(t)$ is infeasible since $h_0(t)$ is an infinite-dimensional parameter while we only have a finite number of observations. We will once again adopt the commonly used strategy adopted in Chapters 2–5, where $h_0(t)$ is approximated using a *finite-dimensional* subspace with its dimension $m$ that can grow with the sample size $n$ but at a slower rate, such that $m/n \to 0$ as $n \to \infty$. We also require the assumption that as $n \to \infty$, the approximate $h_0(t)$ converges to the true $h_0(t)$, which is true under certain regularity conditions as demonstrated in Wong & Severini (1991).

Specifically, the finite-dimensional subspace we use is spanned by non-negative basis functions $\psi_u(t)$ (where $u = 1, \ldots, m$) such that $h_0(t)$ ($\geq 0$) is approximated as

$$h_0(t) = \sum_{u=1}^{m} \theta_u \psi_u(t). \quad (6.7)$$

The basis functions discussed in Chapter 2 can be used here as well, including: indicator functions (resulting in a piecewise constant approximation to $h_0(t)$); M-splines as per Ramsay (1988); and Gaussian basis functions. For details on these basis functions, please refer to Chapter 2.

Corresponding to (6.7), the cumulative baseline hazard is given by

$$H_0(t) = \sum_{u=1}^{m} \theta_u \Psi_u(t), \quad (6.8)$$

where $\Psi_u(t) = \int_0^t \psi_u(s) ds$ is the cumulative basis function.

We explained in Chapter 2 that one major benefit of directly approximating $h_0(t)$ using (6.7) is that its cumulative hazard $H_0(t)$ can be easily calculated (as $\Psi_u(t)$ is readily available in most cases). If a functional transformation of $h_0(t)$ were adopted, such as $\eta(t) = \log h_0(t)$, to avoid the non-negativity constraint for $h_0(t)$, calculating the cumulative baseline hazard $H_0(t)$ would usually be more challenging. Alternatively, working with the transformation $\eta(t) = \log H_0(t)$ is another popular option. Although this approach helps to avoid the non-negative requirement on $H_0(t)$ and the corresponding $h_0(t)$ can be easily worked out, the monotonic increasing property of $H_0(t)$ (equivalent to $h_0(t) \geq 0$) can be difficult to impose.

Substituting the approximation (6.7) into the Cox model (6.5), the cumulative hazard $H_i(t) = \int_0^t h_i(s) ds$ can be written as

$$H_i(t) = H_{0i}^*(t) e^{\mathbf{x}_i^\top \boldsymbol{\beta}}, \quad (6.9)$$

where $H^*_{0i}(t) = \sum_u \theta_u \Psi^*_{ui}(t)$ with

$$\Psi^*_{ui}(t) = \int_0^t \psi_u(s) e^{\mathbf{z}_i(s)^\mathsf{T}\gamma} ds. \tag{6.10}$$

Since $h_0(t)$ is now fully represented by the vector $\boldsymbol{\theta} = (\theta_1, \ldots, \theta_m)^\mathsf{T}$, in the following discussions, we will replace $h_0(t)$ with $\boldsymbol{\theta}$ if there is no confusion. The parameter vector we wish to estimate is $\boldsymbol{\eta} = (\boldsymbol{\theta}^\mathsf{T}, \boldsymbol{\beta}^\mathsf{T}, \gamma^\mathsf{T})^\mathsf{T}$.

Let $J(\boldsymbol{\theta})$ be the penalty function used to smooth the $h_0(t)$ estimate. Then, the maximum penalized likelihood estimates of $\boldsymbol{\eta}$ are given by:

$$(\widehat{\boldsymbol{\beta}}, \widehat{\gamma}, \widehat{\boldsymbol{\theta}}) = \underset{\boldsymbol{\beta},\gamma,\boldsymbol{\theta}}{\operatorname{argmax}} \{\Phi(\boldsymbol{\eta}) = l(\boldsymbol{\eta}) - \lambda J(\boldsymbol{\theta})\},$$

subject to the constraint $\boldsymbol{\theta} \geq 0$. A computational algorithm for this optimization problem is provided in Section 6.4.

Comparing full likelihood-based estimation with the Breslow method for estimating $h_0(t)$, the latter requires that estimates of $\boldsymbol{\beta}$ and $\gamma$ must be obtained first and then used as inputs when estimating $h_0(t)$. This highlights a significant benefit of the full likelihood method: it simplifies the establishment of the corresponding asymptotic covariance matrix for all the estimated parameters as it estimates all the parameters simultaneously. Consequently, inferences on quantities of interest can be obtained more easily.

The log-likelihood in (6.6) demands $H^*_{0i}(t_i)$ which involves $\Psi^*_{ui}(t)$ for different $u$. We will discuss in the next section how to compute the quantities $\Psi^*_{ui}(t)$.

## 6.4 Computation of maximum penalized likelihood estimates

Next, we will discuss how to obtain the maximum penalized likelihood estimates of $\boldsymbol{\beta}, \gamma$ and $\boldsymbol{\theta}$, where a penalty function is selected to regularize the $\boldsymbol{\theta}$ (i.e., $h_0(t)$) estimate. Computational challenges in obtaining these estimates arise from two aspects: (i) the non-negativity constraint $\boldsymbol{\theta} \geq 0$ and (ii) the time-varying covariates $\mathbf{z}_i(t)$.

To ensure $\boldsymbol{\theta} \geq 0$, we will again utilize the MI algorithm. For the regression coefficients of $\mathbf{x}_i$ and $\mathbf{z}_i(t)$, we will implement Newton algorithms, with computation details of their gradients and Hessian matrices provided in Section 6.6.

### 6.4.1 Computational algorithm

Because the data structures of the time-fixed covariates $\mathbf{x}_i$ and the time-varying covariates $\mathbf{z}_i(t)$ are different, it is more efficient to update $\boldsymbol{\beta}$ and $\gamma$ separately in an alternating strategy within the algorithm.

The algorithm we will explain next is similar to the one presented in Chapter 2. It is essentially a Newton-MI algorithm, but with separate Newton steps used to update $\boldsymbol{\beta}$ and $\boldsymbol{\gamma}$. Afterward, the MI step, as described in Chapter 2, is employed to update $\boldsymbol{\theta}$, which ensures that $\boldsymbol{\theta} \geq 0$. This Newton-MI algorithm is straightforward to implement and remains efficient even for large values of $m$.

The Karush–Kuhn–Tucker (KKT) first-order necessary conditions for the constrained optimal solution of $\boldsymbol{\beta}$, $\boldsymbol{\gamma}$ and $\boldsymbol{\theta}$ are

$$\frac{\partial \Phi}{\partial \beta_j} = 0, \; j = 1, \ldots, p; \tag{6.11}$$

$$\frac{\partial \Phi}{\partial \gamma_b} = 0, \; b = 1, \ldots, q; \tag{6.12}$$

$$\frac{\partial \Phi}{\partial \theta_u} = 0 \text{ if } \theta_u > 0 \text{ or } \frac{\partial \Phi}{\partial \theta_u} < 0 \text{ if } \theta_u = 0, \; u = 1, \ldots, m. \tag{6.13}$$

Equations (6.11) to (6.13) are similar to the KKT equations in Chapter 2, and they can be solved using the Newton-MI algorithm. This algorithm is an iterative process that alternately solves the equations in (6.11), (6.12) and (6.13).

Each iteration follows an alternating updating scheme. At iteration $k+1$, it starts with estimates $\boldsymbol{\beta}^{(k+1)}$, followed by $\boldsymbol{\gamma}^{(k+1)}$ and then $\boldsymbol{\theta}^{(k+1)}$. More specifically, in iteration $k+1$, the following alternating steps are performed:

**Step 1:** From $\boldsymbol{\beta}^{(k)}, \boldsymbol{\gamma}^{(k)}$ and $\boldsymbol{\theta}^{(k)}$, compute $\boldsymbol{\beta}^{(k+1)}$ such that $\Phi(\boldsymbol{\beta}^{(k+1)}, \boldsymbol{\gamma}^{(k)}, \boldsymbol{\theta}^{(k)}) \geq \Phi(\boldsymbol{\beta}^{(k)}, \boldsymbol{\gamma}^{(k)}, \boldsymbol{\theta}^{(k)})$.

**Step 2:** Then, we compute $\boldsymbol{\gamma}^{(k+1)}$ from $\boldsymbol{\beta}^{(k+1)}$, $\boldsymbol{\gamma}^{(k)}$ and $\boldsymbol{\theta}^{(k)}$, such that $\Phi(\boldsymbol{\beta}^{(k+1)}, \boldsymbol{\gamma}^{(k+1)}, \boldsymbol{\theta}^{(k)}) \geq \Phi(\boldsymbol{\beta}^{(k+1)}, \boldsymbol{\gamma}^{(k)}, \boldsymbol{\theta}^{(k)})$.

**Step 3:** Finally, we compute $\boldsymbol{\theta}^{(k+1)} \geq 0$ from $\boldsymbol{\beta}^{(k+1)}$, $\boldsymbol{\gamma}^{(k+1)}$ and $\boldsymbol{\theta}^{(k)}$, such that $\Phi(\boldsymbol{\beta}^{(k+1)}, \boldsymbol{\gamma}^{(k+1)}, \boldsymbol{\theta}^{(k+1)}) \geq \Phi(\boldsymbol{\beta}^{(k+1)}, \boldsymbol{\gamma}^{(k+1)}, \boldsymbol{\theta}^{(k)})$.

The above incremental penalized log-likelihood conditions in the steps ensure that $\Phi(\boldsymbol{\beta}^{(k+1)}, \boldsymbol{\gamma}^{(k+1)}, \boldsymbol{\theta}^{(k+1)}) \geq \Phi(\boldsymbol{\beta}^{(k)}, \boldsymbol{\gamma}^{(k)}, \boldsymbol{\theta}^{(k)})$ by the end of iteration $k+1$. This is a key requirement for algorithm convergence; see Chan & Ma (2012) or Chapter 2.

Steps 1 and 2 are accomplished by the Newton algorithm with line search steps, while Step 3 is carried out by the MI algorithm, ensuring compliance with the non-negative constraints on $\boldsymbol{\theta}$. Next, we will derive the necessary first and second derivatives of $\Phi$ for the implementation of Steps 1 to 3.

Firstly, we define $\mathbf{1}_n$ to be a vector of 1's of length $n$, $\boldsymbol{\delta}$ to be a vector for all $\delta_i$, $\mathbf{Z}_t$ to be an $n \times q$ matrix whose $i$-th row is $\mathbf{z}_i(t_i)^\top = (z_{i1}(t_i), \ldots, z_{iq}(t_i))$ and $\mathbf{X}$ to be the model matrix with dimension $n \times p$, and whose $i$-th row is $\mathbf{x}_i^\top$.

It is easy to derive the first two derivatives of $\Phi$ with respect to $\boldsymbol{\beta}$ and $\boldsymbol{\gamma}$. They are (see (6.9) for the definition of $H_{0i}^*(t)$):

$$\frac{\partial \Phi}{\partial \boldsymbol{\beta}} = \mathbf{X}^\top(-\mathbf{A}\mathbf{1}_n + \boldsymbol{\delta}), \tag{6.14}$$

$$\frac{\partial \Phi}{\partial \boldsymbol{\gamma}} = -\mathbf{B}_t^\top e^{\mathbf{X}\boldsymbol{\beta}} + \mathbf{Z}_t^\top \boldsymbol{\delta}, \tag{6.15}$$

$$\frac{\partial^2 \Phi}{\partial \boldsymbol{\beta} \partial \boldsymbol{\beta}^\mathsf{T}} = -(\mathbf{X}^\mathsf{T} \mathbf{A} \mathbf{X}), \tag{6.16}$$

$$\frac{\partial^2 \Phi}{\partial \boldsymbol{\gamma} \partial \boldsymbol{\gamma}^\mathsf{T}} = -\mathbf{M}, \tag{6.17}$$

where $\mathbf{A}$ is a diagonal matrix, given by:

$$\mathbf{A} = \text{diag}\Big(H_{01}^*(t_1)e^{\mathbf{x}_1^\mathsf{T}\boldsymbol{\beta}}, \dots, H_{0n}^*(t_n)e^{\mathbf{x}_n^\mathsf{T}\boldsymbol{\beta}}\Big), \tag{6.18}$$

$\mathbf{B}_t$ is an $n \times q$ matrix with its $i$-th row given by

$$\left(\frac{\partial H_{0i}^*(t_i)}{\partial \gamma_1}, \dots, \frac{\partial H_{0i}^*(t_i)}{\partial \gamma_q}\right),$$

and $\mathbf{M}$ is a $q \times q$ matrix with the $(r, s)$-th element $m_{rs}$ given by

$$m_{rs} = \sum_{i=1}^{n} \frac{\partial^2 H_{0i}^*(t_i)}{\partial \gamma_r \partial \gamma_s} e^{\mathbf{x}_i^\mathsf{T}\boldsymbol{\beta}}. \tag{6.19}$$

These expressions involve the first two derivatives of $H_{0i}^*(t_i)$, and they are

$$\frac{\partial H_{0i}^*(t_i)}{\partial \gamma_r} = \sum_{u=1}^{m} \theta_u \frac{\partial \Psi_{ui}^*(t_i)}{\partial \gamma_r}, \tag{6.20}$$

$$\frac{\partial^2 H_{0i}^*(t_i)}{\partial \gamma_r \partial \gamma_s} = \sum_{u=1}^{m} \theta_u \frac{\partial^2 \Psi_{ui}^*(t_i)}{\partial \gamma_r \partial \gamma_s}, \tag{6.21}$$

where the derivatives of $\Psi_{ui}^*(t_i)$ are

$$\frac{\partial \Psi_{ui}^*(t_i)}{\partial \gamma_r} = \int_0^{t_i} \psi_u(w) e^{\mathbf{z}_i(w)^\mathsf{T}\boldsymbol{\gamma}} z_{ir}(w) dw, \tag{6.22}$$

$$\frac{\partial^2 \Psi_{ui}^*(t_i)}{\partial \gamma_r \partial \gamma_s} = \int_0^{t_i} \psi_u(w) e^{\mathbf{z}_i(w)^\mathsf{T}\boldsymbol{\gamma}} z_{ir}(w) z_{is}(w) dw. \tag{6.23}$$

From the above first and second derivatives, a Newton algorithm (incorporating line search) can be applied to update $\boldsymbol{\beta}$, and then $\boldsymbol{\gamma}$.

To update $\boldsymbol{\beta}$, we employ one iteration of the Newton algorithm with line search. At iteration $k+1$, starting with $\boldsymbol{\beta}^{(k)}$, $\boldsymbol{\gamma}^{(k)}$ and $\boldsymbol{\theta}^{(k)}$ and employing a line-search step size $\omega_1^{(k)} \in (0, 1]$, we have

$$\boldsymbol{\beta}^{(k+1)} = \boldsymbol{\beta}^{(k)} + \omega_1^{(k)} (\mathbf{X}^\mathsf{T} \mathbf{A}^{(k)} \mathbf{X})^{-1} \mathbf{X}^\mathsf{T} (-\mathbf{A}^{(k)} \mathbf{1}_n + \boldsymbol{\delta}), \tag{6.24}$$

where matrix $\mathbf{A}^{(k)}$ denotes matrix $\mathbf{A}$ of (6.18) but with $\boldsymbol{\beta} = \boldsymbol{\beta}^{(k)}$, $\boldsymbol{\gamma} = \boldsymbol{\gamma}^{(k)}$ and $\boldsymbol{\theta} = \boldsymbol{\theta}^{(k)}$. The line search parameter $\omega_1^{(k)}$ helps to achieve $l(\boldsymbol{\beta}^{(k+1)}, \boldsymbol{\gamma}^{(k)}, \boldsymbol{\theta}^{(k)}) \geq l(\boldsymbol{\beta}^{(k)}, \boldsymbol{\gamma}^{(k)}, \boldsymbol{\theta}^{(k)})$ (the penalty function is not involved when updating $\boldsymbol{\beta}$). The line search can be efficiently conducted using Armijo's rule; see, for example, Luenberger

& Ye (2008). Note that when $[\mathbf{A}^{(k)}]^{1/2}\mathbf{X}$ has full column rank, $\mathbf{X}^\top\mathbf{A}^{(k)}\mathbf{X}$ is positive definite, and hence it guarantees the line search will find an $\omega_1^{(k)} \in (0,1]$ satisfying the log-likelihood increment requirement.

Next, with $\boldsymbol{\beta}^{(k+1)}$, $\gamma^{(k)}$ and $\boldsymbol{\theta}^{(k)}$, we can update $\gamma$ by again implementing one iteration of the Newton algorithm with line search. Let $\mathbf{M}^{(k)}$ denote matrix $\mathbf{M}$ give by (6.19), where its parameters are now set to: $\boldsymbol{\beta} = \boldsymbol{\beta}^{(k+1)}, \gamma = \gamma^{(k)}$ and $\boldsymbol{\theta} = \boldsymbol{\theta}^{(k)}$. One iteration of a Newton algorithm gives

$$\gamma^{(k+1)} = \gamma^{(k)} + \omega_2^{(k)}[\mathbf{M}^{(k)}]^{-1}\left(-(\mathbf{B}_t^{(k)})^\top e^{\mathbf{X}\boldsymbol{\beta}^{(k+1)}} + \mathbf{Z}_t^\top\boldsymbol{\delta}\right), \tag{6.25}$$

where $\omega_2^{(k)} \in (0,1]$ is a line search step size ensuring $l(\boldsymbol{\beta}^{(k+1)}, \gamma^{(k+1)}, \boldsymbol{\theta}^{(k)}) \geq l(\boldsymbol{\beta}^{(k+1)}, \gamma^{(k)}, \boldsymbol{\theta}^{(k)})$ (note again the penalty function is not involved when updating $\boldsymbol{\beta}$). Again, Armijo's rule can be used to compute the step size $\omega_2^{(k)}$. From (6.19) and (6.23), and particularly the structure displayed in (6.23), we can see that $\mathbf{M}$ is a positive definite matrix under certain regular conditions. This positive definiteness will be more obvious, as explained below, when elements of the vector $\mathbf{z}_i(t)$ are discrete functions. A positive definite $\mathbf{M}$ ascertains existence of an $\omega_2^{(k)} \in (0,1]$ for achieving increment of the log-likelihood function.

The formula (6.24) for updating $\boldsymbol{\beta}$ is straightforward to implement. However, the iterative formula (6.25) for $\gamma$ can be computationally intensive as it requires repeated evaluations of the integrals in (6.10), (6.22) and (6.23). It is important to note that this process involves computing $3 \times m \times n$ integrals in each iteration, which can be time-consuming when $n$ is large. In such cases, adopting an efficient approximation method to evaluate these integrals, such as the Gaussian quadrature method, can provide a balance between efficiency, accuracy and ease of implementation.

In many practical situations, however, the time-varying covariates $z_{ib}(t)$ are often piecewise constant (i.e. discrete) functions of $t$. This is especially common when $z_{ib}(t)$ values are measured intermittently rather than continuously. For example, in a clinical trial evaluating the effects of a new medicine for myocardia ischemia, patients' cholesterol levels may be measured only at certain time points. When $z_{ib}(t)$ are discrete functions, the integrals mentioned earlier become easily computable, significantly reducing the computational efforts required for updating $\gamma$. In fact, each integral can be expressed as a summation of a finite number of terms involving differences of cumulative basis functions. This simplifies the updating formula, as shown in (6.31) below.

Next, we explain how to compute these integrals when $z_{ib}(t)$ are discrete functions. Let us first introduce some notations necessitated for the discussions below.

For individual $i$, recall its observed survival time is $t_i$. Suppose there are $n_i$ intermittent measurement points up to (and may include) $t_i$ for values of the time-varying covariates vector $\mathbf{z}_i(t)$. These measurement points are denoted by $t_{i1}, \ldots, t_{in_i}$. A piecewise constant $z_{ib}(t)$ means its value remains constant between any two consecutive measurement points. This means, for $b = 1, \ldots, q$, we have

$$z_{ib}(t) = z_{iab} I(t_{ia} \leq t < t_{i,a+1}), \tag{6.26}$$

where $I$ is an indicator function, and $a = 1, \ldots, n_i$. To simplify notations, we assume, without loss of generality, that $t_{i1} = 0$ and $t_{i,n_i+1} = t_i$. Such a $z_{ib}(t)$ is piecewise constant with $n_i$ pieces over $[0, t_i]$.

The $z_{iab}$ values in (6.26) need to be entered into a data frame before running the R program created to fit a time-varying covariates Cox model using our R function "`tvc_mpl.r`". Table 6.2 displays the required data frame format for time-varying covariates, which is identical to the conventional "long-format" data frame used to accommodate time-varying covariates in the R "`survival`" package.

TABLE 6.2: Example of time-varying covariates long-format data frame.

| subject | start | end | status | $z_1(t)$ | $\cdots$ | $z_q(t)$ |
|---|---|---|---|---|---|---|
| 1 | 0 | $t_{12}$ | 0 | $z_{111}$ | $\cdots$ | $z_{11q}$ |
| | $t_{12}$ | $t_{13}$ | 0 | $z_{121}$ | $\cdots$ | $z_{12q}$ |
| | $t_{13}$ | $t_1$ | 1 | $z_{131}$ | $\cdots$ | $z_{13q}$ |
| 2 | 0 | $t_{22}$ | 0 | $z_{211}$ | $\cdots$ | $z_{21q}$ |
| | $t_{22}$ | $t_2$ | 0 | $z_{221}$ | $\cdots$ | $z_{22q}$ |
| 3 | $\cdots$ | $\cdots$ | $\cdots$ | $\cdots$ | $\cdots$ | $\cdots$ |

This table provides a simple example to illustrate this data frame. In the table, "status" of 0 indicates no event, while 1 indicates an event. Subject 1 experienced an event at time $t_1$, while subject 2 was censored at time $t_2$. Subject 1 has three recordings of time-varying covariates at times $t_{11} = 0$, $t_{12}$ and $t_{13}$, while subject 2 has two recordings at times $t_{21} = 0$ and $t_{22}$. These time points are listed as "start" and "end" in the table.

Table 6.2 illustrates that the $z_{iab}$ values for subject $i$ form an $n_i \times q$ matrix denoted as $\mathbf{Z}_i$. Specifically, the $d$-th row of $\mathbf{Z}_i$ is represented as $\mathbf{z}_{id}^\top = (z_{id1}, \ldots, z_{idq})$, which means that $\mathbf{Z}_i$ can be expressed as $\mathbf{Z}_i = (\mathbf{z}_{i1}, \ldots, \mathbf{z}_{in_i})^\top$.

Then, for $t \in [t_{ia}, t_{i,a+1})$, the integration used to define $\Psi_{ui}^*(t)$ in (6.10) is replaced with a summation as follows:

$$
\begin{aligned}
\Psi_{ui}^*(t) &= \sum_{d=1}^{a-1} \left( \int_{t_{id}}^{t_{i,d+1}} \psi_u(s)ds \right) e^{\mathbf{z}_{id}^\top \gamma} \\
&= \sum_{d=1}^{a-1} [\Psi_u(t_{i,d+1}) - \Psi_u(t_{id})] e^{\mathbf{z}_{id}^\top \gamma},
\end{aligned}
\tag{6.27}
$$

where $\Psi_u(t) = \int_0^t \psi_u(s)ds$. Depending on the selected basis functions, computing $\Psi_u(t)$ is typically straightforward in R. For instance, if M-spline basis functions are employed, then $\Psi_u(t)$ can be represented as I-splines. Alternatively, when Gaussian bases are utilized, $\Psi_u(t)$ corresponds to the normal cumulative distribution functions.

Similarly, computations of $H^*_{0i}(t)$ are also simplified as a summation:

$$H^*_{0i}(t) = \sum_{d=1}^{a-1} [H_0(t_{i,d+1}) - H_0(t_{i,d})] e^{\mathbf{z}_{id}^\mathsf{T}\gamma}, \tag{6.28}$$

where $t \in [t_{ia}, t_{i,a+1})$ and $H_0(t) = \sum_{u=1}^{m} \theta_u \Psi_u(t)$. The evaluation of $H_0(t)$ is again typically straightforward in R.

For discrete time-varying covariates, once the integrals in (6.10), (6.22) and (6.23) are represented as summations, the Newton iterative formula for $\gamma$ can be expressed using a "model matrix" for the time-varying covariates, which is created from the $z_{iab}$ values. Details are provided next.

Let $N = \sum_{i=1}^{n} n_i$. Denote a vector of 1's of length $N$ as $\mathbf{1}_N$. Define the status vector as follows:

$$\boldsymbol{\varrho} = [\varrho_{11}, \ldots, \varrho_{1n_1}, \ldots, \varrho_{n1}, \ldots, \varrho_{nn_n}]^\mathsf{T},$$

Here, $\varrho_{ia} = 1$ only if $a = n_i$ and $\delta_i = 1$; otherwise, $\varrho_{ia} = 0$. Essentially, $\boldsymbol{\varrho}$ corresponds to the "status" column in Table 6.2. Now, let $\mathbf{Z}$ be:

$$\mathbf{Z} = [\mathbf{Z}_1^\mathsf{T}, \ldots, \mathbf{Z}_n^\mathsf{T}]^\mathsf{T}.$$

The matrix $\mathbf{Z}$ has dimensions of $N \times q$ and contains the values of all the time-varying covariates.

Using the matrices defined above, we can express the first two derivatives of the penalized log-likelihood $\Phi$ with respect to $\gamma$, as given in equations (6.15) and (6.17), using matrices as follows:

$$\frac{\partial \Phi}{\partial \gamma} = \mathbf{Z}^\mathsf{T}(-\mathbf{B}\mathbf{1}_N + \boldsymbol{\zeta}), \tag{6.29}$$

$$\frac{\partial^2 \Phi}{\partial \gamma \partial \gamma^\mathsf{T}} = -(\mathbf{Z}^\mathsf{T}\mathbf{B}\mathbf{Z}), \tag{6.30}$$

Here, $\mathbf{B}$ is a block-diagonal matrix defined as:

$$\mathbf{B} = \text{diag}(e^{\mathbf{x}_1^\mathsf{T}\boldsymbol{\beta}}\mathbf{B}_1, \ldots, e^{\mathbf{x}_n^\mathsf{T}\boldsymbol{\beta}}\mathbf{B}_n),$$

where each $\mathbf{B}_i$ is a diagonal matrix of size $n_i \times n_i$, given by:

$$\mathbf{B}_i = \text{diag}\Big([H_0(t_{i2}) - H_0(t_{i1})]e^{\mathbf{z}_{i1}^\mathsf{T}\gamma}, \ldots, [H_0(y_i) - H_0(t_{in_i})]e^{\mathbf{z}_{in_i}^\mathsf{T}\gamma}\Big).$$

Accordingly, the iterative formula (6.25) for $\gamma$ becomes

$$\gamma^{(k+1)} = \gamma^{(k)} + \omega_2^{(k)}(\mathbf{Z}^\mathsf{T}\mathbf{B}^{(k)}\mathbf{Z})^{-1}\mathbf{Z}^\mathsf{T}(-\mathbf{B}^{(k)}\mathbf{1}_N + \boldsymbol{\zeta}). \tag{6.31}$$

This formula appears almost identical to the $\boldsymbol{\beta}$ updating formula given in (6.24) and thus is very easy to implement once the matrix $\mathbf{Z}$ is constructed. The computational burden for updating $\gamma$ is similar to that of $\boldsymbol{\beta}$.

Finally, we will explain how to update $\boldsymbol{\theta}$. Once again, we will adopt the MI algorithm, as we did in the previous chapters. There are two primary reasons for choosing the MI algorithm:

(i) It efficiently enforces the non-negative constraint on $\boldsymbol{\theta}$.

(ii) It is straightforward to implement, even when the length of $\boldsymbol{\theta}$ is large.

These reasons make the MI algorithm a suitable choice for updating $\boldsymbol{\theta}$.

The MI algorithm only demands the first derivative of $\Phi$ with respect to $\boldsymbol{\theta}$, which can be expressed as:

$$\frac{\partial \Phi}{\partial \boldsymbol{\theta}} = \mathbf{C}^{\mathsf{T}}\mathbf{E}^{-1}\boldsymbol{\delta} - \mathbf{C}^{*\mathsf{T}}\mathbf{f}. \tag{6.32}$$

Here, $\mathbf{f}$ is an $n$-vector with elements $e^{\mathbf{x}_i^{\mathsf{T}}\boldsymbol{\beta}}$, $\mathbf{C}$ and $\mathbf{C}^*$ are both $n \times m$ matrices with $(i, u)$-th elements given by $\psi_u(t_i)$ and $\Psi^*_{ui}(t_i)$, respectively. Additionally, $\mathbf{E}$ is a diagonal matrix with diagonal elements $h_0(t_i)$.

The MI algorithm updates $\boldsymbol{\theta}$ using the following iterative scheme:

$$\boldsymbol{\theta}^{(k+1)} = \boldsymbol{\theta}^{(k)} + \omega_3^{(k)}\mathbf{D}^{(k)}\left(\mathbf{C}^{\mathsf{T}}[\mathbf{E}^{(k)}]^{-1}\boldsymbol{\delta} - [\mathbf{C}^{*(k)}]^{\mathsf{T}}\mathbf{f}^{(k)}\right), \tag{6.33}$$

where $\mathbf{D}$ is a diagonal matrix:

$$\mathbf{D} = \operatorname{diag}\left(\frac{\theta_u}{\sum_i \Psi^*_{ui}(y_i)e^{\mathbf{x}_i^{\mathsf{T}}\boldsymbol{\beta}} + \epsilon}\right).$$

In matrix $\mathbf{D}$, the constant $\epsilon$ is a small threshold used to prevent zero denominators. It is worth noting that the choice of $\epsilon$ does not impact the final solution but can affect the convergence speed of the algorithm.

The step size $\omega_3^{(k)} \in (0, 1]$, and it guarantees that

$$l(\boldsymbol{\beta}^{(k+1)}, \boldsymbol{\gamma}^{(k+1)}, \boldsymbol{\theta}^{(k+1)}) \geq l(\boldsymbol{\beta}^{(k+1)}, \boldsymbol{\gamma}^{(k+1)}, \boldsymbol{\theta}^{(k)}).$$

This step size can be efficiently determined in each iteration using the Armijo rule.

### 6.4.2 Algorithm convergence properties

When $\mathbf{A}^{1/2}\mathbf{X}$ and $\mathbf{B}^{1/2}\mathbf{Z}$ have full column rank, matrices $\mathbf{X}^{\mathsf{T}}\mathbf{A}\mathbf{X}$ and $\mathbf{Z}^{\mathsf{T}}\mathbf{B}\mathbf{Z}$ are positive definite, ensuring well-defined updates for $\boldsymbol{\beta}$ and $\boldsymbol{\gamma}$. Following the same argument as in Chan & Ma (2012) and making assumptions similar to Assumptions C1 and C2 in Chapter 2, Section 2.5.2, we can establish similar convergence results to Theorems 2.1 through 2.3. In summary, the following two results hold:

(1) If $\boldsymbol{\theta}^{(k)}$ is non-negative, then $\boldsymbol{\theta}^{(k+1)}$ is also non-negative.

(2) Under certain regularity conditions, this algorithm converges to a solution that satisfies the KKT conditions.

## 6.5 Asymptotic results

The asymptotic properties of the maximum penalized likelihood estimators presented in this section is similar to Chapters 2–5, so their detailed derivations are omitted and we just provide the main results.

### 6.5.1 Asymptotic consistency and asymptotic normality

Let $a = \min_i\{t_i\}$ and $b = \max_i\{t_i\}$, where $t_i \neq \infty$. Let $\boldsymbol{\beta}_0$, $\gamma_0$ and $h_0(t)$ denote the true $\boldsymbol{\beta}$, $\gamma$ and the true baseline hazard. In this section, we let $\widehat{h}_n(t)$ represent the basis function-based baseline hazard MPL estimate. First, we need to amend the Assumption A2.1 with the following assumption:

**Assumption A6.1**

**A6.1:** Assume matrix $\mathbf{X}$ is bounded, and also assume matrix $\mathbf{Z}(t)$ is bounded for any $t \in [a, b]$. Assume $E(\mathbf{X}\mathbf{X}^\top)$ and $E(\mathbf{Z}(t)\mathbf{Z}^\top(t))$ are non-singular for any $t \in [a, b]$.

The asymptotic consistency results are given below.

**Theorem 6.1.** Assume $h_0(t)$ has up to $c$ $(\geq 1)$ derivatives for $t \in [a, b]$. Suppose the number of basis functions $m$ is selected according to $m = n^v$, where $0 < v < 1$, and assume $\mu_n = \lambda/n \to 0$ when $n \to \infty$. Then, as $n \to \infty$, and under Assumption 6.1 and Assumptions 2.2–2.4, we have

(1) $||\widehat{\boldsymbol{\beta}} - \boldsymbol{\beta}^0||_2 \to 0$ (a.s.),

(2) $||\widehat{\gamma} - \gamma^0||_2 \to 0$ (a.s.), and

(3) $\sup_{t\in[a,b]} |\widehat{h}_n(t) - h_0(t)| \to 0$ (a.s.).

□

We denote the vector for the MPL estimates of $\boldsymbol{\beta}, \gamma$ and $\boldsymbol{\theta}$ as $\widehat{\boldsymbol{\eta}} = (\widehat{\boldsymbol{\theta}}^\top \widehat{\boldsymbol{\beta}}^\top, \widehat{\gamma}^\top)^\top$. The regularity assumptions we need for the asymptotic normality results are similar to B2.1–B2.4 except B2.1 needs to be replaced by Assumption B6.1 stated below.

**Assumption B6.1**

**B6.1:** Assume random vectors: $\mathbf{W}_i = (T_i, \mathbf{x}_i^\top, \widetilde{\mathbf{z}}_i(T_i)^\top)^\top$, where $i = 1, \ldots, n$, are independent and identically distributed and the distributions of $\mathbf{x}_i$ and $\mathbf{z}_i(t)$, where $t < T_i$, are independent of the parameter vector $\boldsymbol{\eta}$.

**Theorem 6.2.** Assuming Assumptions B6.1 and B2.2–B2.4 hold. Assume the scaled smoothing value $\mu_n = o(n^{-1/2})$. If there are active constraints in the MPL estimate of $\boldsymbol{\theta}$, the corresponding $\mathbf{U}$ matrix can be constructed in a similar way as (2.54); otherwise, $\mathbf{U}$ can be selected as an identity matrix with appropriate dimensions. Let $\mathbf{F}(\boldsymbol{\eta}) = -n^{-1}E\big(\partial^2 l(\boldsymbol{\eta})/\partial\boldsymbol{\eta}\partial\boldsymbol{\eta}^{\mathsf{T}}\big)$. Then, when $n \to \infty$,

(1) the constrained MPL estimate $\widehat{\boldsymbol{\eta}}$ is consistent for $\boldsymbol{\eta}_0$, and

(2) $\sqrt{n}(\widehat{\boldsymbol{\eta}} - \boldsymbol{\eta}_0)$ converges in distribution to a multivariate normal distribution $N(\mathbf{0}, \widetilde{\mathbf{F}}(\boldsymbol{\eta}_0)^{-1}\mathbf{F}(\boldsymbol{\eta}_0)[\widetilde{\mathbf{F}}(\boldsymbol{\eta}_0)^{-1}]^{\mathsf{T}})$, where $\widetilde{\mathbf{F}}(\boldsymbol{\eta})^{-1} = \mathbf{U}(\mathbf{U}^{\mathsf{T}}\mathbf{F}(\boldsymbol{\eta})\mathbf{U})^{-1}\mathbf{U}^{\mathsf{T}}$.

□

Inference problems we encounter in practice will always have finite sample sizes $n$. When $n$ is large, and assuming the regularity conditions stated in Theorem 6.2 hold, the approximate distribution of $\widehat{\boldsymbol{\eta}} - \boldsymbol{\eta}_0$ is still a multivariate normal distribution with the mean vector zero and the covariance matrix:

$$\widehat{\mathrm{Var}}(\widehat{\boldsymbol{\eta}}) = \mathbf{A}(\widehat{\boldsymbol{\eta}})^{-1}\frac{\partial^2 l(\widehat{\boldsymbol{\eta}})}{\partial\boldsymbol{\eta}\partial\boldsymbol{\eta}^{\mathsf{T}}}\mathbf{A}(\widehat{\boldsymbol{\eta}})^{-1}, \tag{6.34}$$

where

$$\mathbf{A}(\widehat{\boldsymbol{\eta}})^{-1} = \mathbf{U}\left(\mathbf{U}^{\mathsf{T}}\left(-\frac{\partial^2 l(\widehat{\boldsymbol{\eta}})}{\partial\boldsymbol{\eta}\partial\boldsymbol{\eta}^{\mathsf{T}}} + \lambda\frac{\partial^2 J(\widehat{\boldsymbol{\theta}})}{\partial\boldsymbol{\theta}\partial\boldsymbol{\theta}}\right)\mathbf{U}\right)^{-1}\mathbf{U}^{\mathsf{T}}.$$

## 6.5.2 Inferences on survival function $S(t|\mathbf{x}_i, \widetilde{\mathbf{z}}_i(t))$

One important application of a Cox model with time-varying covariates is on dynamic prediction. The prediction accuracy demands the standard deviation of the estimated survival function $\widehat{S}(t|\mathbf{x}_i, \widetilde{\mathbf{z}}_i(t))$. We denote $S(t|\mathbf{x}_i, \widetilde{\mathbf{z}}_i(t))$ by $S_i(t)$.

Since

$$\widehat{S}_i(t) = \exp\{-\widehat{H}_i(t)\},$$

where $\widehat{H}_i(t) = \int_0^t \widehat{h}_0(s)e^{\mathbf{x}_i^{\mathsf{T}}\widehat{\boldsymbol{\beta}} + \mathbf{z}_i^{\mathsf{T}}(s)\widehat{\boldsymbol{\gamma}}}ds$, we have, by repeatedly employing the Delta method, that

$$\begin{aligned}\mathrm{Var}\left(\widehat{S}_i(t)\right) &\approx \widehat{S}_i(t)^2\,\mathrm{Var}\left(\widehat{H}_i(t)\right)\\ &\approx \widehat{S}_i(t)^2\left(\frac{\partial\widehat{H}_i(t)}{\partial\widehat{\boldsymbol{\eta}}}\right)^{\mathsf{T}}\mathrm{Var}(\widehat{\boldsymbol{\eta}})\frac{\partial\widehat{H}_i(t)}{\partial\widehat{\boldsymbol{\eta}}}.\end{aligned} \tag{6.35}$$

Here, the approximation in (6.35) is obtained by treating $\widehat{H}_i(t)$ as a function of $\widehat{\boldsymbol{\eta}}$. The elements of the derivative vector $\widehat{H}_i(t)/\partial\widehat{\boldsymbol{\eta}}$ are given by:

$$\frac{\partial\widehat{H}_i(t)}{\partial\widehat{\beta}_j} = x_{ij}\widehat{H}_i(t), \tag{6.36}$$

$$\frac{\partial\widehat{H}_i(t)}{\partial\widehat{\gamma}_b} = e^{\mathbf{x}_i^{\mathsf{T}}\widehat{\boldsymbol{\beta}}}\int_0^t z_{ib}(s)\widehat{h}_0(s)e^{\mathbf{z}_i^{\mathsf{T}}(s)\widehat{\boldsymbol{\gamma}}}ds, \tag{6.37}$$

$$\frac{\partial\widehat{H}_i(t)}{\partial\widehat{\theta}_u} = e^{\mathbf{x}_i^{\mathsf{T}}\widehat{\boldsymbol{\beta}}}\int_0^t \psi_u(s)e^{\mathbf{z}_i^{\mathsf{T}}(s)\widehat{\boldsymbol{\gamma}}}ds. \tag{6.38}$$

It is not ideal to construct an $100(1-\alpha)\%$ CI of $S_i(t)$ at a time point $t$ by: $\widehat{S}_i(t) \pm z_{\alpha/2}\text{std}(\widehat{S}_i(t))$ as such a CI cannot be ensured to be restrained within $[0,1]$. Similar to the process described in Section 2.9, we use a logit transformation and construct a CI for the transformed $S_i(t)$, and then apply a backward transformation to change this CI back to a CI of $S_i(t)$. The final CI for $S_i(t)$ is given by:

$$\frac{\widehat{S}_i(t)e^{\pm z_{\alpha/2}\frac{1}{\widehat{S}_i(t)(1-\widehat{S}_i(t))}\text{std}(\widehat{S}_i(t))}}{(1-\widehat{S}_i(t))+\widehat{S}_i(t)e^{\pm z_{\alpha/2}\frac{1}{\widehat{S}_i(t)(1-\widehat{S}_i(t))}\text{std}(\widehat{S}_i(t))}}. \tag{6.39}$$

Example 6.6 implements this CI formula to obtain a CI for a predictive survival curve.

## 6.6 Hessian matrix

In Section 6.4, we have obtained the second derivatives $\partial^2 l/\partial\boldsymbol{\beta}\partial\boldsymbol{\beta}^\mathsf{T}$ and $\partial^2 l/\partial\boldsymbol{\gamma}\partial\boldsymbol{\gamma}^\mathsf{T}$. In this section, we will provide the other second derivatives as they are required by the large sample covariance matrix in (6.34).

First, recall that

$$\frac{\partial l}{\partial \beta_j} = \sum_{i=1}^n x_{ij}\Big(-H_{0i}^*(t_i)e^{\mathbf{x}_i^\mathsf{T}\boldsymbol{\beta}}+\delta_i\Big),$$

$$\frac{\partial l}{\partial \theta_u} = \sum_{i=1}^n \left(\delta_i\frac{\psi_i(t_i)}{h_0(t_i)} - \Psi_{ui}^*(t_i)e^{\mathbf{x}_i^\mathsf{T}\boldsymbol{\beta}}\right).$$

From these first derivative results we can obtain the required second derivatives as given below.

$$\frac{\partial^2 l}{\partial\beta_j\partial\gamma_r} = -\sum_{i=1}^n x_{ij}\frac{\partial H_{0i}^*(t_i)}{\partial\gamma_r}e^{\mathbf{x}_i^\mathsf{T}\boldsymbol{\beta}}, \tag{6.40}$$

$$\frac{\partial^2 l}{\partial\beta_j\partial\theta_u} = -\sum_{i=1}^n x_{ij}\Psi_{ui}^*(t_i)e^{\mathbf{x}_i^\mathsf{T}\boldsymbol{\beta}}, \tag{6.41}$$

$$\frac{\partial^2 l}{\partial\gamma_r\partial\theta_u} = -\sum_{i=1}^n \frac{\partial\Psi_{ui}^*(t_i)}{\partial\gamma_r}e^{\mathbf{x}_i^\mathsf{T}\boldsymbol{\beta}}, \tag{6.42}$$

$$\frac{\partial^2 l}{\partial\theta_u\partial\theta_v} = -\sum_{i=1}^n \delta_i\frac{\psi_u(t_i)\psi_v(t_i)}{h_0^2(t_i)}, \tag{6.43}$$

where $\partial H_{0i}^*(t)/\partial\gamma_r$ is given in (6.20) and $\partial\Psi_{ui}^*(t)/\partial\gamma_r$ is given in (6.22).

## 6.7 Smoothing parameter selection

The automatic selection of the smoothing parameter plays a crucial role in the success of the penalized likelihood method discussed in this chapter. This statement holds true for other penalized likelihood methods as well, as demonstrated in Dettoni et al. (2020). A poorly chosen smoothing parameter can significantly affect the quality of the estimates and is likely to result in a sub-optimal solution.

Once again, we employ a marginal likelihood method for automatic smoothing parameter selection, similar to the approach in previous chapters. This method first view the penalty function of $\boldsymbol{\theta}$ as a multivariate normal prior distribution through

$$\boldsymbol{\theta} \sim \mathrm{N}(0_{m\times 1}, \sigma^2\mathbf{R}^{-1}),$$

where the variance $\sigma^2$ is related to the smoothing parameter $\lambda$ through $\sigma^2 = 1/(2\lambda)$. Then, the og-posterior is

$$l_p(\boldsymbol{\beta}, \boldsymbol{\gamma}, \boldsymbol{\theta}) = l(\boldsymbol{\beta}, \boldsymbol{\gamma}, \boldsymbol{\theta}) - \frac{m}{2}\ln\sigma^2 - \frac{1}{2\sigma^2}\boldsymbol{\theta}^\mathsf{T}\mathbf{R}\boldsymbol{\theta}. \tag{6.44}$$

where $l(\boldsymbol{\beta}, \boldsymbol{\gamma}, \boldsymbol{\theta})$ is the log-likelihood given in (6.6), but with $h_0(t)$ replaced by $\boldsymbol{\theta}$.

An estimate of $\lambda$, or equivalently an estimate of $\sigma^2$, can be achieved by maximizing the marginal likelihood, obtained through integrating out of $\boldsymbol{\beta}, \boldsymbol{\gamma}$ and $\boldsymbol{\theta}$ from the posterior given in (6.44). As in the previous chapters, we again approximate this integral using Laplace's method, leading to the following approximated log-marginal likelihood:

$$l_m(\sigma^2) \approx -\frac{m}{2}\log\sigma^2 + l(\widehat{\boldsymbol{\beta}}, \widehat{\boldsymbol{\gamma}}, \widehat{\boldsymbol{\theta}}) - \frac{1}{2\sigma^2}\widehat{\boldsymbol{\theta}}^\mathsf{T}\mathbf{R}\widehat{\boldsymbol{\theta}} - \frac{1}{2}\log|-\widehat{\mathbf{H}} + \mathbf{Q}|, \tag{6.45}$$

where $\widehat{\boldsymbol{\beta}}, \widehat{\boldsymbol{\gamma}}$ and $\widehat{\boldsymbol{\theta}}$ denote the MPL estimates when $\sigma^2$ is given, $\widehat{\mathbf{H}}$ is the Hessian matrix from $l(\boldsymbol{\beta}, \boldsymbol{\gamma}, \boldsymbol{\theta})$ evaluated at the MPL estimates $\widehat{\boldsymbol{\beta}}$, $\widehat{\boldsymbol{\gamma}}$ and $\widehat{\boldsymbol{\theta}}$, and

$$\mathbf{Q} = \begin{pmatrix} \mathbf{0}_{p\times p} & \mathbf{0}_{p\times q} & \mathbf{0}_{p\times m} \\ \mathbf{0}_{q\times p} & \mathbf{0}_{q\times q} & \mathbf{0}_{q\times m} \\ \mathbf{0}_{m\times p} & \mathbf{0}_{m\times q} & \frac{1}{\sigma^2}\mathbf{R} \end{pmatrix},$$

where $\mathbf{0}$ denotes a matrix with all its elements being zero. We should note that although the problem we consider in this chapter is different from the previous chapters, their marginal log-likelihood functional forms (i.e. comparing (6.45) with the previous marginal log-marginal likelihoods) are identical.

The solution to $\sigma^2$ maximizing (6.45), denoted by $\widehat{\sigma}^2$, satisfies

$$\widehat{\sigma}^2 = \frac{\widehat{\boldsymbol{\theta}}^\mathsf{T}\mathbf{R}\widehat{\boldsymbol{\theta}}}{m-\nu}, \tag{6.46}$$

where $\nu = \mathrm{tr}\{(-\widehat{\mathbf{H}} + \widehat{\mathbf{Q}})^{-1}\widehat{\mathbf{Q}}\}$. Here, $\widehat{\mathbf{Q}}$ denotes matrix $\mathbf{Q}$ but with $\sigma^2$ replaced by its current value, and $\nu$ has been considered as equivalent to the "model degrees of freedom" throughout this book.

The expression in (6.46) for $\sigma^2$ allows for an iterative procedure for estimating $\boldsymbol{\beta}$, $\boldsymbol{\gamma}$, $\boldsymbol{\theta}$ and $\sigma^2$ with two steps:

(1) In the first step, with a current estimate of $\sigma^2$, we obtain the corresponding MPL estimates for $\boldsymbol{\beta}$, $\boldsymbol{\gamma}$ and $\boldsymbol{\theta}$.

(2) Then, in the second step, we update $\sigma^2$ using these obtained MPL estimates of $\boldsymbol{\beta}$, $\boldsymbol{\gamma}$ and $\boldsymbol{\theta}$ on the right-hand side of (6.46).

These two steps are repeated until $\nu$ stabilizes, for instance, when the difference between two consecutive $\nu$ values is less than 1 or less than 0.5.

## 6.8 R examples

In this section, we provide two R examples: one example explains how to generate survival data from an extended Cox model that contains a time-varying covariate, and the other example is a simple simulation study comparing the MPL estimates of an extended Cox against the partial likelihood method.

**Example 6.3** (Generate extended Cox model data).
In this example, we explain how to simulate independent survival times $T_i = \min(Y_i, C_i)$. Here, the time to an event $Y_i$ is assumed to follow a distribution where its hazard function is given by an extended Cox model:

$$h_i(t) = 1.5t^{0.5}e^{\beta_1 x_{1i}+\beta_2 x_{2i}+\gamma_1 z_{1i}(t)}, \tag{6.47}$$

and the right censoring time $C_i$, which is independent of $Y_i$, is generated from a uniform distribution $unif(0, \mu_c = 1.05)$ for each $i$, where $\mu_c$ is chosen to give, on average, a desired censoring proportion $\pi_c$ for use in a simulation study.

Thus, independent observations $(t_i, \delta_i)$ for $i = 1, \dots, n$ are generated, where $\delta_i$ is the event time indicator. In this example, we set the regression coefficients to $\beta_1 = 1$, $\beta_2 = 1$ and $\gamma_1 = 1$. The covariates $x_1$ and $x_2$ are time-fixed (baseline) covariates. Values for $x_1$ are randomly drawn from a normal distribution $N(0, 1)$, and values for $x_2$ are randomly drawn from a Bernoulli distribution with parameter $p = 0.5$.

The time-dependent covariate $z_1(t)$ is a discrete variable that starts at zero for all subjects at baseline but changes to one at a randomly selected time point for a random 50% of subjects, after which it remains at one. According to (6.47), the baseline hazard is defined as

$$h_0(t) = \nu t^{\nu-1}/\lambda^{\nu},$$

where $\lambda = 1$ and $\nu = 1.5$.

The method of generating survival times is similar to the one outlined in Austin (2012). When the model has a single time-dependent covariate that begins with zero and changes to the value of one at most once, the hazard for the $i$-th subject is partitioned using two intervals along time $t$. Denote by $t_{0i}$ the time at which the value of the time-dependent covariate switches from zero to one. From this hazard, we can work out the corresponding survival function, and from which a random number generation formula can be obtained as given below:

$$t_i = \begin{cases} \left( \frac{-\log(u_i)}{\lambda e^{\beta_1 x_{1i} + \beta_2 x_{2i}}} \right)^{1/\nu} & \text{if } -\log(u_i) < \lambda e^{\beta_1 x_{1i} + \beta_2 x_{2i}} t_{0i}^{\nu} \\ \left( \frac{-\log(u_i)}{\lambda e^{\beta_1 x_{1i} + \beta_2 x_{2i} + \gamma}} + t_{0i}^{\nu}(1 - e^{-\gamma}) \right)^{1/\nu} & \text{if } -\log(u_i) \geq \lambda e^{\beta_1 x_{1i} + \beta_2 x_{2i}} t_{0i}^{\nu} \end{cases} \tag{6.48}$$

After generating a uniform random number $u_i \sim U(0, 1)$, we can then simulate the required $t_i$ according to (6.48), where $\lambda$ and $\nu$ values have already given before. The R code for generating random numbers for this example is given below. These R code will be employed to generate $n = 1000$ survival times, which include about 50% right-censored times.

```
tvc_simulation <- function(n, beta1, beta2, gamma,
 t_0i_par1, t_0i_par2, c_i_par){

  # draw time-fixed covariate values and regression term
  x1 <- rnorm(n)
  x2 <- rbinom(n, 1, 0.5)
  mu_term <- exp(beta1 * x1 + beta2 * x2)

  # draw t_0i values
  t_0i <- runif(n, t_0i_par1, t_0i_par2)

  # fixed lambda and nu
  lambda <- 1
  nu <- 1.5

  # draw -log(u)
  neg_log_u = -log(runif(n))

  # draw event times
  t <- NULL
  for(i in 1:n){
    if(neg_log_u[i] < lambda * mu_term[i] * t_0i[i]^nu){
      t_i <- (neg_log_u[i]/( mu_term[i]))^(1/nu)
      t <- c(t, t_i)
    }else{
      t_i <- ((neg_log_u[i] - mu_term[i] * t_0i[i]^nu
       + (mu_term[i] * exp(gamma)) * t_0i[i]^nu)/
```

```
      (mu_term[i] * exp(gamma)))^(1/nu)
    t <- c(t, t_i)
  }
}

c_i <- runif(n, 0, c_i_par)
delta <- as.numeric(t < c_i)
TL <- TR <- rep(0,n)
TL[which(delta == 1)] <- t[which(delta == 1)]
TR[which(delta == 1)] <- t[which(delta == 1)]
TL[which(delta == 0)] <- c_i[which(delta == 0)]
TR[which(delta == 0)] <- Inf

# create "long" dataset
i_long <- TL_long <- TR_long <- delta_long <- x1_long
  <- x2_long <- Z <- start <- end <- last_record <- NULL

for(i in 1:n){
  if(TL[i] < t_0i[i]){
    # no change in z observed
    i_long <- c(i_long, i)
    TL_long <- c(TL_long, TL[i])
    TR_long <- c(TR_long, TR[i])
    delta_long <- c(delta_long, delta[i])
    x1_long <- c(x1_long, x1[i])
    x2_long <- c(x2_long, x2[i])
    Z <- c(Z, 0)
    start <- c(start, 0)
    end <- c(end, TL[i])
    last_record <- c(last_record, 1)
  }else{
    # change in z observed
    i_long <- c(i_long, rep(i,2))
    TL_long <- c(TL_long, rep(TL[i],2))
    TR_long <- c(TR_long, rep(TR[i],2))
    delta_long <- c(delta_long, rep(delta[i],2))
    x1_long <- c(x1_long, rep(x1[i],2))
    x2_long <- c(x2_long, rep(x2[i],2))
    Z <- c(Z, 0, 1)
    start <- c(start, 0, t_0i[i])
    end <- c(end, t_0i[i], TL[i])
    last_record <- c(last_record, 0, 1)

  }
}

dat_long <- data.frame(i_long, TL_long, TR_long, delta_long,
x1_long, x2_long, Z, start, end, last_record)
```

```
  # create baseline dataset
  dat_baseline <- data.frame(i = c(1:n), TL = TL, TR = TR,
  delta = delta, x1 = x1, x2 = x2, rep = c(table(i_long)))

  out = list(dat_baseline = dat_baseline, dat_long = dat_long)
  return(out)

}
```

The above R code can be used to generate right-censored survival data from a Cox model time-varying covariates. An example is given below.

```
> # generate datasets
> data <- tvc_simulation(1000, 1, 1, 1, 0.1, 0.5, 1.05)
> dat_baseline <- data$dat_baseline
> dat_long <- data$dat_long
>
> # approx 50% right censoring
> table(dat_baseline$delta)

  0   1
424 576
>
> # approx 50% have change in z observed
> table(dat_baseline$rep)

  1   2
494 506
```

This generated dataset can be used to fit an extended Cox model, and its R code is given below. The reader is encouraged to run this code themselves. The required R function "tvc_fit.R" is available in the Cox model time-varying covariates R package at the GitHub address for this book.

```
# fit using partial likelihood


# fit using MPL method access code for
# tvc_fit() function at <<<<GITHUB>>>>>
dat_long$left_position <- dat_long$tau = rep(0,
    nrow(dat_long))
dat_left <- dat_long
dat_left[which(dat_left$tau != 1), -1] <- 0

ctrl <- tvc_mpl_control(lambda = 0, iter = c(10,
    1000), n_knots = 7, par_initial = c(0, 0, 0.1),
    range = c(0.1, 0.9), line_search = c(1, 1, 1),
    reg_conv = 1e-05)
```

```
test <- tvc_fit((dat_long), (dat_baseline), (dat_left),
    ctrl)
```

□

In the next example, we will conduct a simple simulation study to compare the maximum likelihood estimates of an extended Cox model parameters with the estimates obtained from the partial likelihood method.

**Example 6.4** (A simulation study).
In this example, we provide R code for conducting a Monte Carlo simulation study aiming to compare the maximum likelihood (ML) and the partial likelihood (PL) estimates of extended Cox model regression coefficients. An ML estimate is obtained using the R MPL function simply by setting the smoothing value =0.

This simulation contains $M = 200$ repetitions of samples with two different sample sizes $n = 100$ and $n = 2000$, and two approximate censoring proportions: $\pi_c = 20\%$ and $\pi_c = 80\%$. The Cox model we adopt for this simulation is given in (6.47) of Example 6.3. For each combination of $\pi_c$ and $n$, we estimate the baseline and time-dependent effects using the ML method (estimates denoted by $\widehat{\beta}_1^{ML}, \widehat{\beta}_2^{ML}, \widehat{\gamma}_1^{ML}$) and compare them with the PL estimates (estimates denoted by $\widehat{\beta}_1^{PL}, \widehat{\beta}_2^{PL}, \widehat{\gamma}_1^{PL}$). Here, M-splines are adopted for the basis functions.

Table 6.3 summarizes the bias, standard deviation (SD), and mean-square error (MSE) for the regression coefficients calculated from the simulation study. The following R code is for $n = 2000$, and the R code for $n = 100$ is obtained by simply replacing 2000 with 100 in the '`tvc_simulation`' function. Note that the '`tvc_simulation`' function is given in Example 6.3 above.

```
save_ch6 <- matrix(0, nrow = 200, ncol = 12)

for (s in 1:200) {
    data <- tvc_simulation(2000, 0.5, -0.2, 0.3,
        0.1, 0.7, 3.5)
    dat_baseline <- data$dat_baseline
    dat_long <- data$dat_long

    # fit using mpl
    dat_long$left_position <- dat_long$tau = rep(0,
        nrow(dat_long))
    dat_left <- dat_long
    dat_left[which(dat_left$tau != 1), -1] <- 0
    ctrl <- tvc_mpl_control(lambda = 0, iter = c(1,
        2000), n_knots = 10, par_initial = c(0,
        0, 0.1), range = c(0.1, 0.9), line_search = c(1,
        1, 1), reg_conv = 1e-05)
    test <- tvc_fit((dat_long), (dat_baseline),
        (dat_left), ctrl)
```

```
    save_ch6[s, 1] <- test$parameters$beta[1]
    save_ch6[s, 2] <- test$parameters$beta[2]
    save_ch6[s, 3] <- test$parameters$gamma[1]

    save_ch6[s, 4] <- test$se_Q[1]
    save_ch6[s, 5] <- test$se_Q[2]
    save_ch6[s, 6] <- test$se_H[3]

    # fit using pl
    dat_long$event_long <- dat_long$delta_long *
        dat_long$last_record

    surv.obj <- Surv(time = dat_long$start, time2 = dat_long$end,
        type = "counting", event = dat_long$event_long)
    pl.fit <- coxph(surv.obj ~ dat_long$x1_long +
        dat_long$x2_long + dat_long$Z, data = dat_long,
        id = dat_long$i_long)


    save_ch6[s, 7] <- pl.fit$coefficients[1]
    save_ch6[s, 8] <- pl.fit$coefficients[2]
    save_ch6[s, 9] <- pl.fit$coefficients[3]

    save_ch6[s, 10] <- sqrt(diag(pl.fit$var))[1]
    save_ch6[s, 11] <- sqrt(diag(pl.fit$var))[2]
    save_ch6[s, 12] <- sqrt(diag(pl.fit$var))[3]

}
```

The results of this simulation are summarized in Table 6.3:

These results show that for both $n = 100$ and $n = 2000$, the ML estimates for this example perform similar to PL estimates on both bias and standard deviations for all the regression coefficients.

The ML method for the extended Cox model also provides an estimate of the baseline hazard. Plot in Figure 6.2 compares the true baseline hazard with the average baseline hazard estimate for each of the four $n$ and $\pi$ combinations, together with point-wise 95% Monte Carlo CIs. The results demonstrate that the baseline hazard can be recovered to a close degree of accuracy by the ML method.

□

TABLE 6.3: Comparing estimates $\widehat{\gamma}_1$, $\widehat{\beta}_1$ and $\widehat{\beta}_2$ obtained from the ML and the PL methods.

| | | | $n = 100$ $\pi_c = 20\%$ | $n = 100$ $\pi_c = 80\%$ | $n = 2000$ $\pi_c = 20\%$ | $n = 2000$ $\pi_c = 80\%$ |
|---|---|---|---|---|---|---|
| $\widehat{\gamma}_1$ | PL | Bias | 0.0347 | 0.0397 | 0.0046 | 0.0064 |
| | | SD | 0.3928 | 0.4673 | 0.0735 | 0.0854 |
| | | MSE | 0.2540 | 0.3237 | 0.0946 | 0.1007 |
| | ML | Bias | 0.0201 | 0.0414 | 0.0023 | 0.0075 |
| | | SD | 0.2607 | 0.2927 | 0.0465 | 0.0569 |
| | | MSE | 0.2618 | 0.3191 | 0.1011 | 0.1028 |
| $\widehat{\beta}_1$ | PL | Bias | 0.0201 | 0.0139 | 0.0003 | 0.0034 |
| | | SD | 0.1319 | 0.1813 | 0.0513 | 0.0512 |
| | | MSE | 0.2156 | 0.2110 | 0.2107 | 0.2304 |
| | ML | Bias | 0.0329 | 0.0202 | 0.0005 | 0.0072 |
| | | SD | 0.1233 | 0.1664 | 0.0318 | 0.0478 |
| | | MSE | 0.2242 | 0.2160 | 0.0516 | 0.0712 |
| $\widehat{\beta}_2$ | PL | Bias | -0.0035 | -0.0445 | 0.0013 | 0.0065 |
| | | SD | 0.2415 | 0.3414 | 0.0270 | 0.0532 |
| | | MSE | 0.2912 | 0.3918 | 0.1245 | 0.1713 |
| | ML | Bias | -0.0076 | -0.0534 | 0.0015 | 0.0035 |
| | | SD | 0.2172 | 0.2943 | 0.0269 | 0.0435 |
| | | MSE | 0.2943 | 0.4028 | 0.0298 | 0.0266 |

## 6.9 Application to breast cancer recurrence data

In this section, we implement the MPL method to data from a study of breast cancer recurrence, and this dataset is available at "`https://github.com/MPL-book`". Clinical and demographic characteristics and time to cancer recurrence were recorded for $n = 902$ breast cancer patients at Duke Hospital, USA, between 2000 and 2014. The full dataset and information on variables collected are available at Saha et al. (2018); in this example, we apply the MPL method to only a selection of the available variables. This dataset contains a high percentage of right censoring (approximately $87\%$) of the event of interest (diagnosis of a cancer recurrence).

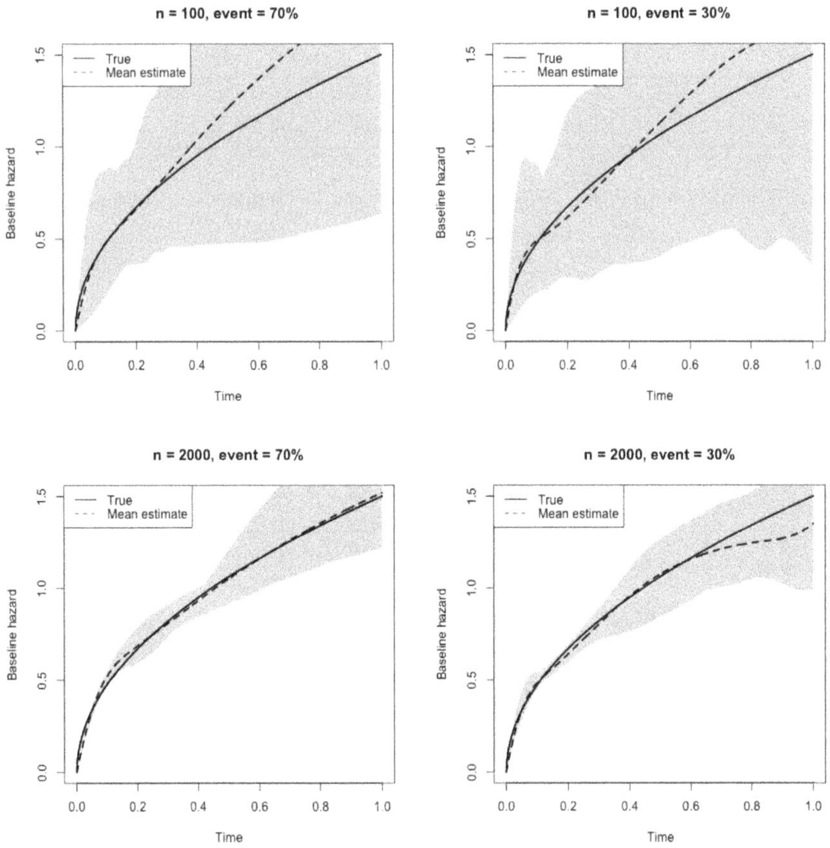


FIGURE 6.2: Estimated baseline hazard functions with 95% Monte Carlo coverage probabilities.

**Example 6.5** (Breast cancer recurrence analysis)**.**
In this example, we select four time-fixed covariates to include in the extended MPL Cox model:

“age at diagnosis", a binary variable indicated whether a patient was under the age of 35 years at the time of the original cancer diagnosis;
“spread to lymph nodes": a binary variable indicated whether the cancer had spread to the lymph nodes at the time of the original diagnosis;
“breast cancer sub-type": a binary variable indicating whether an individual’s breast cancer was of the triple negative sub-type; and
“tumour stage": a binary variable indicating whether the tumour was Stage 4 at the time of the original diagnosis.

TABLE 6.4: Sample observations of the dataset.

| ID | Start | Stop | Status | Age | Lymph Node | Sub-type | Stage | BCS | Mastectomy |
|---|---|---|---|---|---|---|---|---|---|
| 9 | 0 | 98 | 0 | 1 | 1 | 1 | 1 | 0 | 0 |
| 9 | 98 | 407 | 1 | 1 | 1 | 1 | 1 | 0 | 1 |
| 13 | 0 | 29 | 0 | 1 | 0 | 0 | 0 | 0 | 0 |
| 13 | 29 | 1064 | 0 | 1 | 0 | 0 | 0 | 1 | 0 |

Additionally, we include a time-varying covariate indicating whether a patient had undergone one of two possible types of surgical intervention: breast conservation surgery [BCS] or mastectomy at time $t$. Note that each patient could only receive one of the two types of surgery, and that the nature of this time-varying covariate is comparable to the one used in the simulation study in Example 6.4, i.e. each of the dummy variables for the two surgery types start at 0 and may change to 1, after which they will not change again.

Table 6.4 below gives an example of two patients included in this dataset. The first, with patient ID 9, was under 35 years of age and had lymph node involvement and a Stage 4 tumour at diagnosis, and was diagnosed with triple-negative breast cancer. At $t = 98$ days after diagnosis, this patient underwent a mastectomy, and at $t = 407$ days after original diagnosis was diagnosed with a cancer recurrence. The second, with patient ID 13, was under 35 years of age at diagnosis, but did not have lymph node involvement, had a tumour below Stage 4 and did not have triple negative breast cancer. At $t = 29$ days after diagnosis, this patient underwent a breast conservation surgery, and by the end of follow-up for this patient at $t = 1064$ days, there had been no recurrence diagnosed.

We estimate the model

$$h_i(t) = h_0(t)e^{\beta_1 x_{i1}+\beta_2+x_{i2}+\beta_3 x_{i3}+\beta_4 x_{i4}+\gamma_1 z_{i1}(t)+\gamma_2 z_{i2}(t)}$$

where

$x_{i1}$ gives the age at baseline,
$x_{i2}$ gives the lymph node involvement,
$x_{i3}$ gives the triple negative status, and
$x_{i4}$ gives the tumour stage.

The time-varying covariates are $z_{i1}(t)$, which is 0 if participant $i$ had not received a breast conservation surgery at or before time $t$ and 1 if they had received this surgery at or before this time, and $z_{i2}(t)$, which is 0 if the participant had not received a mastectomy at or before time $t$ and 1 if they had. We estimate all of the $\beta$ and $\gamma$ regression parameters using the MPL method outlined in this chapter and approximate the baseline hazard function using M-splines. R code for fitting this model is displayed below.

```
# Read in long format and baseline datasets
# available on Github
bc_longformat <- read.table("bc_longformat.txt",
    header = TRUE, row.names = NULL)
```

```
bc_baseline <- read.table("bc_baseline.txt", header = TRUE,
    row.names = NULL)

# Need to create a dataset that is cut-off at
# the left end of any interval censoring
# intervals

# Note that creating bc_left requires the
# bc_longformat dataset to have the columns
# tau and left_position

bc_longformat$left_position <- bc_longformat$tau <- rep(0,
    nrow(bc_longformat))

for (i in 1:length(unique(bc_longformat$i_long))) {

    # identify rows in bc_longformat
    # associated with individual i
    ind_long <- which(bc_longformat$i_long == i)

    if (bc_longformat$delta_long[ind_long[1]] ==
        3) {
        # if individual i has delta == 3 if
        # z(t) changed for individual i during
        # follow-up
        if (length(ind_long) == 2) {
            if (as.numeric(bc_longformat$start)[ind_long[2]] <
                as.numeric(bc_longformat$TL_long)[ind_long[1]]) {
                # if the change time of z(t)
                # for individual i was before
                # t_i^L
                bc_longformat$left_position[ind_long[2]] <- 1
                bc_longformat$tau[ind_long] <- 1
            } else {
                # if the change time of z(t)
                # for individual i was after
                # t_i^L
                bc_longformat$left_position[ind_long[1]] <- 1
                bc_longformat$tau[ind_long[1]] <- 1
            }

        } else if (length(ind_long == 1)) {
            # if z(t) did not change for
            # individual i during follow-up
            bc_longformat$left_position[ind_long] <- 1
            bc_longformat$tau[ind_long] <- 1
        }
    }
```

```
}

bc_left <- bc_longformat
bc_left[which(bc_left$tau != 1), -1] <- 0

left.i <- which(bc_baseline$delta == 3)
for (li in left.i) {
    # replace 'end' time with left interval
    # time make everything after left interval
    # time 0
    ind <- which(bc_longformat$i_long == li)
    left.pos.ind <- ind[which(bc_left$left_position[ind] ==
        1)]
    bc_left$end[left.pos.ind] <- bc_left$TL[left.pos.ind]
}


# Now we can fit the model

ctrl <- tvc_mpl_control(lambda = 0, iter = c(10,
    3000), n_knots = 6, par_initial = c(0, 0, 1),
    range = c(0.1, 0.9), line_search = c(1, 1, 1),
    reg_conv = 1e-05)
breastcancer_mpl <- tvc_fit(bc_longformat, bc_baseline,
    bc_left, ctrl)
```

TABLE 6.5: Cox model regression parameters and hazard ratio for the breast cancer example.

| Covariate | Parameter | SE | p-value | HR | HR 95% CI |
|---|---|---|---|---|---|
| Age $< 35$ at diagnosis | 0.704 | 0.306 | 0.022 | 2.023 | 1.109, 3.688 |
| Lymph node involvement | 0.791 | 0.274 | 0.004 | 2.207 | 1.289, 3.778 |
| Triple negative sub-type | 1.277 | 0.429 | 0.003 | 3.586 | 1.548, 8.306 |
| Stage 4 tumor at diagnosis | 0.589 | 0.414 | 0.155 | 1.802 | 0.801, 4.060 |
| BCS (vs. no surgery) | -0.986 | 0.311 | 0.002 | 0.373 | 0.203, 0.686 |
| Mastectomy (vs. no surgery) | -1.190 | 0.224 | $< 0.001$ | 0.304 | 0.196, 0.472 |

In Table 6.5, we display the regression parameter results from fitting this model. The time-fixed covariates of age at diagnosis, lymph node involvement and triple negative cancer sub-type all significantly increase the risk of recurrence diagnosis. A Stage 4 tumour at diagnosis was not a statistically significant predictor of risk of recurrence diagnosis. Additionally, both surgery types significantly decrease the risk of recurrence diagnosis compared to having no surgical intervention.

Additionally, from the fitted MPL model, we can obtain estimates of the baseline survival quantities, and predicted survival quantities for any combination of covariates, along with their 95% point-wise confidence intervals. The estimated baseline hazard function and associated point-wise confidence interval is plotted in Figure 6.3. □

## 6.10 Stanford Heart Transplant example

In this section we implement the MPL method to the Stanford Heart Transplant data. This dataset has been studied several times, notably by Crowley & Hu (1977) who study the effects of a number of covariates as to whether heart transplant prolongs survival for patients with heart disease. The data follows 103 patients admitted to the Stanford Heart Transplant Program during which time a search for a suitable donor heart is undertaken. This search takes from between a few days to up to a year. Seventy-five patients undergo a transplant, however some patients die prior to a suitable donor heart is found.

This dataset has been studied in Example 6.2 where the PL method was demonstrated. Here, we study it again using the MPL method.

**Example 6.6** (Stanford heart transplant data analysis).
In this example, we select three covariates:

“age" – age in years at acceptance minus 45;
“year" – acceptance time point after the study began;
“surgery" – an indicator if the patient has previously had bypass surgery.

A single time-dependent covariate called “transplant" is created to indicate whether the patient has a heart transplant at a particular time.

Table 6.6 below display two typical patients: patients 4 and 5, with long-format data. Patient 4 is accepted 0.49 years after the study began at the age of 37.26 years, and has no previous bypass surgery. At day 35 the subject receives a transplant but die 3 days later. Patient 5 is accepted 0.61 years after the study began at the age of 17.8 years, and has no previous bypass surgery. The patient never receives a transplant during the study and dies after 17 days.

Using this data, we estimate the model

$$h_i(t) = h_0(t)e^{\beta_{age}x_{age}+\beta_{year}x_{year}+\beta_{surg}x_{surg}+\gamma_{tran}z_{tran}(t)} \tag{6.49}$$

where:

$x_{age}$ is patient age in years at acceptance;
$x_{year}$ is the year of patient acceptance;
$x_{surg}$ is an indicator if the patient has previously had bypass surgery (1=yes, 0=no); and
$z_{trans}(t)$ is a time-dependent covariate whose values change from 0 prior to a patient reviving a transplant to 1 post a patient receiving a transplant.

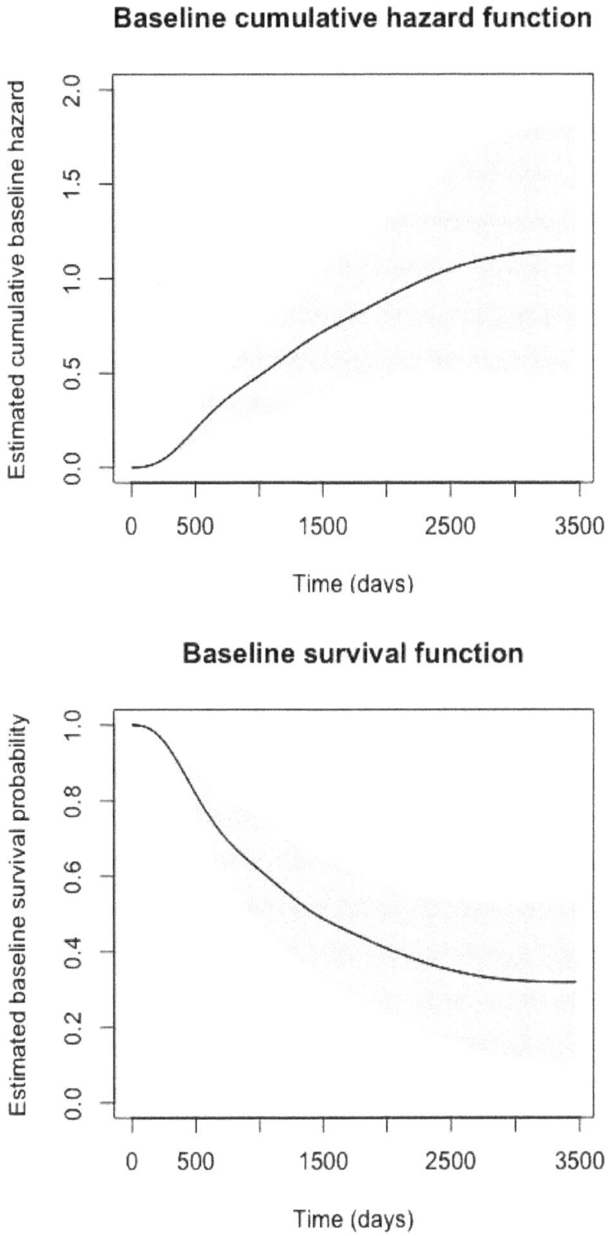


FIGURE 6.3: Baseline cumulative hazard and baseline survival functions for time to breast cancer recurrence from the fitted MPL model, with 95% pointwise confidence intervals.

TABLE 6.6: Patients 4 and 5 from the Stanford Heart Transplant data, in long (or counting process) format.

| Patient Id | start | stop | status | transplant | age | year | surgery |
|---|---|---|---|---|---|---|---|
| 4 | 0 | 35 | 0 | 0 | -7.74 | 0.49 | 0 |
| 4 | 35 | 38 | 1 | 1 | -7.7 | 0.49 | 0 |
| 5 | 0 | 17 | 1 | 0 | -27.2 | 0.61 | 0 |

The regression parameters $(\beta_{age}, \beta_{year}, \beta_{surg}, \gamma_{trans})$ are unknown and require estimation. For maximum penalized likelihood estimation, we approximate the baseline hazard by M-spline functions. R code for this example is displayed below.

```
library(survival)
data("heart")
# We need to do some data
# processing to use the MPL
# package
n <- 103

# Create baseline data set
stnfrd_hrt_baseline <- NULL

for (i in 1:n) {
    ind_long <- which(heart$id ==
        i)
    new_row <- c(heart$id[ind_long[1]],
        heart$stop[ind_long[length(ind_long)]],
        ifelse(heart$event[ind_long[length(ind_long)]] ==
            1, heart$stop[ind_long[length(ind_long)]],
            Inf), heart$event[ind_long[length(ind_long)]],
        as.numeric(heart$age)[ind_long[1]],
        as.numeric(heart$year)[ind_long[1]],
        as.numeric(heart$surgery)[ind_long[1]],
        length(ind_long))
    stnfrd_hrt_baseline <- rbind(stnfrd_hrt_baseline,
        new_row)
}

row.names(stnfrd_hrt_baseline) <- NULL
stnfrd_hrt_baseline <- as.data.frame(stnfrd_hrt_baseline)
colnames(stnfrd_hrt_baseline) <- c("i",
    "TL", "TR", "delta", "age",
    "year", "surgery", "rep")
table(stnfrd_hrt_baseline$rep)

# Create long data set
```

```
stnfrd_hrt_long <- data.frame(i_long = heart$id,
    TL_long = rep(stnfrd_hrt_baseline$TL,
        stnfrd_hrt_baseline$rep),
    TR_long = rep(stnfrd_hrt_baseline$TR,
        stnfrd_hrt_baseline$rep),
    delta_long = rep(stnfrd_hrt_baseline$delta,
        stnfrd_hrt_baseline$rep),
    age_long = rep(stnfrd_hrt_baseline$age,
        stnfrd_hrt_baseline$rep),
    year_long = rep(stnfrd_hrt_baseline$year,
        stnfrd_hrt_baseline$rep),
    surgery_long = rep(stnfrd_hrt_baseline$surgery,
        stnfrd_hrt_baseline$rep),
    transplant = as.numeric(heart$transplant ==
        1), start = heart$start,
    end = heart$stop)

stnfrd_hrt_long$last_record <- 0
for (i in 1:n) {
    ind_long <- which(stnfrd_hrt_long$i_long ==
        i)
    stnfrd_hrt_long$last_record[ind_long[length(ind_long)]] <- 1
}

stnfrd_hrt_long$left_position <- stnfrd_hrt_long$tau <- rep(0,
    nrow(stnfrd_hrt_long))
stnfrd_hrt_left <- stnfrd_hrt_long
stnfrd_hrt_left[which(stnfrd_hrt_left$tau !=
    1), -1] <- 0

# Fit the MPL model
library(splines2)
library(dplyr)
ctrl <- tvc_mpl_control(lambda = 0,
    iter = c(10, 1000), n_knots = 4,
    par_initial = c(0, 0, 0.1),
    range = c(0.1, 0.9), line_search = c(1,
        1, 1), reg_conv = 1e-06)
test <- tvc_fit((stnfrd_hrt_long),
    (stnfrd_hrt_baseline), (stnfrd_hrt_left),
    ctrl)

# Plots of baseline survival
# quantities estimates
plot(test$func_est$h0_est ~ test$func_est$v,
    type = "l")
plot(test$func_est$H0_est ~ test$func_est$v,
    type = "l")
```

```
plot(test$func_est$S0_est ~ test$func_est$v,
    type = "l")

# Predictive survival plots
# year = age = 0, no surgery
# vs. year = age = 0, surgery
# at t = 20
St_surg0 <- LL <- UL <- NULL

for (t in 1:1000) {
    Ht <- (test$func_est$H0_est[t])
    St <- exp(-Ht)
    Psi <- Psi_f(events = test$func_est$v[t],
        knts = test$kn$int_knots,
        Boundary.knots = test$kn$bound_knots)

    dSt_beta <- -St * Ht * as.vector(rep(0,
        3))
    dSt_gamma <- -St * Ht * as.vector(0)
    dSt_theta <- -St * Psi
    dSt_eta <- matrix(c(dSt_beta,
        dSt_gamma, dSt_theta))
    cov <- test$covar_H

    se <- sqrt(t(dSt_eta) %*% cov %*%
        (dSt_eta))
    St_surg0 <- c(St_surg0, St)
    LL <- c(LL, (St - 1.96 * se))
    UL <- c(UL, (St + 1.96 * se))

}

plot(test$func_est$v, St_surg0,
    type = "l", ylim = c(0, 1),
    main = "Survival probability by transplant: none",
    ylab = "Estimated survival probability",
    xlab = "Time (days)")
polygon(x = c(test$func_est$v,
    rev(test$func_est$v)), y = c(LL,
    rev(UL)), border = NA, col = adjustcolor("grey60",
    alpha.f = 0.5))

lines(test$func_est$v, St_surg0,
    lwd = 2)

St_surg_t10 <- LL1 <- UL1 <- NULL

for (t in 1:1000) {
```

```
    theta <- test$parameters$theta
    if (test$func_est$v[t] < 10) {
        Ht <- (test$func_est$H0_est[t])
        St <- exp(-Ht)
        Psi <- Psi_f(events = test$func_est$v[t],
            knts = test$kn$int_knots,
            Boundary.knots = test$kn$bound_knots)

        dSt_beta <- -St * Ht *
            as.vector(rep(0, 3))
        dSt_gamma <- -St * Ht *
            as.vector(0)
        dSt_theta <- -St * Psi
        dSt_eta <- matrix(c(dSt_beta,
            dSt_gamma, dSt_theta))
        cov <- test$covar_H
        se <- sqrt(t(dSt_eta) %*%
            cov %*% (dSt_eta))

        St_surg_t10 <- c(St_surg_t10,
            St)
        LL1 <- c(LL1, (St - 1.96 *
            se))
        UL1 <- c(UL1, (St + 1.96 *
            se))
    } else {
        tMinus1 <- min(which(test$func_est$v >=
            10))

        Psi100 <- Psi_f(events = test$func_est$v[tMinus1],
            knts = test$kn$int_knots,
            Boundary.knots = test$kn$bound_knots)
        Ht100 <- Psi100 %*% theta
        Ht100_g <- Psi100 %*% theta *
            exp(test$parameters$gamma[1])

        Ht <- Ht100 + (test$func_est$H0_est[t] *
            exp(test$parameters$gamma[1])) -
            Ht100_g
        St <- exp(-Ht)
        Psi <- Psi_f(events = test$func_est$v[t],
            knts = test$kn$int_knots,
            Boundary.knots = test$kn$bound_knots)

        dSt_beta <- -St * Ht *
            as.vector(rep(0, 3))
        dSt_gamma <- -St * (test$func_est$H0_est[t] *
            exp(test$parameters$gamma[1])) +
```

```
            St * Ht100_g
        dSt_gamma <- as.vector(c(dSt_gamma))
        dSt_theta <- as.vector(-St *
            exp(test$parameters$gamma[1])) *
            Psi - as.vector(St) *
            Psi100 + as.vector(St *
            exp(test$parameters$gamma[1])) *
            Psi100
        dSt_eta <- matrix(c(dSt_beta,
            dSt_gamma, dSt_theta))
        cov <- test$covar_H
        se <- sqrt(t(dSt_eta) %*%
            cov %*% (dSt_eta))

        St_surg_t10 <- c(St_surg_t10,
            St)
        LL1 <- c(LL1, (St - 1.96 *
            se))
        UL1 <- c(UL1, (St + 1.96 *
            se))
    }
}

LL1[which(LL1 < 0)] = 0

plot(test$func_est$v, St_surg_t10,
    type = "l", ylim = c(0, 1),
    main = "Survival probability by transplant: @ t=10",
    ylab = "Estimated survival probability",
    xlab = "Time (days)")
polygon(x = c(test$func_est$v,
    rev(test$func_est$v)), y = c(LL1,
    rev(UL1)), border = NA, col = adjustcolor("grey60",
    alpha.f = 0.5))
lines(test$func_est$v, St_surg_t10,
    lwd = 2)
```

The results of the MPL and the PL estimation are detailed in Table 6.7, where the PL results are copied from Example 6.2. Each method returns similar accuracy for the regression coefficients and associated standard errors however the MPL method also returns a further twelve parameter estimates for the baseline hazard function displayed in Figure 6.4.

We plot the MPL baseline hazard, baseline cumulative hazard and baseline survival results in Figure 6.4. In Figure 6.5, we plot the predicted survival plots for two individuals, both with `age`, `year` and `surgery` set to 0, and one who never had a transplant and one who had a transplant at $t = 10$. The estimated survival probability over time is slightly higher for the individual who underwent a transplant compared

TABLE 6.7: Parameter estimates for the Stanford Heart Transplant Data.

| | MPL Estimation | | | | PL Estimation | | | |
|---|---|---|---|---|---|---|---|---|
| Parameter | Value | SE | z | pval | Value | SE | z | pval |
| $\widehat{\beta}_{age}$ | 0.028 | 0.013 | 2.101 | 0.047 | 0.027 | 0.014 | 1.981 | 0.024 |
| $\widehat{\beta}_{surg}$ | -0.653 | 0.366 | -1.783 | 0.075 | -0.636 | 0.367 | -1.731 | 0.042 |
| $\widehat{\beta}_{year}$ | -0.124 | 0.063 | -1.984 | 0.018 | -0.146 | 0.070 | -2.074 | 0.019 |
| $\widehat{\gamma}_{tran}$ | -0.065 | 0.239 | -0.274 | 0.785 | -0.015 | 0.313 | -0.046 | 0.482 |

to the individual who did not. However, the confidence intervals indicate that the two survival probabilities are not significantly different over time.

This example demonstrates that the likelihood-based estimate of survival quantities allows for the calculation of confidence intervals for the baseline hazard as well as for predictive survival probabilities. □

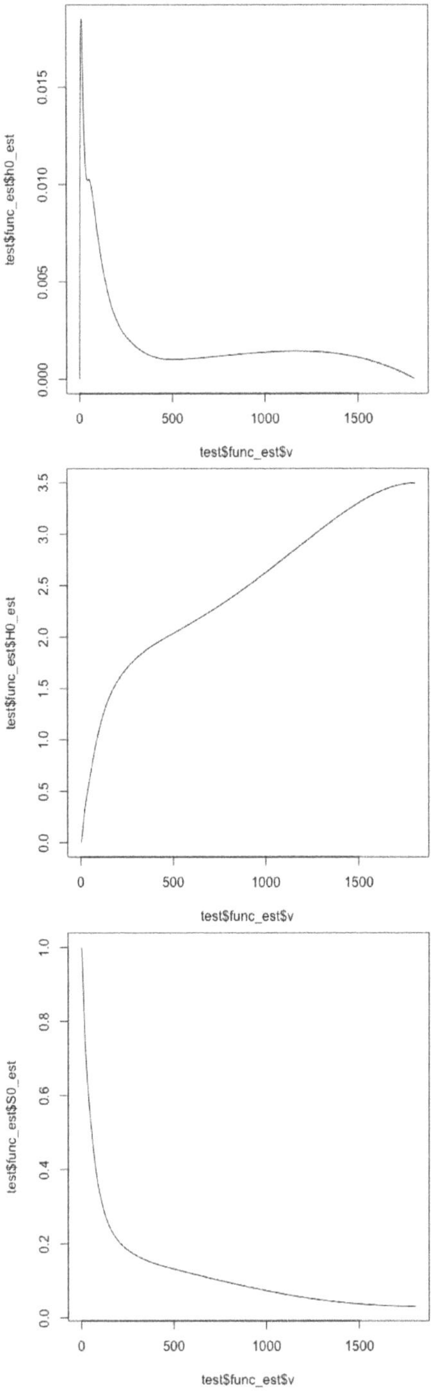


FIGURE 6.4: Baseline hazard, baseline cumulative hazard and baseline survival estimates from the fitted MPL model.

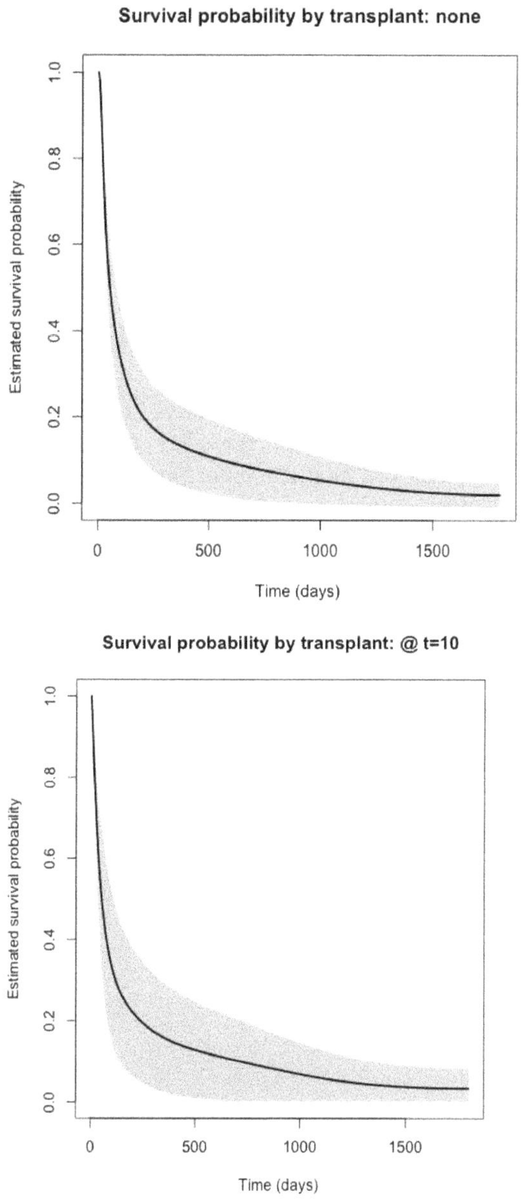


FIGURE 6.5: Predicted survival plots with point-wise 95% confidence intervals comparing survival probability over time between an individual with no transplant and an individual who had a transplant at $t = 10$.

## 6.11 Summary

The PL method used to estimate the extended Cox models (i.e. Cox model with time varying covariates) has two distinct short comings:

(1) The baseline hazard is not estimated by the model, so inferences such as the recovery of survival probabilities require a separate estimation step for the baseline hazard.

(2) The partial likelihood does not produce a covariance matrix for both the regression coefficients and the baseline hazard. This makes joint inferences of the model parameters difficult, except when using computationally intensive methods, such as bootstrapping.

This chapter focuses on the MPL method for extended Cox models. The MPL method allows for the estimation of the regression coefficients for time-fixed and time-varying covariates, as well as the baseline hazard. The MPL estimate of the baseline hazard is developed using non-negative basis functions. Its main computational challenge is ensuring that the coefficients of the basis functions remain non-negative. The asymptotic normality result for the MPL estimates accounts for the possibility that some basis function coefficients can be actively constrained.

The results reported in Thackham & Ma (2020) and Webb & Ma (2023) demonstrate the superior performance of the MPL method over the PL method in recovering the true regression coefficients for time-dependent covariates when sample sizes are not large but censoring proportions are large.

# 7

# *Copula Cox models for dependent right censoring*

Dependent censoring refers to the situation when the event time and censoring time are dependent, and this context contradicts the assumption of independent censoring we have made so far.

One problem that arises with dependent censoring is that some of the parameters of the event time can be influenced by the censoring time. This phenomenon is also known as "informative censoring". This means that the censoring time should also be taken into account in the likelihood function. Ignoring dependent censoring can lead to biases in the estimated parameters, as shown in comments appeared in Xu et al. (2018), Dettoni et al. (2020).

Suppose there is only right censoring (a situation we consider in this chapter). Since the event time and the censoring times cannot be observed simultaneously, an identifiability issue arises regarding the dependence structure of the event and censoring times. Therefore, assumptions about the dependence structure between event and censoring times *must be made* before incorporating dependent censoring into the parameter estimation process.

In the literature, three types of methods for dependent censoring exist: (i) copula methods, (ii) shared frailty methods and (iii) shared predictor methods. In this chapter, we will focus on copula methods.

## 7.1 Introduction

In this chapter, we focus on right-censored survival data. Discussions in the previous chapters assume an important assumption, namely that the censoring time is independent of the event time given the covariates. This assumption usually also implies that the parameters of the censoring time are non-informative about the model parameters of the event time; e.g. Kalbfleisch & Prentice (2002).

The independent censoring assumption underlies many popular methods in traditional survival analysis, such as the Kaplan-Meier estimate (Kaplan & Meier 1958) for survival functions, the Nelson-Aalen estimate (Nelson 1972) for cumulative hazards, and the maximum partial likelihood estimate (Cox 1972) for Cox model

DOI: 10.1201/9781351109710-7

regression coefficients. This assumption is also employed in the full likelihood-based estimation methods described in Chapters 2 to 6.

It is important to note that, in practical scenarios, event times and censoring times can sometimes be correlated. For instance, in a clinical trial, patients may withdraw from the study due to deteriorating health conditions after the trial has begun. This withdrawal often signifies a higher risk of adverse outcomes and mortality for the withdrawn patients compared to those who remain in the study. Ignoring this correlation can lead to biased estimates and possibly misleading conclusions.

It is common for observations to involve a mix of dependent and independent censoring times. In a clinical trial, for example, some participants may exit due to health deterioration, resulting in dependent censoring, while others may experience censoring due to the end of the study, which is considered independent censoring. The latter type of censoring, often referred to as "administrative censoring" (Joffe 2001), will be treated interchangeably with independent censoring in this chapter.

We will incorporate both dependent and independent right censoring times into the survival time observations. Additionally, we require that the independent censoring times be *non-informative*, meaning they do not convey information about the parameters of interest, specifically the parameters related to the event and dependent censoring times. We assume that we can distinguish in advance whether each observed censoring time is dependent or independent.

We will continue to use the survival data notation adopted in the previous chapters. We let $Y_i$ represent the event time of interest and $C_i$ denote the right censoring time, where $i = 1, \dots, n$. Let $A_i$ be the independent right censoring time. The time we can observe is thus $T_i = \min\{Y_i, C_i, A_i\}$. We consider dependence between $Y_i$ and $C_i$ in this chapter.

The model we wish to fit is the Cox model for the event time:

$$h_Y(t|\mathbf{x}_i) = h_{0Y}(t)e^{\mathbf{x}_i^\top \boldsymbol{\beta}}. \tag{7.1}$$

Here, the subscript $Y$ for the baseline hazard indicates that this is a model for the event time $Y$, distinguishing it from the Cox model introduced later in (7.2) for the dependent censoring time $C$. In (7.1), $\mathbf{x}_i$ denotes a $p_1 \times 1$ vector representing the covariate values of individual $i$, and the dimension of $\boldsymbol{\beta}$ in this chapter is $p_1 \times 1$.

We only consider time-fixed covariates in this chapter. We will discuss how to find the MPL estimates of the regression coefficient $\boldsymbol{\beta}$ and the baseline hazard $h_{0Y}(t)$ for this model. Furthermore, we will provide the asymptotic properties of these estimates to enable inferences on quantities of interest.

The Cox model (7.1) is a *marginal* model of the event time only. Later, we will also introduce a marginal model for the dependent right censoring time. There is no need to introduce a model for the independent right censoring time, as we have assumed they are non-informative to the parameters of interest.

There exist three groups of methods to fit the Cox regression models (7.1) when dependent right censoring presents:

(a) Copula methods (for example, Xu et al. (2018), Huang & Zhang (2008), Chen (2010), Chen et al. (2017)).

(b) Frailty methods (for example, Huang & Wolfe (2002)).

(c) Shared predictors methods (for example, Dettoni et al. (2020)).

The basic idea behind the copula method is to use a copula function to approximate the joint survival function for the event time and dependent censoring time. This *approximate* joint survival function can then be used to estimate parameters of the marginal hazard functions, such as the model (7.1). Rich materials on copula methods for dependent right censoring can be found in, for example, Emura & Chen (2018).

The copula method of Huang & Zhang (2008) is in fact a type of partial likelihood approach where the dependence structure between event and censoring times is modeled by an assumed copula function. It employs a self-consistent estimator, combined with the partial likelihood concept, to create an objective function akin to partial likelihood. This function is iteratively optimized, with the baseline cumulative hazard estimate updated in each iteration using the Breslow (1972) method.

The frailty method differs from the copula approach in its underlying idea. In the frailty approach, conditional independence is assumed between the event and censoring times, given the frailty (a type of latent variable). From this assumption and the assumed distribution of the frailty variable, the joint distribution for the event and censoring times can be obtained by integrating out the frailty variable. The frailty approach develops *conditional* hazard models for both the event and censoring times, given the frailty variable. However, interpreting the model parameters in frailty conditional models can be challenging. In contrast, copula marginal models do not suffer from this interpretation issue and are generally preferred in practice.

The idea behind the shared predictor approach is very simple. The approach assumes that if censoring is informative, some covariates can be common for both the event time and censoring time models. More details about this approach can be found in Dettoni et al. (2020).

In this chapter, parameters of model (7.1), that is the regression coefficients and the baseline hazard, are estimated by maximizing the penalized likelihood, where the distribution of the dependent censoring must be included into the likelihood function. The contribution of a copula function is that it provides an assumed joint distribution between the event and right censoring times. A penalty function is used to smooth (or regularize) the baseline hazard estimate. A detailed description of this method is available in Xu et al. (2018). When there are no covariates, Xu et al. (2016) suggest a non-parametric hazard estimation under dependent right censoring using copulas.

A penalty function is not absolutely necessary, so model (7.1) can alternatively be estimated by the maximum likelihood method. Examples of MLE include Chen (2010), Emura & Chen (2018) and Chen et al. (2017).

However, the roughness penalty function we adopt has two benefits: (i) it smoothes the baseline hazard estimate, and (ii) it reduces the sensitivity due to the location and number of knots used to approximate the baseline hazard function.

This chapter provides a detailed explanation of the MPL method for fitting model (7.1) in the presence of dependent right censoring, where this dependence is modeled using a copula function. When there is no dependent censoring, this MPL method naturally reduces to the MPL method for survival data with independent right censoring, as discussed in, for example, Ma et al. (2014).

## 7.2 Dependent right censoring

Recall that the dependent right censoring time for individual $i$ is denoted by $C_i$, and the independent right censoring time by $A_i$. The observed event time is denoted as $T_i$ and is the minimum of the triplet $Y_i$, $C_i$ and $A_i$. Traditional measures of dependence such as covariance between $Y_i$ and $C_i$ cannot be directly obtained from the observed $T_i$ values as for each $i$, at most only one of $Y_i$ or $C_i$ can be observed. Therefore, assumption(s) must be made about the dependence structure between $Y_i$ and $C_i$. In this chapter, to model the dependence structure between $Y_i$ and $C_i$, a copula function is utilized. We will summarize copula functions in the next section; detailed accounts of copulas can be found in, for example, Nelsen (2006).

Fundamentally, a copula function is used to formulate a joint survival function between $Y_i$ and $C_i$ based on the assumed marginal survival functions of $Y_i$ and $C_i$. We have already assumed a marginal Cox hazard model for $Y_i$ given in (7.1), and this leads to the corresponding marginal survival function for $Y_i$.

Similar to $Y_i$, we also assume a Cox model to be the working marginal model for $C_i$. Specifically, we adopt:

$$h_C(t|\mathbf{z}_i) = h_{0C}(t)e^{\mathbf{z}_i^\top \boldsymbol{\phi}}. \tag{7.2}$$

Here, $\mathbf{z}_i$ represents the individual $i$ covariate vector for $C_i$ and it contains $p_2$ covariates, and therefore $\boldsymbol{\phi}$ is a $p_2 \times 1$ vector of regression coefficients. There is flexibility in choosing the covariates for models (7.1) and (7.2). They can be entirely different or even the same.

We denote the marginal survival functions of $Y_i$ and $C_i$ by $S_Y(t|\mathbf{x}_i)$ and $S_C(t|\mathbf{z}_i)$, respectively. From models (7.1) and (7.2), the marginal survival functions are:

$$S_Y(t|\mathbf{x}_i) = e^{-H_Y(t|\mathbf{x}_i)}; \tag{7.3}$$

$$S_C(t|\mathbf{z}_i) = e^{-H_C(t|\mathbf{z}_i)}, \tag{7.4}$$

where $H_Y(t|\mathbf{x}_i)$ and $H_C(t|\mathbf{x}_i)$ are the corresponding cumulative hazards, given by

$$H_Y(t|\mathbf{x}_i) = H_{0Y}(t)e^{\mathbf{x}_i^\top \boldsymbol{\beta}}, \tag{7.5}$$

$$H_C(t|\mathbf{z}_i) = H_{0C}(t)e^{\mathbf{z}_i^\top \boldsymbol{\phi}}. \tag{7.6}$$

In these expressions, $H_{0Y}(t)$ and $H_{0C}(t)$ are the cumulative baseline hazard functions, namely: $H_{0Y}(t) = \int_0^t h_{0Y}(s)ds$ and $H_{0C}(t) = \int_0^t h_{0C}(s)ds$.

Note that there is no need to assume any models for the independent censoring times $A_i$ as they are non-informative for $Y_i$ or $C_i$. This chapter does not consider time-varying covariates, and thus the covariates used in the above models are all time-fixed.

Next, we will begin by explaining copula functions and then demonstrate how to use them to construct the joint survival function of $Y$ and $C$ from their marginal survival functions $S_Y(t)$ and $S_C(t)$. Our focus is on Archimedean copulas.

## 7.3 Copulas

We denote a copula function by $K(a, b; \alpha)$. Here $a$ and $b$, both in $[0, 1]$, are variables of this function. The parameter $\alpha$ is referred to as the degree of association parameter. This parameter $\alpha$ can be converted into another parameter called Kendall's correlation coefficient ($\tau$), as introduced by Kendall (1970). Kendall's $\tau$ quantifies the rank correlation between variables $a$ and $b$ and its range is from $-1$ to 1.

We can construct a joint survival function for both $Y$ and $C$ using the copula function $K(a, b; \alpha)$. For time $t$, this joint survival function is defined as:

$$S_{Y,C}(t, t) = \Pr(Y > t, C > t) = K(S_Y(t), S_C(t); \alpha), \tag{7.7}$$

where $S_Y(\cdot)$ and $S_C(\cdot)$ represent the respective marginal survival functions of $Y$ and $C$. A simple example of the copula function $K$ is given by $K(a, b) = ab$, which is referred to as the independent copula as it assumes no dependence between $Y$ and $C$.

We have mentioned above that we will only consider the Archimedean copulas in this chapter. The Archimedean copulas form an important class of copulas; they are very easy to work with and possess nice properties. An Archimedean copula $K(a, b; \alpha)$ adopts the following general functional expression:

$$K(a, b; \alpha) = \phi^{-1}(\phi(a; \alpha) + \phi(b; \alpha)), \tag{7.8}$$

where $\phi$ is called the generator of copula function $K$. It requires that $\phi(u)$ satisfies:

(i) $\phi(1) = 0$ and,

(ii) $\phi$ is a convex and decreasing function with its domain (0, 1] and range $[0, \infty)$.

The following are some examples of function $\phi$, corresponding to several commonly used Archimedean copulas.

(a) **Clayton copula:**
The generator function for the Clayton copula (Clayton 1978) is

$$\phi(a) = a^{-\alpha} - 1,$$

where $\alpha > 1$. From this $\phi$ function, we can obtain the Clayton copula function for variables $a$ and $b$ as follows:

$$K(a, b; \alpha) = \left(a^{-\alpha} + b^{-\alpha} - 1\right)^{-1/\alpha}. \tag{7.9}$$

This Clayton copula is suitable when there exists positive dependence between $Y$ and $C$, with $\alpha > 1$ representing the strength of the dependence, and the corresponding Kendall's rank correlation coefficient is $\tau = \alpha/(\alpha + 2)$.

(b) **Gumbel-Hougaard copula:**
The generator for the Gumbel-Hougaard copula (Gumbel 1961, Hougaard 1986) is

$$\phi(a) = (-\log a)^{\alpha},$$

where $\alpha \geq 1$. From this $\phi$ function, the Gumbel-Hougaard copula is

$$K(a, b; \alpha) = e^{((-\log a)^{\alpha} + (-\log b)^{\alpha})^{1/\alpha}}. \tag{7.10}$$

Here, the associated Kendall's rank correlations is $\tau = 1 - \alpha^{-1}$. Again, since $\tau > 0$, this copula is suitable for modelling positively correlated $Y$ and $C$.

(c) **Frank copula:**
The generator for the Frank copula (Frank 1979) is

$$\phi(a) = \log \frac{e^{\alpha a} - 1}{e^{\alpha} - 1},$$

where $-\infty < \alpha < \infty$. This $\phi$ function generates the following Frank copula:

$$K(a, b; \alpha) = \frac{1}{\alpha} \log \left( \frac{(e^{\alpha a} - 1)(e^{\alpha b} - 1)}{e^{\alpha} - 1} + 1 \right). \tag{7.11}$$

The corresponding Kendall's $\tau$ is:

$$\tau = 1 - 4\left\{D_1(-\alpha) - 1\right\}/\alpha,$$

with $D_1(\alpha) = \int_0^{\alpha} t/(e^t - 1)dt/\alpha$. Since the Kendall's $\tau$ for the Frank copula can take either negative or positive values, this copula can be used to model either positive or negative dependence between $Y$ and $C$.

**Example 7.1** (Generation of dependent censoring survival data**).**
Generation of dependent censoring survival data can be easily achieved using copula functions in R. In this example, we provide R code for generating dependent right-censored survival data. Particular, the following activities are demonstrated in this example:

(i) Generating survival times with dependent censoring, where both marginal $Y$ and $C$ distributions follow Cox models with Weibull baseline hazards. More specifically,

$$h_Y(t) = h_{0Y}(t)e^{Z1-0.5Z2},$$
$$h_C(t) = h_{0C}(t)e^{0.2Z1-Z2},$$

where both baseline hazards are Weibull hazards with shape parameter $a$ and scale parameter $\lambda$, given by $h_{0Y}(t) = h_{0C}(t) = \lambda^{-a}at^{a-1}$. Data generation from a Cox model with Weibull baseline hazard has already been discussed in, for example, Example 2.6. Here, we explain another approach to generate random survival times. Firstly, for a Cox model with such a Weibull baseline, namely

$$h(t) = \lambda^{-a}at^{a-1}e^{\mathbf{x}^\mathsf{T}\boldsymbol{\beta}},$$

we can rewrite this hazard as

$$h(t) = \left(\lambda\left(e^{-\mathbf{x}^\mathsf{T}\boldsymbol{\beta}}\right)^{\frac{1}{a}}\right)^{-a} at^{a-1}.$$

Therefore, this hazard $h(t)$ can be conceived as a Weilbull hazard with the shape parameter $a$ and scale parameter $\lambda\left(e^{-\mathbf{x}^\mathsf{T}\boldsymbol{\beta}}\right)^{\frac{1}{a}}$. Thus, random numbers for marginal $Y_i$ can be generated from

$$Y_i \sim Weibull(a, \lambda\left(e^{-\mathbf{x}^\mathsf{T}\boldsymbol{\beta}}\right)^{\frac{1}{a}})$$

and random numbers for marginal $C_i$ can be generated from

$$C_i \sim Weibull(a, \lambda\left(e^{-\mathbf{z}^\mathsf{T}\boldsymbol{\phi}}\right)^{\frac{1}{a}}).$$

Then, using the R "`copula`" package, where the **Clayton** copula is adopted for this example with Kendall's correlation coefficient $\tau = 0.7$, we can generate a pair of dependent uniform random numbers representing marginal survival function values. From these dependent uniform random numbers, we can generate their corresponding $Y_i$ and $C_i$ random numbers based on the Weibull distributions as specified above. The sample size we select for this simulation is $n = 2000$.
(ii) Do a scatter plot of $Y_i$ (y-axis) and $C_i$ (x-axis) values to visualize dependence.
(iii) Fit a Cox model using PL by ignoring dependent censoring and comment on the accuracy of the estimated coefficients.

```
library(copula)

n = 2000
# define the copula
tau_ec = 0.7
alpha <- iTau(claytonCopula(100), tau = tau_ec)
cop <- claytonCopula(alpha, dim = 2)
```

```
x <- rep(0, n)  #observed time
del <- rep(0, n)  #event status
eta <- rep(0, n)  #(dependent) censoring status
y <- NULL
c <- NULL

# generate covariates
Z <- cbind(rbinom(n, 1, 0.5), runif(n, -1, 1))

# set true parameter values
betas <- c(1, -0.5)
phis <- c(0.2, -1)

# set weibull distribution parameters
lambda <- 2
a <- 14/3
# lamda and a can be selected differently by
# the reader

# exp(xTb) term for event time
expbetT <- exp(Z %*% betas)
# scale of weibull baseline hazard for event
lamT <- lambda * (exp(-Z %*% betas))^(1/a)
# scale of weibull baseilne hazard for
# censoring
lamCw <- lambda * (exp(-Z %*% phis))^(1/a)

for (i in 1:n) {
    S_tc <- rCopula(1, cop)  #S_event and S_cens probabilities

    # draw random event time from weibull dist
    ti <- qweibull((1 - S_tc[1, 1]), a, lamT[i])
    # draw random censoring time from weibull
    # dist
    ci <- qweibull((1 - S_tc[1, 2]), a, lamCw[i])
    # save event and censoring times
    y <- c(y, ti)
    c <- c(c, ci)

    if (ti < ci) {
        # event observed
        del[i] <- 1
        x[i] <- ti
    } else if (ci < ti) {
        # dependent censoring time observed
        eta[i] <- 1
        x[i] <- ci
    }
```

```
}
```

From this simulated data, we next create a scatter plot of the event and censoring times to visualize their dependence. Also, we fit a Cox model using the PL method, thereby ignoring dependent censoring, to examine the biases in the estimated regression coefficients.

```
> dat<-cbind(x, del, eta)
> head(dat)
            x del eta
[1,] 1.941319   0   1
[2,] 1.140722   0   1
[3,] 1.522070   0   1
[4,] 1.041665   1   0
[5,] 1.303935   0   1
[6,] 1.872389   1   0
>
> #plot
> plot(y ~ c, xlab = "Censoring time", ylab = "Event time")
>
> #fit normal Cox model (ignoring dependent censoring)
> library(survival)
> surv.obj <- Surv(time = x, event = del)
> coxph(surv.obj ~ Z[,1] + Z[,2])
Call:
coxph(formula = surv.obj ~ Z[, 1] + Z[, 2])

            coef exp(coef) se(coef)      z        p
Z[, 1]   1.56313   4.77374  0.06300 24.810  < 2e-16
Z[, 2] -0.20933   0.81112  0.04982 -4.202 2.65e-05

Likelihood ratio test=692.7  on 2 df, p=< 2.2e-16
n= 2000, number of events= 1399
```

There are clear biases (in this case, both biases are positive) in the estimates obtained in the above output compared to the true values of beta (1, -0.5) set in the data simulation, despite the large sample size of $n = 2000$.

The scatter plot in Figure 7.1 displays a strong positive dependence between the event and censoring times. □

## 7.4 Likelihood function

Our main target is to fit the Cox model (7.1) when dependent right censoring presents. Ignoring this dependence when fitting the model is likely to create biased estimates of the model parameters. To accommodate this dependence, as explained in the previous sections, one approach is to employ a copula function to construct a joint survival function between the event time $Y$ and the dependent right censoring time $C$ based on the marginal survival functions of $Y$ and $C$. It is important to note that in this approach, the parameters of both $Y$ and $C$ must be estimated jointly, even though the primary interest lies in the parameters of $Y$ alone.

The observed values for individual $i$, where $i = 1, \ldots, n$, are denoted as $(t_i, \delta_{iY}, \delta_{iC}, \mathbf{x}_i, \mathbf{z}_i)$. Here, $t_i$ represents an observed value of the random variable $T_i$, $\mathbf{x}_i$ is a $p_1 \times 1$ vector for covariate values in model (7.1), and $\mathbf{z}_i$ is a $p_2 \times 1$ vector for covariate values in model (7.2). The indicator $\delta_{iY}$ denotes the event indicator, while $\delta_{iC}$ represents the dependent censoring indicator. Clearly, the indicator for $A_i$, the independent censoring, is $\delta_{iA} = 1 - \delta_{iY} - \delta_{iC}$.

For individual $i$, we denote the density and survival functions for $A_i$ as $f_{iA}(t)$ and $S_{iA}(t)$, respectively. Note that the parameters in $f_{iA}(t)$ and $S_{iA}(t)$ do not overlap with the parameters of $Y_i$ and $C_i$ due to the non-informative censoring assumption of $A_i$. Recall that $T_i$ represents the right-censored event time for individual $i$. To construct the likelihood function, we need the density function of $T_i$ in cases where $\delta_{iY} = 1$, $\delta_{iC} = 1$, or $\delta_{iA} = 1$, i.e. we will derive three different functions under these three conditions.

We denote the density of $T_i$ by $f_T(t|\mathbf{x}_i, \mathbf{z}_i)$. If $\delta_{iY} = 1$, this density function can be written as

$$
\begin{aligned}
f_T(t|\mathbf{x}_i, \mathbf{z}_i; \delta_{iY} = 1) &= \lim_{\Delta t \to 0} \frac{Pr(t \le Y_i < t + \Delta t, C_i > t, A_i > t|\mathbf{x}_i, \mathbf{z}_i)}{\Delta t} \\
&= \lim_{\Delta t \to 0} \frac{S_{Y,C}(t, t|\mathbf{x}_i, \mathbf{z}_i) - S_{Y,C}(t + \Delta t, t|\mathbf{x}_i, \mathbf{z}_i)}{\Delta t} S_{iA}(t) \\
&= - \left. \frac{\partial S_{Y,C}(t_1, t_2|\mathbf{x}_i, \mathbf{z}_i)}{\partial t_1} \right|_{t_1 = t, t_2 = t} S_{iA}(t) \qquad (7.12) \\
&= f_Y(t|\mathbf{x}_i) K_1 \left( S_Y(t|\mathbf{x}_i), S_C(t|\mathbf{z}_i) \right) S_{iA}(t), \qquad (7.13)
\end{aligned}
$$

where $S_{Y,C}(\cdot, \cdot)$ in (7.12) is replaced by the copula joint survival function model (7.7) and $K_1(a, b) = \partial K(a, b)/\partial a$, the derivative of the copula function with respect to its first argument.

Similarly, the density for $T_i$ associated with $\delta_{iC} = 1$ is given by

$$
\begin{aligned}
f_T(t|\mathbf{x}_i, \mathbf{z}_i; \delta_{iC} = 1) &= \lim_{\Delta t \to 0} \frac{S_{Y,C}(t, t|\mathbf{x}_i, \mathbf{z}_i) - S_{Y,C}(t, t + \Delta t|\mathbf{x}_i, \mathbf{z}_i)}{\Delta t} S_{iA}(t) \\
&= f_C(t|\mathbf{z}_i) K_2 \left( S_Y(t|\mathbf{x}_i), S_C(t|\mathbf{z}_i) \right) S_{iA}(t), \qquad (7.14)
\end{aligned}
$$

where $K_2(a, b) = \partial K(a, b)/\partial b$.

Finally, when $\delta_{iA} = 1$ the density for $T_i$ is

$$
\begin{aligned}
f_T(t|\mathbf{x}_i, \mathbf{z}_i; \delta_{iA} = 1) &= S_{Y,C}(t, t|\mathbf{x}_i, \mathbf{z}_i) \lim_{\Delta t \to 0} \frac{S_{iA}(t) - S_{iA}(t + \Delta t)}{\Delta t} \\
&= K \left( S_Y(t|\mathbf{x}_i), S_C(t|\mathbf{z}_i) \right) f_{iA}(t). \qquad (7.15)
\end{aligned}
$$

Using the results from (7.13), (7.14), (7.15), and the marginal hazard models (7.1)

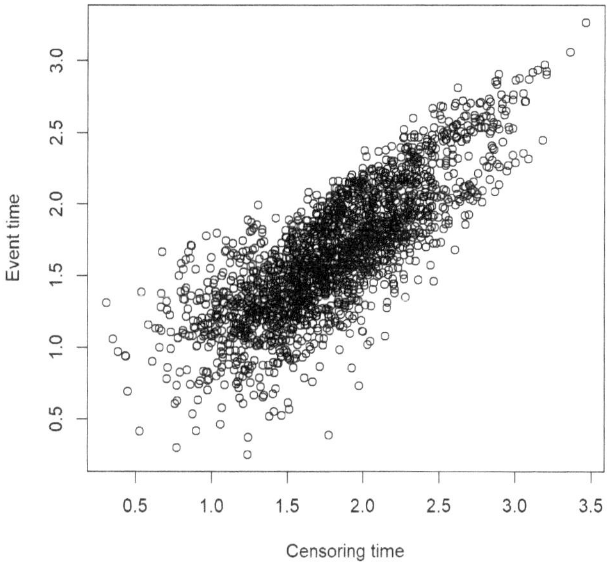


FIGURE 7.1: Scatter plot shows clear dependence between the simulated event and censoring times.

and (7.2), we can easily write down the log-likelihood for independent observations $(t_i, \delta_{iY}, \delta_{iC}, \mathbf{x}_i, \mathbf{z}_i)$, where $i = 1, \ldots, n$:

$$l = \sum_{i=1}^{n} \{\delta_{iY} l_{iY} + \delta_{iC} l_{iC} + (1 - \delta_{iY} - \delta_{iC}) l_{iA}\}, \tag{7.16}$$

where

$$l_{iY} = \log h_{0Y}(t_i) + \mathbf{x}_i^\mathsf{T} \boldsymbol{\beta} - H_Y(t_i|\mathbf{x}_i) + \log K_1(S_Y(t|\mathbf{x}_i), S_C(t|\mathbf{z}_i)), \tag{7.17}$$

$$l_{iC} = \log h_{0C}(t_i) + \mathbf{z}_i^\mathsf{T} \boldsymbol{\phi} - H_C(t_i|\mathbf{z}_i) + \log K_2(S_Y(t|\mathbf{x}_i), S_C(t|\mathbf{z}_i)), \tag{7.18}$$

$$l_{iA} = \log K(S_Y(t|\mathbf{x}_i), S_C(t|\mathbf{z}_i)). \tag{7.19}$$

Here, the cumulative hazards $H_Y(\cdot)$ and $H_C(\cdot)$ are defined in (7.5) and (7.6), respectively. It is worth noting that since neither $f_{iA}(t)$ nor $S_{iA}(t)$ contain parameters for the $Y_i$ and $C_i$ models, they are omitted from the log-likelihood function.

While the primary focus is on estimating $\boldsymbol{\beta}$ and $h_{0Y}(t)$, it is important to note that maximizing the log-likelihood $l$ of (7.16) also involves estimating $\boldsymbol{\phi}$ and $h_{0C}(t)$. Since $h_{0Y}(t)$ and $h_{0C}(t)$ are non-parametric, this estimation task can be challenging. To address this challenge, we will approximate $h_{0Y}(t)$ and $h_{0C}(t)$ using finite-dimensional spaces, as explained in the next section.

## 7.5 Approximating baseline hazards using basis functions

Directly estimating $h_{0Y}(t)$ or $h_{0C}(t)$ from the log-likelihood (7.16) is ill-posed because these parameters are infinite-dimensional, while we have only a finite number of observations. To address this issue, we follow the approach used previously and approximate $h_{0Y}(t)$ and $h_{0C}(t)$ as follows:

$$h_{0Y}(t) = \sum_{u=1}^{m_1} \theta_u \psi_{u1}(t), \tag{7.20}$$

$$h_{0C}(t) = \sum_{u=1}^{m_2} \gamma_u \psi_{u2}(t), \tag{7.21}$$

where $\psi_{u1}(t)$ and $\psi_{u2}(t)$ are non-negative basis functions. It is possible that different sets of basis functions may be adopted for the baseline hazards of $Y$ and $C$.

Since non-negative basis functions are employed, it is sufficient to require $\theta_u \geq 0$ for $1 \leq u \leq m_1$ and $\gamma_u \geq 0$ for $1 \leq u \leq m_2$ in order to achieve the constraints $h_{0Y}(t) \geq 0$ and $h_{0C}(t) \geq 0$.

Corresponding to the approximations (7.20) and (7.21), the approximate cumulative baseline hazards are

$$H_{0Y}(t) = \sum_{u=1}^{m_1} \theta_u \Psi_{u1}(t), \tag{7.22}$$

$$H_{0C}(t) = \sum_{u=1}^{m_2} \gamma_u \Psi_{u2}(t). \tag{7.23}$$

The corresponding baseline survival functions are easily obtained from $S_{0Y}(t) = \exp\{-H_{0Y}(t)\}$ and $S_{0C}(t) = \exp\{-H_{0C}(t)\}$.

Different choices of non-negative basis functions are available. For example, Section 2.3 contains details of piecewise constant, M-spline and Gaussian basis functions as a few examples.

Despite $h_{0Y}(t)$ and $h_{0C}(t)$ being expressed using a finite number of basis functions, they remain flexible due to the flexibility in selecting $m_1$ and $m_2$. In particular, we allow $m_1$ and $m_2$ to approach infinity, but at a slow rate, as $n \to \infty$, ensuring the non-parametric nature of the baseline hazards is preserved.

Let $\boldsymbol{\theta}$ and $\boldsymbol{\gamma}$ be the vectors for coefficients $\theta_u$ and $\gamma_u$, respectively. We aim to estimate the parameter vector $\boldsymbol{\eta} = (\boldsymbol{\theta}^\mathsf{T}, \boldsymbol{\gamma}^\mathsf{T}, \boldsymbol{\beta}^\mathsf{T}, \boldsymbol{\phi}^\mathsf{T})^\mathsf{T}$. As explained in Chapter 2, Section 2.3, direct maximization of the log-likelihood for this estimation problem is not ideal, as the results can be non-smooth and sensitive to the number and location of knots used to define the basis functions. To address these issues, we incorporate a penalty function into the log-likelihood and seek the maximum penalized likelihood estimates, which are typically numerically stable.

Roughness penalties are popular choices; they regularize $h_{0Y}(t)$ and $h_{0C}(t)$ using their derivatives. For example, if we adopt the second derivative-based roughness penalties, the penalty functions are given as follows:

$$J_1(\boldsymbol{\theta}) = \int_t h''_{0Y}(t)^2 dt = \boldsymbol{\theta}^\mathsf{T} \mathbf{R}_1 \boldsymbol{\theta}, \tag{7.24}$$

$$J_2(\boldsymbol{\gamma}) = \int_t h''_{0C}(t)^2 dt = \boldsymbol{\gamma}^\mathsf{T} \mathbf{R}_2 \boldsymbol{\gamma}. \tag{7.25}$$

Here, matrices $\mathbf{R}_1$ and $\mathbf{R}_2$ have dimensions of $m_1 \times m_1$ and $m_2 \times m_2$, respectively. The $(u, v)$th element of $\mathbf{R}_1$ is $\int_t \psi''_{u1}(t)\psi''_{v1}(t)dt$, and the $(u, v)$th element of $\mathbf{R}_2$ is $\int_t \psi''_{u2}(t)\psi''_{v2}(t)dt$. For piecewise constant basis functions, we need to replace the derivatives with differences; see Section 2.3.

## 7.6 Maximum penalized likelihood estimation

An efficient and stable algorithm is key to the successful implementation of MPL estimation, especially given the computational challenges of this constrained optimization problem.

We wish to estimate the parameter vector $\boldsymbol{\eta}$ by maximizing the following penalized log-likelihood:

$$\Phi(\boldsymbol{\eta}) = l(\boldsymbol{\eta}) - \lambda_1 J_1(\boldsymbol{\theta}) - \lambda_2 J_2(\boldsymbol{\gamma}), \tag{7.26}$$

where the log-likelihood $l(\boldsymbol{\eta})$ is provided in (7.16). Here, $\lambda_1 \geq 0$ and $\lambda_2 \geq 0$ are the smoothing parameters, and the functions $J_1(\boldsymbol{\theta})$ and $J_2(\boldsymbol{\gamma})$ are penalty functions. These penalties are employed to constrain the estimates of $\boldsymbol{\theta}$ and $\boldsymbol{\gamma}$ so that their corresponding baseline hazards are smooth. Examples of penalty functions are the quadratic penalties outlined in (7.24) and (7.25). These quadratic penalty functions are commonly used for $\boldsymbol{\theta}$ and $\boldsymbol{\gamma}$; see Chapters 2–6.

In addition to smoothing the baseline hazard estimates, penalty functions also play a crucial role in stabilizing the estimates of $\boldsymbol{\theta}$ and $\boldsymbol{\gamma}$, providing more numerically stable parameter estimates. These penalties help to reduce the influence of non-important basis functions by driving their coefficients (i.e., the $\theta_u$'s) towards zero. As a result, the MPL method exhibits reduced sensitivity to the number and location of knots, a distinct advantage over maximum likelihood methods (without penalties) such as the conventional method of sieves. Further discussions on this point can be found in Section 2.4.

In this section, we focus on the computational aspects of the MPL method when dealing with dependent right censoring. As in previous chapters, we recommend a tactical approach of alternating updating of the parameters, where blocks of parameters are updated sequentially in each iteration. For the problem discussed in this chapter, the natural blocks of parameters are: $\boldsymbol{\beta}$, $\boldsymbol{\phi}$, $\boldsymbol{\theta}$ and $\boldsymbol{\gamma}$. Alternatively, parameters can be divided into two blocks: $\boldsymbol{\beta}, \boldsymbol{\phi}$ and $\boldsymbol{\theta}, \boldsymbol{\gamma}$ based on their types. In the following discussions, we adopt the former way of defining the blocks. Algorithms for the latter parameter separation strategy can be developed in a similar manner and so will not be explained further.

Therefore, we wish to solve the following constrained optimization problem:

$$(\widehat{\boldsymbol{\beta}}, \widehat{\boldsymbol{\phi}}, \widehat{\boldsymbol{\theta}}, \widehat{\gamma}) = \underset{\boldsymbol{\beta},\boldsymbol{\phi},\boldsymbol{\theta},\gamma}{\text{argmax}}\, \Phi(\boldsymbol{\eta}), \tag{7.27}$$

subject to $\boldsymbol{\theta} \geq 0$, and $\gamma \geq 0$, and here the inequalities are interpreted element-wisely. Note that function $\Phi$ is defined in equation (7.26).

The KKT necessary conditions for this constrained optimization are

$$\frac{\partial \Phi}{\partial \beta_{j_1}} = 0 \text{ and } \frac{\partial \Phi}{\partial \phi_{j_2}} = 0, \tag{7.28}$$

$$\frac{\partial \Phi}{\partial \theta_{u_1}} = 0 \text{ if } \theta_{u_1} > 0 \text{ and } \frac{\partial \Phi}{\partial \theta_{u_1}} < 0 \text{ if } \theta_{u_1} = 0, \tag{7.29}$$

$$\frac{\partial \Phi}{\partial \gamma_{u_2}} = 0 \text{ if } \gamma_{u_2} > 0 \text{ and } \frac{\partial \Phi}{\partial \gamma_{u_2}} < 0 \text{ if } \gamma_{u_2} = 0, \tag{7.30}$$

where $j_1 = 1, \ldots, p_1$, $j_2 = 1, \ldots, p_2$, $u_1 = 1, \ldots, m_1$ and $u_2 = 1, \ldots, m_2$. As explicated before, we adopt an alternating algorithm similar to Chapter 2 to solve equations (7.28), (7.29) and (7.30); see also Ma et al. (2014) for similar ideas. The rationale of this algorithm is that it solves the KKT conditions on $\boldsymbol{\beta}, \boldsymbol{\phi}, \boldsymbol{\theta}$ and $\gamma$ efficiently, even when $m_1$ or $m_2$ is large. The latter can happen when, for example, event times are used to define a piecewise constant approximation to $h_{0Y}(t)$ and $h_{0C}(t)$.

For our computational approach, the basic strategy is: attempt to solve manageable sub-problems in each iteration. More specifically, in iteration $k + 1$, we wish to solve the following alternating sub-optimization problems:

$$\boldsymbol{\beta}^{(k+1)} = \underset{\boldsymbol{\beta}}{\text{argmax}}\, \Phi(\boldsymbol{\beta}, \boldsymbol{\phi}^{(k)}, \boldsymbol{\theta}^{(k)}, \gamma^{(k)}), \tag{7.31}$$

$$\boldsymbol{\phi}^{(k+1)} = \underset{\boldsymbol{\phi}}{\text{argmax}}\, \Phi(\boldsymbol{\beta}^{(k+1)}, \boldsymbol{\phi}, \boldsymbol{\theta}^{(k)}, \gamma^{(k)}), \tag{7.32}$$

$$\boldsymbol{\theta}^{(k+1)} = \underset{\boldsymbol{\theta} \geq 0}{\text{argmax}}\, \Phi(\boldsymbol{\beta}^{(k+1)}, \boldsymbol{\phi}^{(k+1)}, \boldsymbol{\theta}, \gamma^{(k)}), \tag{7.33}$$

$$\gamma^{(k+1)} = \underset{\gamma \geq 0}{\text{argmax}}\, \Phi(\boldsymbol{\beta}^{(k+1)}, \boldsymbol{\phi}^{(k+1)}, \boldsymbol{\theta}^{(k+1)}, \gamma). \tag{7.34}$$

Note that problems in (7.31) and (7.32) have no constraints, while (7.33) and (7.34) involve the non-negativity constraints.

It is important to realize that solving the above sub-optimization problems *exactly* in each iteration can be too costly. We recommend solving each sub-optimization problem only partially using a single step of an algorithm. More specifically, we solve (7.31) and (7.32) using a one-step Newton (or quasi-Newton) algorithm and solve (7.33) and (7.34) using a one-step MI algorithm due to the constraints. One important condition for the convergence of this algorithm is that the objective function $\Phi$ must be non-decreasing after updating each parameter vector. Therefore, at iteration $k+1$, we demand the following conditions hold after updating each parameter:

$$\Phi(\boldsymbol{\beta}^{(k)}, \boldsymbol{\phi}^{(k)}, \boldsymbol{\theta}^{(k)}, \gamma^{(k)}) \leq \Phi(\boldsymbol{\beta}^{(k+1)}, \boldsymbol{\phi}^{(k)}, \boldsymbol{\theta}^{(k)}, \gamma^{(k)}) \tag{7.35}$$

after updating $\boldsymbol{\beta}$,

$$\Phi(\boldsymbol{\beta}^{(k+1)}, \boldsymbol{\phi}^{(k)}, \boldsymbol{\theta}^{(k)}, \gamma^{(k)}) \leq \Phi(\boldsymbol{\beta}^{(k+1)}, \boldsymbol{\phi}^{(k+1)}, \boldsymbol{\theta}^{(k)}, \gamma^{(k)}) \tag{7.36}$$

after updating $\boldsymbol{\phi}$,

$$\Phi(\boldsymbol{\beta}^{(k+1)}, \boldsymbol{\phi}^{(k+1)}, \boldsymbol{\theta}^{(k)}, \gamma^{(k)}) \leq \Phi(\boldsymbol{\beta}^{(k+1)}, \boldsymbol{\phi}^{(k+1)}, \boldsymbol{\theta}^{(k+1)}, \gamma^{(k)}) \tag{7.37}$$

after updating $\boldsymbol{\theta}$, and

$$\Phi(\boldsymbol{\beta}^{(k+1)}, \boldsymbol{\phi}^{(k+1)}, \boldsymbol{\theta}^{(k+1)}, \gamma^{(k)}) \leq \Phi(\boldsymbol{\beta}^{(k+1)}, \boldsymbol{\phi}^{(k+1)}, \boldsymbol{\theta}^{(k+1)}, \gamma^{(k+1)}) \tag{7.38}$$

after updating $\gamma$.

At iteration $k+1$, if the Hessian matrix of $\Phi$ with respect to $\boldsymbol{\beta}$ and the Hessian matrix of $\Phi$ with respect to $\boldsymbol{\phi}$ are guaranteed to be negative definite, then we can update $\boldsymbol{\beta}$ and $\boldsymbol{\phi}$ using the Newton algorithm; otherwise, we can use the Gauss-Newton or quasi-Newton algorithm. In either case, we can express the updating formulas for $\boldsymbol{\beta}$ and $\boldsymbol{\phi}$ as follows:

$$\boldsymbol{\beta}^{(k+1)} = \boldsymbol{\beta}^{(k)} + \omega_1^{(k)} \mathbf{S}_{\boldsymbol{\beta}}^{-1}(\boldsymbol{\beta}^{(k)}, \boldsymbol{\phi}^{(k)}, \boldsymbol{\theta}^{(k)}, \gamma^{(k)}) \frac{\partial \Phi(\boldsymbol{\beta}^{(k)}, \boldsymbol{\phi}^{(k)}, \boldsymbol{\theta}^{(k)}, \gamma^{(k)})}{\partial \boldsymbol{\beta}}, \tag{7.39}$$

$$\boldsymbol{\phi}^{(k+1)} = \boldsymbol{\phi}^{(k)} + \omega_2^{(k)} \mathbf{S}_{\boldsymbol{\phi}}^{-1}(\boldsymbol{\beta}^{(k+1)}, \boldsymbol{\phi}^{(k)}, \boldsymbol{\theta}^{(k)}, \gamma^{(k)}) \frac{\partial \Phi(\boldsymbol{\beta}^{(k+1)}, \boldsymbol{\phi}^{(k)}, \boldsymbol{\theta}^{(k)}, \gamma^{(k)})}{\partial \boldsymbol{\phi}}, \tag{7.40}$$

where $\omega_1^{(k)}$ and $\omega_2^{(k)}$, both in $(0, 1]$, are line search step sizes, and they can ensure that (7.35) and (7.36) are achieved, respectively. Values for $\omega_1^{(k)}$ and $\omega_2^{(k)}$ can be determined from, for example, the Armijo's line search method as described in Luenberger (1984). Matrices $\mathbf{S}_{\boldsymbol{\beta}}(\boldsymbol{\beta}, \boldsymbol{\phi}, \boldsymbol{\theta}, \gamma)$ and $\mathbf{S}_{\boldsymbol{\phi}}(\boldsymbol{\beta}, \boldsymbol{\phi}, \boldsymbol{\theta}, \gamma)$ are required to be positive definite, and both matrices depend on $\boldsymbol{\beta}, \boldsymbol{\phi}, \boldsymbol{\theta}, \gamma$. There are several options for the selection of the $\mathbf{S}_{\boldsymbol{\beta}}(\boldsymbol{\beta}, \boldsymbol{\phi}, \boldsymbol{\theta}, \gamma)$ and $\mathbf{S}_{\boldsymbol{\phi}}(\boldsymbol{\beta}, \boldsymbol{\phi}, \boldsymbol{\theta}, \gamma)$ matrices, and they include:

(1) Fisher information matrices given by, respectively,

$$\mathbf{S}_{\boldsymbol{\beta}} = -E\left(\frac{\partial^2 \Phi}{\partial \boldsymbol{\beta} \partial \boldsymbol{\beta}^{\mathsf{T}}}\right), \quad \mathbf{S}_{\boldsymbol{\phi}} = -E\left(\frac{\partial^2 \Phi}{\partial \boldsymbol{\phi} \partial \boldsymbol{\phi}^{\mathsf{T}}}\right).$$

Although the Fisher information matrices can assure that $\mathbf{S}_{\boldsymbol{\beta}}$ and $\mathbf{S}_{\boldsymbol{\phi}}$ are non-negative definite, a problem with this option is that these matrices are difficult to compute for the problem we consider because they demand intractable expectation calculations.

(2) Negative Hessian matrices are given by:

$$\mathbf{S}_{\beta} = -\frac{\partial^2 \Phi}{\partial \boldsymbol{\beta} \partial \boldsymbol{\beta}^{\mathsf{T}}}, \ \mathbf{S}_{\phi} = -\frac{\partial^2 \Phi}{\partial \boldsymbol{\phi} \partial \boldsymbol{\phi}^{\mathsf{T}}}.$$

These matrices are easy to compute, but their drawback is that they may not always be non-negative definite, particularly before the maximum is reached, when $\Phi$ is not a concave function. In this case, a possible remedy is to adopt quasi-Newton algorithms where, for example, positive terms of the Hessian matrix expression are removed; see the algorithms for $\boldsymbol{\beta}$ and $\boldsymbol{\phi}$ we discuss later.

(3) In the case that the negative Hessian matrices are not positive definite, another option is the Gauss-Newton algorithm, where $\mathbf{S}_{\beta}$ and $\mathbf{S}_{\phi}$ become:

$$\mathbf{S}_{\beta} = \frac{\partial \Phi}{\partial \boldsymbol{\beta}} \frac{\partial \Phi}{\partial \boldsymbol{\beta}^{\mathsf{T}}}, \ \mathbf{S}_{\phi} = \frac{\partial \Phi}{\partial \boldsymbol{\phi}} \frac{\partial \Phi}{\partial \boldsymbol{\phi}^{\mathsf{T}}}.$$

The Gauss-Newton method is particularly attractive as it only involves the first derivative. However, a drawback of this method is that it may be numerically unstable, particularly when iterations on $\boldsymbol{\beta}$ and $\boldsymbol{\phi}$ are close to the stationary points of $\Phi$.

The strategies for defining iterations as mentioned above require the calculation of first and, in some cases, second derivatives of $\Phi$ with respect to $\boldsymbol{\beta}$ and $\boldsymbol{\phi}$. Details of these derivatives will be furnished next. Since the notations involved in these derivative formulae can be tedious, for the purpose of simplifying the notation, we let $S_{iY} = S_Y(t_i|\mathbf{x}_i)$, $S_{iC} = S_C(t_i|\mathbf{z}_i)$, $H_{iY} = H_Y(t_i|\mathbf{x}_i)$ and $H_{iC} = H_C(t_i|\mathbf{z}_i)$.

From the log-likelihood expression given in (7.16), we can easily obtain the first derivatives (gradients) for $\boldsymbol{\beta}$ and $\boldsymbol{\phi}$; they are:

$$\frac{\partial \Phi}{\partial \beta_j} = \sum_{i=1}^{n} \{\delta_{iY} - \delta_{iY} H_{iY} - \Lambda_{i1} S_{iY} H_{iY}\} x_{ij}, \tag{7.41}$$

for $j = 1, \ldots, p_1$, and

$$\frac{\partial \Phi}{\partial \phi_j} = \sum_{i=1}^{n} \{\delta_{iC} - \delta_{iC} H_{iC} - \Lambda_{i2} S_{iC} H_{iC}\} z_{ij}, \tag{7.42}$$

for $j = 1, \ldots, p_2$. The above gradient formulae depend on $\Lambda_{i1}$ and $\Lambda_{i2}$, and they are given by

$$\Lambda_{i1} = \delta_{iY} \frac{K_{11}(S_{iY}, S_{iC})}{K_1(S_{iY}, S_{iC})} + \delta_{iC} \frac{K_{21}(S_{iY}, S_{iC})}{K_2(S_{iY}, S_{iC})} + (1 - \delta_{iY} - \delta_{iC}) \frac{K_1(S_{iY}, S_{iC})}{K(S_{iY}, S_{iC})},$$

$$\Lambda_{i2} = \delta_{iC} \frac{K_{22}(S_{iY}, S_{iC})}{K_2(S_{iY}, S_{iC})} + \delta_{iY} \frac{K_{12}(S_{iY}, S_{iC})}{K_1(S_{iY}, S_{iC})} + (1 - \delta_{iY} - \delta_{iC}) \frac{K_2(S_{iY}, S_{iC})}{K(S_{iY}, S_{iC})}.$$

Here, $K_1$ and $K_2$ have already been defined in Section 7.4, and

$$K_{11}(a,b) = \frac{\partial^2 K}{\partial a^2}, \; K_{22}(a,b) = \frac{\partial^2 K}{\partial b^2},$$
$$K_{12}(a,b) = \frac{\partial^2 K}{\partial a \partial b}, \; K_{21}(a,b) = \frac{\partial^2 K}{\partial b \partial a} = K_{12}(a,b).$$

The second derivatives are readily obtained from the gradients in (7.41) and (7.42), but their expressions are tedious. Elements of the Hessian (i.e. second derivative) matrix with respect to $\boldsymbol{\beta}$ are given by

$$\frac{\partial^2 \Phi}{\partial \beta_j \partial \beta_r} = -\sum_{i=1}^{n} \left\{ \delta_{iY} H_{iY} - \frac{\partial \Lambda_{i1}}{\partial S_{iY}} H_{iY}^2 S_{iY}^2 - \Lambda_{i1} H_{iY}^2 S_{iY} + \Lambda_{i1} H_{iY} S_{iY} \right\} x_{ij} x_{ir}, \tag{7.43}$$

where

$$\begin{aligned} \frac{\partial \Lambda_{i1}}{\partial S_{iY}} =& \delta_{iY} \frac{K_{111}(S_{iY}, S_{iC}) K_1(S_{iY}, S_{iC}) - K_{11}^2(S_{iY}, S_{iC})}{K_1^2(S_{iY}, S_{iC})} \\ &+ \delta_{iC} \frac{K_{211}(S_{iY}, S_{iC}) K_2(S_{iY}, S_{iC}) - K_{21}^2(S_{iY}, S_{iC})}{K_2^2(S_{iY}, S_{iC})} \\ &+ (1 - \delta_{iY} - \delta_{iC}) \frac{K_{11}(S_{iY}, S_{iC}) K(S_{iY}, S_{iC}) - K_1^2(S_{iY}, S_{iC})}{K^2(S_{iY}, S_{iC})}. \end{aligned} \tag{7.44}$$

Similarly, elements of the Hessian with respect to $\boldsymbol{\phi}$ are:

$$\frac{\partial^2 \Phi}{\partial \phi_j \partial \phi_r} = -\sum_{i=1}^{n} \left\{ \delta_{iC} H_{iC} - \frac{\partial \Lambda_{i2}}{\partial S_{iC}} H_{iC}^2 S_{iC}^2 - \Lambda_{i2} H_{iC}^2 S_{iC} + \Lambda_{i2} H_{iC} S_{iC} \right\} z_{ij} z_{ir}, \tag{7.45}$$

where

$$\begin{aligned} \frac{\partial \Lambda_{i2}}{\partial S_{iC}} =& \delta_{iC} \frac{K_{222}(S_{iY}, S_{iC}) K_2(S_{iY}, S_{iC}) - K_{22}^2(S_{iY}, S_{iC})}{K_2^2(S_{iY}, S_{iC})} \\ &+ \delta_{iY} \frac{K_{122}(S_{iY}, S_{iC}) K_1(S_{iY}, S_{iC}) - K_{12}^2(S_{iY}, S_{iC})}{K_1^2(S_{iY}, S_{iC})} \\ &+ (1 - \delta_{iY} - \delta_{iC}) \frac{K_{22}(S_{iY}, S_{iC}) K(S_{iY}, S_{iC}) - K_2^2(S_{iY}, S_{iC})}{K^2(S_{iY}, S_{iC})}. \end{aligned} \tag{7.46}$$

In the above expressions for $\partial \Lambda_{i1} / \partial S_{iY}$ and $\partial \Lambda_{i2} / \partial S_{iC}$, we have

$$K_{111}(a,b) = \frac{\partial^3 K(a,b)}{\partial a^3}, \; K_{222}(a,b) = \frac{\partial^3 K(a,b)}{\partial b^3},$$
$$K_{122}(a,b) = \frac{\partial^3 K(a,b)}{\partial a \partial b^2}, \; K_{211}(a,b) = \frac{\partial^3 K(a,b)}{\partial b \partial a^2}$$

Let $\mathbf{X}$ be the model matrix for model (7.1) whose $i$-th row is $\mathbf{x}_i^{\mathsf{T}}$, and $\mathbf{Z}$ be the model matrix for model (7.2) whose $i$-th row is $\mathbf{z}_i^{\mathsf{T}}$. The above second derivative formulae of $\Phi$ with respect to $\boldsymbol{\beta}$ and $\boldsymbol{\phi}$ indicate the corresponding Hessian matrices can be expressed in quadratic forms such that

$$\frac{\partial^2 \Phi}{\partial \boldsymbol{\beta} \partial \boldsymbol{\beta}^T} = -\mathbf{X}^{\mathsf{T}} \mathbf{D}_{\boldsymbol{\beta}} \mathbf{X}, \tag{7.47}$$

$$\frac{\partial^2 \Phi}{\partial \boldsymbol{\phi} \partial \boldsymbol{\phi}^T} = -\mathbf{Z}^{\mathsf{T}} \mathbf{D}_{\boldsymbol{\phi}} \mathbf{Z}. \tag{7.48}$$

Here, $\mathbf{D}_{\boldsymbol{\beta}}$ is a diagonal matrix whose diagonal elements are given by:

$$d_{i\boldsymbol{\beta}} = \delta_{iY} H_{iY} - \frac{\partial \Lambda_{i1}}{\partial S_{iY}} H_{iY}^2 S_{iY}^2 - \Lambda_{i1} H_{iY}^2 S_{iY} + \Lambda_{i1} H_{iY} S_{iY}, \tag{7.49}$$

and $\mathbf{D}_{\boldsymbol{\phi}}$ is also a diagonal matrix with diagonal elements

$$d_{i\boldsymbol{\phi}} = \delta_{iC} H_{iC} - \frac{\partial \Lambda_{i2}}{\partial S_{iC}} H_{iC}^2 S_{iC}^2 - \Lambda_{i2} H_{iC}^2 S_{iC} + \Lambda_{i2} H_{iC} S_{iC}. \tag{7.50}$$

The final matrix expressions in (7.47) and (7.48) may not be negative definite when $\boldsymbol{\beta}$ and $\boldsymbol{\phi}$ are not at the maximum (global or local) of the penalized likelihood function. This is because the diagonal elements $d_{i\boldsymbol{\beta}}$ and $d_{i\boldsymbol{\phi}}$ may be negative. A simple modification to avoid this issue is to replace $d_{i\boldsymbol{\beta}}$ with $\delta_{iY} H_{iY} + \Lambda_{i1} H_{iY} S_{iY}$ and $d_{i\boldsymbol{\phi}}$ with $\delta_{iC} H_{iC} + \Lambda_{i2} H_{iC} S_{iC}$. These modifications result in the so-called *quasi-Newton* algorithms. These modified second derivative matrices in the quasi-Newton algorithms ensure the convergence of these algorithms.

Next, we explain how to update $\boldsymbol{\theta}$ and $\boldsymbol{\gamma}$ while respecting the non-negativity constraints $\boldsymbol{\theta} \geq 0$ and $\boldsymbol{\gamma} \geq 0$. We again adopt the MI algorithm for its ease of implementation.

Recall the MI algorithm only requires the first derivatives of $\Phi$ with respect to $\boldsymbol{\theta}$ and $\boldsymbol{\gamma}$, and thus we first formulate these derivatives below:

$$\frac{\partial \Phi}{\partial \theta_u} = \sum_{i=1}^{n} \left( \delta_{iY} \frac{\psi_{u1}(t_i)}{h_{0Y}(t_i)} - \{\delta_{iY} + \Lambda_{i1} S_{iY}\} \Psi_{u1}(t_i) e^{\mathbf{x}_i^{\mathsf{T}} \boldsymbol{\beta}} \right) - \lambda_1 \frac{\partial J_1(\boldsymbol{\theta})}{\partial \theta_u}, \tag{7.51}$$

for $u = 1, \ldots, m_1$, and

$$\frac{\partial \Phi}{\partial \gamma_u} = \sum_{i=1}^{n} \left( \delta_{iC} \frac{\psi_{u2}(t_i)}{h_{0C}(t_i)} - \{\delta_{iC} + \Lambda_{i2} S_{iC}\} \Psi_{u2}(t_i) e^{\mathbf{z}_i^{\mathsf{T}} \boldsymbol{\phi}} \right) - \lambda_2 \frac{\partial J_2(\boldsymbol{\gamma})}{\partial \gamma_u}. \tag{7.52}$$

for $u = 1, \ldots, m_2$. Note that $\Lambda_{i1}$ and $\Lambda_{i2}$ in (7.51) and (7.52) have already been defined before.

Following Chapter 2, we can easily develop the MI algorithms for estimation of $\boldsymbol{\theta} \geq 0$ and $\boldsymbol{\gamma} \geq 0$. More specifically, with the initial values $\boldsymbol{\theta}^{(0)} > 0$ and $\boldsymbol{\phi}^{(0)} > 0$, $\boldsymbol{\theta}$ is updated according to

$$\boldsymbol{\theta}^{(k+1)} = \boldsymbol{\theta}^{(k)} + \omega_3^{(k)} \mathbf{V}_{\boldsymbol{\theta}}(\boldsymbol{\beta}^{(k+1)}, \boldsymbol{\phi}^{(k+1)}, \boldsymbol{\theta}^{(k)}, \boldsymbol{\gamma}^{(k)}) \frac{\partial \Phi(\boldsymbol{\beta}^{(k+1)}, \boldsymbol{\phi}^{(k+1)}, \boldsymbol{\theta}^{(k)}, \boldsymbol{\gamma}^{(k)})}{\partial \boldsymbol{\theta}}, \tag{7.53}$$

and $\phi$ is updated by

$$\boldsymbol{\phi}^{(k+1)} = \boldsymbol{\phi}^{(k)} + \omega_4^{(k)} \mathbf{V}_\gamma(\boldsymbol{\beta}^{(k+1)}, \boldsymbol{\phi}^{(k+1)}, \boldsymbol{\theta}^{(k+1)}, \gamma^{(k)}) \frac{\partial \Phi(\boldsymbol{\beta}^{(k+1)}, \boldsymbol{\phi}^{(k+1)}, \boldsymbol{\theta}^{(k+1)}, \gamma^{(k)})}{\partial \theta}, \tag{7.54}$$

where $\mathbf{V}_{\boldsymbol{\theta}}$ is a diagonal matrix given by

$$\mathbf{V}_{\boldsymbol{\theta}} = \text{diag}(\theta_1/\xi_{11}, \ldots, \theta_{m_1}/\xi_{1m_1}) \tag{7.55}$$

and $\mathbf{V}_{\boldsymbol{\phi}}$ a diagonal matrix given by

$$\mathbf{V}_\gamma = \text{diag}(\phi_1/\xi_{21}, \ldots, \phi_{m_2}/\xi_{2m_2}). \tag{7.56}$$

Here,

$$\xi_{1u} = \sum_{i=1}^{n} \left(\delta_{iY} + [\Lambda_{i1}]^+ S_{iY} \Psi_{1u}(t_i)\right) e^{\mathbf{x}_i^\top \boldsymbol{\beta}} + \lambda_1 \left[\frac{\partial J_1(\boldsymbol{\theta})}{\partial \theta_u}\right]^+ + \epsilon, \tag{7.57}$$

$$\xi_{2u} = \sum_{i=1}^{n} \left(\delta_{iC} + [\Lambda_{i2}]^+ S_{iC} \Psi_{2u}(t_i)\right) e^{\mathbf{z}_i^\top \boldsymbol{\phi}} + \lambda_2 \left[\frac{\partial J_2(\boldsymbol{\phi})}{\partial \phi_u}\right]^+ + \epsilon, \tag{7.58}$$

where the notation $[c]^+$ is defined the same as in other chapters, i.e. $[c]^+ = \max(c, 0)$, and $\epsilon$ is a small non-zero constant (i.e $10^{-3}$) used to avoid $\xi_{1u}$ or $\xi_{2u}$ to be zero. It is easy to verify that both $\boldsymbol{\theta}^{(k+1)}$ and $\gamma^{(k+1)}$ are non-negative given $\boldsymbol{\theta}^{(k)} \geq 0$ and $\gamma^{(k)} \geq 0$, and that both step sizes $\omega_3^{(k)}, \omega_4^{(k)} \in (0, 1]$. The step sizes $\omega_3^{(k)}$ and $\omega_4^{(k)}$ used in formulas (7.53) and (7.54) can be determined by the Armijo rule. Step size $\omega_3^{(k)}$ ensures that

$$\Phi(\boldsymbol{\beta}^{(k+1)}, \boldsymbol{\phi}^{(k+1)}, \boldsymbol{\theta}^{(k+1)}, \gamma^{(k)}) \geq \Phi(\boldsymbol{\beta}^{(k+1)}, \boldsymbol{\phi}^{(k+1)}, \boldsymbol{\theta}^{(k)}, \gamma^{(k)})$$

and step size $\omega_4^{(k)}$ assures

$$\Phi(\boldsymbol{\beta}^{(k+1)}, \boldsymbol{\phi}^{(k+1)}, \boldsymbol{\theta}^{(k+1)}, \gamma^{(k+1)}) \geq \Phi(\boldsymbol{\beta}^{(k+1)}, \boldsymbol{\phi}^{(k+1)}, \boldsymbol{\theta}^{(k+1)}, \gamma^{(k)}).$$

The convergence properties of the above algorithm for estimation of $(\boldsymbol{\beta}, \boldsymbol{\phi}, \boldsymbol{\theta}, \gamma)$ can be obtained following proofs given in Chan & Ma (2012).

We wish to emphasize again that the MI algorithms above are very efficient even when the number of knots used to approximate the baseline hazards is large. This can be the case if one adopts a piecewise constant approximation to the baseline hazards such as the conventional non-parametric estimates. MI is easy to derive and implement since only the first derivative of the penalized likelihood function is required, and there is no need to solve a large linear system of equations in each iteration. In the context of dependent censoring, our experience with this algorithm is that it performs well for different datasets and generally possesses a good convergence speed.

## 7.7 Other second derivatives of Hessian

The large sample normal distribution developed in the next section requires a full Hessian matrix, so we need to derive the second derivatives that have not been provided previously, particularly: $\partial^2\Phi/\partial\beta_j\partial\phi_r$ $\partial^2\Phi/\partial\theta_u\partial\beta_j$, $\partial^2\Phi/\partial\theta_u\partial\phi_j$, $\partial^2\Phi/\partial\gamma_u\partial\beta_j$ and $\partial^2\Phi/\partial\gamma_u\partial\phi_j$.

These second derivatives are given below.

$$\frac{\partial^2\Phi}{\partial\beta_j\partial\phi_r} = \sum_{i=1}^{n} \frac{\partial\Lambda_{i1}}{\partial S_{iC}} H_{iY} H_{iC} S_{iY} S_{iC} z_{ir} x_{ij}, \tag{7.59}$$

$$\frac{\partial^2\Phi}{\partial\theta_u\partial\beta_j} = -\sum_{i=1}^{n} \left\{\delta_{iY} - \frac{\partial\Lambda_{i1}}{\partial S_{iY}} H_{iY} S_{iY}^2 - \Lambda_{i1} H_{iY} S_{iY} + \Lambda_{i1} S_{iY}\right\} \Psi_{u1}(t_i) e^{\mathbf{x}_i^\top \boldsymbol{\beta}} x_{ij}, \tag{7.60}$$

$$\frac{\partial^2\Phi}{\partial\theta_u\partial\phi_j} = \sum_{i=1}^{n} \frac{\partial\Lambda_{i1}}{\partial S_{iC}} S_{iC} S_{iY} H_{iC} \Psi_{u1}(t_i) e^{\mathbf{x}_i^\top \boldsymbol{\beta}} z_{ij}, \tag{7.61}$$

$$\frac{\partial^2\Phi}{\partial\gamma_u\partial\theta_j} = \sum_{i=1}^{n} \frac{\partial\Lambda_{i2}}{\partial S_{iY}} S_{iY} S_{iC} H_{iY} \Psi_{u2}(t_i) e^{\mathbf{z}_i^\top \boldsymbol{\beta}} x_{ij}, \tag{7.62}$$

$$\frac{\partial^2\Phi}{\partial\gamma_u\partial\phi_j} = -\sum_{i=1}^{n} \left\{\delta_{iC} - \frac{\partial\Lambda_{i2}}{\partial S_{iC}} H_{iC} S_{iC}^2 - \Lambda_{i2} H_{iC} S_{iC} + \Lambda_{i2} S_{iC}\right\} \Psi_{u2}(t_i) e^{\mathbf{z}_i^\top \boldsymbol{\phi}} z_{ij}. \tag{7.63}$$

Here, $\partial\Lambda_{i1}/\partial S_{iY}$ and $\partial\Lambda_{i2}/\partial S_{iC}$ have already been given in (7.44) and (7.46) respectively, and

$$\begin{aligned}\frac{\partial\Lambda_{i2}}{\partial S_{iY}} &= \delta_{iC}\frac{K_{122}(S_{iY},S_{iC})K_2(S_{iY},S_{iC}) - K_{22}(S_{iY},S_{iC})K_{12}(S_{iY},S_{iC})}{K_2^2(S_{iY},S_{iC})}\\ &+ \delta_{iY}\frac{K_{112}(S_{iY},S_{iC})K_1(S_{iY},S_{iC}) - K_{12}(S_{iY},S_{iC})K_1^2(S_{iY},S_{iC})}{K_1^2(S_{iY},S_{iC})}\\ &+ (1-\delta_{iY}-\delta_{iC})\frac{K_{12}(S_{iY},S_{iC})K(S_{iY},S_{iC}) - K_2(S_{iY},S_{iC})K_1(S_{iY},S_{iC})}{K^2(S_{iY},S_{iC})},\end{aligned} \tag{7.64}$$

$$\begin{aligned}\frac{\partial\Lambda_{i1}}{\partial S_{iC}} &= \delta_{iY}\frac{K_{211}(S_{iY},S_{iC})K_1(S_{iY},S_{iC}) - K_{11}(S_{iY},S_{iC})K_{21}(S_{iY},S_{iC})}{K_1^2(S_{iY},S_{iC})}\\ &+ \delta_{iC}\frac{K_{212}(S_{iY},S_{iC})K_2(S_{iY},S_{iC}) - K_{12}(S_{iY},S_{iC})K_{22}(S_{iY},S_{iC})}{K_2^2(S_{iY},S_{iC})}\\ &+ (1-\delta_{iY}-\delta_{iC})\frac{K_{21}(S_{iY},S_{iC})K(S_{iY},S_{iC}) - K_1(S_{iY},S_{iC})K_2(S_{iY},S_{iC})}{K^2(S_{iY},S_{iC})}.\end{aligned} \tag{7.65}$$

In (7.65), $K_{212}(a, b)$ is define as

$$K_{212}(a, b) = \frac{\partial^3 K(a, b)}{\partial b \partial a \partial b}.$$

## 7.8 Asymptotic properties

The asymptotic results for the dependent right censoring problem in this chapter are developed similarly to previous chapters, and therefore we will omit their proofs. We will now summarize the asymptotic consistency and asymptotic normality results in Theorems 7.1 and 7.2.

Let $\beta_0, \phi_0, h_{0Y}(t), h_{0C}(t)$ denote the true parameters, and $\widehat{\beta}, \widehat{\phi}$, $\widehat{h}_{nY}(t)$, and $\widehat{h}_{nC}(t)$ denote their MPL estimates as described previously, where each unknown baseline hazard is approximated using a finite number, denoted by $m_1$ and $m_2$ respectively, of basis functions. Let $\boldsymbol{\theta}$ and $\gamma$ represent the vectors for the coefficients of the basis functions adopted to approximate $h_{0Y}(t)$ and $h_{0C}(t)$, respectively. Let $\tilde{a} = \min_i\{t_i\}$ and $\tilde{b} = \max_i\{t_i\}$.

The results in Theorem 7.1 state that the MPL estimates converge to their true values when the number of basis functions $m_1 \to \infty$ and $m_2 \to \infty$, but they converge to infinity slower than $n \to \infty$, i.e., $m_1/n \to 0$ and $m_2/n \to 0$. Let the scaled smoothing values be defined as $\mu_{1n} = \lambda_1/n$ and $\mu_{2n} = \lambda_2/n$. We assume that both $\mu_{1n}$ and $\mu_{2n}$ tend to zero as $n \to \infty$.

The proofs of the theorems in this section, namely Theorems 7.1 and 7.2, closely resemble those of Theorems 2.4 and 2.5. Therefore, they will not be provided in this section. Theorem 7.1 requires the following assumptions.

**Assumptions:**

A7.1 Matrices $\mathbf{X}$ and $\mathbf{Z}$ are both bounded and matrices $E(\mathbf{X}\mathbf{X}^\top)$ and $E(\mathbf{Z}\mathbf{Z}^\top)$ are non-singular.

A7.2 The penalty functions $J_1(\cdot)$ and $J_2(\cdot)$ are bounded.

A7.3 The coefficients vector $\boldsymbol{\theta}$ is in a compact subset of $R^{m_1}$ and the coefficients vector $\gamma$ is in a compact subset of $R^{m_2}$. Moreover, the basis functions $\psi_{u1}(t)$ and $\psi_{u2}(t)$ are all bounded.

A7.4 Assume for any $h_{0Y}(x)$ and $h_{0C}(x)$, there exist $h_{nY}(t)$ and $h_{nC}(t)$ such that $\max_{t\in[\tilde{a},\tilde{b}]} |h_{nY}(t) - h_{0Y}(t)| \to 0$ and $\max_{t\in[a,b]} |h_{nC}(t) - h_{0C}(t)| \to 0$ as $m_1, m_2 \to \infty$ and $n \to \infty$ but $m_1/n \to 0$ and $m_2/n \to 0$.

A7.5 The copula function $K(S_Y(t), S_C(t))$ correctly models the joint survival function between $Y$ and $C$ for $t \in [a, b]$.

**Theorem 7.1.** Assume that Assumptions A7.1–A7.4 hold and $h_{0Y}(t)$ and $h_{0C}(t)$

have up to $r \geq 1$ derivatives. Assume $m_1 = n^{\upsilon_1}$ and $m_2 = n^{\upsilon_2}$, where $0 < \upsilon_1, \upsilon_2 < 1$, and $\mu_{1n}$ and $\mu_{2n} \to 0$ as $n \to \infty$. Then, when $n \to \infty$:

(1) $\|\widehat{\boldsymbol{\beta}} - \boldsymbol{\beta}_0\|_2 \to 0$ (a.s.),

(2) $\|\widehat{\boldsymbol{\phi}} - \boldsymbol{\phi}_0\|_2 \to 0$ (a.s.).

(3) $\sup_{t \in [\tilde{a}, \tilde{b}]} |\widehat{h}_{nY}(t) - h_{0Y}(t)| \to 0$ (a.s.).

(4) $\sup_{t \in [\tilde{a}, \tilde{b}]} |\widehat{h}_{nC}(t) - h_{0C}(t)| \to 0$ (a.s).

□

The results in Theorem 7.1 explain the MPL estimates of the baseline hazards $h_{0Y}^0(t)$ and $h_{0C}^0(t)$ are consistent under certain regularity conditions. Next, we summarize the asymptotic normality results for the estimated parameters of $\boldsymbol{\beta}$, $\boldsymbol{\phi}$, $\boldsymbol{\theta}$ and $\boldsymbol{\gamma}$ where $m_1$ of $\boldsymbol{\theta}$ and $m_2$ of $\boldsymbol{\gamma}$ are fixed. While having fixed values for $m_1$ and $m_2$ might suggest that the results are parametric, the fact that $m_1$ and $m_2$ can vary with the sample size $n$ deviates this MPL method away from conventional parametric methods.

Let vector $\boldsymbol{\eta} = (\boldsymbol{\theta}^\top, \boldsymbol{\gamma}^\top, \boldsymbol{\beta}^\top, \boldsymbol{\phi}^\top)^\top$, where $m_1$ and $m_2$ are fixed. Let $\widehat{\boldsymbol{\eta}}$ denote the constrained MPL estimate of $\boldsymbol{\eta}$ where $\boldsymbol{\theta} \geq 0$ and $\boldsymbol{\gamma} \geq 0$ are respected. Let $\boldsymbol{\eta}_0$ be the true parameter vector. Assume both smoothing parameters $\lambda_1$ and $\lambda_2$ are $O(\sqrt{n})$, and therefore both $\mu_{1n}$ and $\mu_{2n}$ are $o(n^{-1/2})$.

We have emphasized in the previous chapters that it is extremely important for the basis function approximation methods that some constraints may be active; if this fact is ignored, it may cause unpleasant consequences such as the covariance matrix of the estimates may not be non-positive definite. Detailed information on how to handle active constraints can be found in Chapter 2.

Assume there are $q$ active constraints from $\boldsymbol{\theta}$ and $s$ actives constraints from $\boldsymbol{\gamma}$. Let matrix $\mathbf{U}$ be defined similarly as (2.54). This matrix has the dimension of $(m_1 + m_2 + p_1 + p_2) \times (m_1 + m_2 + p_1 + p_2 - q - s)$, and its rows corresponding to the active constraints are set to zero while the other rows form an identity matrix. Let $\mathbf{F}(\boldsymbol{\eta}) = -n^{-1}E(\partial^2 l(\boldsymbol{\eta})/\partial\boldsymbol{\eta}\partial\boldsymbol{\eta}^\top)$. Assume the matrix $\mathbf{U}^\top\mathbf{F}(\boldsymbol{\eta})\mathbf{U}$ is invertible in a neighborhood of $\boldsymbol{\eta}_0$. The matrix $\mathbf{U}$ defined in Assumption B7.5 is used to indicate the active constraints. Note that $\mathbf{U}^T\mathbf{U} = \mathbf{I}$ which has dimension $(m_1 + m_2 + p_1 + p_2 - q - s) \times (m_1 + m_2 + p_1 + p_2 - q - s)$, where $q$ and $r$ are the number of active constraints from $\boldsymbol{\theta}$ and $\boldsymbol{\gamma}$ respectively.

The asymptotic normality result in Theorem 7.2 requires the following assumptions.

**Assumptions:**

B7.1 Random vectors $\mathbf{W}_i = (T_i, \mathbf{x}_i^\top, \mathbf{z}_i^\top)^\top$, $1 \leq i \leq n$, are independent and identically distributed (although $Y_i$ and $C_i$ are dependent), and the distributions of $\boldsymbol{x}_i$ and $\boldsymbol{z}_i$ are independent of $\boldsymbol{\eta}$.

B7.2 Let $\Omega$ be the parameter space for $\boldsymbol{\eta}$ where $\Omega$ is a compact subset of $R^{p_1+p_2+m_1+m_2}$. Assume $E(n^{-1}l(\boldsymbol{\eta}))$ exists and has a unique maximum at $\boldsymbol{\eta}_0 \in \Omega$.

B7.3 For the penalty functions $J_1(\cdot)$ and $J_2(\cdot)$, both $J_1(\boldsymbol{\eta})$ and $J_2(\boldsymbol{\eta})$ are continuous and bounded over $\Omega$, and their first two derivatives exist with respect to all $\boldsymbol{\eta} \in \Omega$. Moreover, their second derivatives are bounded in a neigbouthood of $\boldsymbol{\eta}_0$.

B7.4 Assume $l(\boldsymbol{\eta})$ is bounded and is twice continuously differentiable in a neighbourhood of $\boldsymbol{\eta}_0$, and the matrix

$$n^{-1}E\left(\frac{\partial^2 l(\boldsymbol{\eta})}{\partial\boldsymbol{\eta}\partial\boldsymbol{\eta}^\mathsf{T}}\right)$$

exist.

**Theorem 7.2.** Assume the Assumptions B7.1–B7.4 hold and both $\mu_{1n}$ and $\mu_{2n}$ are $o(n^{-1/2})$. Assume there are $q$ and $s$ active constraints in the MPL estimates of $\boldsymbol{\theta}$ and $\boldsymbol{\gamma}$ respectively. Let $\mathbf{F}(\boldsymbol{\eta}) = -n^{-1}E\big(\partial^2 l(\boldsymbol{\eta})/\partial\boldsymbol{\eta}\partial\boldsymbol{\eta}^\mathsf{T}\big)$. Then, we have:

(1) The constrained MPL estimate $\hat{\boldsymbol{\eta}}$ is consistent for $\boldsymbol{\eta}_0$.

(2) $n^{1/2}(\hat{\boldsymbol{\eta}} - \boldsymbol{\eta}_0)$ converges in distribution to $N(\mathbf{0}, \widetilde{\mathbf{F}}(\boldsymbol{\eta}_0)^{-1}\mathbf{F}(\boldsymbol{\eta}_0)\widetilde{\mathbf{F}}(\boldsymbol{\eta}_0)^{-1})$ when $n \to \infty$, where $\widetilde{\mathbf{F}}(\boldsymbol{\eta}_0)^{-1} = \mathbf{U}(\mathbf{U}^\mathsf{T}\mathbf{F}(\boldsymbol{\eta}_0)\mathbf{U})^{-1}\mathbf{U}^\mathsf{T}$.

□

We have already explained an efficient way to compute the matrix $\widetilde{\mathbf{F}}(\boldsymbol{\eta}_0)^{-1}$ in Chapter 2 without using the matrices multiplications as defined above. In practice, $\boldsymbol{\eta}_0$ is unavailable. A common strategy is to replace it with $\hat{\boldsymbol{\eta}}$ due to the strong consistency result. On the other hand, computing the expectation for the information matrix $\mathbf{F}(\boldsymbol{\eta})$ can be challenging. As a result, this matrix is typically replaced by the negative Hessian matrix (i.e., the observed information matrix) $-\partial^2 l(\boldsymbol{\eta})/\partial\boldsymbol{\eta}\partial\boldsymbol{\eta}^\mathsf{T}$ or its approximation.

The asymptotic normality given in Theorem 7.2 requires $n \to \infty$, which is never true in practice. Therefore, this asymptotic result needs to be modified to provide a large sample approximate normal distribution. In particular, the variance formula needs to be altered to accommodate for a large $n$, meaning that the smoothing parameters should be retained in the covariance matrix formula.

Similar to Corollary 2.1, we can state the following large sample normality result. Assume the Assumptions B7.1–B7.4 hold. Then, when $n$ is large, the distribution of $\hat{\boldsymbol{\eta}} - \boldsymbol{\eta}_0$ is approximately multivariate normal with mean $\mathbf{0}$ and the covariance matrix:

$$\widehat{\mathrm{Var}}(\widehat{\boldsymbol{\eta}}) = \mathbf{A}(\widehat{\boldsymbol{\eta}})^{-1}\left(-\frac{\partial l(\widehat{\boldsymbol{\eta}})}{\partial\boldsymbol{\eta}\partial\boldsymbol{\eta}^\mathsf{T}}\right)\mathbf{A}(\widehat{\boldsymbol{\eta}})^{-1}, \tag{7.66}$$

where

$$\mathbf{A}(\widehat{\boldsymbol{\eta}})^{-1} = \mathbf{U}\left\{\mathbf{U}^\mathsf{T}\left[-\frac{\partial^2 l(\widehat{\boldsymbol{\eta}})}{\partial\boldsymbol{\eta}\partial\boldsymbol{\eta}^\mathsf{T}} + \lambda_1\frac{\partial^2 J_1(\widehat{\boldsymbol{\theta}})}{\partial\boldsymbol{\theta}\partial\boldsymbol{\theta}^\mathsf{T}} + \lambda_2\frac{\partial^2 J_2(\widehat{\boldsymbol{\gamma}})}{\partial\boldsymbol{\gamma}\partial\boldsymbol{\gamma}^\mathsf{T}}\right]\mathbf{U}\right\}^{-1}\mathbf{U}.$$

## 7.9 Automated smoothing parameter calculation

The penalized likelihood method discussed in this section involves two smoothing parameters. While a similar marginal likelihood-based approach can be developed to estimate these parameters, the method we will explain here has its unique characteristics, specifically designed to handle both smoothing parameters.

The quadratic penalty functions for $\boldsymbol{\theta}$ and $\boldsymbol{\gamma}$ means they can be viewed as having prior distributions:

$$\boldsymbol{\theta} \sim N(\mathbf{0}_{m_1 \times 1},\ \sigma_1^2 \mathbf{R}_1^{-1}),$$
$$\boldsymbol{\gamma} \sim N(\mathbf{0}_{m_2 \times 1},\ \sigma_2^2 \mathbf{R}_2^{-1}).$$

The smoothing parameters are $\lambda_1 = 1/(2\sigma_1^2)$ and $\lambda_2 = 1/(2\sigma_2^2)$. Using these prior distributions, the penalized log-likelihood (or log-posterior) is equivalent to the following expression

$$\Phi(\boldsymbol{\beta}, \boldsymbol{\phi}, \boldsymbol{\theta}, \boldsymbol{\gamma}) \propto -\frac{m_1}{2} \log \sigma_1^2 - \frac{m_2}{2} \log \sigma_2^2 + l(\boldsymbol{\beta}, \boldsymbol{\phi}, \boldsymbol{\theta}, \boldsymbol{\gamma}) - \frac{1}{2\sigma_1^2} \boldsymbol{\theta}^\mathsf{T} \mathbf{R}_1 \boldsymbol{\theta} - \frac{1}{2\sigma_2^2} \boldsymbol{\gamma}^\mathsf{T} \mathbf{R}_2 \boldsymbol{\gamma}. \tag{7.67}$$

After integrating out the parameters $\boldsymbol{\beta}, \boldsymbol{\phi}, \boldsymbol{\theta}, \boldsymbol{\gamma}$ from the posterior density $\exp\{\Phi(\boldsymbol{\beta}, \boldsymbol{\phi}, \boldsymbol{\theta}, \boldsymbol{\gamma})\}$, and then apply the Laplace's approximation, we have the following approximated log-marginal density function:

$$l_m(\sigma_1^2, \sigma_2^2) \approx -\frac{m_1}{2} \log \sigma_1^2 - \frac{m_2}{2} \log \sigma_2^2 + l(\widehat{\boldsymbol{\beta}}, \widehat{\boldsymbol{\phi}}, \widehat{\boldsymbol{\theta}}, \widehat{\boldsymbol{\gamma}}) - \frac{1}{2\sigma_1^2} \boldsymbol{\theta}^\mathsf{T} \mathbf{R}_1 \boldsymbol{\theta} - \frac{1}{2\sigma_2^2} \boldsymbol{\gamma}^\mathsf{T} \mathbf{R}_2 \boldsymbol{\gamma} - \frac{1}{2} \log |-\widehat{\mathbf{H}} + \mathbf{Q}(\sigma_1^2, \sigma_2^2)|, \tag{7.68}$$

where $\widehat{\boldsymbol{\beta}}, \widehat{\boldsymbol{\phi}}, \widehat{\boldsymbol{\theta}}$ and $\widehat{\boldsymbol{\gamma}}$ are the MPL estimates and $\widehat{\mathbf{H}}$ is the Hessian from the log-likelihood $l(\boldsymbol{\beta}, \boldsymbol{\phi}, \boldsymbol{\theta}, \boldsymbol{\gamma})$ where $\boldsymbol{\beta}, \boldsymbol{\phi}, \boldsymbol{\theta}, \boldsymbol{\gamma}$ are replaced by their MPL estimates, and $\mathbf{Q}$ is a block diagonal matrix given by

$$\mathbf{Q}(\sigma_1^2, \sigma_2^2) = \text{diag}(\mathbf{0}_{p_1 \times p_1}, \mathbf{0}_{p_2 \times p_2}, \mathbf{R}_1/\sigma_1^2, \mathbf{R}_2/\sigma_2^2).$$

Setting the differentiation of $l_m$ with respect to $\sigma_1^2$ to 0 will yield that the solution of $\sigma_1^2$ satisfies

$$\sigma_1^2 = \frac{\boldsymbol{\theta}^\mathsf{T} \mathbf{R}_1 \boldsymbol{\theta}}{m_1 - \nu_1}, \tag{7.69}$$

where

$$\nu_1 = \text{tr}\left( (-\widehat{\mathbf{H}} + \mathbf{Q})^{-1} \begin{pmatrix} \mathbf{0}_{p_1 \times p_1} & \mathbf{0}_{p_1 \times p_2} & \mathbf{0}_{p_1 \times m_1} & \mathbf{0}_{p_1 \times m_2} \\ \mathbf{0}_{p_2 \times p_1} & \mathbf{0}_{p_2 \times p_2} & \mathbf{0}_{p_2 \times m_1} & \mathbf{0}_{p_2 \times m_2} \\ \mathbf{0}_{m_1 \times p_1} & \mathbf{0}_{m_1 \times p_2} & \mathbf{R}_1/\sigma_1^2 & \mathbf{0}_{m_1 \times m_2} \\ \mathbf{0}_{m_2 \times p_1} & \mathbf{0}_{m_2 \times p_2} & \mathbf{0}_{m_2 \times m_1} & \mathbf{0}_{m_2 \times m_2} \end{pmatrix} \right). \tag{7.70}$$

Because it involves many zero matrices, we can simplify the expression for $\nu_1$. Let us partition the $(\widehat{-\mathbf{H}+\mathbf{Q}})^{-1}$ matrix according to $\boldsymbol{\beta}, \boldsymbol{\phi}, \boldsymbol{\theta}$ and $\boldsymbol{\gamma}$:

$$(-\widehat{\mathbf{H}}+\mathbf{Q})^{-1} = \begin{pmatrix} \mathbf{A}_{11} & \mathbf{A}_{12} & \mathbf{A}_{13} & \mathbf{A}_{14} \\ \mathbf{A}_{21} & \mathbf{A}_{22} & \mathbf{A}_{23} & \mathbf{A}_{24} \\ \mathbf{A}_{31} & \mathbf{A}_{32} & \mathbf{A}_{33} & \mathbf{A}_{34} \\ \mathbf{A}_{41} & \mathbf{A}_{42} & \mathbf{A}_{43} & \mathbf{A}_{44} \end{pmatrix}.$$

Then, $\nu_1$ becomes:

$$\nu_1 = \text{tr}(\mathbf{A}_{33}\mathbf{R}_1)/\sigma_1^2. \tag{7.71}$$

Similarly, by setting the differentiation of $l_m$ with respect to $\sigma_2^2$ to 0 will give us that the solution of $\sigma_2^2$ satisfies

$$\sigma_2^2 = \frac{\boldsymbol{\gamma}^\mathsf{T}\mathbf{R}_2\boldsymbol{\gamma}}{m_2 - \nu_2}, \tag{7.72}$$

where

$$\nu_2 = \text{tr}(\mathbf{A}_{44}\mathbf{R}_2)/\sigma_2^2. \tag{7.73}$$

$\nu_1$ and $\nu_2$ are usually referred to as the degrees of freedom.

These results provide us with a procedure to estimate the model parameters $\boldsymbol{\beta}$, $\boldsymbol{\phi}$, $\boldsymbol{\theta}$, $\boldsymbol{\gamma}$, and also the smoothing parameters $\lambda_1 = 1/(2\sigma_1^2)$ and $\lambda_2 = 1/(2\sigma_2^2)$. It involves "inner" and "outer" iterations. During the inner iterations, we keep the values of $\sigma_1^2$ and $\sigma_2^2$ fixed, and we obtain corresponding estimates for $\boldsymbol{\beta}$, $\boldsymbol{\phi}$, $\boldsymbol{\theta}$, $\boldsymbol{\gamma}$, using the algorithm described in this chapter, running it until convergence. Then, we update $\sigma_1^2$ and $\sigma_2^2$ using the formulas (7.69) and (7.72), respectively, where the $\sigma_1^2$ and $\sigma_2^2$ values on the right-hand side of these formulas are set to their current values. This process continues until both $\text{df}_1 = \nu_1$ and $\text{df}_2 = \nu_2$ are stabilized, which typically means the changes in the "df" values between two consecutive outer iterations are less than 1 or even less than 0.5.

## 7.10 Sensitivity analysis

We have adopted the Archimedean copula function (see (7.8)) to model dependence, which includes the parameter $\alpha$, often referred to as the degree of association parameter. Parameter $\alpha$ can be transformed into the Kendall's rank correlation coefficient parameter, commonly denoted as Kendall's $\tau$.

In practice, Kendall's $\tau$ value is generally unknown. Therefore, it is essential to perform a sensitivity analysis to evaluate how the estimates, especially the regression coefficients of vector $\boldsymbol{\beta}$ and the baseline hazard, baseline cumulative hazard, or baseline survival, are affected with varying values of $\tau$. Some examples of sensitivity analysis in the context of copula can be found in Chen (2010), Huang & Zhang (2008) and Xu et al. (2016).

A sensitivity analysis can be conducted by selecting a sequence of possible $\tau$ values within the range appropriate for $\tau$. Typically, we have prior knowledge of whether $\tau$ is positive or negative. For example, in a clinical trial assessing the effectiveness of a treatment to prevent cancer recurrence, it may be known that early dropouts from the trial were mainly due to adverse effects of the treatment. In such cases, there is likely a positive dependence between cancer recurrence time and censoring time. Therefore, we might choose a sequence of non-negative $\tau$ values for the sensitivity analysis, such as $\tau = \{0, 0.2, 0.4, 0.6, 0.8\}$, where $\tau = 0$ suggests independence and $\tau = 0.8$ suggests strong dependence. Then, estimates of the parameters of interest, such as $\beta$ and $h_{0Y}(t)$ are obtained corresponding to these selected $\tau$ values. Plots of the CIs for all the $\beta_j$'s, where each CI plot also displays a $p$-value for the significant test, will clearly show how the $\beta_j$s and their $p$-values change with $\tau$. Plots of the baseline hazard estimates for the selected $\tau$ also illustrate their sensitivity to the choice of $\tau$. Example 7.3 below contains a sensitivity analysis procedure.

## 7.11 R '`survivalMPLdc`' package and examples

The R package '`survivalMPLdc`' can be used to fit Cox models under dependent right censoring. This package can be installed either from R CRAN or GitHub.

For CRAN, you may try:

```
install.packages("survivalMPLdc")
library("survivalMPLdc")
```

For GitHub, you may try:

```
# install.packages("devtools")
devtools::install_github("Kenny-Jing-Xu/survivalMPLdc")
library(survivalMPLdc)
```

In this section, we will present two R examples using this package. The first is a simple simulation example, while the second example demonstrates how to utilize this package with a dementia dataset included in the package.

**Example 7.2** (An example with simulated data)**.**
This example conducts a simple simulation study to evaluate the Cox hazard model estimation method when dependent censoring exists. We adopt sample sizes $n = 100$ and 500 with the number of replications $N = 300$. We aim to:

(i) Investigate the effects of the copula type and $\tau$ value on the MPL estimates of $\beta$ and $\phi$ (see Tables 7.1 and 7.2).

(ii) Compare the asymptotic standard deviation with the Monte Carlo standard deviation for the estimates of regression coefficients and baseline hazard (see Tables 7.1–7.4).

Data are generated as the following. Two covariates are from two distributions: $X_1 \sim \text{Bernoulli}(0.5)$ and $X_2 \sim \text{uniform}(-10, 10)$. The Cox model marginal hazards for $Y$ and $C$ are, respectively,

$$h_Y(t) = h_{0Y}(t)\exp(-0.5X_1 + 0.1X_2)$$
$$h_C(t) = h_{0C}(t)\exp(0.3X_1 + 0.2X_2),$$

where the baseline hazards are: $h_{0Y}(t) = h_{0C}(t) = \lambda^{-a}at^{a-1}$, where we adopted $a = 2$ and $\lambda = 2$ in this simulation. Here, the true $\boldsymbol{\beta} = (-0.5, 0.1)^{\mathsf{T}}$ and the true $\boldsymbol{\phi} = (0.3, 0.2)^{\mathsf{T}}$.

Next, we briefly explain how to generate dependent survival and censoring times using the Frank copula (the reader can check Example 7.1 to understand this random number generation process). The pair of dependent random numbers $(S_{Y_i}, S_{C_i})$ is firstly generated using the R function "frankCopula", and then the inversion method is used to generate $(Y_i, C_i)$ where $Y_i$ is generated from $S_{T_i}$ and $C_i$ from $S_{C_i}$. More details on how to generate dependent random numbers using copulas in references such as Clayton (1978).

For each replication sample, we compute the MPL estimates $\widehat{\boldsymbol{\beta}}$, $\widehat{\boldsymbol{\phi}}$, $\widehat{\boldsymbol{\theta}}$ and $\widehat{\gamma}$ with an assumed $\tau$, where both $h_{0Y}(t)$ and $h_{0C}(t)$ are approximated using M-splines with the smoothing values estimated using the marginal likelihood method as described in Section 7.9. The code for this simulation study is given below.

```
# function for simulating data

gen_depcen <- function(n, tau) {

    # set up copula
    alpha <- iTau(frankCopula(100),
        tau = tau)
    cop <- frankCopula(alpha, dim = 2)

    x <- rep(0, n)  #observed time
    del <- rep(0, n)  #event status
    eta <- rep(0, n)  #(dependent) censoring status
    y <- NULL
    c <- NULL

    # generate covariates
    Z <- cbind(rbinom(n, 1, 0.5), runif(n,
        -10, 10))

    # set true parameter values
    betas <- c(-0.5, 0.1)
    phis <- c(0.3, 0.2)
```

```
    # set weibull distribution
    # parameters
    lambda <- 2
    a <- 2

    for (i in 1:n) {
        S_tc <- rCopula(1, cop)
        # S_event and S_cens
        # probabilities

        # draw random event time
        ti <- (-log(S_tc[1])/((lambda^(-a)) *
            exp(Z[i, ] %*% betas)))^(1/a)
        # draw random censoring
        # time
        ci <- (-log(S_tc[2])/((lambda^(-a)) *
            exp(Z[i, ] %*% phis)))^(1/a)
        # save event and censoring
        # times
        y <- c(y, ti)
        c <- c(c, ci)

        if (ti < ci) {
            # event observed
            del[i] <- 1
            x[i] <- ti
        } else if (ci < ti) {
            # dependent censoring
            # time observed
            eta[i] <- 1
            x[i] <- ci
        }
    }
    dat <- cbind(x, del, eta)

    out <- list(dat = dat, Z = Z)
    return(out)

}

# function for running simulation
# study with n=500
ch7_save <- matrix(0, nrow = 300, ncol = 48)


for (s in 1:300) {

    dat <- gen_depcen(n = 500, tau = 0.8)
```

```
surv.dat <- dat$dat
Z.dat <- dat$Z

# fit independent censoring
ctrl1 <- coxph_mpl_dc.control(ordSp = 4,
    binCount = 30, tau = 0, copula = "independent",
    pent = "penalty_mspl", smpart = "REML",
    penc = "penalty_mspl", smparc = "REML",
    mid = 1, asy = 1, ac = 1, cv = 1,
    cat.smpar = "No")
coxMPLests_tau0 <- coxph_mpl_dc(surv = surv.dat,
    cova = Z.dat, control = ctrl1)

ch7_save[s, 1] <- coxMPLests_tau0$mpl_beta[1]
ch7_save[s, 2] <- coxMPLests_tau0$mpl_beta[2]
ch7_save[s, 3] <- coxMPLests_tau0$mpl_beta_sd[1]
ch7_save[s, 4] <- coxMPLests_tau0$mpl_beta_sd[2]

ch7_save[s, 5] <- coxMPLests_tau0$mpl_phi[1]
ch7_save[s, 6] <- coxMPLests_tau0$mpl_phi[2]
ch7_save[s, 7] <- coxMPLests_tau0$mpl_phi_sd[1]
ch7_save[s, 8] <- coxMPLests_tau0$mpl_phi_sd[2]

# fit frank cop and tau = 0.8
ctrl2 <- coxph_mpl_dc.control(ordSp = 4,
    binCount = 30, tau = 0.8, copula = "frank",
    pent = "penalty_mspl", smpart = "REML",
    penc = "penalty_mspl", smparc = "REML",
    mid = 1, asy = 1, ac = 1, cv = 1,
    cat.smpar = "No")
coxMPLests_tau0.8_frank <- coxph_mpl_dc(surv = surv.dat,
    cova = Z.dat, control = ctrl2)

ch7_save[s, 9] <- coxMPLests_tau0.8_frank$mpl_beta[1]
ch7_save[s, 10] <- coxMPLests_tau0.8_frank$mpl_beta[2]
ch7_save[s, 11] <- coxMPLests_tau0.8_frank$mpl_beta_sd[1]
ch7_save[s, 12] <- coxMPLests_tau0.8_frank$mpl_beta_sd[2]

ch7_save[s, 13] <- coxMPLests_tau0.8_frank$mpl_phi[1]
ch7_save[s, 14] <- coxMPLests_tau0.8_frank$mpl_phi[2]
ch7_save[s, 15] <- coxMPLests_tau0.8_frank$mpl_phi_sd[1]
ch7_save[s, 16] <- coxMPLests_tau0.8_frank$mpl_phi_sd[2]

# fit frank cop and tau = 0.4
ctrl3 <- coxph_mpl_dc.control(ordSp = 4,
    binCount = 30, tau = 0.4, copula = "frank",
    pent = "penalty_mspl", smpart = "REML",
    penc = "penalty_mspl", smparc = "REML",
```

```
    mid = 1, asy = 1, ac = 1, cv = 1,
    cat.smpar = "No")
coxMPLests_tau0.4_frank <- coxph_mpl_dc(surv = surv.dat,
    cova = Z.dat, control = ctrl3)

ch7_save[s, 17] <- coxMPLests_tau0.4_frank$mpl_beta[1]
ch7_save[s, 18] <- coxMPLests_tau0.4_frank$mpl_beta[2]
ch7_save[s, 19] <- coxMPLests_tau0.4_frank$mpl_beta_sd[1]
ch7_save[s, 20] <- coxMPLests_tau0.4_frank$mpl_beta_sd[2]

ch7_save[s, 21] <- coxMPLests_tau0.4_frank$mpl_phi[1]
ch7_save[s, 22] <- coxMPLests_tau0.4_frank$mpl_phi[2]
ch7_save[s, 23] <- coxMPLests_tau0.4_frank$mpl_phi_sd[1]
ch7_save[s, 24] <- coxMPLests_tau0.4_frank$mpl_phi_sd[2]

# fit clayton cop and tau =
# 0.8
ctrl4 <- coxph_mpl_dc.control(ordSp = 4,
    binCount = 30, tau = 0.8, copula = "clayton",
    pent = "penalty_mspl", smpart = "REML",
    penc = "penalty_mspl", smparc = "REML",
    mid = 1, asy = 1, ac = 1, cv = 1,
    cat.smpar = "No")
coxMPLests_tau0.8_clay <- coxph_mpl_dc(surv = surv.dat,
    cova = Z.dat, control = ctrl4)

ch7_save[s, 25] <- coxMPLests_tau0.8_clay$mpl_beta[1]
ch7_save[s, 26] <- coxMPLests_tau0.8_clay$mpl_beta[2]
ch7_save[s, 27] <- coxMPLests_tau0.8_clay$mpl_beta_sd[1]
ch7_save[s, 28] <- coxMPLests_tau0.8_clay$mpl_beta_sd[2]

ch7_save[s, 29] <- coxMPLests_tau0.8_clay$mpl_phi[1]
ch7_save[s, 30] <- coxMPLests_tau0.8_clay$mpl_phi[2]
ch7_save[s, 31] <- coxMPLests_tau0.8_clay$mpl_phi_sd[1]
ch7_save[s, 32] <- coxMPLests_tau0.8_clay$mpl_phi_sd[2]

# fit clayton cop and tau =
# 0.4
ctrl5 <- coxph_mpl_dc.control(ordSp = 4,
    binCount = 30, tau = 0.4, copula = "clayton",
    pent = "penalty_mspl", smpart = "REML",
    penc = "penalty_mspl", smparc = "REML",
    mid = 1, asy = 1, ac = 1, cv = 1,
    cat.smpar = "No")
coxMPLests_tau0.4_clay <- coxph_mpl_dc(surv = surv.dat,
    cova = Z.dat, control = ctrl5)

ch7_save[s, 33] <- coxMPLests_tau0.4_clay$mpl_beta[1]
```

```
    ch7_save[s, 34] <- coxMPLests_tau0.4_clay$mpl_beta[2]
    ch7_save[s, 35] <- coxMPLests_tau0.4_clay$mpl_beta_sd[1]
    ch7_save[s, 36] <- coxMPLests_tau0.4_clay$mpl_beta_sd[2]

    ch7_save[s, 37] <- coxMPLests_tau0.4_clay$mpl_phi[1]
    ch7_save[s, 38] <- coxMPLests_tau0.4_clay$mpl_phi[2]
    ch7_save[s, 39] <- coxMPLests_tau0.4_clay$mpl_phi_sd[1]
    ch7_save[s, 40] <- coxMPLests_tau0.4_clay$mpl_phi_sd[2]

    # fit frank cop and tau = 0.8,
    # maximum likelihood (no
    # smoothing)
    ctrl6 <- coxph_mpl_dc.control(ordSp = 4,
        binCount = 10, tau = 0.8, copula = "frank",
        pent = "penalty_mspl", smpart = 0,
        penc = "penalty_mspl", smparc = 0,
        mid = 1, asy = 1, ac = 0, cv = 0,
        cat.smpar = "No", maxit2 = 1)
    coxests_tau0.8_frank_ML <- coxph_mpl_dc(surv = surv.dat,
        cova = Z.dat, control = ctrl6)

    ch7_save[s, 41] <- coxests_tau0.8_frank_ML$mpl_beta[1]
    ch7_save[s, 42] <- coxests_tau0.8_frank_ML$mpl_beta[2]
    ch7_save[s, 43] <- coxests_tau0.8_frank_ML$mpl_beta_sd[1]
    ch7_save[s, 44] <- coxests_tau0.8_frank_ML$mpl_beta_sd[2]

    ch7_save[s, 45] <- coxests_tau0.8_frank_ML$mpl_phi[1]
    ch7_save[s, 46] <- coxests_tau0.8_frank_ML$mpl_phi[2]
    ch7_save[s, 47] <- coxests_tau0.8_frank_ML$mpl_phi_sd[1]
    ch7_save[s, 48] <- coxests_tau0.8_frank_ML$mpl_phi_sd[2]


}
```

The simulation results are presented in Tables 7.1, 7.2, 7.3 and 7.4. Specifically, Tables 7.1 and 7.2 summarize the simulation results for the MPL estimates of $\boldsymbol{\beta}$ and $\boldsymbol{\phi}$ with sample sizes $n = 100$ and $n = 500$ respectively, while Tables 7.3 and 7.4 provide the MPL and ML results for the estimation of the baseline hazard $h_{0Y}(t)$ with sample sizes $n = 100$ and $n = 500$ respectively.

For all the tables, the true $\tau = 0.8$ and the true $\lambda_t = 2$, resulting in an event proportion $\pi_T = 46\%$ and a censoring proportion $\pi_C = 35\%$). Various copulas and $\tau$ values (denoted as $\widetilde{\tau}$ in these tables) are assumed when performing MPL estimates.

All the tables include bias (BIAS), Monte Carlo standard deviation (MCSD), average of asymptotic standard deviations (ASD) and coverage probability (CP).

To evaluate $\widehat{\boldsymbol{\theta}}$ and the baseline hazard estimate $\widehat{h}_{0Y}(t)$, we compare $\widehat{h}_{0Y}(t)$ with

TABLE 7.1: Summary of simulation results for the MPL estimates of $\beta$ and $\phi$ with sample size $n = 100$ and true $\tau = 0.8$.

| | BIAS | ASD | MCSD | CP(95%) | BIAS | ASD | MCSD | CP(95%) |
|---|---|---|---|---|---|---|---|---|
| | [Frank copula and $\widetilde{\tau} = 0.8$] | | | | [Frank copula and $\widetilde{\tau} = 0.4$] | | | |
| $\beta_1$ | -0.040 | 0.179 | 0.186 | 0.962 | -0.149 | 0.260 | 0.287 | 0.894 |
| $\beta_2$ | 0.001 | 0.020 | 0.018 | 0.958 | -0.070 | 0.028 | 0.026 | 0.556 |
| $\phi_1$ | 0.016 | 0.147 | 0.136 | 0.984 | 0.282 | 0.169 | 0.152 | 0.583 |
| $\phi_2$ | 0.004 | 0.018 | 0.022 | 0.918 | 0.041 | 0.021 | 0.021 | 0.572 |
| | [Clayton copula and $\widetilde{\tau} = 0.8$]] | | | | [Independent censoring] | | | |
| $\beta_1$ | 0.049 | 0.154 | 0.216 | 0.863 | -0.968 | 0.315 | 0.355 | 0.122 |
| $\beta_2$ | -0.024 | 0.012 | 0.054 | 0.836 | -0.112 | 0.032 | 0.030 | 0.087 |
| $\phi_1$ | 0.140 | 0.210 | 0.289 | 0.847 | 0.421 | 0.184 | 0.168 | 0.368 |
| $\phi_2$ | -0.078 | 0.011 | 0.028 | 0.842 | 0.056 | 0.022 | 0.026 | 0.314 |
| | [Clayton copula and $\widetilde{\tau} = 0.4$] | | | | [ ] | | | |
| $\beta_1$ | -0.064 | 0.278 | 0.315 | 0.735 | | | | |
| $\beta_2$ | -0.059 | 0.031 | 0.031 | 0.845 | | | | |
| $\phi_1$ | 0.033 | 0.178 | 0.169 | 0.868 | | | | |
| $\phi_2$ | 0.042 | 0.021 | 0.026 | 0.854 | | | | |

$h_{0Y}(t)$ at the selected points $(0.5, 1, 1.5, 2, 2.5, 3, 3.5)$ using BIAS, MCSD and ASD these points. We do not evaluate the estimates of $h_{0C}(t)$ as this function is not our primary focus.

Tables 7.1 and 7.2 show that the average asymptotic standard deviations are very close to their Monte Carlo standard deviations, indicating the asymptotic covariance formulae for $\widehat{\beta}$ and $\widehat{\phi}$ are in general accurate. For data generated using $\tau = 0.8$ (i.e. strong dependence between $Y$ and $C$) and $\lambda_t = 2$, we observe that when assuming the Frank copula (i.e. the true copula) and $\tau = 0.8$ (i.e, the true $\tau$ value), the MPL method is successful for both sample sizes in obtaining quality estimates judged by their small biases and close to 95% coverage probabilities (CPs). Estimates obtained from independent censoring have the largest biases and poor coverage probabilities. Assuming a $\tau$ value with incorrect magnitude but correct sign (e.g. $\widetilde{\tau} = 0.4$) still improves the bias over the independent censoring case. We also observe that estimates using an incorrect copula but correct $\tau$ give lower biases than estimates from an incorrect $\tau$ but correct copula, indicating the importance of $\tau$ in the MPL estimation. Comparing the results for $n = 200$ we can see that they mainly differ over BIAS while their SD are similar.

Based on the data generated with $\tau = 0.8$ and Frank copula, assessments of the baseline hazard MPL and ML estimates, obtained using $\widetilde{\tau} = 0.8$ and Frank copula, are provided in Figure 7.2 and Table 7.4, where the ML estimates were acquired

TABLE 7.2: Summary of simulation results for the MPL estimates of $\beta$ and $\phi$ with sample size $n = 500$ and true $\tau = 0.8$.

| | BIAS | ASD | MCSD | CP(95%) | BIAS | ASD | MCSD | CP(95%) |
|---|---|---|---|---|---|---|---|---|
| | [Frank copula and $\widetilde{\tau} = 0.8$] | | | | [Frank copula and $\widetilde{\tau} = 0.4$] | | | |
| $\beta_1$ | 0.008 | 0.090 | 0.088 | 0.959 | -0.195 | 0.129 | 0.123 | 0.786 |
| $\beta_2$ | 0.001 | 0.010 | 0.011 | 0.951 | -0.071 | 0.015 | 0.015 | 0.488 |
| $\phi_1$ | 0.021 | 0.075 | 0.081 | 0.912 | 0.269 | 0.087 | 0.096 | 0.697 |
| $\phi_2$ | 0.004 | 0.009 | 0.009 | 0.968 | 0.039 | 0.011 | 0.009 | 0.783 |
| | [Clayton copula and $\widetilde{\tau} = 0.8$]] | | | | [Independent censoring] | | | |
| $\beta_1$ | 0.052 | 0.189 | 0.176 | 0.893 | -0.906 | 0.159 | 0.159 | 0.000 |
| $\beta_2$ | -0.080 | 0.041 | 0.017 | 0.873 | -0.115 | 0.017 | 0.016 | 0.000 |
| $\phi_1$ | -0.116 | 0.221 | 0.199 | 0.784 | 0.405 | 0.095 | 0.105 | 0.051 |
| $\phi_2$ | -0.017 | 0.055 | 0.024 | 0.776 | 0.055 | 0.012 | 0.011 | 0.000 |
| | [Clayton copula and $\widetilde{\tau} = 0.4$] | | | | [ ] | | | |
| $\beta_1$ | 0.264 | 0.139 | 0.145 | 0.838 | | | | |
| $\beta_2$ | -0.059 | 0.016 | 0.017 | 0.918 | | | | |
| $\phi_1$ | 0.236 | 0.092 | 0.101 | 0.683 | | | | |
| $\phi_2$ | 0.040 | 0.011 | 0.010 | 0.835 | | | | |

TABLE 7.3: Results for the MPL and ML estimates of $h_{0Y}(t)$ assuming Frank copula with $\widetilde{\tau} = 0.8$. The sample size $n = 100$.

| $t$ | 0.4 | 0.8 | 1.2 | 1.6 | 2.0 |
|---|---|---|---|---|---|
| | | MPL | | | |
| BIAS | -0.016 | -0.006 | 0.014 | -0.034 | -0.025 |
| MCSD | 0.067 | 0.090 | 0.183 | 0.208 | 0.249 |
| MC CP | 0.70 | 0.70 | 0.80 | 0.63 | 0.70 |
| | | ML | | | |
| BIAS | -0.034 | 0.008 | 0.008 | -0.039 | 0.005 |
| MCSD | 0.092 | 0.192 | 0.302 | 0.315 | 0.424 |
| MC CP | 0.60 | 0.77 | 0.67 | 0.70 | 0.63 |

simply by letting $h_1 = h_2 = 0$. Clearly, the smoothing in the MPL method allows for more accurate estimation of the true baseline hazard function, particularly in small sample sizes or where the number of basis functions selected may be too large.

□

TABLE 7.4: Results for the MPL and ML estimates of $h_{0Y}(t)$ assuming Frank copula with $\widetilde{\tau} = 0.8$. The sample size $n = 500$.

| $t$ | 0.4 | 0.8 | 1.2 | 1.6 | 2.0 |
|---|---|---|---|---|---|
| | | MPL | | | |
| BIAS | -0.005 | 0.004 | 0.018 | -0.012 | 0.008 |
| MCSD | 0.058 | 0.113 | 0.141 | 0.168 | 0.166 |
| MC CP | 0.73 | 0.75 | 0.80 | 0.78 | 0.73 |
| | | ML | | | |
| BIAS | -0.008 | 0.004 | 0.026 | -0.028 | 0.007 |
| MCSD | 0.071 | 0.157 | 0.197 | 0.236 | 0.239 |
| MC CP | 0.66 | 0.73 | 0.74 | 0.65 | 0.62 |

**Example 7.3** (Application to a dementia data).
In this example, we apply the MPL method from this chapter to fit a proportional hazards model using the dementia dataset from the PRIME study (Brodaty et al. (2011)). This dataset is available in the `survivalMPLdc` package.

The longitudinal study involved 970 patients with either dementia or mild cognitive impairment living in the community and attending one of nine memory clinics in Australia. Patients were assessed annually with additional visits at 3 months and 6 months. Demographics (e.g., age, sex, education level, and who the patient lived with) and diagnosis (i.e., type of dementia) were collected at baseline. At baseline and each visit, a research nurse or specialist clinician administered measures of cognition. For more details about this study, see Brodaty et al. (2011). Dementia involves progressive decline in cognition and function, and as a result, patients often require institutionalization.

In this example, we are interested in the time to institutionalization for potential dementia patients and we wish to identify important predictors for this response variable. Predicting institutionalization is important clinically when planning patients' management and at a social level. The following predictors are considered: age, sex (female), educational level (high school or above), living alone, dementia type (Alzheimer disease), baseline cognitive ability (MMSE), baseline functional ability (SMAF), baseline neuropsychiatric symptoms (total NPI), baseline dementia severity (CDR), baseline caregiver burden (ZBI), medication types (benzodiazepines, anti-psychotics), change in cognitive ability at three months (i.e., 3 months from the baseline assessment), change in functional ability at three months and change in neuropsychiatric symptoms at three months.

We focused on the patients with complete data, which contains 583 patients. Among them, 156 patients (26.8%) were institutionalized during the study; 146 (25.0%) withdrew from the study before the three-year period; and 281 (48.2%) were followed for the full 3-year period but were not institutionalized. Since patients who withdrew from the studies are older and have lower cognitive and functional abilities, they appear more likely to be institutionalized than the patients who remained in the

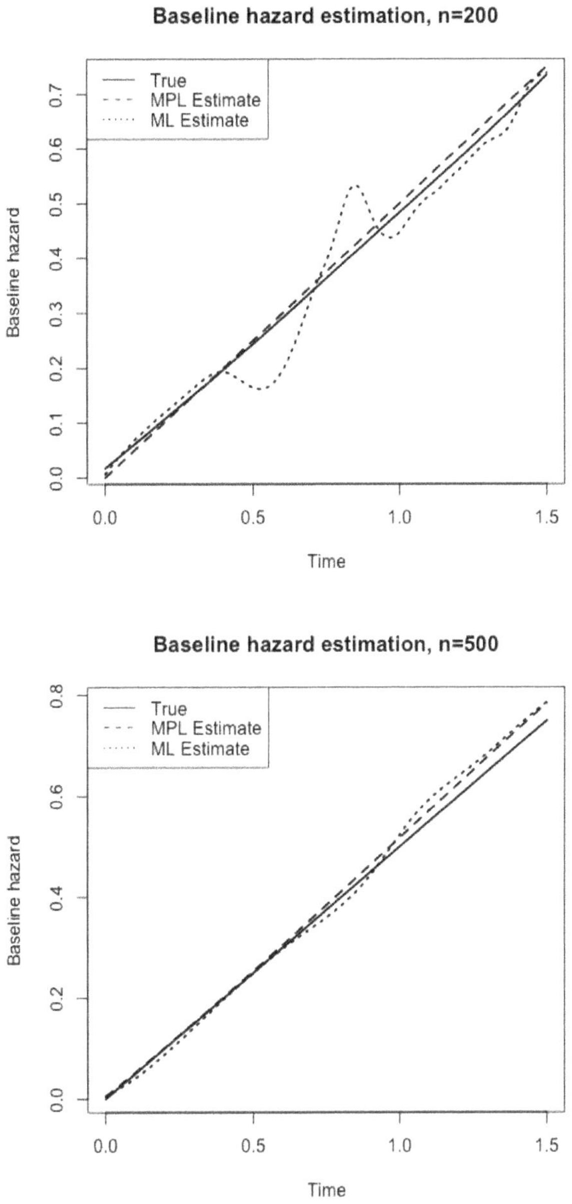


FIGURE 7.2: Baseline hazard function estimation using a Frank copula and $\tau = 0.8$, comparing the MPL estimates (with smoothing) to the ML estimates (without smoothing).

studies. It is reasonable to consider withdrawal from studies as dependent censoring, and we believe that Kendall's $\tau > 0$ for this example.

Here, we analyze this dataset by assuming a Frank copula model for dependent censoring and perform a sensitivity analysis similar to Chen (2010) and Huang and Zhang Huang & Zhang (2008). This means that we fit the model repeatedly using Kendall's $\tau = 0$ (i.e., independent censoring), 0.2 (weak association between event and censoring), 0.5 (medium association) and 0.8 (strong association). Note that the results based on other copulas, such as Clayton or Gumbel copulas, can also be used to fit the model; however, the results do not differ substantially. We use M-splines to approximate the baseline hazard, and the smoothing parameter value is determined by the marginal likelihood-based automatic smoothing parameter selection procedure.

R code for this example is displayed next.

```
# set up PRIME data set
data(PRIME)
names(PRIME) <- c("Time", "Institutionalized",
    "Drop Out", "Age", "Gender",
    "High Education", "Alzheimer Disease",
    "Baseline CDR", "Baseline MMSE",
    "Baseline SMAF", "Baseline ZBI",
    "Baseline NPI", "Benzodiazepine",
    "Antipsychotic", "Living Alone",
    "3-month Change MMSE", "3-month Change SMAF",
    "3-month Change NPI")

# time observed, event and
# dependent censoring
# indicators
surv <- as.matrix(PRIME[, 1:3])

# covariates
cova <- as.matrix(PRIME[, -c(1:3)])

# the number covariates
n <- dim(PRIME)[1]
p <- dim(PRIME)[2] - 3

# MPL under independent
# censoring
binCount <- 30
control <- coxph_mpl_dc.control(ordSp = 4,
   binCount = binCount, tie = "Yes",
   tau = 0, copula = "independent",
   pent = "penalty_mspl", smpart = "REML",
   penc = "penalty_mspl", smparc = "REML",
   mid = 1, asy = 1, ac = 1, cv = 1,
   cat.smpar = "No")
```

```
coxMPLests_tau0 <- coxph_mpl_dc(surv = surv,
    cova = cova, control = control,
    )

MPL_beta_tau0 <- coef(object = coxMPLests_tau0,
    parameter = "beta")
MPL_phi_tau0 <- coef(object = coxMPLests_tau0,
    parameter = "phi")
mpl_h0t_tau0 <- coxMPLests_tau0$mpl_h0t
mpl_h0Ti_tau0 <- approx(surv[,
    1], mpl_h0t_tau0, xout = seq(0,
    max(surv[, 1]), 0.01), method = "constant",
    rule = 2, ties = mean)$y

# MPL estimate under tau=0.2
control <- coxph_mpl_dc.control(ordSp = 4,
    binCount = binCount, tie = "Yes",
    tau = 0.2, copula = "frank",
    pent = "penalty_mspl", smpart = "REML",
    penc = "penalty_mspl", smparc = "REML",
    mid = 1, asy = 1, ac = 1, cv = 1,
    cat.smpar = "No")
coxMPLests_tau0.2 <- coxph_mpl_dc(surv = surv,
    cova = cova, control = control,
    )

MPL_beta_tau0.2 <- coef(object = coxMPLests_tau0.2,
    parameter = "beta")
MPL_phi_tau0.2 <- coef(object = coxMPLests_tau0.2,
    parameter = "phi")

mpl_h0t_tau0.2 <- coxMPLests_tau0.2$mpl_h0t
mpl_h0Ti_tau0.2 <- approx(surv[,
    1], mpl_h0t_tau0.2, xout = seq(0,
    max(surv[, 1]), 0.01), method = "constant",
    rule = 2, ties = mean)$y

# MPL estimate under tau=0.5
control <- coxph_mpl_dc.control(ordSp = 4,
    binCount = binCount, tie = "Yes",
    tau = 0.5, copula = "frank",
    pent = "penalty_mspl", smpart = "REML",
    penc = "penalty_mspl", smparc = "REML",
    mid = 1, asy = 1, ac = 1, cv = 1,
    cat.smpar = "No")
coxMPLests_tau0.5 <- coxph_mpl_dc(surv = surv,
    cova = cova, control = control,
    )
```

```
mpl_beta_phi_zp_tau0.5 <- coxMPLests_tau0.5$mpl_beta_phi_zp
MPL_beta_tau0.5 <- coef(object = coxMPLests_tau0.5,
    parameter = "beta")
MPL_phi_tau0.5 <- coef(object = coxMPLests_tau0.5,
    parameter = "phi")
mpl_h0t_tau0.5 <- coxMPLests_tau0.5$mpl_h0t
mpl_h0Ti_tau0.5 <- approx(surv[,
    1], mpl_h0t_tau0.5, xout = seq(0,
    max(surv[, 1]), 0.01), method = "constant",
    rule = 2, ties = mean)$y

# MPL estimate under tau=0.8
control <- coxph_mpl_dc.control(ordSp = 4,
    binCount = binCount, tie = "Yes",
    tau = 0.8, copula = "frank",
    pent = "penalty_mspl", smpart = "REML",
    penc = "penalty_mspl", smparc = "REML",
    mid = 1, asy = 1, ac = 1, cv = 1,
    cat.smpar = "No")
coxMPLests_tau0.8 <- coxph_mpl_dc(surv = surv,
    cova = cova, control = control,
    )
mpl_beta_phi_zp_tau0.8 <- coxMPLests_tau0.8$mpl_beta_phi_zp
MPL_beta_tau0.8 <- coef(object = coxMPLests_tau0.8,
    parameter = "beta")
MPL_phi_tau0.8 <- coef(object = coxMPLests_tau0.8,
    parameter = "phi")
mpl_h0t_tau0.8 <- coxMPLests_tau0.8$mpl_h0t
mpl_h0Ti_tau0.8 <- approx(surv[,
    1], mpl_h0t_tau0.8, xout = seq(0,
    max(surv[, 1]), 0.01), method = "constant",
    rule = 2, ties = mean)$y
```

Figure 7.3 displays the estimated regression coefficient versus $\tau$ value plots for a selection of the covariates. The plots also include their 95% confidence intervals (CIs). Clearly, the following predictors are significant for all $\tau$ values: baseline MMSE, MMSE over three months; baseline SMAF, SMAF over three months (marginally significant); baseline NPI and NPI over three months. Other predictors varied in statistical significance across different $\tau$ values: Age becomes marginally significant when $\tau = 0.8$; both benzodiazepines and antipsychotic medication classes are significant only when $\tau < 0.2$; Living alone is significant as long as $\tau > 0.2$.

Figure 7.4 contains plots of baseline hazard estimates. Clearly, the MPL baseline hazard estimates not only display clear trends for the risk of institutionalization over time after adjusting the covariates, they also provide easy comparisons of baseline hazards for different $\tau$ values.

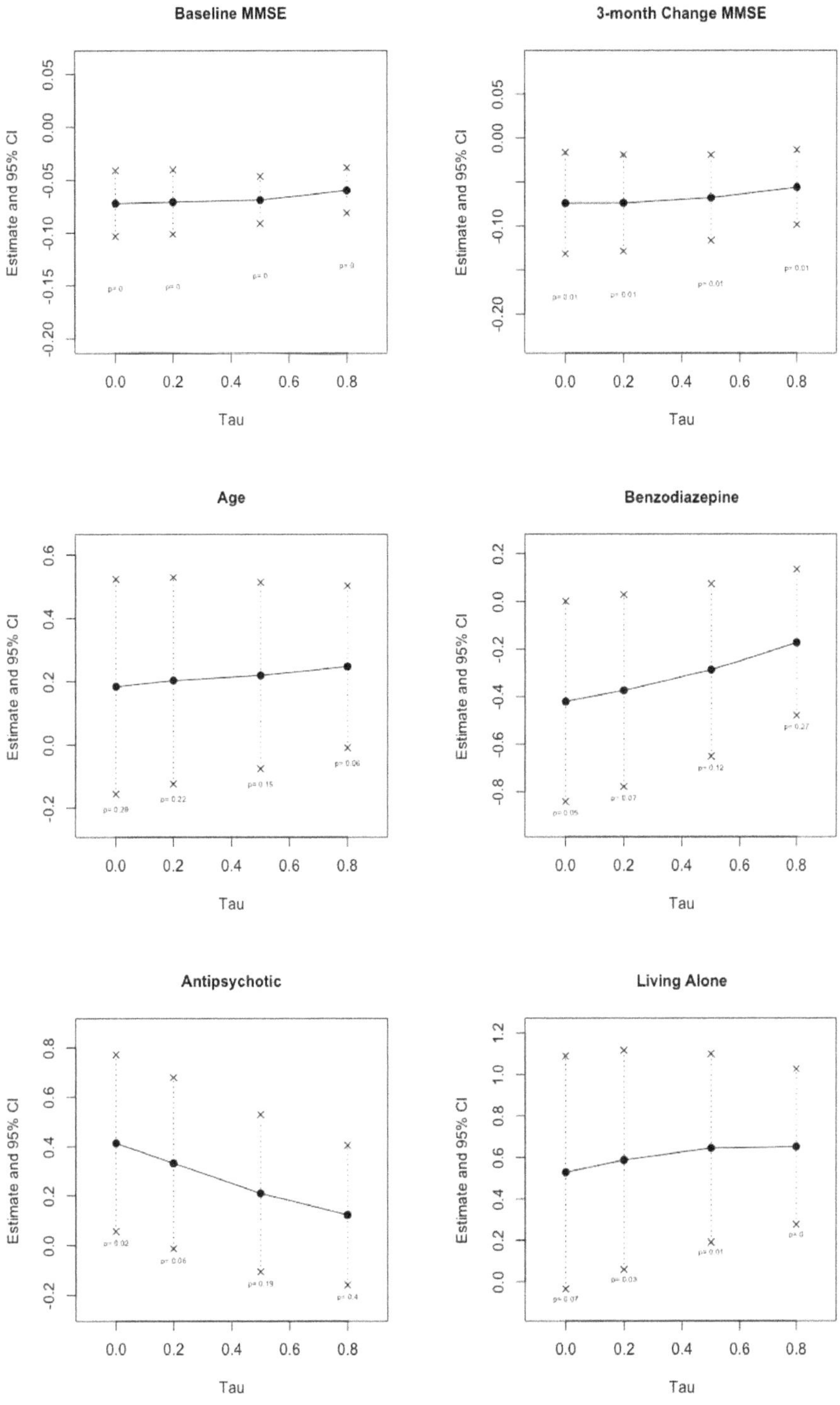
Baseline MMSE
3-month Change MMSE
Age
Benzodiazepine
Antipsychotic
Living Alone
Estimate and 95% CI
Tau
0.0
0.2
0.4
0.6
0.8

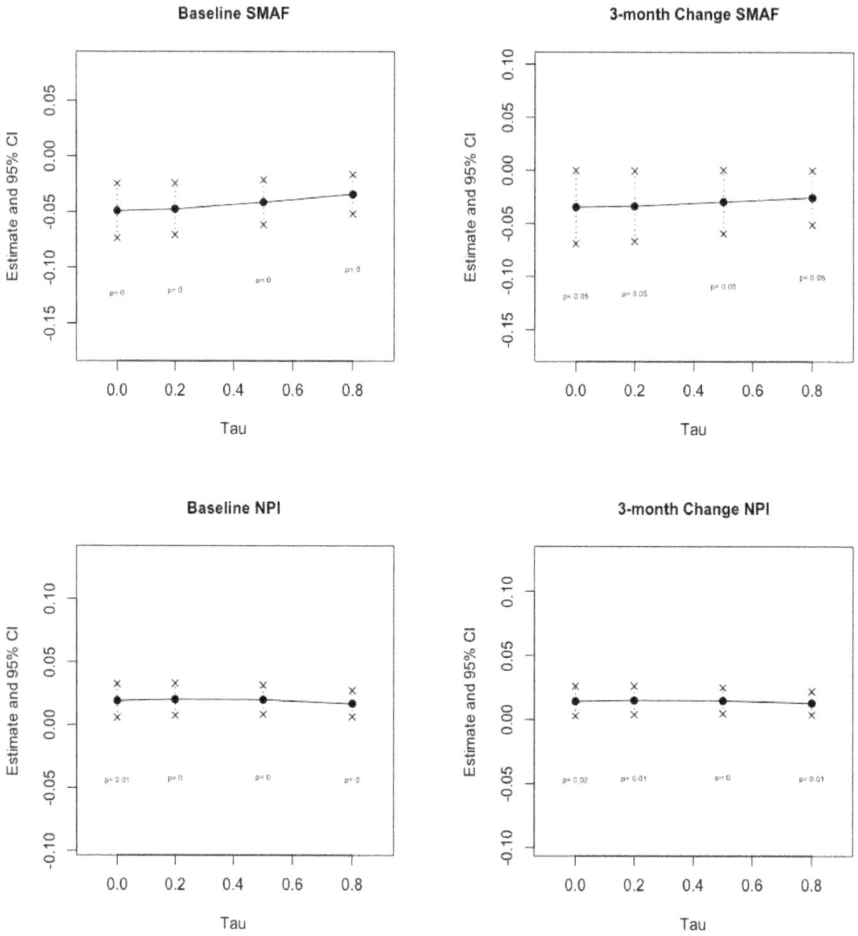


FIGURE 7.3: Plots of the estimated $\beta$ against $\tau$ and their 95% CIs.

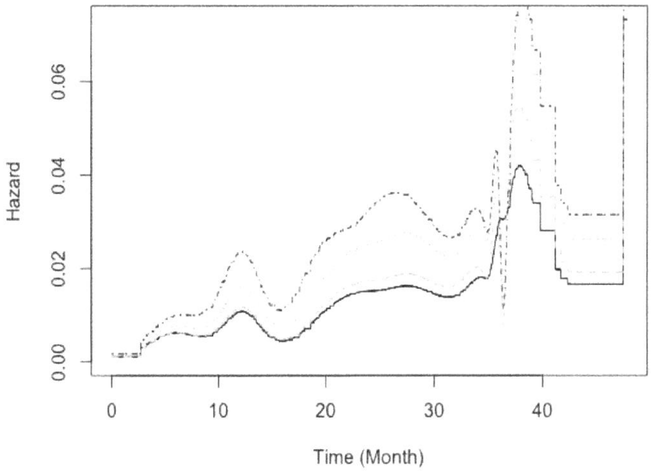


FIGURE 7.4: Plots of the baseline hazard function estimates corresponding to different $\tau$ values.

□

## 7.12 Summary

This chapter explains estimation methods for Cox proportional hazard models under dependent censoring using copulas and penalized likelihood. This approach is capable of providing high-quality MPL estimates for regression coefficients and the baseline hazard, especially when the selected copula function and Kendall's $\tau$ accurately model the dependence between failure and censoring times. An incorrect copula has less effect on parameter estimates than an incorrect $\tau$. The MPL estimate of the baseline hazard typically offers smaller biases and much-improved interpretations compared to its maximum likelihood counterpart. The large sample variance formula for the MPL estimates, given by (7.66), is generally accurate, as shown by the simple simulation study in Example 7.2.

For practical data analysis, estimating Kendall's $\tau$ value exactly can be challenging. However, we can gauge the range of $\tau$ using expert knowledge or experience Huang & Zhang (2008), Chen (2010). Then, a sensitivity analysis can be performed using this range of $\tau$ values, resulting in various regression coefficients and baseline hazard estimates. The copula method of this chapter can be extended to other semi-parametric hazard models where informative censoring is present, such as proportional odds, additive hazards, or accelerated failure time models.

# 8

# *Additive hazards model*

The semi-parametric Cox model is the most widely used regression model for analyzing survival data. However, meeting the proportional hazards assumption (see Chapter 2) can be challenging in practice. Therefore, an alternative semi-parametric regression model to the Cox model can be appealing. The accelerated failure time (AFT) model is a popular alternative, and the semi-parametric additive hazards model offers another important option.

Another important reason for the additive hazard model is that it enables more direct interpretation of the regression coefficients as they reflect directly a risk increase or decrease when a covariate changes.

In this chapter, we will delve into the semi-parametric additive hazards model within the broader framework of partly interval-censored survival data. This context encompasses right-censored survival data as a special case. Our primary emphasis in this chapter lies in the computational aspects of fitting such a model, addressing the non-negativity constraint on both the baseline and individual hazards – constraints that have largely been overlooked in the current literature.

## 8.1 Introduction

The additive hazards model, as introduced by Aalen (1989) and Lin & Ying (1994), offers an important alternative to the conventional proportional hazards model of Cox (1972) for the analysis of time-to-event data. There are two reasons for why people select an additive hazards model to analyze their survival data.

(i) In cases where the hazards (or risks) change, instead of hazards ratio, is the primary interest.

(ii) In cases where the proportional hazards assumption of the Cox model does not hold, the additive hazards model can serve as an alternative approach to analyze the provided survival data.

Although inclusion of *time-dependent covariate(s)* in a Cox model can help to deal with the issue of failing the proportional hazards assumption, the additive hazard model provides another approach to deal with this problem.

For the full likelihood-based method discussed in this chapter, we again consider partly interval-censored (e.g. Kim (2003)) survival (or event) data. Thus, they can

DOI: 10.1201/9781351109710-8

contain event times as well as left, right and interval censoring times. Let us start with defining the semi-parametric additive hazard model that we will consider in this chapter.

Recall that in the previous chapters we have used $Y_i$ to denote the event time of interest for subject $i$. Due to censoring, $Y_i$ may not be observed exactly; in this chapter, we allow $Y_i$ to be partly interval-censored. In order to accommodate interval-censored observations, the observed survival data for subject $i$ is denoted by $[t_i^L, t_i^R]$ like before. Specifically, if $t_i^L = t_i^R = t_i$ then $Y_i = t_i$ and $t_i$ is an event time; if $t_i^L = 0$ then $Y_i$ is left-censored at $t_i^R < \infty$ so that $0 < Y_i \le t_i^R$; if $t_i^R = \infty$ and $t_i^L \neq 0$ then $Y_i$ is right-censored at $t_i^L$ so that $Y_i > t_i^L$; otherwise, $Y_i$ is interval censored with $t_i^L \le Y_i \le t_i^R$.

As before, we define $\delta_i$, $\delta_i^L$, $\delta_i^R$, and $\delta_i^I$ as indicators for event, left, right, and interval censoring, respectively. It is clear that $\delta_i = 1 - \delta_i^L - \delta_i^R - \delta_i^I$.

Starting from Section 8.4, we will express the additive hazards model where time-fixed and time-varying covariates are separated, similar to Chapter 6. However, in this section and Sections 8.2 and 8.3, to maintain consistency with the model specifications in Aalen (1989) and Lin & Ying (1994), we will present the additive hazards model with all covariates expressed as functions of time $t$. In fact, with this expression, any time-fixed covariate simply becomes a constant for all $t$.

Let the covariate vector be $\mathbf{x}_i(t) = (x_{i1}(t), \ldots, x_{ip}(t))^\top$ as a vector of $p$ covariates for subject $i$. If a covariate $x_j$ is time-fixed then the corresponding $x_{ij}(t)$ remains a constant for all $t$.

The observations, including the observed survival times and the *covariates history*, are denoted by $(t_i^L, t_i^R, \widetilde{\mathbf{x}}_i(t_i^R))$ for $i = 1, \ldots, n$, where $\widetilde{\mathbf{x}}_i(t) = \{\mathbf{x}_i(v) : 0 \le v \le t\}$ represents the covariates history up to time $t$. In practice, $\mathbf{x}_i(t)$ are usually only observed at a finite number of time points, so they are generally not continuous functions of $t$; for more discussions on this, please see Chapter 6 and also Section 8.4 of this chapter.

There are two additive hazards model, given by Lin & Ying (1994) and Aalen (1989) respectively. They are described below.

**Lin & Ying's additive hazards model:** Lin & Ying's additive hazards model (Lin & Ying 1994) is a semi-parametric model that describes the hazard function of subject $i$ as the sum of a non-parametric baseline hazard and a linear predictor $\mathbf{x}_i^\top \boldsymbol{\beta}$. It can be expressed as follows:

$$h_Y(t|\widetilde{\mathbf{x}}_i(t)) = h_0(t) + \mathbf{x}_i(t)^\top \boldsymbol{\beta}, \tag{8.1}$$

where $\boldsymbol{\beta} = (\beta_1, \cdots, \beta_p)^\top$ is a regression coefficient vector and $h_0(t)$ is an unspecified baseline hazard. Clearly, the hazard model (8.1) indicates, for each subject, the covariates act additively on the baseline hazard, while for the Cox model, its covariates act multiplicatively on the baseline hazard. Therefore, for the additive hazard model, we can comprehend $\boldsymbol{\beta}$ as the covariates' effects on hazards difference (or risk difference), namely the difference between the hazard of subject $i$ and the baseline hazard.

**Allen's additive hazards model:** The Aalen's additive hazards model is very similar to that of Lin & Ying except it accommodates more general time-dependent regression coefficients. Therefore, Allen's additive hazards model appears as

$$h_Y(t|\widetilde{\mathbf{x}}_i(t)) = h_0(t) + \mathbf{x}_i(t)^\top \boldsymbol{\beta}(t), \tag{8.2}$$

where $\boldsymbol{\beta}(t) = (\beta_1(t), \ldots, \beta_p(t))^\top$ is a time dependent regression coefficient vector.

The above additive hazards model are established based on the assumption of an additive effect of the covariates on hazards. This assumption needs to be checked when applying such a model, similar to checking the proportional hazards assumption for the Cox model. To verify the appropriateness of such a model, a goodness-of-fit procedure, such as the one provided in Martinussen & Scheike (2006), can be implemented.

Similar to the Cox regression model, the additive hazard model in (8.1) (or (8.2)) also needs some constraints in order for $h_Y(t|\widetilde{\mathbf{x}}_i(t))$ to become a valid hazard function. For this, the necessary constraints on an additive hazards model are:

$$h_0(t) \geq 0 \text{ and } h_Y(t|\widetilde{\mathbf{x}}_i(t)) \geq 0, \; \forall t, \tag{8.3}$$

where the first constraint is on the baseline hazard and the second on the hazard of each individual. Since it is computationally difficult to request $h_Y(t|\widetilde{\mathbf{x}}_i(t)) \geq 0$ for all $t$, we will later impose this constraint only on certain observation time points. Such a simplification to the functional constraint also appear in Ghosh (2001). These constraints in (8.3) solidify one important reason for why the estimation problem we are facing now is more challenge than the Cox model as the latter only involves the $h_0(t) \geq 0$ constraint.

From the additive hazard model in (8.1), the cumulative hazard for subject $i$ also appears in an additive form, given by:

$$H_Y(t|\widetilde{\mathbf{x}}_i(t)) = H_0(t) + \mathbf{X}_i(t)^\top \boldsymbol{\beta}, \tag{8.4}$$

where $H_0(t)$ is the cumulative baseline hazard, $\mathbf{X}_i(t) = \int_0^t \mathbf{x}_i(s)ds$ is the cumulative covariate. The process $\{\mathbf{X}_i(t) : \; t \geq 0\}$ is called the cumulative covariate process.

From the cumulative hazard of (8.4), we can derive the corresponding survival function by

$$S_Y(t|\widetilde{\mathbf{x}}_i(t)) = e^{-H_Y(t|\widetilde{\mathbf{x}}_i(t))}. \tag{8.5}$$

For the purpose of simplification of notations, in the following discussions we may express $h_Y(t|\widetilde{\mathbf{x}}_i(t))$ by $h_i(t)$, $H_Y(t|\widetilde{\mathbf{x}}_i(t))$ by $H_i(t)$ and $S(t|\widetilde{\mathbf{x}}_i(t))$ by $S_i(t)$ when there are no confusions.

**Example 8.1** (Generate random numbers from an additive hazards model).
In this example, we explain how to generate random numbers from the following additive hazards model:

$$h_i(t) = h_0(t) + \beta_1 x_{i1} + \beta_2 x_{i2},$$

where $h_0(t) = 3t^2$ (a Weibull baseline hazard), $x_{i1} \sim$ Bernoulli$(0.6)$, $x_{i2} \sim$ unif$(1, 2)$ and $\beta_1 = -0.5$. Possible values for coefficient $\beta_2$ can be determined from the requirement that $h_i(t) \geq 0$. Since

$$h_i(t) \geq \min_t h_0(t) + \beta_1 x_{i1} + \beta_2 x_{i2},$$

and $\min_t h_0(t) = 0$ for $h_0(t) = 3t^2$, we have $h_i(t) \geq 0$ if $\beta_2$ satisfies the condition:

$$-0.5x_{i1} + \beta_2 x_{i2} \geq 0.$$

Note that $x_{i1}$ takes a value of either 0 or 1. When $x_{i1} = 0$, then any $\beta_2 > 0$ will be fine. When $x_{i1} = 1$, we want $\beta_2 x_{i2} > 0.5$. This can be guaranteed if $\beta_2 > 0.5$ since the minimum of $x_{i2}$ is 1 for this example.

We select $\beta_2 = 1.5$ to generate random numbers in this example. Based on this, the random number generation steps are:
(i) Generate $x1 \sim$ Bernoulli$(0.6)$ and $x2 \sim$ unif$(1, 2)$.
(ii) Generate $U \sim$ unif$(0, 1)$ and then solve for $t$ from

$$t^3 + (-0.5x1 + 1.5x2)t + \log U = 0.$$

Note that only positive solutions will be retained.

R code for generating 100 random numbers from the model, 60% events and 40% right censoring is given below.

```
# Function we need to solve for t
target_func <- function(t, x1, x2, logU) {
    t^3 + (-0.5 * x1 + 1.5 * x2) * t + logU
}

# Loop to generate data
n <- 0
y <- delta <- x1 <- x2 <- NULL
while (n < 100) {
    x1_i <- rbinom(1, 1, 0.6)
    x2_i <- runif(1, 1, 2)
    logU <- log(runif(1))
    t_i <- uniroot(target_func, c(0, 10), x1 = x1_i,
        x2 = x2_i, logU = logU)$root

    if (t_i > 0) {
        c_i <- rexp(1, 1.3)
        x1 <- c(x1, x1_i)
        x2 <- c(x2, x2_i)
        n <- n + 1
        if (c_i < t_i) {
            delta <- c(delta, 0)
            y <- c(y, c_i)
        } else if (t_i < c_i) {
```

```
            delta <- c(delta, 1)
            y <- c(y, t_i)
        }
    }
}

table(delta)  #check censoring proportion

## delta
##  0  1
## 36 64

ah.data <- data.frame(y = y, delta = delta, X1 = x1,
    X2 = x2)
```

□

Several existing methods can be employed to fit the additive hazards model (8.1) or (8.2), such as the least-squares approach of Aalen (1989), the counting process based estimating equation approaches of Lin & Ying (1994), Lin et al. (1998) and Wang et al. (2010), the GLM approach of Farrington (1996) and the constrained optimization methods of Zeng et al. (2006), Ghosh (2001) and Li & Ma (2019). However, we need to be aware of the following issues associated with some of these methods:

(i) Some approaches do not appropriately consider the non-negativity constraints on the baseline and individual hazards given in (8.3). For example, Aalen (1989), Lin & Ying (1994) and Farrington (1996) totally ignore such constraints, while Zeng et al. (2006) only consider constraining the baseline hazard. Ghosh (2001) addresses both the baseline and individual hazards, but it can only handle current status data, which is a special case of the partly interval-censored data considered in this chapter.

(ii) Most of the existing methods do not directly provide a smoothed estimate of the baseline hazard. In some of the previous chapters, we have explained that detecting a changing pattern in the baseline hazard is more easily done with a smoothed estimate of the baseline hazard.

This chapter will focus on the penalized log-likelihood approach to fit the additive hazards model, considering constraints on both the baseline and individual hazards. The penalty function on the baseline hazard $h_0(t)$ helps to obtain a smooth estimate of $h_0(t)$ and also reduces numerical instabilities caused by improperly selected knots locations and the number of knots. Note that knots are used to construct an approximated $h_0(t)$ function.

Before discussing the constrained penalized likelihood method, we first summarize several existing approaches for additive hazards model. This includes the method of Aalen (1989) in Section 8.2, and some counting process methods, specifically the method of Lin & Ying (1994), the method of Lin et al. (1998) and the method of Wang et al. (2010) in Section 8.3.

## 8.2 Aalen's method

Aalen (1989) considered the following general additive hazards model:

$$\begin{aligned} h_{Y_i}(t|\widetilde{\mathbf{x}}_i(t)) &= h_0(t) + \mathbf{x}_i(t)^\top \boldsymbol{\beta}(t), \\ &= (1, \mathbf{x}_i(t)^\top)\mathbf{b}(t). \end{aligned} \tag{8.6}$$

Here, $\mathbf{b}(t) = (h_0(t), \boldsymbol{\beta}(t)^\top)^\top$ represents a vector of the baseline hazard and $p$ regression coefficient functions and $\mathbf{x}_i(t) = (x_{i1}(t), \ldots, x_{ip}(t))^\top$ denotes a vector of covariates. It should be noted that equation (8.6) is Aalen's original additive hazard model specification, where the baseline hazard is incorporated into the coefficient function vector $\mathbf{b}(t)$, and the associated covariate vector becomes $(1, \mathbf{x}_i(t)^\top)$.

Comparing with the Lin&Ying model in (8.1), the model (8.6) is more general as its regression coefficients can be time-dependent or time-fix (corresponding to a constant function), while for (8.1), the regression coefficients are time-fixed. We will refer to (8.6) as the generalized additive hazards model.

In (8.6), neither $h_0(t)$ nor $\boldsymbol{\beta}(t)$ assumes a particular functional form, making them both non-parametric. Due to this characteristic, (8.6) is also referred to as a non-parametric model by some researchers (Martinussen & Scheike 2006).

For the model (8.6), estimating the cumulative regression coefficient function is easier than estimating the regression coefficient itself, similar to the fact that estimating a cumulative distribution function is easier than estimating a density function. This is primarily because a cumulative function is always a smooth function, which implies that they are self-regularized. Unlike estimating the hazard function, no additional regularization constraints are required when estimating the cumulative coefficient functions.

Aalen (1989) proposed a least squares approach (see below) to estimate the following cumulative $\mathbf{b}(t)$ function:

$$\mathbf{B}(t) = \int_0^t \mathbf{b}(s)ds, \tag{8.7}$$

where $\mathbf{b}(t)$ is a $(p+1)$-vector with its first element being the baseline hazard. Let $\widehat{\mathbf{B}}(t)$ be the estimated $\mathbf{B}(t)$. Aalen (1989) also developed the covariance matrix of $\widehat{\mathbf{B}}(t)$; see (8.9) below.

From the estimate $\widehat{\mathbf{B}}(t)$, the corresponding estimate of $\mathbf{b}(t)$ can be obtained by taking the first-order difference of $\widehat{\mathbf{B}}(t)$. However, such an estimate can be noisy. Better estimates for $\mathbf{b}(t)$ can be obtained through smoothing techniques, such as the kernel smoothing method given in Huffer & McKeague (1991).

To explain how to estimate $\mathbf{B}(t)$, we first construct a matrix $\mathbf{Z}(t)$ as follows. If subject $i$ is in the risk set at time $t^-$, meaning that at $t^-$ the event of interest has not occurred for subject $i$ and this subject is not right-censored, then the $i$th row of $\mathbf{Z}(t)$ is $(1, \mathbf{x}_i(t)^\mathsf{T})$. If subject $i$ is not in the risk set at time $t$, then the corresponding row of $\mathbf{Z}(t)$ contains zero values. Let $\mathbf{e}(t)$ be an $n$-vector with the $i$th element equal to 1 if subject $i$ experiences the event at time $t$, and equal to 0 for otherwise. Let $\tau_1 < \tau_2 < \cdots$ be the ordered event times (note that censoring times are not included). Then, the least squares estimate of $\mathbf{B}(t)$ is given by:

$$\widehat{\mathbf{B}}(t) = \sum_{\tau_i \leq t} [\mathbf{Z}(\tau_i)^\mathsf{T}\mathbf{Z}(\tau_i)]^{-1}\mathbf{Z}(\tau_i)^\mathsf{T}\mathbf{e}(\tau_i). \tag{8.8}$$

From the theory of martingales and stochastic integrals, Aalen (1989) also devised the variance-covariance matrix of $\widehat{\mathbf{B}}(t)$, give by:

$$\mathrm{Var}(\widehat{\mathbf{B}}(t)) = \sum_{\tau_i \leq t} [\mathbf{Z}(\tau_i)^\mathsf{T}\mathbf{Z}(\tau_i)]^{-1}\mathbf{Z}(\tau_i)^\mathsf{T}\mathcal{D}(\mathbf{e}(\tau_i))\mathbf{Z}(\tau_i)[\mathbf{Z}(\tau_i)^\mathsf{T}\mathbf{Z}(\tau_i)]^{-1} \tag{8.9}$$

where $\mathcal{D}(\mathbf{e}(\tau_i))$ denotes a diagonal matrix with its diagonal elements given by $\mathbf{e}(\tau_i)$. The covariance matrix $\mathrm{Var}(\widehat{\mathbf{B}}(t))$ only exists up to the smallest time at which the matrix $\mathbf{Z}(\tau_i)^\mathsf{T}\mathbf{Z}(\tau_i)$ becomes singular.

It is clear that this least squares estimation procedure does not respect the non-negativity constraints specified in (8.3). This is a disadvantage of this method.

**Example 8.2** (R example for fitting Aalen model).
In this example, we adopt the generated random numbers from Example 8.1 (60% event and 40% right censoring times) to demonstrate how to fit an Aalen model using the `aalen` function in the R package "`timereg`".

(i) Fit a model.
(ii) Plot the estimated cumulative $\beta$ function.

```
# Load library and fit the model
library(timereg)
aalen.surv <- Surv(time = y, event = delta)
aalen.fit <- aalen(aalen.surv ~ X1 + X2, data = ah.data)

# Plot estimated cumulative beta functions
# (true in red)
plot(aalen.fit)
```

Figure 8.1 presents plots of cumulative regression coefficients along with their associated 95% point-wise confidence intervals. The linearity observed in both plots suggests that both $\beta_1$ and $\beta_2$ are likely constants. As $t$ varies from 0 to 1, the cumulative coefficient decreases to -0.5 in the top panel and increases to about 1.5 in the bottom panel. Therefore, reasonable estimates would be $\beta_1 = -0.5$ and $\beta_2 = 1.5$. □

## 8.3 Some counting process methods

In this section, we will summarize several counting process methods for the additive hazards model (8.1) where regression coefficients are not time-dependent, particularly the method of Lin & Ying (1994), the method of Lin et al. (1998) and the method of Wang et al. (2010). We mainly explain their computational details.

### 8.3.1 The method of Lin-Ying

Lin & Ying (1994) proposed an estimation method for fitting the additive hazards model (8.1). They developed their method using estimating equations for the model parameters, which are constructed similarly to the partial likelihood estimating equations for the Cox model. However, their estimating equations only estimate the regression coefficient vector $\boldsymbol{\beta}$ and do not directly provide an estimate for $h_0(t)$, which is treated as a nuisance parameter. Furthermore, this estimating equation-based approach do not consider the non-negativity constraints in (8.3).

The method proposed by Lin & Ying (1994) only accommodates event and right-censoring times in the observed survival data. To simplify our discussions, we assume no overlapping among these events or right-censoring times. Denoting the observed survival times as $(t_i, \delta_i)$ for $i = 1, \ldots, n$, where $\delta_i = 1$ if $t_i$ is an event time and 0 for otherwise, we can define the event counting process $\{N_i(t) : t \geq 0\}$ for subject $i$. This process starts at 0 and jumps to 1 at the observed event time. Corresponding to $N_i(t)$, there exists an at-risk process $\{D_i(t) : t \geq 0\}$ – also a 0-1 process indicating, with the value 1, whether subject $i$ is at risk at time $t$. For any subject $i$, the counting process function $N_i(t)$ can be uniquely decomposed (e.g., Andersen & Gill (1982)) as

$$N_i(t) = M_i(t) + \int_0^t D_i(s) dH_i(s|\widetilde{\mathbf{x}}_i(t)), \tag{8.10}$$

where $M_i(t)$ is a local square integrable martingale. Hence, we have

$$\sum_{i=1}^n dN_i(t) = \sum_{i=1}^n D_i(t) dH_i(t|\widetilde{\mathbf{x}}_i(t)). \tag{8.11}$$

Equation (8.11) can be employed to estimate the baseline or cumulative baseline hazard for any given $\boldsymbol{\beta}$. For instance, in the context of Cox models, an estimator for the cumulative baseline hazard $H_0(t)$ from (8.11) is given by

$$\widehat{H}_0(t; \boldsymbol{\beta}) = \int_0^t \frac{\sum_{i=1}^n dN_i(s)}{\sum_{i=1}^n D_i(s) \exp\{\mathbf{x}_i(s)^\top \boldsymbol{\beta}\}}, \tag{8.12}$$

and this is the well-known Breslow estimator (Breslow 1972).

Under the additive hazards model (8.1), equation (8.11) gives the following $H_0(t)$ estimator :

$$\widehat{H}_0(t; \boldsymbol{\beta}) = \int_0^t \frac{\sum_{i=1}^n \{dN_i(s) - D_i(s)\mathbf{x}_i(s)^\top \boldsymbol{\beta} ds\}}{\sum_{i=1}^n D_i(s)}. \tag{8.13}$$

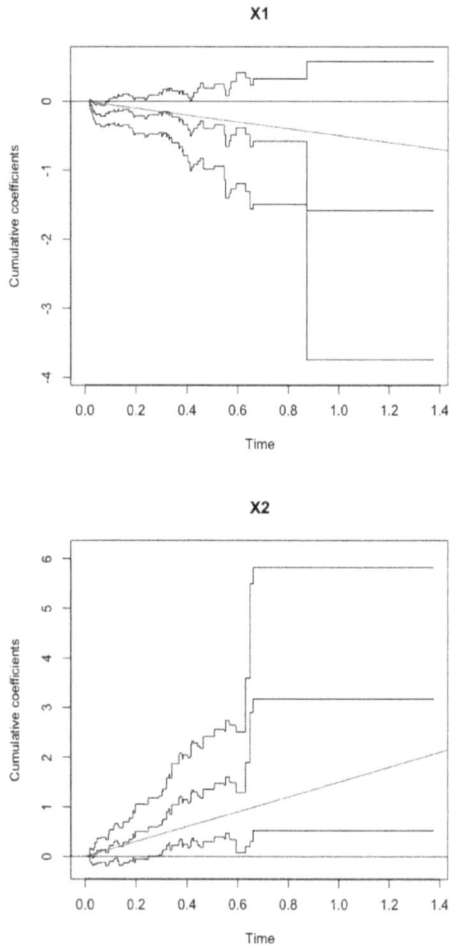


FIGURE 8.1: Cumulative coefficient estimates and their 95% point-wise CIs

Using this formula, we can derive an estimator for the baseline hazard $h_0(t)$, given by

$$\widehat{h}_0(t;\boldsymbol{\beta}) = \frac{\sum_{i=1}^n \{dN_i(t) - D_i(t)\mathbf{x}_i(t)^\top \boldsymbol{\beta}\}}{\sum_{i=1}^n D_i(t)}. \tag{8.14}$$

However, this estimator is unsatisfactory since it cannot guarantee $\widehat{h}_0(t;\boldsymbol{\beta}) \geq 0$, while the Breslow estimator (8.12) does not suffer from this issue.

Mimicking the partial likelihood score function for the Cox model, which is given by,

$$U_P(\boldsymbol{\beta}) = \sum_{i=1}^n \mathbf{x}_i(t)\{dN_i(t) - D_i(t)e^{\mathbf{X}_i(t)^\top \boldsymbol{\beta}} dH_0(t)\},$$

Lin & Ying (1994) proposed following "score" function for the additive hazards model:

$$U(\boldsymbol{\beta}) = \sum_{i=1}^{n} \int_0^{\infty} \mathbf{x}_i(t)\{dN_i(t) - D_i(t)(dH_0(t) + \mathbf{x}_i(t)^{\mathsf{T}}\boldsymbol{\beta}dt)\},$$

and this function is equivalent to

$$U(\boldsymbol{\beta}) = \sum_{i=1}^{n} \int_0^{\infty} \{\mathbf{x}_i(t) - \overline{\mathbf{x}}(t)\}\{dN_i(t) - D_i(t)\mathbf{x}_i(t)^{\mathsf{T}}\boldsymbol{\beta}dt\}, \tag{8.15}$$

where

$$\overline{\mathbf{x}}(t) = \frac{\sum_{i=1}^{n} D_i(t)\mathbf{x}_i(t)}{\sum_{i=1}^{n} D_i(t)},$$

representing the average covariates values for all subjects at time $t$.

Therefore, we can estimate $\boldsymbol{\beta}$ by solving the estimating equation

$$U(\boldsymbol{\beta}) = 0,$$

where $U(\boldsymbol{\beta})$ is given by (8.15), to give

$$\widehat{\boldsymbol{\beta}} = \left[\sum_{i=1}^{n} \int_0^{\infty} D_i(t)\{\mathbf{x}_i(t) - \overline{\mathbf{x}}(t)\}^{\otimes 2}dt\right]^{-1} \left[\int_0^{\infty} \sum_{i=1}^{n} D_i(t)\{\mathbf{x}_i(t) - \overline{\mathbf{x}}(t)\}dN_i(t)\right], \tag{8.16}$$

where for vector $\mathbf{a}$, $\mathbf{a}^{\otimes 2} = \mathbf{a}\mathbf{a}^{\mathsf{T}}$.

The asymptotic distribution of $\widehat{\boldsymbol{\beta}}$ is also given in Lin & Ying (1994). Let $\boldsymbol{\beta}_0$ denote the true regression coefficient vector. Following the counting process theory (e.g., Andersen et al. (1993)), Lin and Ying (1994) developed an asymptotic distribution for $\sqrt{n}(\widehat{\boldsymbol{\beta}} - \boldsymbol{\beta}_0)$, and this is a multivariate normal distribution with mean vector $\mathbf{0}_{p\times 1}$ and covariance matrix $\mathbf{A}^{-1}\mathbf{B}\mathbf{A}^{-1}$, where

$$\mathbf{A} = \frac{1}{n}\sum_{i=1}^{n} \int_0^{\infty} D_i(t)\{\mathbf{x}_i(t) - \overline{\mathbf{x}}(t)\}^{\otimes 2}dt \tag{8.17}$$

$$\mathbf{B} = \frac{1}{n}\sum_{i=1}^{n} \int_0^{\infty} \{\mathbf{x}_i(t) - \overline{\mathbf{x}}(t)\}^{\otimes 2}dN_i(t) \tag{8.18}$$

Plugging in the estimate of $\boldsymbol{\beta}$ into (8.13) yields an estimator for the cumulative baseline hazard denoted as $\widehat{H}_0(t; \widehat{\boldsymbol{\beta}})$. This, in turn, leads to the estimator for the survival function:

$$\widehat{S}(t; \widehat{\boldsymbol{\beta}}|\widetilde{\mathbf{x}}_i(t)) = e^{-\widehat{H}_0(t;\widehat{\boldsymbol{\beta}}) - \mathbf{X}_i(t)^{\mathsf{T}}\widehat{\boldsymbol{\beta}}}. \tag{8.19}$$

Using standard counting process techniques, one may derive that, under certain regularity conditions, quantities $\sqrt{n}(\widehat{H}_0(t; \widehat{\boldsymbol{\beta}}) - H_0(t))$ and $\sqrt{n}(\widehat{S}_0(t; \widehat{\boldsymbol{\beta}}) - S_0(t))$ both converge weakly to zero mean Gaussian processes and their covariance matrices are available in Lin & Ying (1994).

**Example 8.3** (Lin & Ying method for an additive hazards model).
In this example, we use the ah function in the R package addhazard to fit an additive hazards model, where the dataset is from Example 8.1.
(i) Fit a model.
(ii) Plot the estimated cumulative baseline hazard function.

```
# Load library and fit the model
library(addhazard)
aalen.surv <- Surv(time = y, event = delta)
ah.fit <- ah(aalen.surv ~ X1 + X2, data = ah.data,
    ties = FALSE)

# Predict baseline hazard function
newdata <- data.frame(X1 = 0, X2 = 0)
time <- predict(ah.fit, newdata, newtime = seq(from = 0.1,
    to = 2, by = 0.1))$time
baseline_haz <- predict(ah.fit, newdata, newtime = seq(from = 0.1,
    to = 2, by = 0.1))$L
plot(baseline_haz ~ time, type = "s", ylab = "baseline hazard")
```

The outputs of this AH model fitting using the Lin & Ying method are displayed below. It is clear that the $\beta_1$ estimate is not very accurate in this example.

```
> summary(ah.fit)
Call:
ah(formula = aalen.surv ~ X1 + X2, data = ah.data,
  ties = FALSE)

      coef      se lower.95 upper.95      z p.value
X1 -2.3810  0.8267  -4.0013  -0.7607 -2.880 0.00397 **
X2  1.4859  1.2981  -1.0583   4.0301  1.145 0.25234
---
Signif. codes:  0 '***' 0.001 '**' 0.01 '*' 0.05 '.' 0.1 ' ' 1
```

□

### 8.3.2 The method of Lin, Oakes and Ying

Recall that the method of Lin and Ying is applicable only to right-censored survival data, meaning they include only event or right-censoring times. Lin et al. (1998) extended the counting process idea to facilitate a straightforward approach for estimating the regression coefficient vector $\beta$ in the additive hazards model under current-status data, also known as the type I interval-censored data. Subsequently, Wang et al. (2010) further extended this method to study type II interval-censored survival data, which may contain left-, right- or interval-censored survival times, and we will summarize their work in the next subsection.

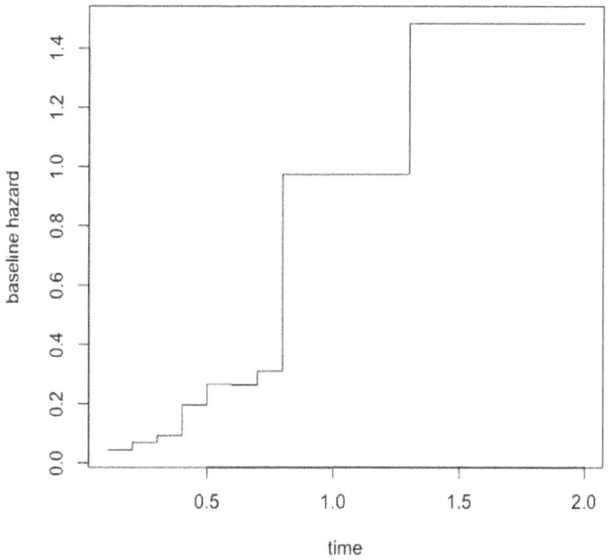


FIGURE 8.2: Plot of the baseline hazard estimate obtained from the Lin & Ying method.

Current status data is a type of survival data exclusively composed of left- or right-censored survival data, making it a special case of the partly interval-censored data discussed in Chapter 2. This type of survival data typically arises from a follow-up study where each individual is monitored at a single time point (which can be random). At this time point, the event status is determined: either the event has already occurred, resulting in left-censored time or the event is yet to occur, resulting in right-censored time. As each subject $i$ has only a single time point $c_i$ for the status of the event-of-interest, we can assume that all covariates have no values except at 0 and $c_i$, namely, $\mathbf{x}_i(t)$ is available for $t = 0$ and $t = c_i$.

In current-status data, the exact event times are unknown, but it is known whether the event time $Y_i$ is below or above a single monitoring time $c_i$ (which can be random). For current status observations, alongside the covariate values, we have the monitoring time $c_i$ and the indicator value $\delta_i^R = I(Y_i \geq c_i)$ for subject $i$. Note $\delta_i^R$ is a right-censoring indicator.

Let the observation from subject $i$ be denoted as $(c_i, \delta_i^R, \widetilde{\mathbf{x}}_i(c_i))$, where $\widetilde{\mathbf{x}}_i(c_i)$ denotes the history of $\mathbf{x}_i(t)$ up to $c_i$. To apply the counting process method, it is crucial to define the event counting process $N_i(t)$ for the current-status observation of subject $i$. Lin et al. (1998) define $N_i(t)$ $(t \geq 0)$ for subject $i$ as follows:

$$N_i(t) = \delta_i^R I(t \geq c_i). \tag{8.20}$$

This $N_i(t)$ is a 0-1 process for $t \geq 0$: it is 0 when $t < c_i$ and jumps to 1 at $c_i$ if $c_i$ is a right censoring time, remaining at 1 thereafter.

Corresponding to the counting process $N_i(t)$, they also define the at-risk process $\{D_i(t) : t \geq 0\}$ as:

$$D_i(t) = I(t \leq c_i).$$

The value of $D_i(t)$ is 1 when $t \leq c_i$ (associated with $N_i(t) = 0$) and 0 when $t > c_i$, meeting the definition of an at-risk process.

We denote the hazard associated with $dN_i(t) = 1$ as $\lambda_i(t)$. Thus, the quantity $\lambda_i(t)dt$ represents the probability of a jump (i.e., $dN_i(t) = 1$) at time $t$ for subject $i$. Note that this hazard $\lambda_i(t)$ is distinct from the event hazard $h_i(t)$.

For subject $i$, $dN_i(t) = 1$ occurs at $t$ only if both of the following conditions are true: (i) $t = c_i$, and (ii) subject $i$ is event-free up to this time $t$. Therefore, the probability of $dN_i(t) = 1$ is given by the product of the probabilities of these two conditions.

Let $\lambda_c(t)$ be the hazard function for the monitoring times $(c_1, \ldots, c_n)$. Assuming $\lambda_c(t)$ does not depend on the covariate vector $\mathbf{x}_i$. In this context, by utilizing $\lambda_c(t)$ and the survival function (8.5), we can express the probability of $dN_i(t) = 1$ as:

$$\begin{aligned}\lambda_i(t)dt &= \lambda_c(t)dt\, e^{-H_0(t) - \mathbf{X}_i(t)^{\mathsf{T}}\boldsymbol{\beta}} \\ &= \lambda_0(t)e^{-\mathbf{X}_i(t)^{\mathsf{T}}\boldsymbol{\beta}}dt, \end{aligned} \tag{8.21}$$

where $\mathbf{X}_i(t)$ denotes the cumulative $\mathbf{x}_i(t)$, namely $\mathbf{X}_i(t) = \int_0^t \mathbf{x}_i(s)ds$, and

$$\lambda_0(t) = \lambda_c(t)e^{-H_0(t)}. \tag{8.22}$$

The expression in (8.21) indicates that:

(1) $\lambda_i(t)$ is the hazard function associated with the data $(c_i, \delta_i^R)$, $i = 1, \ldots, n$, where $\delta_i^R = I(Y_i \geq c_i)$.

(2) $\lambda_i(t)$ essentially adopts a Cox model with time-varying covariates $-\mathbf{X}_i(t)$.

Therefore, $\boldsymbol{\beta}$ can be estimated using the partial likelihood method for a Cox model with time-varying covariates, where the survival dataset is: $(c_i, \delta_i^R, -\widetilde{\mathbf{X}}_i(c_i))$, where $\widetilde{\mathbf{X}}_i(c_i)$ is the history of $\mathbf{X}_i(t)$ up to $c_i$.

A straightforward scenario is when all $\mathbf{x}_i$ are time-fixed. In this case, $\mathbf{X}_i(t) = \mathbf{x}_i t$. With this, the estimation of $\boldsymbol{\beta}$ can be achieved simply by employing the partial likelihood method on the survival dataset $(c_i, \delta_i^R, -\mathbf{x}_i t)$ for $t \leq c_i$, where $i = 1, \ldots, n$.

The R package `survival` can be utilized to fit such a Cox model and obtain an estimate of $\boldsymbol{\beta}$.

**Example 8.4** (Additive model with current-status data).
In this example we will demonstrate how to implement the method of Lin, Oakes & Ying discussed in this subsection. We will perform the following activities:

(i) Generate 100 current status survival data observations, where the true survival times are obtained similar to Example 8.1.
(ii) Prepare data as described above for estimating $\beta$ using a Cox model.
(iii) Estimate $\beta$ using the R `survival` package.

The R function for simulate a current-status dataset is given below, where the R function `target_func` was defined in Example 8.1.

```
# Generate current status data
n <- 0
y <- t <- c <- delta <- x1 <- x2 <- NULL

while (n < 100) {
    x1_i <- rbinom(1, 1, 0.2)
    x2_i <- runif(1, 0, 0.5)
    logU <- log(runif(1))
    t_i <- uniroot(target_func, c(0, 10),
        x1 = x1_i, x2 = x2_i, logU = logU)$root
    if (t_i > 0) {
        c_i <- rexp(1, 1)
        x1 <- c(x1, x1_i)
        x2 <- c(x2, x2_i)
        n <- n + 1
        if (c_i < t_i) {
            delta <- c(delta, 1)
            y <- c(y, c_i)
        } else if (t_i < c_i) {
            delta <- c(delta, 0)
            y <- c(y, c_i)
        }
        t <- c(t, t_i)
        c <- c(c, c_i)
    }
}
```

```
> cs.data <- data.frame(y = y, delta = delta, X1 = x1,
  X2 = x2)
> table(cs.data$delta)

 0  1
42 58
>
> # Fit model to current status data
> library(survival)
> cs.data <- data.frame(y = y, delta = delta, X1 = x1,
  X2 = x2)
> cs.surv <- Surv(time = cs.data$y, event = cs.data$delta)
> fit.tvc = coxph(cs.surv ~ tt(X1) + tt(X2), data = cs.data,
+             tt = function(x, t,...) -t*x)
```

```
> summary(fit.tvc)
Call:
coxph(formula = cs.surv ~ tt(X1) + tt(X2), data = cs.data,
  tt = function(x, t, ...) -t * x)

  n= 100, number of events= 58

          coef exp(coef) se(coef)      z Pr(>|z|)
tt(X1) -2.139     0.118    0.827 -2.59   0.0097 **
tt(X2)  3.727    41.564    2.782  1.34   0.1803
---
Signif. codes:  0 '***' 0.001 '**' 0.01 '*' 0.05 '.' 0.1 ' ' 1

        exp(coef) exp(-coef) lower .95 upper .95
tt(X1)      0.118     8.4944    0.0233     0.595
tt(X2)     41.564     0.0241    0.1781  9701.496

Concordance= 0.542  (se = 0.041 )
Likelihood ratio test= 6.49  on 2 df,   p=0.04
Wald test            = 6.69  on 2 df,   p=0.04
Score (logrank) test = 7.07  on 2 df,   p=0.03
```

Comparing with the true values: $\beta_1 = -0.5$ and $\beta_2 = 1.5$ (see Example 8.1), we can see the above estimates are not very accurate. The modification described below may improve the estimates. □

Since the process $N_i(t)$ is considered censored (its censoring indicator $\delta_i^R = 0$) when subject $i$ experiences the event before $c_i$, the "censoring times" are also dependent on $\boldsymbol{\beta}$ – a type of informative censoring. This information is not utilized in the above estimation process, which might lead to biased and less efficient estimates. Lin et al. (1998) suggest the following modification to improve efficiency.

Suppose the monitoring times $c_i$ are informative about $\boldsymbol{\beta}$, such as when $c_i$ is influenced by the covariate vector $\mathbf{x}_i(t)$. We can modify the above approach for estimating $\boldsymbol{\beta}$ by incorporating this information. More specifically, we assume $c_i$ is related to $\mathbf{x}_i(t)$ through a proportional hazards model:

$$\lambda_c(t) = \lambda_{0c}(t)e^{\mathbf{x}_i(t)^\mathsf{T}\boldsymbol{\gamma}}, \tag{8.23}$$

where $\lambda_{0c}(t)$ is its baseline hazard. Under this assumption, the expression in (8.21) is now modified to:

$$\lambda_i(t) = \lambda_0(t)e^{-\mathbf{X}_i(t)^\mathsf{T}\boldsymbol{\beta}+\mathbf{x}_i(t)^\mathsf{T}\boldsymbol{\gamma}}, \tag{8.24}$$

where $\lambda_0(t) = \lambda_{0c}(t)e^{-H_0(t)}$. Thus, $\boldsymbol{\beta}$ (and $\boldsymbol{\gamma}$) can be estimated by utilizing the R `survival` package with the data $(c_i, \delta_i^R, -\widetilde{\mathbf{X}}_i(c_i), \widetilde{\mathbf{x}}_i(c_i))$, where $\widetilde{\mathbf{X}}_i(c_i)$ and $\widetilde{\mathbf{x}}_i(c_i)$ have been defined previously. The regression coefficient vector of $-\mathbf{X}_i(t)$ gives the $\boldsymbol{\beta}$ estimate and the coefficient vector of $\mathbf{x}_i(t)$ gives the $\boldsymbol{\gamma}$ estimate.

**Example 8.5** (Redo Example 8.4).
In this example, we re-attempt Example 8.4 using the above-modified approach to exam particularly the changes in bias and variance.

```
# Generate current status data
n <- 0
y <- t <- c <- delta <- x1 <- x2 <- NULL

while (n < 100) {
    x1_i <- rbinom(1, 1, 0.2)
    x2_i <- runif(1, 0, 0.5)
    logU <- log(runif(1))
    t_i <- uniroot(target_func, c(0, 10),
        x1 = x1_i, x2 = x2_i, logU = logU)$root
    if (t_i > 0) {
        c_i <- rexp(1, 1)
        x1 <- c(x1, x1_i)
        x2 <- c(x2, x2_i)
        n <- n + 1
        if (c_i < t_i) {
            delta <- c(delta, 1)
            y <- c(y, c_i)
        } else if (t_i < c_i) {
            delta <- c(delta, 0)
            y <- c(y, c_i)
        }
        t <- c(t, t_i)
        c <- c(c, c_i)
    }
}
```

```
> cs.data1 <- data.frame(y = y, delta = delta, X1 = x1,
  X2 = x2)
> # Fit this modified model to current status data
> cs.surv1 <- Surv(time = cs.data1$y, event = cs.data1$delta)
> fit.tvc1 = coxph(cs.surv ~ tt(X1) + tt(X2) + X1 + X2,
   data = cs.data1, tt = function(x, t,...) -t*x)
> summary(fit.tvc1)
Call:
coxph(formula = cs.surv ~ tt(X1) + tt(X2) + X1 + X2,
  data = cs.data1, tt = function(x, t, ...) -t * x)

  n= 100, number of events= 58

           coef exp(coef) se(coef)      z Pr(>|z|)
tt(X1) -3.6657    0.0256   1.5993 -2.29    0.022 *
tt(X2)  0.0550    1.0565   5.0455  0.01    0.991
X1     -0.7541    0.4704   0.6253 -1.21    0.228
```

```
X2        -1.7706     0.1702    1.7800 -0.99     0.320
---
Signif. codes:  0 '***' 0.001 '**' 0.01 '*' 0.05 '.' 0.1 ' ' 1

        exp(coef) exp(-coef) lower .95 upper .95
tt(X1)     0.0256     39.085  1.11e-03  5.88e-01
tt(X2)     1.0565      0.947  5.36e-05  2.08e+04
X1         0.4704      2.126  1.38e-01  1.60e+00
X2         0.1702      5.874  5.20e-03  5.57e+00
```

□

### 8.3.3 The method of Wang, Sun and Tong

Wang et al. (2010) extended the current status data approach proposed by Lin et al. (1998) to Type II interval-censored survival data, which encompasses observations that involve left-, right- and interval-censored survival times. It is important to note that exact survival times are not permitted within this extension, which contrasts with the likelihood-based method discussed in the subsequent sections.

Before introducing this method, let us establish some notations. For subject $i$, define two inspection time points $U_i$ and $V_i$, with $U_i < V_i$ strictly, such that the event may occur before $U_i$, between $U_i$ and $V_i$ or after $V_i$. We use indicators $\delta_{i1}$, $\delta_{i2}$ and $\delta_{i3} = 1-\delta_{i1}-\delta_{i2}$ to represent these three possibilities, respectively. The observation for subject $i$ can then be represented as $(U_i, V_i, \delta_{1i}, \delta_{2i}, \delta_{3i}, \widetilde{\mathbf{x}}_i(V_i))$, where $\widetilde{\mathbf{x}}_i(V_i))$ denotes the covariates process (history) up to $V_i$. The restriction to type II interval censoring ensures that $V_i > U_i$ strictly for each $i$, and this means there will be no event times in the observed survival times.

For subject $i$, it is necessary to define two event-counting processes to accommodate interval censoring: $\{N_{1i}(t) : t \geq 0\}$ and $\{N_{2i}(t \,|\, U_i) : t \geq 0\}$, where the second process depends on a given $U_i$ $(< V_i)$. These processes $N_{1i}(t)$ and $N_{2i}(t \,|\, U_i)$ are defined as

$$N_{1i}(t) = (1 - \delta_{1i})I(t \geq U_i), \tag{8.25}$$

$$N_{2i}(t \,|\, U_i) = \begin{cases} 0 & \text{if } t < U_i, \\ \delta_{3i}I(t \geq V_i) & \text{if } t \geq U_i. \end{cases} \tag{8.26}$$

Thus, for subject $i$, $N_{1i}(t)$ equals 0 for $t < U_i$ and will have a jump to the value of 1 at $t = U_i$ if the event for subject $i$ does not occur before $U_i$. For a given $U_i$, $N_{2i}(t \,|\, U_i)$ equals 0 for $t < U_i$ and will have a jump at $t = V_i$ if the event time of this subject is right-censored at $V_i$. The at-risk process functions corresponding to $N_{1i}(t)$ and $N_{2i}(t \,|\, U_i)$ are, respectively,

$$D_{1i}(t) = I(t < U_i), \tag{8.27}$$

$$D_{2i}(t \,|\, U_i) = I(t < V_i | t \geq U_i) = I(U_i \leq t < V_i). \tag{8.28}$$

The approach of Wang et al. (2010) assumed models for the random inspection times $\{U_i\}$ and $\{V_i\}$. More specifically, they adopted Cox models similar to the method of Lin et al. (1998):

$$\lambda_i^U(t|\mathbf{x}_i) = \lambda_u(t)e^{\mathbf{x}_i(t)^\top\gamma}, \tag{8.29}$$

$$\lambda_i^V(t|U_i,\mathbf{x}_i) = I(t > U_i)\lambda_v(t)e^{\mathbf{x}_i(t)^\top\gamma}. \tag{8.30}$$

These two models share the same regression coefficients because $U_i$ and $V_i$ are paired together, implying identical coefficient effects for both.

Following the same arguments as Lin et al. (1998) (see also the last section), we can derive the hazard functions $\lambda_{i1}(t)$ and $\lambda_{i2}(t)$ for respectively $N_{1i}(t)$ and $N_{2i}(t \mid U_i)$. They are:

$$\lambda_{1i}(t) = \lambda_{01}(t)e^{-\mathbf{X}_i(t)^\top\boldsymbol{\beta}+\mathbf{x}_i(t)^\top\gamma}, \tag{8.31}$$

$$\lambda_{2i}(t) = I(t > t_i^L)\lambda_{02}(t)e^{-\mathbf{X}_i(t)^\top\boldsymbol{\beta}+\mathbf{x}_i(t)^\top\gamma}, \tag{8.32}$$

where $\lambda_{01}(t) = \lambda_u(t)e^{-H_0(t)}$ and $\lambda_{02}(t) = \lambda_v(t)e^{-H_0(t)}$. Since both (8.31) and (8.32) involve $\boldsymbol{\beta}$ and $\gamma$, estimating these two parameter vectors becomes non-trivial. Wang et al. (2010) proposed a method as summarized next.

For a given $\gamma = \widehat{\gamma}$, Wang et al. (2010) proposed to estimate $\boldsymbol{\beta}$ by solving the following estimating equation

$$U_{\boldsymbol{\beta}}(\boldsymbol{\beta},\widehat{\gamma}) = 0, \tag{8.33}$$

where

$$U_{\boldsymbol{\beta}}(\boldsymbol{\beta},\widehat{\gamma}) = \sum_{i=1}^n (1-\delta_{1i})\left[\mathbf{X}_i(U_i) - \frac{v_{1\boldsymbol{\beta}}^{(1)}(U_i;\boldsymbol{\beta},\widehat{\gamma})}{v_{1\boldsymbol{\beta}}^{(0)}(U_i;\boldsymbol{\beta},\widehat{\gamma})}\right] + \sum_{i=1}^n \delta_{3i}\left[\mathbf{X}_i(V_i) - \frac{v_{2\boldsymbol{\beta}}^{(1)}(V_i;\boldsymbol{\beta},\widehat{\gamma})}{v_{2\boldsymbol{\beta}}^{(0)}(V_i;\boldsymbol{\beta},\widehat{\gamma})}\right], \tag{8.34}$$

where, for $j = 0, 1$ and $k = 1, 2$,

$$v_{k\boldsymbol{\beta}}^{(j)}(t;\boldsymbol{\beta},\gamma) = n^{-1}\sum_{i=1}^n D_{ki}(t)e^{-\mathbf{X}_i(t)^\top\boldsymbol{\beta}+\mathbf{x}_i(t)^\top\gamma}\mathbf{X}_i^{(j)}(t).$$

Here, $\mathbf{a}^{(0)} = 1$ and $\mathbf{a}^{(1)} = \mathbf{a}$.

For estimating $\gamma$, we leverage the assumption that both $U$ and $V$ follow a Cox model, as specified in (8.29) and (8.30). This provides a method to estimate $\gamma$ by solving the following estimating equation:

$$U_\gamma(\gamma) = 0, \tag{8.35}$$

where

$$U_\gamma(\gamma) = \sum_{i=1}^n \left[\mathbf{x}_i(U_i) - \frac{v_{1\gamma}^{(1)}(U_i;\gamma)}{v_{1\gamma}^{(0)}(U_i;\gamma)}\right] + \sum_{i=1}^n \left[\mathbf{x}_i(V_i) - \frac{v_{2\gamma}^{(1)}(V_i;\gamma)}{v_{2\gamma}^{(0)}(V_i;\gamma)}\right],$$

and here $v_{1\gamma}$ and $v_{2\gamma}$ are given by

$$v_{k\gamma}^{(j)}(t;\gamma) = n^{-1}\sum_{i=1}^{n} D_{ki}(t)e^{-\mathbf{x}_i(t)^\mathsf{T}\gamma}\,\mathbf{x}_i^{(j)}(t),$$

where $k = 1, 2$ and $j = 0, 1$.

## 8.4 Model setup for full likelihood-based methods

For the rest of this chapter, we will concentrate on Lin & Ying's additive hazard model (8.1). However, we will separate time-fixed and time-varying covariates and assume all regression coefficients remain constant over time. Specifically, the model we will consider can be expressed as the following:

$$h_{Y_i}(t|\mathbf{x}_i, \widetilde{\mathbf{z}}_i(t)) = h_0(t) + \mathbf{x}_i^\mathsf{T}\boldsymbol{\beta} + \mathbf{z}_i(t)^\mathsf{T}\boldsymbol{\gamma}. \tag{8.36}$$

Here, $\mathbf{x}_i = (x_{i1}, \ldots, x_{ip})^\mathsf{T}$ is a vector representing $p$ time-fixed covariate values of subject $i$, $\mathbf{z}_i(t) = (z_{i1}(t), \ldots, z_{iq}(t))^\mathsf{T}$ is a vector for $q$ time-varying covariate values of $i$, and $\widetilde{\mathbf{z}}_i(t)$ represents the history of $\mathbf{z}_i(t)$ up to time $t$. The definition of the time-varying covariate history $\widetilde{\mathbf{z}}_i(t)$ can be found in Section 6.1. Coefficient vectors $\boldsymbol{\beta}$ and $\boldsymbol{\gamma}$ are time-fixed, as opposed to Aalen's additive hazards model.

Similar to Chapter 6, we assume that $\mathbf{z}_i(t)$ is not continuously available as a function of $t$; instead, values of $\mathbf{z}_i(t)$ are only available at a finite number of intermittent time points. For subject $i$, let $t_{i1}, \ldots, t_{in_i}$ be the intermittent time points, so that when $t_{i,a-1} < t \leq t_{ia}$,

$$z_{ib}(t) = z_{ib}(t_{ia}) \equiv z_{iab}, \tag{8.37}$$

where $a = 1, \ldots, n_i$ and $b = 1, \ldots, q$. Here, the values $t_{i0}$ and $\mathbf{z}(t_{i0})$ (corresponding to $a = 1$) need to be defined. Specifically, we assume that $t_{i0} = 0$ and $z_{ib}(t_{i0}) = z_{ib}(t_{i1})$. With these piecewise constant values for $z_{ib}(t)$, the model matrix derived from the time-varying covariates appears as a long-format matrix, given by:

$$\mathbf{Z} = \begin{pmatrix} z_{111} & \cdots & z_{11q} \\ \vdots & \vdots & \vdots \\ z_{1n_1 1} & \cdots & z_{1n_1 q} \\ \vdots & \vdots & \vdots \\ z_{n11} & \cdots & z_{n1q} \\ \vdots & \vdots & \vdots \\ z_{1n_n 1} & \cdots & z_{1n_n q} \end{pmatrix}. \tag{8.38}$$

Under this setting, the cumulative hazard derived from (8.36) is

$$H(t|\mathbf{x}_i, \widetilde{\mathbf{z}}_i(t)) = H_0(t) + \mathbf{x}_i^\mathsf{T}\boldsymbol{\beta}t + \mathbf{Z}_i(t)^\mathsf{T}\boldsymbol{\gamma}, \tag{8.39}$$

where $\mathbf{Z}_i(t) = (Z_{i1}(t), \dots, Z_{iq}(t))^\top$ denotes the cumulative function of $\mathbf{z}_i(t)$, and here $Z_{ib}(t) = \int_0^t z_{ib}(s)ds$. When $z_{ib}(t)$ is piecewise constant, the integral in $Z_{ib}(t)$ can be simplified. In fact, for $t_{i,a-1} < t \leq t_{ia}$, $Z_{ib}(t)$ can be expressed as:

$$\begin{aligned} Z_{ib}(t) &= \int_0^t z_{ib}(s)ds \\ &= \sum_{d=1}^{a} z_{idb}(t_{id}^* - t_{i,d-1}), \end{aligned}$$

where $t_{id}^* = t_{id}I(d < a) + tI(d = a)$ with $I(\mathcal{A})$ being the indicator function for event $\mathcal{A}$. In fact, $t_{id}^* = t_{id}$ if $d < a$ and $t_{id}^* = t$ if $d = a$. Let the vector $\mathbf{z}_{id} = (z_{id1}, \dots, z_{idq})^\top$, the last term of (8.39) can be expressed as

$$\mathbf{Z}_i(t)^\top \gamma = \sum_{d=1}^{a} (t_{id}^* - t_{i,d-1})\, \mathbf{z}_{id}^\top \gamma, \tag{8.40}$$

when $t_{i,a-1} < t \leq t_{ia}$.

As before, we denote the partly interval-censored survival data using $[t_i^L, t_i^R]$. Without loss of generality, we assume the last (and also the second last when considering an interval-censored time) intermittent point is defined as follows:

(1) For an event time (i.e. $t_i^L = t_i^R = t_i$), we have $t_{in_i} = t_i$.
(2) For a left-censored time (i.e. $t_i^L = 0$ and $t_i^R$ finite), we have $t_{in_i} = t_i^R$.
(3) For a right-censored time (i.e. $t_i^L$ is finite but $t_i^R = \infty$), we have $t_{in_i} = t_i^L$.
(4) For an interval-censored time (i.e. $t_i^L \neq t_i^R$, where $t_i^L \neq 0$ and $t_i^R \neq \infty$), we have $t_{i,n_i-1} = t_i^L$ and $t_{in_i} = t_i^R$.

In the discussion that follows, we will use the set of all intermittent points, denoted as

$$\mathcal{C} = \{t_{11}, \dots, t_{1n_1}, \dots, t_{n1}, \dots, t_{nn_n}\}, \tag{8.41}$$

to enforce the constraint $h(t|\mathbf{x}_i, \widetilde{\mathbf{z}}_i(t)) \geq 0$ for each $i$.

A common characteristic of the additive hazards model methods discussed in Sections 8.2 and 8.3 is their omission of the non-negativity constraints outlined in (8.3). This omission can result in sub-optimal estimates.

In the Cox model, the sole constraint required is $h_0(t) \geq 0$. As discussed in Chapter 2, the MI algorithm (or other constrained optimization algorithms) can be employed to enforce this constraint. However, in the additive hazards model, both $h_0(t) \geq 0$ and $h(t|\mathbf{x}_i, \widetilde{\mathbf{z}}_i(t)) \geq 0$ constraints are necessary, with the latter proving particularly challenging to impose. The likelihood-based method explored in this chapter can enforce these non-negativity constraints and is sufficiently flexible to handle partly interval-censored survival times.

Due to the challenges associated with imposing the $h(t|\mathbf{x}_i, \widetilde{\mathbf{z}}_i(t)) \geq 0$ constraints for additive hazards model, there is a limited availability of constrained optimization algorithms for optimizing the log-likelihood or penalized log-likelihood function.

Notably, the Alternating Direction Maximization Method (ADMM) and the Primal-Dual Interior Point Method are efficient algorithms that can be used with additive hazards model; see Rathnayake (2019), Ghosh (2001) and Li & Ma (2019). This chapter will primarily focus on the latter method.

Hence, we will utilize the primal-dual interior point algorithm to facilitate the constrained maximum penalized likelihood estimation of both the regression coefficients and the baseline hazard in the additive hazards model, with the likelihood constructed from general partly interval-censored survival times. The rationale behind employing a penalty function has already been discussed in the preceding chapters, primarily aiding in the regularization of the non-parametric baseline hazard estimate.

Detailed information about the primal-dual interior point algorithm can be found in references such as Wright (1997). This algorithm is effective in enforcing the constraints on the additive hazards model. In fact, it ensures that each subject's hazard remains non-negative at the intermittent points set $\mathcal{C}$ and also ensures the baseline hazard is non-negative.

**Example 8.6** (Random numbers with time-fixed and time-varying covariates).
In this example, we generate random survival data from an additive hazards model that includes both time-fixed and time-varying covariates. The hazard function for the event time $Y_i$ is given by:

$$h_{Y_i}(t) = h_0(t) + 0.5x_{i1} - 3z_{i1}(t) + \gamma_2 z_{i2}(t),$$

where $h_0(t) = 3t^2$, $x_{i1} \sim$ Bernoulli$(0.5)$, $z_{i1}(t) = t - 1$ and $z_{i2}(t)$ is defined as follows: it equals 0 if $t < \tau_i$ and 1 if $t \geq \tau_i$, with $\tau_i$ sampled from a uniform distribution unif$(0.5, 2.5)$. This hazard function can be re-written as:

$$h_{Y_i}(t) = \begin{cases} 3t^2 + 0.5x_{1i} - 3(t-1) & \text{if } t \leq \tau_i, \\ 3t^2 + 0.5x_{1i} - 3(t-1) + \gamma_2 & \text{if } t > \tau_i. \end{cases}$$

We only generate right-censored data, with censoring times $C_i$ sampled from unif$(0.3, 0.7)$.

We need to establish a feasible range for $\gamma_2$ so that

$$\min_t h_{Y_i}(t) \geq 0,$$

representing a valid hazard function. It is easy to verify that the minimum of $3t^2 - 3(t-1)$ occurs at $t = 1/2$ and its minimum value is 2.25. Therefore, $\min_{t\geq 0} h_{Y_i}(t) \geq 0$ when $t \leq \tau_i$, and when $t > \tau_i$ we have $\min_{t\geq 0} h_{Y_i}(t) = 2.25 + \gamma_2$. Hence, the feasible range for $\gamma_2$ is:

$$\gamma_2 > -2.25.$$

In this example, we set $\gamma_2 = -1$.

Random numbers from this hazard can be generated using the cumulative hazard (where $\gamma_2 = -1$), given by

$$H_{Y_i}(t) = t^3 + 0.5x_{i1}t - 1.5(t-1)^2 - \max(0, t - \tau_i).$$

The steps involved in generating random numbers from this distribution are:

(1) Generate $U_i \sim \text{unif}(0,1)$.

(2) Generate $x_{1i} \sim \text{Bernoulli}(0.5)$ and $\tau_i \sim \text{unif}(0.5, 2.5)$.

(3) If $-\log U_i < H_{Y_i}(\tau_i)$, then solve $t^3 + 0.5x_{1i}t - 1.5(t-1)^2 + \log U_i = 0$ for $t$; otherwise, solve $t^3 + 0.5x_{1i}t - 1.5(t-1)^2 - (t - \tau_i) + \log U_i = 0$ for $t$, and we denote the solution by $y_i$.

(4) Generate $c_i \sim \text{unif}(0.3, 0.7)$. A simulated right-censored event time is: $t_i = \min(y_i, c_i)$.

(5) Next, for this individual $i$, we generate intermittent points $t_{i1}, \ldots, t_{i,n_i}$. First, generate an $n_i$ from a discrete uniform distribution between 1, 10. Then, setting $t_{i,n_i} = t_i$ and generate $t_{i1} < t_{i2} < \cdots < t_{i,n_i-1}$ spreading evenly but strictly between 0 and $t_{in_i}$. Also, we insert $\tau_i$ into these points and obtain a total of $n_i + 1$ intermittent points. The two time-varying covariates values for each of these $n_i$ points are calculated and they form a row of the **Z** matrix (see (8.38)).

```
n = 50  #sample size
N_i <- 10  #maximum number of intermittent points
id_long <- x_long <- z1_long <- z2_long <- NULL
start <- end <- event <- NULL

for (i in 1:n) {

    # generate from Unif[0,1]
    u_i <- runif(1)
    neg.log.u <- -log(u_i)

    # generate time fixed
    # covariate
    x <- rbinom(1, 1, 0.5)

    # generate change point in z2

    tau <- runif(1, 0.5, 2.5)
    H_tau <- tau^3 + 0.5 * tau * x -
        1.5 * (tau - 1)^2

    # generate true event time
    if (neg.log.u < H_tau) {
        y_i <- uniroot(function(t) t^3 +
            0.5 * t * x - 1.5 * (t -
            1)^2, c(0, 10))$root
    } else {
        y_i <- uniroot(function(t) t^3 +
            0.5 * t * x - 1.5 * (t -
            1)^2 - (t - tau), c(0, 10))$root
```

```
    }

    # generate censoring time
    c_i <- runif(1, 0.3, 0.8)

    # save event or censoring time
    t_i <- min(c_i, y_i)

    # generate observation times
    n_i <- sample(1:N_i, 1)
    t_ia <- quantile(c(0, t_i), c(seq(0,
        1, length.out = n_i + 1)))[2:(n_i +
        1)]
    if (tau < t_i) {
        t_ia <- sort(c(t_ia, tau))
    }
    n_i = length(t_ia)

    # generate time varying
    # covariates
    z_1 <- t_ia + 1
    z_2 <- rep(0, length(t_ia))
    z_2[which(t_ia > tau)] <- 1

    # create 'long' dataset
    id_long_i <- rep(i, n_i)
    x_long_i <- rep(x, n_i)
    event_i <- rep(0, n_i)
    if (y_i < c_i) {
        event_i[n_i] <- 1
    }

    if (n_i > 1) {
        end_i <- c(t_ia[2:n_i], t_i)
    } else {
        end_i <- (t_i)
    }

    id_long <- c(id_long, id_long_i)
    x_long <- c(x_long, x_long_i)
    z1_long <- c(z1_long, z_1)
    z2_long <- c(z2_long, z_2)
    start <- c(start, t_ia)
    end <- c(end, end_i)
    event <- c(event, event_i)

}
```

```
df <- data.frame(id_long, x_long, z1_long,
    z2_long, start, end, event)
```

□

## 8.5 Penalized likelihood

As discussed in previous chapters, a key advantage of likelihood-based estimation methods lies in the routine derivation of a large-sample approximate normal distribution for estimated parameters, encompassing both parametric and non-parametric components. Section 8.7 will present valuable asymptotic results to guide inferences on regression coefficients, baseline hazards and other quantities, including survival probabilities.

The upcoming method approximates the baseline hazard using a finite number of non-negative basis functions, akin to the approaches in previous chapters. It concurrently estimates the approximated baseline hazard and the regression coefficients by maximizing a penalized log-likelihood function, ensuring non-negativity constraints on both the baseline hazard and the hazard of each subject.

Utilizing basis functions, we can approximate the baseline hazard $h_0(t)$ as:

$$h_0(t) = \sum_{u=1}^{m} \theta_u \psi_u(t), \tag{8.42}$$

where $\psi(t) \geq 0$ represents non-negative basis functions. Various basis functions discussed earlier, such as M-splines, Gaussian density functions and indicator functions, are applicable here. From this approximation, we can derive the cumulative baseline hazard function as:

$$H_0(t) = \sum_{u=1}^{m} \theta_u \Psi_u(t), \tag{8.43}$$

where $\Psi_u(t) = \int_0^t \psi_u(s)ds$, the cumulative basis function.

Accordingly, the constraint $h_0(t) \geq 0$ now simply translates to $\theta_u \geq 0$ for all $u$. This condition is sufficient to ensure the non-negativity of $h_0(t)$. However, if the basis functions $\psi_u(t)$ are indicator functions, the condition $\theta_u \geq 0$ for all $u$ becomes both necessary and sufficient to guarantee a non-negative $h_0(t)$. Indicator basis functions provide a piecewise constant approximation of $h_0(t)$.

The constraints $\boldsymbol{\theta} \geq 0$ can be achieved using the MI algorithm. However, enforcing the constraints $h(t|\mathbf{x}_i, \widetilde{\mathbf{z}}_i(t)) \geq 0, \forall t \in C$ for all $i$ is challenging. To address this challenge, we will implement a primal-dual interior point algorithm to enforce both sets of constraints simultaneously.

In the following sections, we will present a primal-dual interior point algorithm for computing the constrained optimization problem. Subsequently, we will develop asymptotic results, including the large sample normality result. The remainder of this section is dedicated to deriving the penalized likelihood function.

Firstly, under partly interval censoring and the additive hazards model in (8.36), the log-likelihood for subject $i$ is given by:

$$\begin{aligned} l_i(\boldsymbol{\beta}, \boldsymbol{\gamma}, \boldsymbol{\theta}) =& \delta_i\big(\log h_i(t_i) - H_i(t_i)\big) - \delta_i^R H_i(t_i) \\ &+ \delta_i^L \log\big(1 - S_i(t_i)\big) + \delta_i^I \log\big(S_i(t_i^L) - S_i(t_i^R)\big), \end{aligned} \tag{8.44}$$

where $h_i(t)$ is the hazard given by (8.42) with $h_0(t) = \boldsymbol{\theta}^\mathsf{T}\boldsymbol{\psi}(t)$ and here $\boldsymbol{\psi}(t) = (\psi_1(t), \dots, \psi_m(t))^\mathsf{T}$ and $\boldsymbol{\theta} = (\theta_1, \cdots, \theta_m)^\mathsf{T}$, $H_i(t) = H(t|\mathbf{x}_i, \widetilde{\mathbf{z}}_i(t))$ is the cumulative hazard function of $i$ and its formula is given in (8.39), and $S_i(t) = \exp\{-H_i(t)\}$, the survival function for subject $i$.

The log-likelihood from all the observed partly interval-censored survival data is then

$$l(\boldsymbol{\beta}, \boldsymbol{\gamma}, \boldsymbol{\theta}) = \sum_{i=1}^{n} l_i(\boldsymbol{\beta}, \boldsymbol{\gamma}, \boldsymbol{\theta}). \tag{8.45}$$

We aim to estimate $\boldsymbol{\beta}$ and $\boldsymbol{\theta}$ by maximizing the penalized log-likelihood

$$\Phi(\boldsymbol{\beta}, \boldsymbol{\gamma}, \boldsymbol{\theta}) = l(\boldsymbol{\beta}, \boldsymbol{\gamma}, \boldsymbol{\theta}) - \lambda J(\boldsymbol{\theta}), \tag{8.46}$$

subject to $\boldsymbol{\theta} \geq 0$ and $h_i(t) \geq 0, \forall t \in \mathcal{C}$. In (8.46), $J(\boldsymbol{\theta})$ represents a penalty function used to smooth the estimate of $h_0(t)$, and $\lambda \geq 0$ denotes the smoothing parameter, which balances the goodness of fit to the data (reflected by a large value of $l(\boldsymbol{\beta}, \boldsymbol{\theta})$) and the smoothness of the $h_0(t)$ estimate.

Similar to previous chapters, we once again employ the well-known roughness penalty, as discussed in Green & Silverman (1994), to smooth the estimates of $h_0(t)$. This involves the application of the penalty function:

$$J(\boldsymbol{\theta}) = \int h_0''(t)^2 dt = \boldsymbol{\theta}^\mathsf{T}\mathbf{R}\boldsymbol{\theta},$$

where the matrix $\mathbf{R}$ is of dimensions $m \times m$ with the $(u, v)$th element $r_{uv} = \int \psi''u(t)\psi''v(t), dt$. This penalty function discourages rapid changes in $h_0(t)$.

The use of a penalty function serves to smooth the estimate of $h_0(t)$ and reduces sensitivity to the number and location of knots (see, for example, Ruppert et al. 2003, page 66). Consequently, a penalty function helps mitigate numerical instabilities often encountered in basis function-based approximation methods.

## 8.6 Gradient and Hessian

The primal-dual interior-point algorithm we will discuss in the next section demands the first two derivatives of $\Phi(\boldsymbol{\beta}, \boldsymbol{\gamma}, \boldsymbol{\theta})$ with respect to $\boldsymbol{\beta}$, $\boldsymbol{\gamma}$ and $\boldsymbol{\theta}$. These derivatives are given below.

**First derivatives:**
It is easy to derive that

$$\begin{aligned}\frac{\partial\Phi}{\partial\beta_j} =&\sum_{i=1}^{n} x_{ij}\left\{\delta_i\Big(\frac{1}{h_i(t_i)}-t_i\Big)-\delta_i^R t_i^L+\delta_i^L\frac{S_i(t_i^R)t_i^R}{1-S_i(t_i^R)}\right.\\&\left.+\delta_i^I\frac{-S_i(t_i^L)t_i^L+S_i(t_i^R)t_i^R}{S_i(t_i^L)-S_i(t_i^R)}\right\},\end{aligned} \tag{8.47}$$

for $j=1\ldots,p$,

$$\begin{aligned}\frac{\partial\Phi}{\partial\gamma_b} =&\sum_{i=1}^{n}\left\{\delta_i\Big(\frac{z_{ib}(t_i)}{h_i(t_i)}-Z_{ib}(t_i)\Big)-\delta_i^R Z_{ib}(t_i^L)+\delta_i^L\frac{S_i(t_i^R)Z_{ib}(t_i^R)}{1-S_i(t_i^R)}\right.\\&\left.+\delta_i^I\frac{-S_i(t_i^L)Z_{ib}(t_i^L)+S_i(t_i^R)Z_{ib}(t_i^R)}{S_i(t_i^L)-S_i(t_i^R)}\right\},\end{aligned} \tag{8.48}$$

for $b=1,\ldots,q$, and

$$\begin{aligned}\frac{\partial\Phi}{\partial\theta_u} =&\sum_{i=1}^{n}\left\{\delta_i\Big(\frac{\psi_u(t_i)}{h_i(t_i)}-\Psi_u(t_i)\Big)-\delta_i^R \Psi_u(t_i^L)+\delta_i^L\frac{S_i(t_i^R)\Psi_u(t_i^R)}{1-S_i(t_i^R)}\right.\\&\left.+\delta_i^I\frac{-S_i(t_i^L)\Psi_u(t_i^L)+S_i(t_i^R)\Psi_u(t_i^R)}{S_i(t_i^L)-S_i(t_i^R)}\right\}-\lambda\frac{\partial J(\boldsymbol{\theta})}{\partial\theta_u},\end{aligned} \tag{8.49}$$

for $u=1,\ldots,m$.

**Second derivatives:**
The second derivatives are more complicated to derive; they are given by:

$$\begin{aligned}\frac{\partial^2\Phi}{\partial\beta_j\partial\beta_t} =&\sum_{i=1}^{n} x_{ij}x_{it}\left\{-\delta_i\frac{1}{h_i(t_i)^2}-\delta_i^L\frac{S_i(t_i^R)(t_i^R)^2}{(1-S_i(t_i^R))^2}\right.\\&\left.-\delta_i^I\frac{S_i(t_i^L)S_i(t_i^R)(t_i^R-t_i^L)^2}{(S_i(t_i^L)-S_i(t_i^R))^2}\right\},\end{aligned} \tag{8.50}$$

$$\begin{aligned}\frac{\partial^2\Phi}{\partial\gamma_b\partial\gamma_l} =&\sum_{i=1}^{n}\left\{-\delta_i\frac{z_{ib}(t_i)z_{il}(t_i)}{h_i(t_i)^2}-\delta_i^L\frac{S_i(t_i^R)Z_{ib}(t_i^R)Z_{il}(t_i^R)}{(1-S_i(t_i^R))^2}\right.\\&\left.-\delta_i^I\frac{S_i(t_i^L)S_i(t_i^R)(Z_{ib}(t_i^R)-Z_{ib}(t_i^L))(Z_{il}(t_i^R)-Z_{il}(t_i^L))}{(S_i(t_i^L)-S_i(t_i^R))^2}\right\},\end{aligned} \tag{8.51}$$

$$\begin{aligned}\frac{\partial^2\Phi}{\partial\theta_u\partial\theta_v} =&\sum_{i=1}^{n}\left\{-\delta_i\frac{\psi_u(t_i)\psi_v(t_i)}{h_i(t_i)^2}-\delta_i^L\frac{S_i(t_i^R)\Psi_u(t_i^R)\Psi_v(t_i^R)}{(1-S_i(t_i^R))^2}\right.\\&\left.-\delta_i^I\frac{S_i(t_i^L)S_i(t_i^R)(\Psi_u(t_i^R)-\Psi_u(t_i^L))(\Psi_v(t_i^R)-\Psi_v(t_i^L))}{(S_i(t_i^L)-S_i(t_i^R))^2}\right\}\\&-\lambda\frac{\partial^2 J(\boldsymbol{\theta})}{\partial\theta_u\partial\theta_v},\end{aligned} \tag{8.52}$$

$$\frac{\partial^2 \Phi}{\partial \beta_j \partial \gamma_b} = \sum_{i=1}^{n} x_{ij} \Bigg\{ -\delta_i \frac{z_{ib}(t_i)}{h_i(t_i)^2} - \delta_i^L \frac{S_i(t_i^R) Z_{il}(t_i^R) t_i^R}{(1-S_i(t_i^R))^2} - \delta_i^I \frac{S_i(t_i^L) S_i(t_i^R)(Z_{il}(t_i^R) - Z_{il}(t_i^L))(t_i^R - t_i^L)}{(S_i(t_i^L) - S_i(t_i^R))^2} \Bigg\}, \tag{8.53}$$

$$\frac{\partial^2 \Phi}{\partial \beta_j \partial \theta_u} = \sum_{i=1}^{n} x_{ij} \Bigg\{ -\delta_i \frac{\psi_u(t_i)}{h_i(t_i)^2} - \delta_i^L \frac{S_i(t_i^R) \Psi_u(t_i^R) t_i^R}{(1-S_i(t_i^R))^2} - \delta_i^I \frac{S_i(t_i^L) S_i(t_i^R)(\Psi_u(t_i^R) - \Psi_u(t_i^L))(t_i^R - t_i^L)}{(S_i(t_i^L) - S_i(t_i^R))^2} \Bigg\}, \tag{8.54}$$

and

$$\frac{\partial^2 \Phi}{\partial \gamma_b \partial \theta_u} = \sum_{i=1}^{n} \Bigg\{ -\delta_i \frac{z_{ib}(t_i)\psi_u(t_i)}{h_i(t_i)^2} - \delta_i^L \frac{S_i(t_i^R) Z_{ib}(t_i^R) \Psi_u(t_i^R)}{(1-S_i(t_i^R))^2} - \delta_i^I \frac{S_i(t_i^L) S_i(t_i^R)(Z_{ib}(t_i^R) - Z_{ib}(t_i^L))(\Psi_u(t_i^R) - \Psi_u(t_i^L))}{(S_i(t_i^L) - S_i(t_i^R))^2} \Bigg\}. \tag{8.55}$$

## 8.7 A primal-dual interior-point algorithm

An informative introduction to the primal-dual interior-point algorithm can be found, for example, in Wright (1997). In this section, we will utilize this algorithm to solve the constrained optimization problem associated with the additive hazards model given in (8.36).

We begin the discussion by introducing some notations, aimed at simplifying the expressions of the constraints. Let $\boldsymbol{\eta} = (\boldsymbol{\theta}^\top, \boldsymbol{\beta}^\top, \boldsymbol{\gamma}^\top)^\top$, representing a vector of the unknown parameters that we wish to estimate. We define another vector as follows:

$$\mathbf{C}(t) = (\psi_1(t), \ldots, \psi_m(t), x_{i1}, \ldots, x_{ip}, z_{i1}(t), \ldots, z_{iq}(t))^\top.$$

Then, the hazard function in (8.36) can be rewritten as:

$$h(t|\mathbf{x}_i, \widetilde{\mathbf{z}}_i(t)) = \mathbf{C}(t)^\top \boldsymbol{\eta}. \tag{8.56}$$

Recall that the individual hazard constraints $h(t|\mathbf{x}i, \widetilde{\mathbf{z}}i(t)) \geq 0$ are imposed on those $t$ values where $t \in \mathcal{C}$ with $\mathcal{C}$ defined in (8.41). Since there are a total of $N = \sum_{i=1}^{n} n_i$ points in $\mathcal{C}$, we have $N$ constraints associated with $\mathcal{C}$. Additionally, we have $m$ constraints from $\boldsymbol{\theta} \geq 0$. To consolidate all these constraints, we define the following matrices. Let $\mathbf{M}_a$ be an $m \times (m+p+q)$ matrix given by

$$\mathbf{M}_a = \left( \mathbf{I}_{m\times m} \quad \mathbf{0}_{m\times(p+q)} \right),$$

where $\mathbf{I}$ denotes an identity matrix and $\mathbf{0}$ denotes a matrix of zeros. Clearly, $\mathbf{M}_a\boldsymbol{\eta} = \boldsymbol{\theta}$, so $\mathbf{M}_a$ can be used to define the constraint $\boldsymbol{\theta} \geq 0$. On the other hand, let $\mathbf{M}_b$ be an $N \times (m+p+q)$ matrix given by

$$\mathbf{M}_b = \begin{pmatrix} \mathbf{C}(t_{11})^\mathsf{T} \\ \mathbf{C}(t_{12})^\mathsf{T} \\ \vdots \\ \mathbf{C}(t_{1n_1})^\mathsf{T} \\ \vdots \\ \mathbf{C}(t_{n1})^\mathsf{T} \\ \vdots \\ \mathbf{C}(t_{nn_n})^\mathsf{T} \end{pmatrix},$$

then $\mathbf{M}_b\boldsymbol{\eta}$ will give the individual hazard values. Combining $\mathbf{M}_a$ and $\mathbf{M}_b$, we create a new matrix $\mathbf{M}$ given by

$$\mathbf{M} = \begin{pmatrix} \mathbf{M}_a \\ \mathbf{M}_b \end{pmatrix} \tag{8.57}$$

The dimension of $\mathbf{M}$ is $(m+N) \times (m+p+q)$.

Now, we can define the following function

$$f(\boldsymbol{\eta}) = -\mathbf{M}\boldsymbol{\eta}. \tag{8.58}$$

This function represents the constraints we intend to impose when estimating the parameter set $\boldsymbol{\eta}$. The constrained optimization problem can now be reformulated as

$$\widehat{\boldsymbol{\eta}} = \underset{\boldsymbol{\eta}\in\mathcal{F}}{\operatorname{argmin}} \left\{-\Phi(\boldsymbol{\eta})\right\}, \tag{8.59}$$

where $\mathcal{F}$ is a feasible set defined by

$$\mathcal{F} = \left\{\boldsymbol{\eta} \,|\, f_r(\boldsymbol{\eta}) = -\mathbf{M}_r\boldsymbol{\eta} \leq 0,\ r = 1, \ldots, m, m+1, \ldots, m+N\right\},$$

where $\mathbf{M}_r$ denotes the $r$th row of matrix $\mathbf{M}$. Note that here we have converted the maximization problem into a minimization problem and changed the non-negative constraint to a non-positive constraint to align with the traditional constrained optimization setting.

Utilizing the primal-dual interior point algorithm, the solution to the constrained optimization problem in (8.59) can be obtained by solving the following central path equations (see, for example, Chapter 8 of Sun & Yuan (2006)):

$$\frac{\partial\Phi(\boldsymbol{\eta})}{\partial\boldsymbol{\eta}} + \mathbf{M}^\mathsf{T}\boldsymbol{\chi}_{(m+N)\times 1} = \mathbf{0}_{(m+p+q)\times 1}, \tag{8.60}$$

$$f(\boldsymbol{\eta}) + \mathbf{s} = \mathbf{0}_{(m+N)\times 1}, \tag{8.61}$$

$$\chi_r\, s_r = 0, \tag{8.62}$$

$$\chi_r \geq 0, s_r \geq 0. \tag{8.63}$$

Here, $r = 1, \ldots, m, m+1, \ldots, m+N$, $\boldsymbol{\chi}$ is a vector for the Lagrange multipliers, and $\mathbf{s}$ is a vector for the slack variables used to convert the inequality constraints into equality constraints. Note that these central path equations can be regarded as a different expression of the well-known KKT conditions we have used throughout this book.

The system of equations (8.60)–(8.63) can be solved using the primal-dual interior-point algorithm. In fact, this algorithm solves simultaneously the primal problem for $\boldsymbol{\eta}$, and the dual problem for $\boldsymbol{\chi}$ and $\mathbf{s}$.

In each iteration, the algorithm begins by employing the Newton method to solve equations (8.60)–(8.62) and obtain the Newton direction. Subsequently, a line search is conducted along the Newton direction to ensure that the inequality condition (8.63) is satisfied with the new updates. However, this strategy may lead to slow convergence, as discussed in Wright (1997).

A more effective alternative is to modify the Newton procedure by requiring that each pairwise product $\chi_r s_r$ has the same positive value. To achieve this, we set $\chi_r s_r = \xi \varpi$, where $\xi$ is a centering parameter in the range [0, 1], and $\varpi$ is a duality measure defined as:

$$\varpi = \frac{\boldsymbol{\chi}^\mathsf{T} \mathbf{s}}{m+N}. \tag{8.64}$$

Here, $\varpi$ represents the average value of $\chi_r s_r$. The modified Newton step maintains strictly positive pairwise products $\chi_r s_r$, allowing for longer steps before violating the positivity condition. As $\varpi$ approaches zero during each iteration, $\chi_r s_r$ decreases at the same rate, eventually satisfying condition (8.62). This type of algorithm has been extensively studied in works such as Wright (1997) and Sun & Yuan (2006).

More specifically, we define the following functions (called residuals):

$$r_p(\boldsymbol{\eta}, \mathbf{s}) = f(\boldsymbol{\eta}) + \mathbf{s},$$
$$r_d(\boldsymbol{\eta}, \boldsymbol{\chi}) = \frac{\partial \Phi(\boldsymbol{\eta})}{\partial \boldsymbol{\eta}} + \mathbf{M}^\mathsf{T} \boldsymbol{\chi}.$$

For a fixed $\tau > 0$ and for given $\boldsymbol{\eta}, \boldsymbol{\chi}$ and $\mathbf{s}$, let the set (which depends on $\varpi$)

$$\begin{aligned}\mathcal{N}(\varpi) = \{ & (\boldsymbol{\eta}, \boldsymbol{\chi}, \mathbf{s}) : |r_p(\boldsymbol{\eta}, \mathbf{s})| \le \tau\varpi, |r_d(\boldsymbol{\eta}, \boldsymbol{\chi})| \le \tau\varpi, \chi_r s_r = \xi\varpi, \\ & \chi_r \ge 0, s_r \ge 0 \; \forall i \}.\end{aligned}$$

Let $\boldsymbol{\Xi} = \mathrm{diag}(\chi_1, \ldots, \chi_{m+N})$ and $\mathbf{S} = \mathrm{diag}(s_1, \ldots, s_{m+N})$ be diagonal matrices with diagonal elements given by $\chi_r$ and $s_r$ respectively. With these notations, we outline below the computational procedure of the algorithm.

1. Set an initial $\boldsymbol{\eta}^{(0)}$ with its components $\boldsymbol{\theta}^{(0)} = \mathbf{1}_{m\times 1}$, $\boldsymbol{\beta}^{(0)} = \mathbf{0}_{p\times 1}$ and $\boldsymbol{\gamma}^0 = \mathbf{0}_{q\times 1}$, where $\mathbf{1}$ denotes a matrix of 1's and $\mathbf{0}$ a matrix of 0's. These initials satisfy $f(\boldsymbol{\eta}^{(0)}) = (-\mathbf{1}_{m\times 1}^\mathsf{T}, \mathbf{0}_{(m+N)\times 1}^\mathsf{T}, \mathbf{0}_{(m+N)\times 1}^\mathsf{T})^\mathsf{T}$, so that $\boldsymbol{\eta}^0 \in \mathcal{F}$. Accordingly, we can set the initials for $\boldsymbol{\chi}$ and $\mathbf{s}$ as: $\boldsymbol{\chi}^{(0)} = (\mathbf{0}_{m\times 1}^\mathsf{T}, \mathbf{1}_{(m+N)\times 1}^\mathsf{T}, \mathbf{1}_{(m+N)\times 1}^\mathsf{T})^\mathsf{T}$ and $\mathbf{s}^{(0)} = (\mathbf{1}_{m\times 1}^\mathsf{T}, \mathbf{0}_{(m+N)\times 1}^\mathsf{T}, \mathbf{0}_{(m+N)\times 1}^\mathsf{T})^\mathsf{T}$, then the conditions (8.61)–(8.63) are satisfied.

2. After $k$ iterations, we obtained the updates $\boldsymbol{\eta}^{(k)}$, $\boldsymbol{\chi}^{(k)}$ and $\mathbf{s}^{(k)}$, and the the corresponding duality measure is given by

$$\varpi^{(k)} = \frac{(\boldsymbol{\chi}^{(k)})^{\mathsf{T}}\mathbf{s}^{(k)}}{m+M}.$$

Then, at the $(k+1)$th iteration, we first calculate the Newton direction $(d\boldsymbol{\eta}, d\boldsymbol{\chi}, d\mathbf{s})^{\mathsf{T}}$ by solving the following linear system:

$$\begin{pmatrix} \partial^2\Phi(\boldsymbol{\eta}^{(k)})/\partial\boldsymbol{\eta}\partial\boldsymbol{\eta}^{\mathsf{T}} & \mathbf{M}^{\mathsf{T}} & \mathbf{0}_{(m+p+q)\times(m+N)} \\ \mathbf{M} & \mathbf{0}_{(m+N)\times(m+N)} & \mathbf{I}_{(m+N)\times(m+N)} \\ \mathbf{0}_{(m+N)\times(m+N)} & \mathbf{S}^{(k)} & \boldsymbol{\Xi}^{(k)} \end{pmatrix} \begin{pmatrix} d\boldsymbol{\eta}^{(k)} \\ d\boldsymbol{\chi}^{(k)} \\ d\mathbf{s}^{(k)} \end{pmatrix}$$
$$= \begin{pmatrix} -r_d(\boldsymbol{\eta}^{(k)}, \boldsymbol{\chi}^{(k)}) \\ -r_p(\boldsymbol{\eta}^{(k)}, \mathbf{s}^{(k)}) \\ -\boldsymbol{\Xi}^{(k)}\mathbf{S}^{(k)}\mathbf{1}_{(m+N)\times 1} + \xi\mu^{(k)}\mathbf{1}_{(m+N)\times 1} \end{pmatrix}, \tag{8.65}$$

where $\xi$ is fixed and usually is set to $\xi = 0.5$. Then, the update of $\boldsymbol{\eta}, \mathbf{u}, \mathbf{s}$ are obtained according to

$$\begin{pmatrix} \boldsymbol{\eta}^{(k+1)} \\ \boldsymbol{\chi}^{(k+1)} \\ \mathbf{s}^{(k+1)}) \end{pmatrix} = \begin{pmatrix} \boldsymbol{\eta}^{(k)} \\ \boldsymbol{\chi}^{(k)} \\ \mathbf{s}^{(k)} \end{pmatrix} + \omega^{(k)} \begin{pmatrix} d\boldsymbol{\eta}^{(k)} \\ d\boldsymbol{\chi}^{(k)} \\ d\mathbf{s}^{(k)} \end{pmatrix}. \tag{8.66}$$

The step length $\omega^{(k)}$ in (8.66) is chosen to be the first element in the sequence $\{1, \kappa, \kappa^2, \kappa^3, \cdots\}$, where $\kappa \in (0,1)$ (e.g. $\kappa = 0.6$), such that $(\boldsymbol{\eta}^{(k+1)}, \boldsymbol{\chi}^{(k+1)}, \mathbf{s}^{(k+1)}) \in \mathcal{N}(\varpi^{(k)})$ and also the following condition holds:

$$\varpi^{(k+1)} \leq (1 - 0.01\omega^{(k)})\varpi^{(k)}.$$

3. Repeat Step 2 until a convergence criterion is satisfied or the maximum number of iterations is reached, whichever comes first.

A useful convergence criterion is to stop the iterations when $\mu^{(k)}$ is below a certain tolerance criterion, such as $10^{-5}$ or $10^{-10}$. Such a small $\mu^{(k)}$ will ensure that the corresponding $\eta^{(k)}$ approximately satisfies equations (8.60)–(8.63), and thus convergence is achieved when the derivative equation (8.60) is very close to zero.

A global convergence result of the primal-dual interior point algorithm is given in the following theorem.

**Theorem 8.1.** The sequence of iterates $\{\boldsymbol{\eta}^{(k)}, \boldsymbol{\chi}^{(k)}, \mathbf{s}^{(k)}\}$ generated by the primal-dual interior-point algorithm converges to a solution satisfying the KKT conditions (8.60)–(8.63). Furthermore, the sequence of the duality measure $\{\varpi^{(k)}\}$ converges Q-linearly to zero, and the sequences of residual norms $\|r_p(\boldsymbol{\eta}^{(k)}, \mathbf{s}^{(k)})\|$ and $\|r_d(\boldsymbol{\eta}^{(k)}, \boldsymbol{\chi}^{(k)})\|$ converge R-linearly to zero. □

The proof of this theorem can be found in Theorem 6.1 of Wright (1997), so it is omitted here.

## 8.8 Asymptotic properties and inferences

In this section, we provide two asymptotic results which explain theoretical properties of the MPL estimates under the additive hazards model. These results are provided in Theorems 8.2 and 8.3.

Let $a = \min_i\{t_i^L : t_i^L \neq 0\}$ and $b = \max_i\{t_i^R : t_i^R \neq \infty\}$. Let $\boldsymbol{\beta}_0$, $\boldsymbol{\gamma}_0$ and $h_0(t)$ be the true parameters and $\widehat{\boldsymbol{\beta}}$, $\widehat{\boldsymbol{\gamma}}$ and $\widehat{h}_0(t)$ be their MPL estimates, where $t \in [a, b]$. The MPL estimate of $h_0(t)$ is obtained using the approximation given in equation (8.42). The number of basis functions $m$, used in (8.42), is allowed to increase with the sample size $n$ but at a slower pace. That is, $m \to \infty$ when $n \to \infty$, but $m/n \to 0$. Also, the scaled smoothing $\mu_n = \lambda/n$ goes to zero when $n \to \infty$.

Next theorem provides the consistent results for the MPL estimates. The assumptions it requires are identical to those stated in Theorem 6.1.

**Theorem 8.2.** Suppose that Assumption 6.1 and Assumptions 2.2–2.4 hold, where the log-likelihood is replaced with (8.45). Assume that $h_0(t)$ has up to $c \geq 1$ derivatives for $t \in [a, b]$. Additionally, assume that $m = n^{\upsilon}$, where $0 < \upsilon < 1$, and $\mu_n = \lambda/n \to 0$ as $n \to \infty$. Then, as $n \to \infty$,

1. $\|\widehat{\boldsymbol{\beta}} - \boldsymbol{\beta}_0\|_2 \to 0$ (a.s.).
2. $\|\widehat{\boldsymbol{\gamma}} - \boldsymbol{\gamma}_0\|_2 \to 0$ (a.s.).
3. $\sup_{t\in[a,b]} |\widehat{h}_0(t) - h_0(t)| \to 0$ (a.s.).

□

Apart from the above asymptotically consistent results, we need asymptotic normality results for $\widehat{\boldsymbol{\beta}}$, $\widehat{\boldsymbol{\gamma}}$ and $\widehat{\boldsymbol{\theta}}$ in order to make inferences.

For the parameter vector $\boldsymbol{\eta} = (\boldsymbol{\theta}^\top, \boldsymbol{\beta}^\top, \boldsymbol{\gamma}^\top)^\top$, let $\boldsymbol{\eta}_0 = (\boldsymbol{\theta}_0^\top, \boldsymbol{\beta}_0^\top, \boldsymbol{\gamma}_0^\top)^\top$ denote the true value of $\boldsymbol{\eta}$. The constraints for the additive hazards model differ from those in the previous chapters, necessitating special considerations. Recall that the constraints can be expressed as

$$\mathbf{M}\boldsymbol{\eta} \geq 0, \tag{8.67}$$

where matrix $\mathbf{M}$ is defined in (8.57). If a row of $\mathbf{M}$ multiplied by $\boldsymbol{\eta}$ results in 0, it represents an active constraint. The principles for developing the asymptotic covariance matrix for constrained maximum likelihood estimation in the presence of active constraints are discussed in Moore & Sadler (2006). We will adopt this method to handle active constraints in the MPL estimate $\widehat{\boldsymbol{\eta}}$ discussed in this chapter.

As in previous chapters where we addressed active non-negative constraints, we will formulate a matrix $\mathbf{U}$ to accommodate the active constraints discussed in this chapter. However, the construction of this matrix $\mathbf{U}$ differs significantly from our approach in earlier chapters as the constraints in (8.67) are more complicated.

According to equation (8.60), the solution to the constrained MPL must satisfy the following equation:

$$\frac{\partial \Phi}{\partial \eta} + \mathbf{M}^\mathsf{T}\chi = 0, \tag{8.68}$$

where $\chi$ is a vector of Lagrange multipliers. An element $\chi_r$ in $\chi$ is strictly positive if the corresponding constraint is active, and $\chi_r = 0$ otherwise.

Suppose there are $c > 0$ active constraints. With this assumption, we can divide $\chi$ into two components: one for active constraints denoted by a $c$-vector $\chi_A$ and the other for non-active constraints. Correspondingly, we denote the portion of $\mathbf{M}$ that relates to active constraints as $\mathbf{M}_A$. As a result, we have:

$$\mathbf{M}^\mathsf{T}\chi = \mathbf{M}_A^\mathsf{T}\chi_A. \tag{8.69}$$

When the number of active constraints, $c$, is less than the number of all the parameters, the null space of $\mathbf{M}_A$ is non-empty because we can find a non-zero vector $\mathbf{u}$ that satisfies $\mathbf{M}_A\mathbf{u} = \mathbf{0}_{c\times 1}$

Assuming the dimension of this null space is $e$. Let $\mathbf{U}_{(m+p+q)\times e}$ be a matrix whose columns form orthonormal bases of the null space of $\mathbf{M}_A$. Then, matrix $\mathbf{U}$ satisfies:

$$\mathbf{M}_A\mathbf{U} = \mathbf{0}_{c\times e} \text{ and } \mathbf{U}^\mathsf{T}\mathbf{U} = \mathbf{I}_{e\times e}. \tag{8.70}$$

This property, along with equation (8.68), implies

$$\mathbf{U}^\mathsf{T}\frac{\partial \Phi}{\partial \eta} = 0.$$

This equation is equivalent to the one in (2.55). Consequently, it will yield the asymptotic normality results as stated in Theorem 8.3. These results rely on assumptions similar to B6.1 and B2.2 to 2.4.

**Theorem 8.3.** Assuming Assumptions B6.1 and B2.2–B2.4 hold where the log-likelihood is replaced with (8.45), the log-likelihood associated with the additive hazards model. Assume the scaled smoothing parameter $\mu_n = o(n^{-1/2})$. Let $\mathbf{F}(\eta) = -n^{-1}E\big(\partial^2 l(\eta)/\partial\eta\partial\eta^\mathsf{T}\big)$. If there are active constraints, construct matrix $\mathbf{U}$ using the orthonormal bases of the null space of $\mathbf{M}_A$, where $\mathbf{M}_A$ has been defined above. Then, when $n \to \infty$,

1. The constrained MPL estimate $\widehat{\eta}$ is consistent for $\eta_0$, and

2. $n^{1/2}(\widehat{\eta}-\eta_0)$ converges in distribution to $N(\mathbf{0}, \widetilde{\mathbf{F}}(\eta_0)^{-1}\mathbf{F}(\eta_0)\widetilde{\mathbf{F}}(\eta_0)^{-1})$, where the matrix $\widetilde{\mathbf{F}}(\eta)^{-1} = \mathbf{U}(\mathbf{U}^\mathsf{T}\mathbf{F}(\eta)\mathbf{U})^{-1}\mathbf{U}^\mathsf{T}$.

□

Given that $\eta_0$ is typically unavailable in practice, we can readily substitute it with the MPL estimate $\widehat{\eta}$ due to the strong consistency result. Additionally, computing the expected information matrix $\mathbf{F}(\eta)$ can be challenging, and it is possible to replace it with the negative Hessian matrix or an approximation thereof.

Finite sample inferences, i.e. inferences made when $n$ is large but finite, are the only possibility in practice, which means the penalty term should be kept in an approximation to the asymptotic covariance matrix. Results in Theorem 8.3 need to be modified for this purpose.

In fact, when $n$ is large, the distribution of $\widehat{\boldsymbol{\eta}} - \boldsymbol{\eta}_0$ can be approximated by a multivariate normal distribution with the approximate covariance matrix

$$\widehat{\mathrm{Var}}(\widehat{\boldsymbol{\eta}}) = -\mathbf{A}(\widehat{\boldsymbol{\eta}})^{-1}\frac{\partial^2 l(\widehat{\boldsymbol{\eta}})}{\partial\boldsymbol{\eta}\boldsymbol{\eta}^{\mathsf{T}}}\mathbf{A}(\widehat{\boldsymbol{\eta}})^{-1}, \tag{8.71}$$

where

$$\mathbf{A}(\widehat{\boldsymbol{\eta}})^{-1} = \mathbf{U}\Big(\mathbf{U}^{\mathsf{T}}\Big(-\frac{\partial^2 l(\widehat{\boldsymbol{\eta}})}{\partial\boldsymbol{\eta}\partial\boldsymbol{\eta}^{\mathsf{T}}} + \lambda\frac{\partial^2 J(\widehat{\boldsymbol{\eta}})}{\partial\boldsymbol{\eta}\partial\boldsymbol{\eta}^{\mathsf{T}}}\Big)\mathbf{U}\Big)^{-1}\mathbf{U}^{\mathsf{T}}.$$

In the following Example 8.7, we conduct a simple simulation study in which we compare the average of the large sample standard deviation for each estimated parameter, calculated using formula (8.71), with the Monte-Carlo standard deviations to assess the accuracy of (8.71). The results reveal that this asymptotic variance formula is generally accurate for the estimates of $\boldsymbol{\beta}$, $\boldsymbol{\gamma}$ and $h_0(t)$. Furthermore, the study demonstrates that the bias in the MPL estimates is typically negligible when smoothing parameters and $m$ are relatively small compared to $n$.

## 8.9 Smoothing parameter selection

Let $\sigma^2 = 1/(2\lambda)$, and our goal is to estimate $\sigma^2$. The prior distribution corresponding to the penalty function we use is $N(\mathbf{0}_{m\times 1}, \sigma^2\mathbf{R})$. The parameter $\sigma^2$ can be estimated in the same way as in the previous chapters, using the marginal log-likelihood, where the marginal likelihood is obtained by integrating out all the parameters except $\sigma^2$ from the posterior distribution.

Using Laplace's approximation to the marginal likelihood, the solution for $\sigma^2$, denoted by $\widehat{\sigma}^2$, that maximizes this approximate log-marginal likelihood, can be verified to satisfy:

$$\widehat{\sigma}^2 = \frac{\widehat{\boldsymbol{\theta}}^{\mathsf{T}}\mathbf{R}\widehat{\boldsymbol{\theta}}}{m-\nu}, \tag{8.72}$$

where $\nu$ is given by:

$$\nu = \mathrm{tr}\{(\widehat{-\mathbf{H}} + \mathbf{Q}(\widehat{\sigma}^2))^{-1}\mathbf{Q}(\widehat{\sigma}^2)\}.$$

Here, $\widehat{\mathbf{H}}$ is the Hessian matrix from the log-likelihood $l(\boldsymbol{\beta}, \boldsymbol{\gamma}, \boldsymbol{\theta})$, evaluated at $\widehat{\boldsymbol{\beta}}$, $\widehat{\boldsymbol{\gamma}}$ and $\widehat{\boldsymbol{\theta}}$, and

$$\mathbf{Q}(\sigma^2) = \begin{pmatrix} \mathbf{0}_{p\times p} & \mathbf{0}_{p\times q} & \mathbf{0}_{p\times m} \\ \mathbf{0}_{q\times p} & \mathbf{0}_{q\times q} & \mathbf{0}_{q\times m} \\ \mathbf{0}_{m\times p} & \mathbf{0}_{m\times q} & \frac{1}{\sigma^2}\mathbf{R} \end{pmatrix}.$$

The expression (8.72) suggests an iterative procedure. Specifically, with $\sigma^2$ fixed at its current estimate, the corresponding MPL estimates of $\boldsymbol{\beta}, \boldsymbol{\gamma}$ and $\boldsymbol{\theta}$ are obtained. Then, $\sigma^2$ is updated using formula (8.72), where $\widehat{\boldsymbol{\beta}}$, $\widehat{\boldsymbol{\gamma}}$, $\widehat{\boldsymbol{\theta}}$ and $\widehat{\sigma}^2$ on the right-hand side are replaced by their most current estimates. These iterations continue until the degree of freedom, $m - \nu$ stabilizes, i.e. difference between consecutive degrees of freedom is less than 1.

## 8.10 R examples

In this section, we will present two R examples: the first example is a simulation study, and the second example applies the MPL estimation method discussed in this chapter to a Breast Cancer dataset.

**Example 8.7** (A simulation study with right-censored data).
In this R example, we conduct a simulation study to assess the performance of the MPL method. Our objectives are to:

(i) examine the impact of censoring proportion and sample size on MPL estimates, (ii) validate the accuracy of the asymptotic standard deviations provided in (8.71), and (iii) compare the MPL method with the counting process method described by Lin & Ying (1994).

In our simulation, we consider sample sizes of $n = 100$ and $n = 500$, representing small and intermediate sample sizes, respectively. For the simulated datasets, we will have, approximately, $\pi_c = 20\%$ and $80\%$ right censoring proportions.

To simplify the discussion in this example, we use only time-independent covariates. The covariates and survival data are generated in the same manner as in Example 8.1. Therefore, the additive hazards model for the simulation is given by:

$$h_{Y_i}(t) = 3t^2 - 0.5x_{i1} + 1.5x_{i2},$$

where $x_{i1} \sim \text{Bernoulli}(0.6)$, $x_{i2} \sim \text{unif}(1, 2)$ and $Y_i$ represents the event time.

In this simulation, we focus on survival data with right censoring. Interval censoring will be addressed in the next example. Our objective here is to compare the MPL method with the counting process method proposed by Lin & Ying (1994). The general censoring scenario will be presented in the following example, where we aim to evaluate the MPL results for interval-censored event data.

After a $Y_i$ is generated from the additive hazards model above, we next simulate survival times with right censoring. We generate the right censoring time $C_i$ from exponential distributions with a mean of 1, which yields a right censoring proportion of $\pi_c = 20\%$, or a mean of 0.5, resulting in a right censoring proportion of $\pi_c = 80\%$. Then, the right-censored event time is given by $T_i = \min(Y_i, C_i)$.

For each combination of sample size and censoring proportion, we simulate 200 repeated samples.

We approximate the baseline hazard $h_0(t)$ using a piecewise constant function, which is equivalent to employing indicator basis functions. In this context, we represent $h_0(t)$ with a vector $\boldsymbol{\theta} = (\theta_1, \cdots, \theta_m)^\top$, where $m$ is the number of bins. These bins are selected such that each one contains approximately the same number of observations, denoted as $n_c$.

Our experience suggests that $\boldsymbol{\beta}$ estimates are generally not very sensitive to $n_c$ as long as it is not excessively large and the smoothing parameter is appropriately chosen. In this simulation, we set $n_c = 2$ for $n = 100$ and $n_c = 5$ for $n = 500$. The smoothing parameter was automatically selected as described in Section 8.9. For the penalty function, we used a 2nd-order difference penalty:

$$J(\boldsymbol{\theta}) = \sum_{j=2}^{m-1} (\theta_{j-1} - 2\theta_j + \theta_{j+1})^2$$

in the simulation, which penalizes discrepancies between neighboring $\theta_u$'s. The convergence criterion we adopted is $\mu^{(k)} < 10^{-5}$ for the primal-dual interior-point algorithm.

The R code for this simulation is given below.

```
library(RConics)
library(addhazard)

# function for generating data -
# right censoring only
generate_ch8_7 <- function(n) {

    x_1 <- rbinom(n, 1, 0.6)
    x_2 <- runif(n, 1, 2)

    neg.logU <- -log(runif(n, 0.1, 1))

    y <- t <- delta <- rep(0, n)
    for (i in 1:n) {
        y[i] <- as.numeric(cubic(c(1,
            0, -x_1[i] * 0.5 + x_2[i] *
                1.5, -neg.logU[i]))[1])
    }

    c <- rexp(n, 3)

    delta <- as.numeric(y < c)
    t <- c
    t[delta == 1] <- y[delta == 1]


    data = data.frame(id = rep(1:n),
```

```
        t, delta, x_1, x_2)
    return(data)

}

# loop for running simulations
# (Scenario 1)
for (s in 1:200) {
    ah_data = generate_ch8_7(100)

    # fit L&Y model
    aalen.surv <- Surv(time = ah_data$t,
        event = ah_data$delta)
    ah.fit <- ah(aalen.surv ~ x_1 +
        x_2, data = ah_data, ties = FALSE)

    save_87[s, 1] = ah.fit$coef[1]
    save_87[s, 2] = ah.fit$coef[2]
    save_87[s, 3] = ah.fit$se[1]
    save_87[s, 4] = ah.fit$se[1]

    # fit MPL model
    ah_data_new <- cbind(rep(0, 100),
        ah_data$t, ah_data$delta, ah_data$x_1,
        ah_data$x_2, ah_data$id)
    ah_data_new = data.frame(ah_data_new)
    colnames(ah_data_new) = c("start",
        "stop", "event", "x_1", "x_2",
        "id")
    # ah_data_new$stop[which(ah_data_new$event
    # == 0)] <- Inf

    ctrl <- ah_mpl.control(basis = "uniform",
        smooth = NULL, max.iter = c(10,
            1000, 1000), tol = 1e-05,
        n.knots = c(50, 0), range.quant = c(0.1,
            0.9), min.theta = 1e-10,
        penalty = 2L, order = 3L, epsilon = c(1e-16,
            1e-10), ties = "epsilon",
        seed = NULL)


    fit_mpl1 <- ah_survmpl(Surv(ah_data_new$start,
        ah_data_new$stop, ah_data_new$event) ~
        x_1 + x_2, data = ah_data_new,
        control = ctrl)

    save_87[s, 5] = fit_mpl1$coef$Beta[1]
```

```
    save_87[s, 6] = fit_mpl1$coef$Beta[2]
    save_87[s, 7] = fit_mpl1$se$Beta[1,
        2]
    save_87[s, 8] = fit_mpl1$se$Beta[2,
        2]

}
```

TABLE 8.1: BIAS, MCSD, AASD and MSE for comparing MPL with the Lin&Ying's method (right censored data).

| | | | $n = 100$ | | $n = 500$ | |
|---|---|---|---|---|---|---|
| $\pi_c$ | | | 20% | 80% | 20% | 80% |
| $n_c$ | | | 2 | 2 | 5 | 5 |
| $\beta_1 = -0.5$ | MPL | BIAS | 0.034 | 0.071 | -0.008 | 0.058 |
| | | MCSD | 0.627 | 0.793 | 0.377 | 0.454 |
| | | AASD | 0.618 | 0.629 | 0.375 | 0.427 |
| | | MSE | (0.078) | (0.181) | (0.031) | (0.049) |
| | L&Y | BIAS | -0.057 | -0.118 | 0.032 | 0.106 |
| | | MCSD | 0.618 | 0.789 | 0.380 | 0.447 |
| | | AASD | 0.705 | 0.803 | 0.387 | 0.456 |
| | | MSE | (0.381) | (0.663) | (0.145) | (0.215) |
| $\beta_2 = 1.5$ | MPL | BIAS | 0.047 | 0.063 | 0.001 | 0.039 |
| | | MCSD | 0.538 | 0.481 | 0.361 | 0.439 |
| | | AASD | 0.412 | 0.363 | 0.366 | 0.343 |
| | | MSE | (0.124) | (0.1542) | (0.024) | (0.041) |
| | L&Y | BIAS | 1.162 | 0.542 | 0.844 | 0.610 |
| | | MCSD | 1.092 | 1.263 | 0.601 | 0.711 |
| | | AASD | 0.705 | 0.803 | 0.398 | 0.456 |
| | | MSE | (2.539) | (1.880) | (1.071) | (0.865) |

The simulation results presented in Table 8.1 include several metrics: average absolute bias (BIAS), Monte Carlo standard deviation (MCSD), average asymptotic standard deviation (AASD) and mean squared error (MSE).

These results indicate that, when the sample size remains constant, all the reported metrics decrease as the censoring proportion decreases. Likewise, for a fixed censoring proportion, the metrics decrease as the sample size increases. A comparison of MCSD with AASD demonstrates the accuracy of the sandwich formula provided in (8.71), with accuracy improving as sample sizes increase.

The results in Table 8.1 compare the MPL method with the counting process method of Lin & Ying (1994) (referred to as the L&Y method). The L&Y method is specifically designed for estimating regression coefficients with right-censored survival data in the context of the additive hazards model; see Section 8.3.1. The L&Y results were obtained using the R package `addhazard`. Note that in the L&Y

method, the baseline hazard estimate can obtained after estimating $\widehat{\beta}$ using a procedure similar to the Breslow method, which can be highly non-smooth. We do not report L&Y baseline hazard estimation in this simulation example.

It is evident that in general, for both sample sizes we tested, the MPL method can achieve smaller bias for estimating the regression coefficients. Also, the MPL estimates for the three regression parameters have lower MCSDs and MSEs. Comparing MCSDs with AASDs in these two methods separately, we can observe that the asymptotic standard deviation formulas from both methods accurately approximate variances. □

**Example 8.8** (A simulation study with interval censored data).
This example can be seen as a continuation of Example 8.7. However, it delves into more complex interval-censored survival data.

Survival data in this example are generated from the same additive hazard as in Example 8.7 but with a general censoring scheme including left, right and interval censoring.

We first generate an event time from the additive hazard model (see Example 8.7) and this time is denoted by $y_i$. Let $(t_i^L, t_i^R]$ be interval censoring times. We first generate two monitoring times by $c_i^L \sim \text{unif}(0, 1)$ and $c_i^R = L_i + \text{unif}(0, 1)$. The proportion of censoring can be controlled by a uniform random variable $u_i$. If $u_i \geq \pi_c$ ($\pi_c$ is predetermined), $Y_i$ is exactly observed and we set $t_i^L = t_i^R = y_i$. If $u_i < \pi_c$, $Y_i$ is censored and there are three cases: if $y_i \leq c_i^L$, it is left-censored and we set $t_i^L = 0$ and $t_i^R = c_i^L$; if $c_i^L < y_i \leq c_i^R$, $Y_i$ is interval-censored and we set $t_i^L = c_i^L$ and $t_i^R = c_i^R$; if $y_i > c_i^R$, $Y_i$ is right-censored so that $t_i^L = c_i^R$ and $t_i^R = +\infty$.

The R code for data generation and model fitting for this simulation is given below. In this R code, the R function "AH_MPL.R" is available on the GiyHub address for this book.

```
# Function for
# generating interval
# censored data

generate_ch8_7_intcens <- function(n,
    pi_E) {

    x_1 <- rbinom(n, 1, 0.6)
    x_2 <- runif(n, 1, 2)

    neg.logU <- -log(runif(n,
        0.1, 1))

    y <- rep(0, n)
    for (i in 1:n) {
        y[i] <- as.numeric(cubic(c(1,
            0, -x_1[i] * 0.5 +
                x_2[i] * 1.5,
            -neg.logU[i]))[1])
```

```
    }

    L <- runif(n)
    R <- L + runif(n)
    u_i <- runif(n)

    delta <- t_L <- t_R <- rep(0,
        n)

    for (i in 1:n) {
        if (u_i[i] < pi_E) {
            delta[i] <- 1
            t_L[i] <- 0
            t_R[i] <- y[i]
        } else if (y[i] < L[i]) {
            delta[i] <- 2
            t_L[i] <- 0
            t_R[i] <- L[i]
        } else if (L[i] < y[i] &
            y[i] < R[i]) {
            delta[i] <- 3
            t_L[i] <- L[i]
            t_R[i] <- R[i]
        } else if (R[i] < y[i]) {
            delta[i] = 0
            t_L[i] <- 0
            t_R[i] <- R[i]
        }
    }

    data <- data.frame(id = rep(1:n),
        t_L, t_R, delta, x_1,
        x_2)
    return(data)

}

# Set up simulation
# study

save <- matrix(0, ncol = 13,
    nrow = 300)
save_h0 <- NULL

for (s in 1:300) {

    # generate and set
    # up data
    dat <- generate_ch8_7_intcens(300,
        pi_E = 0.8)

    n <- 300
    y <- matrix(0, n, 5)
    colnames(y) <- c("Id",
        "star_time", "stop_time",
        "surtim_id", "cen_id")
```

```
    y[, 1] <- c(1:n)
    y[, 2] <- dat$t_L
    y[, 3] <- dat$t_R
    y[, 4] <- 1
    y[, 5] <- dat$delta

    X <- matrix(0, n, 2)
    X <- dat[, 5:6]

    # fit model
    control <- AH_MPL.control(n,
        smooth = 1000, n.obs_basis = 50,
        max.iter = c(100,
            5000, 50000),
        tol_1 = 1e-05, tol_2 = 1,
        tau = 1000, min.theta = 1e-10)

    fit <- AH_MPL(y, X, control)

    # regression
    # coefficients
    save[s, 1] <- fit$Beta[1]
    save[s, 2] <- fit$Beta[2]

    save[s, 3] <- fit$se.beta[1]
    save[s, 4] <- fit$se.beta[2]

    # baseline hazard
    # estimation
    h0.ind <- c(round(quantile(1:length(fit$bins$Alpha),
        c(0.25, 0.5, 0.75))))

    save[s, 5] <- c(fit$Theta)[h0.ind[1]]
    save[s, 6] <- c(fit$Theta)[h0.ind[2]]
    save[s, 7] <- c(fit$Theta)[h0.ind[3]]

    save[s, 8] <- (fit$bins$Alpha[-length(fit$bins$Alpha)])[h0.ind[1]]
    save[s, 9] <- (fit$bins$Alpha[-length(fit$bins$Alpha)])[h0.ind[2]]
    save[s, 10] <- (fit$bins$Alpha[-length(fit$bins$Alpha)])[h0.ind[3]]

    save[s, 11] <- sqrt(diag(fit$cov.theta))[h0.ind[1]]
    save[s, 12] <- sqrt(diag(fit$cov.theta))[h0.ind[2]]
    save[s, 13] <- sqrt(diag(fit$cov.theta))[h0.ind[3]]

    save_h0 <- cbind(save_h0,
        fit$bins$Alpha[2:21],
        c(fit$Theta)[1:20])
    # select length of
    # alpha and theta to
    # save based on
    # sample size &
    # number of basis
    # functions used

}
```

Summary of the MPL estimation results is given in Table 8.2. It is clear that the bias increases with the censoring proportion.

TABLE 8.2: BIAS, MCSD, AASD and MSE for assessing the MPL method.

| | | | |
|---|---|---|---|
| $\pi_c$ | | 20% | 80% |
| $n_c$ | | 2 | 2 |
| $\beta_1 = 0.5$ | BIAS | -0.336 | -0.662 |
| | MCSD | 0.392 | 0.671 |
| | AASD | 0.379 | 0.611 |
| | MSE | (0.275) | (0.803) |
| $\beta_2 = -1.5$ | BIAS | 0.137 | 0.756 |
| | MCSD | 0.618 | 0.517 |
| | AASD | 0.342 | 1.034 |
| | MSE | (0.134) | (0.834) |

□

**Example 8.9** (Application to an AIDS data).
In this example, we apply MPL to fit an additive hazards model to the AIDS data studied in Lindsay & Ryan (1998). This dataset has also been analyzed in Lindsay & Ryan (1998) using a proportional hazard model. This data set is available on the GitHub web site for this book.

Despite its small sample size ($n = 31$), this dataset contains interval, left and right-censored observations. Its wide censoring intervals pose challenges for model fitting tasks. The dataset pertains to the development of drug resistance, measured using a plaque reduction assay, to zidovudine in patients.

An additive hazard model is appropriate if the study's focus is on assessing covariate effects on changes in the hazard of drug resistance from the baseline. This type of analysis can offer more direct and meaningful insights than the log hazard ratios provided by the proportional hazard model.

The patients were enrolled in four clinical trials for the treatment of AIDS, and blood samples were collected at scheduled visit times dictated by the four protocols. Thus, the time origin of each patient was the starting time of treatment with the drug zidovudine.

These survival data exhibit wide censoring intervals due to the limited number of assessments on each patient, driven by the high costs of the plaque reduction assay. Among the 31 patients, 13 are left-censored, 5 are interval-censored and 13 are right-censored. None of them has an exact event time.

There are four covariates which are believed to be related to development of resistance:

$X_1 =$ stage of disease,
$X_2 =$ dose of zidovudine ,

$X_3$ = CD4 lymphocyte counts at the time of randomization with CD4 $\in [100, 399]$ and
$X_4$ = CD4 $\geq 400$.

For patient $i$, we defined $T_i$ to be the time from starting treatment to development of resistance.

We again used an equal number of observations with $n_c = 2$ in each bin, where for the interval censored times their middle points were used for calculating $n_c$.

We first fit an additive hazard model by the MPL method and then, as a comparison, fit an AH model again using the R "adhazard" package (Lin & Ying's method).

```
# Read in and set up data for
# analysis
Lindsey <- read.csv("Lindsey.csv")

n <- 31
y <- matrix(0, n, 5)
colnames(y) <- c("Id", "star_time",
    "stop_time", "surtim_id", "cen_id")
y[, 1] <- c(1:31)
y[, 2] <- Lindsey[, 1]
y[, 3] <- Lindsey[, 2]
y[, 4] <- 1
y[, 5] <- Lindsey[, 3]
X <- matrix(0, 31, 4)
X <- Lindsey[, 4:7]

# Fit AH model
control <- AH_MPL.control(n, smooth = 1000,
    n.obs_basis = 2, max.iter <- c(100,
        5000, 50000), tol_1 = 1e-05,
    tol_2 = 1, tau = 1000, min.theta = 1e-10)
fit <- AH_MPL(y, X, control)

# Fit L&Y Model
library(addhazard)
head(Lindsey)
Lind.cs <- data.frame(y = Lindsey[,
    2], delta = as.numeric(Lindsey[,
    3] > 0), x1 = Lindsey[, 4], x2 = Lindsey[,
    5], x3 = Lindsey[, 6], x4 = Lindsey[,
    7])


fit.Lind = coxph(Surv(y, delta) ~ tt(x1) +
    tt(x2) + tt(x3) + tt(x4), data = Lind.cs,
    tt = function(x, t, ...) -t * x)
summary(fit.Lind)
```

Table 8.3 reports the MPL estimates and it also contains the results from the "`adhazard`" package.

For the MPL approach, none of the covariates are significant; while for the Lin&Ying method, only "stage" is significant. The additive hazard model model explains that an advanced disease stage at the baseline is likely to increase the risk of drug resistance while a high CD4 count at the baseline indicates a reduced risk.

Figure 8.3 displays the baseline hazard estimate, with its 95% asymptotic piecewise CIs. Clearly, the additive hazard model explains that the baseline hazard gradually reduces with time. This phenomenon matches our expectations as longer a patient stays in the study means the drug is still effective for the patient, and therefore less chance of drug resistance.

TABLE 8.3: Regression coefficient estimates given by the MPL method with asymptotic standard deviations (astd), $p$-values and 95% confidence intervals for $n_c = 2$.

| Method | Effects | $\hat{\beta}$ | astd | $p$-value | 95% C.I |
|---|---|---|---|---|---|
| MPL | stage | 0.02768 | 0.0307 | 0.3666 | (-0.0324, 0.0878) |
| | dose | 0.0000 | 0.0243 | 0.9992 | (-0.0475, 0.0476) |
| | CD4: 100-399 | -0.04472 | 0.0604 | 0.4592 | (-0.1631, 0.0737) |
| | CD4: $\geq$ 400 | -0.04469 | 0.0563 | 0.4273 | (-0.1550, 0.0657) |
| L&Y | stage | 1.5900 | 0.6834 | 0.0200 | (0.2505, 2.9295) |
| | dose | 1.3829 | 0.8060 | 0.0862 | (-0.1968, 2.9626) |
| | CD4: 100-399 | 0.4301 | 0.8901 | 0.6289 | (-1.3145, 2.1748) |
| | CD4: $\geq$ 400 | 0.7080 | 0.8905 | 0.4266 | (-1.0374, 2.4534) |

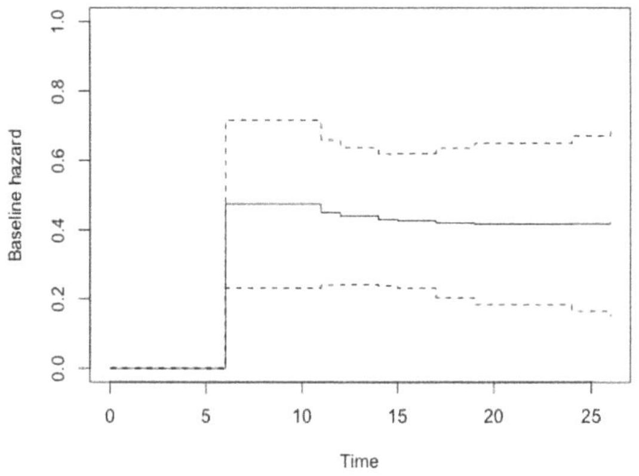


FIGURE 8.3: Plot of baseline hazard estimates from the additive hazard model baseline hazard using the MPL method.

□

## 8.11 Summary

In this chapter, we have discussed the additive hazards model and its fitting process. This model is suitable when the data analyst's focus is on hazards differences rather than hazard ratios. Traditional methods for fitting the additive hazards model, such as Aalen's method Aalen (1989) or counting process methods (e.g., Lin & Ying (1994), Lin et al. (1998), Wang et al. (2010)), typically overlook the constraints required by the additive hazards model. In contrast, the MPL method discussed in this chapter offers greater versatility as it can handle more general partly interval-censored survival data and takes into account all the necessary constraints for the model. Due to the complexity of these constraints, we employ a primal-dual interior point algorithm to solve the constrained optimization problem.

# 9

# *Parametric survival models for competing risks data*

In this chapter, we study some alternatives to semi-parametric proportional hazard models in the context of competing risks.

## 9.1 Introduction to parametric survival models

Common parametric models for univariate survival outcomes are described in Kalbfleisch & Prentice (2002, Chapter 3). Gamma, Weibull, inverse-Gaussian and log-Normal distributions may be specified for the underlying time to event. Multivariate families of distributions are available for use with multivariate survival (Hougaard 2000).

For competing risks, where an outcome event may be of different types (*causes* or *modes*), parametric models allow simple estimation and interpretation including extrapolation of long-term event probabilities, which are of inherent interest and which cannot generally be identified from nonparametric and semiparametric models (Jeong & Fine 2007).

It is sometimes possible that there exists an unobservable correlation between the competing events. This motivated development of survival models inducing correlation between competing risks. Study of the effect of correlation between competing events presupposes the existence of a joint distribution of latent variables. These are latent failure times of each cause. However, this joint density is non-identifiable (Tsiatis 1975) so estimation of correlation from competing events data is not possible in a fully nonparametric context. A common response to ill-posedness of an inverse problem – here due to non-identifiability of the joint distribution – is imposing additional constraints (O'Sullivan 1986). While imposing the assumption of independent competing risks results in the equality of cause-specific hazards (available from cause-specific modelling) and marginal hazards, this constraint is too strong for many applications. Indeed, as noted by Liu (2012), the modelling of dependent failure modes remains a challenging area of importance to survival. A much less restrictive assumption is a parametric model, such as multivariate normal distribution for a flexible monotone power transformation of time to event, as proposed by Deresa & Van Keilegom (2020) ("DVK"). Identifiability is fully restored under mild

DOI: 10.1201/9781351109710-9 

conditions by such an assumption (DVK Theorem 2.1) or by imposing other constraints such as the assumptions of copula models (Emura & Chen 2018). Both approaches therefore offer the pathway to fitting models including correlation.

A useful alternative to proportional hazards assumptions (Cox 1972) in survival analysis is the accelerated failure time (AFT) model (Wei 1992). Indeed Hougaard (1999), in a general review, concludes that a major drawback of the widely used proportional hazards model is that the model and its estimate of relative risk are not robust towards neglected covariates. However, parametric AFT models are robust, motivating the consideration with correlated bivariate survival outcomes of AFT models.

While Dignam & Kocherginsky (2008) and Dignam et al. (2012) study effects of correlation comparing Cox models for cause-specific hazards with Fine-Gray regression models for CIFs by generating simulated competing risks data with constant hazards of event types, similar comparisons of Cox PH models with AFTs have not been conducted until recently. In simulations using Weibull distributions to generate data, both PH assumptions and AFT assumptions hold simultaneously, so estimation methods are directly comparable in this case. The bivariate normal censored (BNC) linear model of Gares et al. (2019) provides another context for comparison of PH and AFT models, with findings we study below.

## 9.2 Direct parametric modelling of the CIF

With competing risks, Jeong & Fine (2006) specify direct parametric models for the cumulative incidence functions (CIFs) $F_r$, suggesting an improper Gompertz distribution suitable for a cure model. This approach can simplify estimation of otherwise unidentifiable asymptotes of event probabilities and reduce parameters over alternatives.

The cumulative incidence function specifies the cumulative probability of an event of interest in the presence of other competing events. A simple Gompertz model or other direct parametric model for the cumulative incidence function simplifies interpretation and is more natural than the usual cause-specific hazard parameterization. Maximum likelihood analysis can be used to estimate simultaneously parametric models for cumulative incidence functions of each event cause.

Jeong and Fine's direct formulation allows choice of parametric model, not restricted to Gompertz, for the CIF $F_r$. Using standard parametric models for survival functions with an additional parameter for the levelling-off point (which may be less than 1) of the (sub)distribution, provides a simpler specification than using a mixture model (Larson & Dinse 1985).

For $K = 2$ competing event causes, the observable data are denoted as $(t_i, \delta_{1i}, \delta_{2i})$ for $i = 1, \ldots, n$, where $t_i$ is a value from the random variable $T_i = \min(Y_{i1}, Y_{i2}, C_i)$ and $\delta_{1i}, \delta_{2i}$ indicate censoring status of each event mode. Parametric inference for the cumulative incidence functions will use the likelihood function

$$L = \prod_{i=1}^{n} f_1(t_i)^{\delta_{1i}}\, f_2(t_i)^{\delta_{2i}}\, S(t_i)^{1-\delta_{1i}-\delta_{2i}},$$

where $S(t)$ is the overall survival function and $f_r(t) = dF_r(t)\,/\,dt$, $r = 1, 2$, is a possibly improper probability density function for the $r$th cause-specific event. Formulating the likelihood in terms of the cumulative incidence functions assumes potential failure times $Y_{ir}$, $r = 1, 2$ for different cause-specific events and differs from the traditional formulation using cause-specific hazard functions in Prentice et al. (1978).

The cumulative incidence function (Kalbfleisch & Prentice 2002) quantifies the cumulative probability of cause-specific failure in the presence of competing events without assumptions about the dependence among the events (Korn & Dorey 1992, Pepe & Mori 1993, Gaynor et al. 1993). In their paper Jeong and Fine established that group effect estimates are consistent and asymptotically unbiased, and their finite properties have been verified regardless of correlation structure.

An alternative to the specification of models of cumulative incidence of outcomes by mode (cause) of outcome is to instead specify the cumulative incidence of time to first event, together with the probability of the mode of that event, conditional on that time. Bryant & Dignam (2004) proposed a semiparametric inference procedure by parameterizing only the cause-specific hazard function $H_r$ in the cumulative incidence function,

$$F_r(t) = \int_0^t S(u)\, dH_r(u),$$

with $S(t)$ estimated nonparametrically.This approach is easy to implement with parametric forms of distributions $H_r$ (Weibull, logistic, ...). The probability of mode $r$, conditional on an event having occurred at time $t$,is calculable from hazards for all modes $r = 1, \ldots, K$ at that time, and the probability of any such event at time $t$ from the multivariate survival distribution. Or each conditional probability of a specified mode can be fitted empirically as a smooth function of time in GLMs involving covariates including time.

## 9.3 Univariate parametric AFT models

In univariate AFT models, use of parametric models in clinical studies is more common than use of semi-parametric approaches. This differs from the situation for proportional hazards model.

Bivariate distributions are available to set parametric models for transition rates in multistate models, or (alternatively) to fully specify latent failure times by mode of event.

A classical approach is that of Kalbfleisch and Prentice, see below.

With covariates $\mathbf{x}$, cause-specific hazard function can be modeled using an AFT model of the form

$$h_r(t; \mathbf{x}) = h_{r0}(t\, e^{-\mathbf{x}^\mathsf{T}\boldsymbol{\beta}_r})\, e^{-\mathbf{x}^\mathsf{T}\boldsymbol{\beta}_r},$$

where $\boldsymbol{\beta}_r$ and $h_{r0}(\cdot)$ are the regression coefficients and baseline hazard function, respectively, for the $r$-th failure type (Kalbfleisch & Prentice 2002, p.256; Kalbfleisch & Prentice 2002, Ch. 7.2.3, equation (7.13)).

In this situation, the covariates directly influence the failure time of the event of type-$r$. This approach can be expressed as the AFT model

$$\log T = \mathbf{x}^\mathsf{T}\boldsymbol{\beta}_r + \epsilon_r,$$

where $T = \min(Y_1, Y_2, \ldots, Y_K)$, with $Y_r$ denoting the latent failure time due to cause $r$, and $\epsilon_r$ is an error term with an unspecified distribution function. In practice, this model can be estimated by treating all observations with failure due to all other causes except for $r$ as censored. Inferences about a particular $\boldsymbol{\beta}_r$ can be obtained using parametric methods with standard statistical packages.

The regression parameter estimates from this cause-specific AFT model are typically interpreted in the same way as conventional AFT models for survival data in the absence of competing risks. However, when there is dependence between the causes of failure, the effects may reflect the influence of competing events, sometimes in a counter-intuitive way, as demonstrated in simulation studies (Choi & Cho 2019). When analyzing competing risks using cause-specific hazard modeling, the covariate effects obtained may not necessarily reflect the cumulative incidence of a given failure type (Dignam et al. 2012, Putter et al. 2007).

### 9.3.1 Weibull marginals

The most convenient parametric models for continuous-time data are ones for which the cause-specific hazard functions $h_k(t)$ are parameterized, with functionally distinct parameters $\theta_k$. The likelihood then factors, allowing individual maximization of each failure mode in which observations of the corresponding cause with censored status accorded to any other causes (Lawless 2003, Chapter 9.3).

The Weibull distribution, with cause-specific hazard functions

$$h_{r0}(t|\alpha_r, \gamma_r) = \frac{\gamma_r}{\alpha_r}\left(\frac{t}{\alpha_r}\right)^{\gamma_r - 1}$$

then provides an AFT regression model in which we specify, by substituting for parameter $\alpha_r$,

$$\alpha_r \leftarrow \alpha_r(\mathbf{x}) = \exp(\mathbf{x}^\mathsf{T}\boldsymbol{\beta}_r).$$

### 9.3.2 Log-normal marginals

AFTs with lognormal marginals offer readily interpretable regression models for cause-specific median times to failure. A univariate normal censored linear regression with failure-censored data using an iterative least square method is implemented by Schmee & Hahn (1979). Let $S_N(z)$ denote the survival function of the univariate standard normal distribution and $f_N(z)$ the standard normal density. An AFT regression model with baseline hazard function

$$h_{r0}(t) = (1/t\sigma_r)\, f_N\left(\frac{\log t - \mu_r}{\sigma_r}\right) / S_N\left(\frac{\log t - \mu_r}{\sigma_r}\right),$$

substituting for the parameter $\mu_k$

$$\mu_r \leftarrow \mu_r(\mathbf{x}) = \exp(\mathbf{x}^\mathsf{T}\boldsymbol{\beta}_r),$$

is available for competing risks using the approach of Kalbfleisch & Prentice (2002). As simulation with hazards specified as above can be effected by independent sampling of events in a succession of small time intervals, this approach does not induce correlation in times to events by cause, though this can be induced when different causes share common covariates.

## 9.4 Linear survival models with dependent censoring

In simulation studies, we may wish to assess the consequences of correlation between event times of different types. The AFT models provide a useful direct approach to do this.

As discussed above, AFTs with lognormal hazards offer readily interpretable regression models for cause-specific median times to failure.

It is convenient in simulating an AFT model to measure times on a log scale, and to restrict attention to the occurrence of an event of interest (event 1) or the occurrence of a competing event (event 2). While more than two events might be considered, interest in many medical trials is focused on a single primary outcome, with any competing events distracting from the measurement of treatment effect, with few observations to individually model some causes. Thus pooling any competing risks together is common. We restrict discussion to two event causes.

Since a competing risk is an event whose occurrence precludes the occurrence of the primary event of interest, only the first-occurring event is observed; the observed time to an event is the minimum of two correlated times. Follow-up of events of types 1 and 2 are both subject to independent censoring at time $C$; both event types are subject to the same censoring. This censoring is assumed to be non-informative.

Therefore, we observe $(T_i, D_i)$ for $i = 1 \ldots n$, with $T_i = \min(Y_{i1}, Y_{i2}, C_i)$, and $D_i$ indicating event type or censoring. Here

$$D_i = \begin{cases} 1: & \text{if the event of interest is observed,} \\ 2: & \text{if an event of a competing risk is observed,} \\ 0: & \text{if no event is observed during follow-up.} \end{cases}$$

Here, $C_i$ denotes the time of censoring.

The observed data is a random sample consisting of observations $(t_i, d_i)$, for $i = 1 \ldots n$, where $t_i$ is the observed $T_i$, and $d_i \in \{0, 1, 2\}$ is an observed $D_i$. Corresponding latent survival times constitute the latent data matrix $\mathbf{Y}$ with $i$th row $(Y_{i1}, Y_{i2})$.

Note that for the rest of this chapter, we will focus on log time scale and the corresponding new notations will be introduced in the next subsection.

### 9.4.1 The BNC model for competing risks

Let $Y_r^* = \log Y_r$ for $r = 1, \ldots, K$ and $C^* = \log C$. For two competing risks ($K = 2$) assume latent event times comprise regression pairs with the usual bivariate normal model assumptions, including a general covariance matrix $\mathbf{\Sigma}$ common to all subjects.

We refer to this as a bivariate normal censored (BNC) linear model for competing risks. The BNC model, EM algorithm, and estimation properties of this model were introduced in a conference paper (Hudson et al. 2018). In this and subsequent sections of this Chapter we present details of the approaches and simulation findings of this conference paper.

Specifically, assume that *latent* observations are an i.i.d. sample of size $n$ from random vector $(Y_1^*, Y_2^*)^\top$ with the conditional distribution, given $X_1 = x_1, \ldots, X_p = x_p$, bivariate normal with mean vector $\boldsymbol{\mu} = (\mathbf{x}^\top \boldsymbol{\beta}_1, \mathbf{x}^\top \boldsymbol{\beta}_2)^\top$ and covariance matrix $\mathbf{\Sigma}$, of dimension $2 \times 2$. Here $\mathbf{x} = (x_1, \ldots, x_p)^\top$. Regression coefficient vectors $\boldsymbol{\beta}_1, \boldsymbol{\beta}_2$ form the columns of a $p \times 2$ matrix $\mathbf{B}$. Assume hereafter that

1. $(Y_1^*, Y_2^*)^\top$ and $C^*$ are conditionally independent given $X = (X_1, \ldots, X_p)^\top$, and that
2. $Y_1^* - \mu_1, Y_2^* - \mu_2$, and $C^*$ are independent of $X$.

We define data matrices $\mathbf{Y}^*, \mathbf{X}$, where the rows of $\mathbf{Y}^*$ are given by latent values $\mathbf{y}_i^{*\top} = (y_{i1}^*, y_{i2}^*)$ and rows of $\mathbf{X}$ are given by observed covariates values $\mathbf{x}_i^\top = (x_{i1}, \ldots, x_{ip})$. Then, in matrix-vector notation, the data matrices $\mathbf{Y}^*, \mathbf{X}$ provide expected values: $E(\mathbf{Y}^*) = \mathbf{M} = \mathbf{XB}$, where $\mathbf{Y}^*$ is of dimension $n \times 2$ and $\mathbf{X}$ is of dimension $n \times p$. Here $\mathbf{M}$ is the matrix of subject-specific means for each cause, so that in row $i$ the means for subject $i$ are the $i$th row of $\mathbf{M}$, denoted as $(m_{i1}, m_{i2})$. The BNC linear model specifies, for known covariate matrix $\mathbf{X}$, the means for subject $i$ as the linear model $(m_{i1}, m_{i2}) = \mathbf{x}_i^\top \mathbf{B}$. The covariance matrix $\mathbf{\Sigma}$, of dimension $2 \times 2$, with correlation coefficient $\rho$, is assumed common to all subjects.

Our R package `bnc` follows this approach to provide simulation data for bivariate normal (BVN) competing risks under the BNC model (Hudson et al. 2019).

### 9.4.2 The DVK model

Deresa & Van Keilegom (2020) introduced a more general model which includes the BNC model above. Again survival is measured on the log scale, e.g. $Y^*, C^*$ take values in $(-\infty, +\infty)$. Firstly, they assume a multivariate normal model, without our restriction to bivariate data. Secondly, a parameterized transformation model is applied to the log survival times of each cause. Finally, covariates are specified to be cause-specific; so for bivariate observations, expected values of subject $i$ with covariate vectors $\mathbf{x}_{i1}$ for cause 1, $\mathbf{x}_{i2}$ for cause 2, are: $m_{ir} = \mathbf{x}_{ir}^{\mathsf{T}} \boldsymbol{\beta}_r$, for $r = 1, 2$. Thus, in general for $K$ risks, the DVK model is

$$\Lambda_\alpha(Y_{ir}^*) = \mathbf{x}_{ir}^{\mathsf{T}} \boldsymbol{\beta}_r + \epsilon_{ir}, \tag{9.1}$$

for causes $r = 1, \ldots, K$, where $\Lambda_\alpha(\cdot)$, parameterized by $\alpha$, is a power transformation family. The error vector $(\epsilon_{i1}, \ldots, \epsilon_{iK})$ is multivariate normal, with mean vector $\mathbf{0}$ and $K \times K$ covariance matrix $\boldsymbol{\Sigma}$.

Deresa & Van Keilegom (2020) demonstrated the identifiability of the DVK model under specified conditions. The BNC model, being a particular case of the DVK model, shares this identifiability.

In order to simplify notation when describing data generation and formulating our EM algorithm we confine discussion to the BNC model ($m = 2$) with all covariates for each cause in common and simple log transformation of survival ($\Lambda(y^*) = y^*$). However, note that generalization of this EM algorithm for the DVK model with cause-specific covariates is possible.

### 9.4.3 Cause-specific hazards of BNC competing events

In the following discussions we adopt a log time scale, so that $Y^*$ means $\log Y$ and $t^*$ means $\log t$. With data on the log scale, the cause-specific hazard function to a first-occurring event of cause 1 at $t^* = \log t$ is denoted by $\lambda_1(t^*)$ and to an event of cause 2 by $\lambda_2(t^*)$. Under the bivariate normal joint distribution of $Y_1^*, Y_2^*$, the function $\lambda_1(t^*)$ is

$$\begin{aligned} \lambda_1(t^* | \mathbf{B}, \boldsymbol{\Sigma}) &= \lim_{\delta t \to 0} \frac{1}{\delta t} P\left[Y_1^* \in (t^*, t^* + \delta t) | Y_1^* > t^*, Y_2^* > t^*\right] \\ &= \lim_{\delta t \to 0} \frac{P\left[Y_1^* \in (t^*, t^* + \delta t), Y_2^* > t^*\right] / \delta t}{P\left[Y_1^* > y, Y_2^* > t^*\right]}. \end{aligned} \tag{9.2}$$

Let $\sigma_1, \sigma_2$ denote standard deviations of $Y_1, Y_2$. Define standardized values $a, b$ of $t^*$ under the two marginals by $a = (t^* - \mu_1)/\sigma_1$, $b = (t^* - \mu_2)/\sigma_2$, where $\mu_1, \mu_2$ are respectively the expectations of $Y_1^*, Y_2^*$, dependent on the realized values of covariates $X_1, \ldots, X_p$. Then the numerator of the above expression for the hazard function becomes

$$\begin{aligned} &\lim_{\delta t \to 0} \frac{1}{\delta t} P\left[Y_1^* \in (t^*, t^* + \delta t), Y_2^* > t^*\right] \\ &= f_{Y_1^*}(t^*)\; P\left[Y_2^* > t^* \,|\, Y_1^* = t^*\right] \\ &= \sigma_1^{-1}\, f_N(a)\, S_N\left(\frac{b - \rho a}{\sqrt{1 - \rho^2}}\right). \end{aligned}$$

Here, $f_{Y_1^*}(\cdot)$ represents the density function of $Y_1^*$, $f_N(\cdot)$ and $S_N(\cdot)$ are respectively the density and survival function of the standard normal distribution $N(0,1)$, and $\rho$ is the correlation between $Y_1^*$ and $Y_2^*$. The final equality for the numerator follows because the conditional distribution of $Y_2^*$ given $Y_1^*$ is univariate normal:

$$[Y_2^*|Y_1^* = t^*] \sim N\left(\mu_2 + \rho\frac{\sigma_2}{\sigma_1}(t^* - \mu_1),\ \sigma_2^2(1-\rho^2)\right).$$

Thus, defining the joint bivariate normal survival function as $S_{12}(t_1^*, t_2^*) = P\left[Y_1^* > t_1^*, Y_2^* > t_2^*\right]$, we have

$$\lambda_1(t^*|\mathbf{B}, \mathbf{\Sigma}) = \sigma_1^{-1} f_N(a)\ S_N\left(\frac{b - \rho a}{\sqrt{1-\rho^2}}\right) / S_{12}(t^*, t^*). \qquad (9.3)$$

The corresponding hazard $\lambda_2(t^*|\mathbf{B}, \mathbf{\Sigma})$ for time to observe the competing risk is readily obtained by exchanging $a$ and $b$ and also $\sigma_1$ and $\sigma_2$.

The sum of these hazards, $\lambda_1(t^*) + \lambda_2(t^*)$, equals the hazard of time to the first event. The ratio $\lambda_r(t^*)/(\lambda_1(t^*) + \lambda_2(t^*))$ determines the conditional probability of event type $r$ $(r = 1, 2)$ for first events at $t^*$, i.e. at log time $t^*$.

### 9.4.4 BNC data generation

Observations censored by a competing risk and following the BVN model for two competing risks can be directly simulated using the cause-specific hazard functions $\lambda_1(\cdot), \lambda_2(\cdot)$ above(Allignol et al. 2011).

Alternatively, using a latent failure approach (see Crowder 1991), first generate a random sample of $n$ bivariate observations $(z_{i1}, z_{i2})^\top$, $i = 1, \ldots, n$, from the BVN distribution with mean vector $\mathbf{0} = (0,0)^\top$, variance matrix $\mathbf{R}$ given by

$$\mathbf{R} = \begin{pmatrix} 1 & \rho \\ \rho & 1 \end{pmatrix}. \qquad (9.4)$$

Actually, this $\mathbf{R}$ is a correlation matrix, and it is related to the covariance matrix $\mathbf{\Sigma}$ via:

$$\mathbf{R} = \mathbf{W\Sigma W},$$

where $\boldsymbol{W}$ is a diagonal matrix, $2 \times 2$, with entries $\sigma_1^{-1}, \sigma_2^{-1}$. Based on this, we can convert $(z_{i1}, z_{i2})^\top$ into a vector $(y_{i1}^*, y_{i2}^*)^\top$ that follows a bivariate normal with mean $\boldsymbol{\mu}_i = (m_{i1}, m_{i2})^\top$ and covariance matrix $\mathbf{\Sigma}$ via

$$(y_{i1}, y_{i2})^\top = (m_{i1}, m_{i2})^\top + \mathbf{W}^{-1}(z_{i1}, z_{i2})^\top.$$

Now we can represent the collection of all $(y_{i1}^*, y_{i2}^*)$ by an $n \times 2$ matrix $\mathbf{Y}^*$ where its $i$th row is given by $(y_{i1}^*, y_{i2}^*)$. In fact, let $\mathbf{Z}$ be a $n \times 2$ matrix whose $i$th row is $(z_{i1}, z_{i2})$. Another way to express $\mathbf{Y}^*$ is as follows. Specifying covariate values in the $n \times p$ matrix $\mathbf{X}$ and regression coefficients $\boldsymbol{\beta}_1, \boldsymbol{\beta}_2$ in a $p \times 2$ matrix $\mathbf{B}$, a bivariate latent data matrix $\mathbf{Y}^*$, is then given as

$$\mathbf{Y}^* = \mathbf{XB} + \mathbf{ZW}^{-1}. \qquad (9.5)$$

The *observed* data consists of $n$ observations $(t_i^*, d_i)$, for $i = 1 \ldots n$, where $t_i^*$ is an observed value of the random variable $T_i^* = \min(Y_{i1}^*, Y_{i2}^*, C_i^*)$, where $C_i$ is a random censoring time (log scale) following a distribution (such as uniform or exponential distribution) that is independent of $(Y_{i1}^*, Y_{i2}^*)$, and $d_i \in \{0, 1, 2\}$ indicates which is the first event (or 0 if no event).

### 9.4.5 BNC likelihood function and inference

In this section, we set a notation for censoring outcomes. We then define a likelihood function for BNC observations censored by a competing risk.

In competing risks data all observations are subject to censoring, not only by end of follow-up but also by competing events. When $d_i = 1$, time to event 2 is censored by an observed event of cause 1; i.e. $Y_{i1}^* = t_i^*$ is observed and $Y_{i2}^* > t_i^*$ is censored. Similarly, when $d_i = 2$, $Y_{i2}^* = t_i^*$ is observed and $Y_{i1}^* > t_i^*$ is censored. Finally, $d_i = 0$ when times to both events exceed the period of follow-up: $Y_{i1} > t_i^*,\ Y_{i2} > t_i^*$ for observed end time of follow-up $T_i^* = t_i^*$. The distribution of $C_1, \ldots, C_n$ need not be included in likelihood calculations when independent censoring is non-informative.

Hence the *Likelihood function* for competing risks observations $(T_i^* = t_i^*, D_i = d_i;\ i = 1, \ldots, n)$ subject to non-informative independent censoring by $C$ may be defined as:

$$\begin{aligned} &L(\mathbf{B}, \boldsymbol{\Sigma};\ \mathbf{t}^*, \mathbf{d}) \\ &\quad = \prod_{i:d_i=1} f_{Y_1^*}(t_i^*)\, P\,[Y_2^* > t_i^* \,|\, Y_1^* = t_i^*] \\ &\qquad \times \prod_{i:d_i=2} f_{Y_2^*}(t_i^*)\, P\,[Y_1^* > t_i^* \,|\, Y_2^* = t_i^*] \times \prod_{i:d_i=0} S_{12}(t_i^*, t_i^*) \\ &\quad = \prod_{i:d_i=1} f_{Y_1^*}(t_i^*)\, S_{2|1}(t_i^* | t_i^*) \\ &\qquad \times \prod_{i:d_i=2} f_{Y_2^*}(t_i^*)\, S_{1|2}(t_i^* | t_i^*) \times \prod_{i:d_i=0} S_{12}(t_i^*, t_i^*), \end{aligned} \tag{9.6}$$

where $\mathbf{t}^*$ and $\mathbf{d}$ are respectively vectors for $\{t_i^*\}$ and $\{d_i\}$,

$$S_{1|2}(a|b) = P\{Y_1^* > a \,|\, Y_2^* = b\}, \tag{9.7}$$

$$S_{2|1}(a|b) = P\{Y_2^* > a \,|\, Y_1^* = b\}, \tag{9.8}$$

and, as before, $S_{12}(a, b) = P\{Y_1^* > a,\ Y_2^* > b\}$.

These probabilities depend on the design matrix $\mathbf{X}$, coefficient matrix $\mathbf{B}$ and covariance matrix $\boldsymbol{\Sigma}$ of the bivariate normal distribution. In the linear model each observation has its own covariate values, a row vector of the covariate matrix $\mathbf{X}$.

Recall that $\mathbf{M} = (\mathbf{m}_1, \mathbf{m}_2)$ be the matrix of expected values of $\mathbf{Y}^* = (\mathbf{Y}_1^*, \mathbf{Y}_2^*)$ where $\mathbf{Y}_r^* = (Y_{1r}^*, \ldots, Y_{nr}^*)^\mathsf{T}$ for $r = 1, 2$ . Then $\mathbf{m}_r = \mathbf{X}\boldsymbol{\beta}_r$ is the vector containing expected times to event $r$. According to (9.5),

$$(\mathbf{Y}^* - \mathbf{M})\mathbf{W} \sim \mathrm{BVN}(\mathbf{0}, \mathbf{R}),$$

where the $2 \times 2$ weight matrix $\mathbf{W}$ is diagonal with entries $\sigma_1^{-1}, \sigma_2^{-1}$, and $\mathbf{R}$ is the correlation matrix specified in (9.4). Thus, given the covariate matrix $\mathbf{X}$, all likelihood terms may be expressed in terms of the standard bivariate normal distribution; the likelihood is a function of $\rho$ and values $\mathbf{z}_1, \mathbf{z}_2, \mathbf{d}$ dependent on the observed random sample. Vectors $\mathbf{z}_1, \mathbf{z}_2$ are themselves functions of $\mathbf{y}^*$ and parameters $\beta_1, \beta_2, \sigma_1, \sigma_2$.

The well-known separability of the likelihood by competing risk (see Ch. 8 of Kalbfleisch & Prentice (2002) and Ch. 9. 1. 2 of Lawless (2003)) in AFT models permits columns of $\mathbf{B}$ to be estimated individually after censoring events of other causes. However, this separability depends on parameters not being common to components of different causes. For the likelihood of the BVN model, some parameters, specifically those of $\mathbf{\Sigma}$, are involved in factors for each cause. Only in the case $\rho = 0$ will the optimization simplify. As noted above, a correlated BVN distribution assumption can be replaced by the lesser assumption of cause-specific hazards $\lambda_1(\cdot), \lambda_2(\cdot)$ defined in equation (9.3) in order to estimate $\mathbf{B}$ by maximum likelihood. However, in the correlated BVN distribution, these cause-specific hazards will be functions of $\mathbf{\Sigma}$, which we treat as unknown. Therefore, the optimization available for $\rho = 0$ is not readily available with correlated competing risks (even when the correlation is known).

## 9.5 An EM algorithm for parametric competing risks models

Experience coding numerical optimization of the likelihood (9.6) to estimate parameters of the BNC model shows lack of convergence is common for routine optimization methods (Newton method, etc). An alternative approach, the EM algorithm, provides assured incremental increases in likelihood during its iterations, leading to satisfactory convergence which can be further accelerated. We describe a novel EM algorithm below for BNC models.

An EM algorithm was introduced for univariate survival by Aitkin (1981) (see also Chapter 1). Our generalization, coded in our R package `bnc`, provides a maximum likelihood (ML) or maximized penalized likelihood (MPL) solution for the BNC linear model. The EM algorithm is used to iterate between imputation of moments of failure times subject to censoring by a competing event or an independent cause (such as loss to follow-up or termination of the study) (E-step) and maximum likelihood estimation (MLE) with these moments (M-step). Complex calculations are evaluated in closed form using a lemma of Stein (Liu 1994).

For the likelihood function (9.6) for *bivariate* observations $(\mathbf{t}^*, \mathbf{d})$, current estimates of the BNC parameters at each step $m$ are $\boldsymbol{\theta}^{(k)} = \left(\mathbf{B}^{(k)}, \mathbf{\Sigma}^{(k)}\right)$. The new estimate $\mathbf{B}^{(k+1)}$ of $\mathbf{B}$ is obtained as

$$\mathbf{B}^{(k+1)} = (\mathbf{X}^\mathsf{T}\mathbf{X})^{-1}\mathbf{X}\mathbf{Y}^{*(k)}$$

with $\mathbf{Y}^{*(k)}$ denoting an imputed $\mathbf{Y}^*$ at iteration $k$. More specifically, for this estimation, for example when $d_i = 2$, so that $Y_{i1}^*$ is censored by the event of cause 2 at $t_i^*$, the censored observation $Y_{i1}^*$ is imputed by

$$Y_{i1}^{*(k)} = E(Y_1^* | Y_1^* > t_i^*,\ Y_2^* = t_i^*;\ \boldsymbol{\mu}_i^{(k)}, \boldsymbol{\Sigma}^{(k)}),$$

where $\boldsymbol{\mu}_i^{(k)} = \mathbf{x}_i \mathbf{B}^{(k)}$. Similarly, when $d_i = 1$, then $Y_{i2}^*$ is imputed by

$$Y_{i2}^{*(k)} = E(Y_2^* | Y_1^* = t_i^*,\ Y_2^* > t_i^*;\ \boldsymbol{\mu}_i^{(k)}, \boldsymbol{\Sigma}^{(k)}).$$

When $d_i = 0$, $Y_{i1}^*$ is imputed by

$$Y_{i1}^{*(k)} = E(Y_1^* | Y_1^* > t_i^*, Y_2^* > t_i^*;\ \boldsymbol{\mu}_i^{(k)}, \boldsymbol{\Sigma}^{(k)}),$$

with a similar expression to impute of $Y_{i2}^*$.

At iteration $k$, a complete data sufficient statistic for the covariance matrix $\boldsymbol{\Sigma}^{(k)}$ is

$$\mathbf{V}^{(k)} = (\mathbf{Y}^{*(k)} - \mathbf{X}\mathbf{B}^{(k)})^{\mathsf{T}}(\mathbf{Y}^{*(k)} - \mathbf{X}\mathbf{B}^{(k)}) = (\mathbf{Y}^{*(k)})^{\mathsf{T}}\mathbf{Q}\mathbf{Y}^{*(k)}$$

for known projection matrix $\mathbf{Q} = \mathbf{I} - \mathbf{X}(\mathbf{X}^{\mathsf{T}}\mathbf{X})^{-1}\mathbf{X}^{\mathsf{T}}$. The EM update to $\boldsymbol{\Sigma}$ therefore includes imputation of quadratic terms (squares and cross-products) in $(Y_1^*, Y_2^*)$. The new estimate is obtained using the censored observation $t_i^*$ by replacing linear terms as above and quadratic terms using appropriate conditional distributions. For $i' \neq i$ the statistical independence of observations reduces calculations to imputation of linear statistics, but for $i' = i$ more complex conditional expectations must be evaluated. For example, when $d_i = 2$, a quadratic term $Y_{i1}^{*2}$ in the censored observation $Y_{i1}$ is imputed using

$$(Y_{i1}^{*2})^{(k)} = E(Y_1^{*2} | Y_1^* > t_i^*, Y_2 = t_i^*;\ \boldsymbol{\mu}_i^{(k)}, \boldsymbol{\Sigma}^{(k)}),$$

and when $d_i = 1$ we need to replace $Y_{i2}^{*2}$ by

$$(Y_{i2}^{*2})^{(k)} = E(Y_2^{*2} | Y_1^* = t_i^*, Y_2 > t_i^*;\ \boldsymbol{\mu}_i^{(k)}, \boldsymbol{\Sigma}^{(k)}).$$

Similarly, when $d_i = 0$, we need to impute $Y_{i1}^{*2}$ and $Y_{i2}^{*2}$ by, respectively,

$$(Y_{i1}^{*2})^{(k)} = E(Y_1^{*2} | Y_1^* > t_i^*, Y_2^* > t_i^*;\ \boldsymbol{\mu}_i^{(k)}, \boldsymbol{\Sigma}^{(k)}),$$
$$(Y_{i2}^{*2})^{(k)} = E(Y_2^{*2} | Y_1^* > t_i^*, Y_2^* > t_i^*;\ \boldsymbol{\mu}_i^{(k)}, \boldsymbol{\Sigma}^{(k)}),$$

and $Y_{i1}^* Y_{i2}^*$ by

$$(Y_{i1}^* Y_{i2}^*)^{(k)} = E(Y_1^* Y_2^* | Y_1 > t_i^*, Y_2 > t_i^*;\ \boldsymbol{\mu}_i^{(k)}, \boldsymbol{\Sigma}^{(k)}).$$

We provide all required moment results for the EM algorithm in a bivariate normal censored linear model in Appendix B, with corresponding R-code in the package `bnc`. Both the EM algorithm and these results for the bivariate normal distribution appear to be new, and may prove useful in contexts other than BNC.

Standard errors of the EM algorithm solution are available using the forward-difference method (FDM) of Jamshidian & Jennrich (2000) and returned by the R package `turboEM`, using function `stderror()` for "turbo" objects (Bobb & Varadhan 2018).

### 9.5.1 Mildly penalized likelihood

As the convergence can be to a boundary $\hat{\rho}_{\text{ML}} = \pm 1$ and consequently the rate of convergence of EM can be very slow, we regularize by penalizing the log-likelihood $\log L = \log L(\mathbf{B}, \mathbf{\Sigma} \mid \mathbf{t}^*, \mathbf{d})$ to replace it as:

$$\log L \leftarrow \log L - (\nu + 3) * \log(\det(\mathbf{\Sigma})/2 - \kappa * \text{tr}(\mathbf{\Sigma}^{-1}))/2,$$

(Green 1990, Barnard et al. 2000).

The `turboEM` algorithm is available for MPL with this penalized objective function. As the convergence of standard EM can be very slow, we use an accelerated EM form (`turboEM` method `squareEM`) The resulting algorithm for bivariate normal MPL estimation is implemented in our R fit using function `bnc::bnc()`.

## 9.6 R packages for parametric survival models

- The `survival::survreg` function fits univariate AFT models for common parametric distributions, including Weibull, lognormal and loglogistic.
- the `aftsrr` function in a contributed R package `aftgee` allows for log-rank and Gehan-type estimations with an extension to multivariate survival data (Chiou, Kang & Yan 2014).
- Our `bnc` function in a contributed R package `bnc` fits the BNC linear model using the EM algorithm to provide an MPL estimator and is available on request from the authors.
- Deresa and Van Keilegom utilize the nonlinear optimization package `nloptr` for likelihood optimization and `numDeriv` for computing the Hessian matrix.

## 9.7 Examples

In this section, we provide two examples where the BNC model can be implemented.

**Example 9.1** (ALT study of time to COVID infection).
Consider two different animal models or human epidemiological studies concerning:

**Study A.** time to infection with the coronavirus (COVID) and its association with influenza infection;

**Study B.** time to COVID reinfection and its association with (time to) a first COVID infection.

Accelerated life tests may be planned in which increasing viral loads of coronavirus are applied in animal subjects. In the first study an illness-death model would provide alternative routes to "COVID" infection either directly from starting “well” state or indirectly after first experiencing influenza (state “flu”). In the second study a three state model would provide pathways to COVID infection and reinfection starting from starting “well” state, first moving to an infected state followed by “recovered” state with final outcome state “reinfected”.

In a multistate model for pathways in the first study the distribution of first event occurring (COVID or flu infection) is that of the minimum of two random variables $T_1, T_2$ with transition rates $h_{13}(t), h_{12}(t)$ specified in parametric form. This model is completed by specifying a third hazard function $h_{23}(t)$ applying for those having experienced flu between infection with influenza and subsequent infection with COVID. In a Markov multi-state model, a hypothesis of interest will be whether there is no residual effect of flu infection: $h_{13}(t) \equiv h_{23}(t)$.

An AFT model simplifies studying effects on times of differing viral loads. Instead of including the covariate (measuring viral load administered to the subject) as a multiplicative factor in the hazard functions, the AFT model introduces a regression coefficient that adjusts the expected times to infection by the covariate.

In order to interpret the effect of flu antibodies, animal studies might consider follow-up on two groups: animals with no pre-existing flu antibodies, and animals who have been vaccinated for flu. Notice that in this context interpretation of the marginal distribution of $T_1$ in a world in which flu has been eliminated is not difficult – it is the distribution of time to COVID infection with an effective flu vaccine. □

**Example 9.2** (BNC for RT benefits in advanced Lung Cancer therapy).
In survival data from a study on oesophogeal cancer, a patient may experience more than one event, including recurrence of the original cancer, new primary cancer and death. Radiation oncologists are often interested in comparing patterns of local or regional recurrences alone as first events to identify a subgroup of patients who need to be treated by radiation therapy after surgery. The cumulative incidence function provides estimates of the cumulative probability of locoregional recurrences in the presence of other competing events. See Jeong & Fine (2006) for illustration of use of direct parametric modelling in Section 9.2. □

## 9.8 Simulation study of the two competing events BNC model

We describe simulation studies conducted to assess the performance of the BNC linear model:

i. to assess the effect of correlation on the estimation of relative increase in survival in two competing events; and
ii. to compare estimates provided by different survival models (BNC linear model and Cox PH).

We study the parametric estimation of treatment benefits in first-event data from correlated bivariate normal competing risks. Performance criteria are sampling distribution (density plots and boxplots displaying quantiles) of ML, MPL, Cox HR and other estimates. In addition, in two-sample (corresponding to two-competing events) datasets of first-event times, we examine sampling distributions of log median ratio (mean difference in log-time to event 1) estimated between treated and control subjects.

## 9.8.1 Simulation design

We generated censored data using the latent variable approach for a BVN linear model. The latent log-survival-times $(Y_1^*, Y_2^*)$ were a random sample of $n$ bivariate normally distributed observations with means $\mathbf{M} = \mathbf{XB}$ determined by the two-sample design, and fixed covariance matrix $\mathbf{\Sigma}$ specified by choices $\sigma_1, \sigma_2$ and $\rho$.

The parameter $\tau$ denotes the maximum length of follow-up and is chosen to fix the censoring proportion $S_1(\tau)$ of events of interest not occurring before end of follow-up, or equivalently the marginal cumulative incidence $F_1(\tau) = 1 - S_1(\tau)$ of events of interest. This proportion $S_1(\tau)$ does not include the censoring of the event of interest by the competing event. Marginal cumulative incidence matches the observed CIF only when the competing event is absent or very late occurring.

In two-sample simulations with design covariates $X_1 = 1$ (constant) and $X_2$ with independent Bernoulli distributed values 0 and 1 (indicating treatment), the randomized trial treatment allocation produces mean vectors that differ only between treatment and control arms. The linear model constant coefficient $\beta_{1r}$ sets baseline survivals while $\beta_{2r}$ sets treatment effects. Then $\boldsymbol{\beta}_r = (\beta_{1r}, \beta_{2r})^\mathsf{T}$, so the mean value matrix $\mathbf{M} = (m_{ir})$ is an $n \times 2$ matrix, with $m_{ir} \in \{\mu_r^C, \mu_r^T\}$ for $r = 1, 2$. Group means $\mu_r^C = \beta_{1r}$ and $\mu_r^T = \beta_{1r} + \beta_{2r}$ were specified, for $r = 1, 2$, as the mean log times to event $r$ applying to all observations in Control and Treatment groups, respectively. We fixed the mean in the Control group to be $\mu_1^C = 0$, and $\sigma_1 = \sigma_2 = 1$. *Treatment effect* is defined as the parameter $\Delta = \mu_1^T - \mu_1^C = \beta_{21}$, the expected difference between Treated and Controls in log-time to the event of interest. Then $\exp(\Delta)$ is the median ratio (MR) on the original time scale. By setting $\beta_{22} = 0$, we specified that the competing event (event 2) has the same risk of occurrence in both arms: $\mu_2^T = \mu_2^C$.

#### 9.8.1.1 Hazard ratios of BVN linear model

Including a treatment indicator in a two-competing risks BVN model provides an estimator of median ratio (MR), the ratio of median survival time in treatment versus control populations. This differs from the hazard ratio. Comparisons are made between the parametric method and standard semi-parametric methods assuming fixed hazard ratios, in particular with Cox's proportional cause-specific hazard model.

As is well known, proportional hazard models differ from AFT models except with Weibull distributional assumptions. Consequently, lognormal modelling implies a time-dependent HR, violating the proportional hazards assumption of cause-specific hazard models. The hazard ratio of Treatment versus Control for time to event 1 (note not log-time) in the case $\rho = 0$ is

$$\mathrm{HR}_1(t) = \frac{\lambda_1^T(t)}{\lambda_1^C(t)} = \frac{\lambda_1(t \mid \Delta, \sigma_1)}{\lambda_1(t \mid 0, \sigma_1)}$$

for hazard function

$$\lambda_1(t|\Delta, \sigma_1) = \frac{1}{t\sigma_1}\lambda_N\left(\frac{\log t - \Delta}{\sigma_1}\right).$$

Here $\Delta$ represents the treatment effect $\beta_{21}$ as defined above and $\lambda_N$ is the hazard function of the (univariate) standard normal distribution (see Appendix B). So, for independent competing causes,

$$\mathrm{HR}_1(t) = \frac{\lambda_N((\log t - \Delta)/\sigma_1)}{\lambda_N(\log t/\sigma_1)}$$

Note this HR is independent of $\beta_{12}$ only when $\rho = 0$.

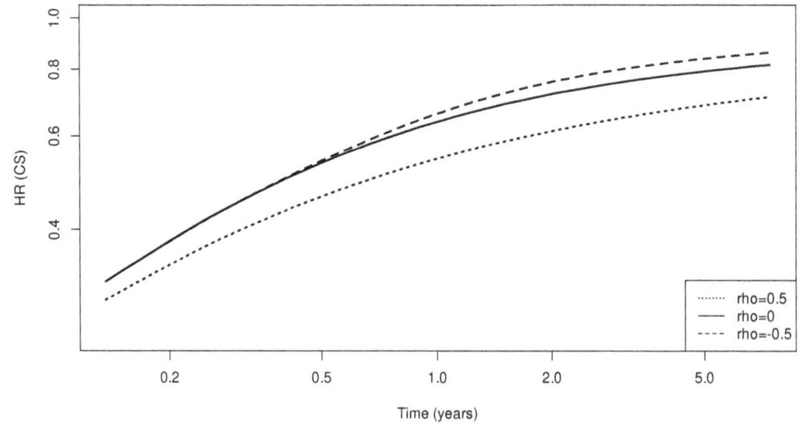


FIGURE 9.1: Hazard ratio of two-sample model with log-normal error distribution varying $\rho \in \{-0.5, 0, 0.5\}$. Hazard ratio is for treatment versus control group with treatment benefit $\beta_{21} = \Delta = 0.5$, with $\sigma_{11} = \sigma_{22} = 1$. Other parameters are $\beta_{11} = \beta_{21} = \beta_{22} = 0$.

For dependent competing risks, the corresponding HR depends also on parameters $\rho$, $\mu_2$ and $\sigma_2$. Figure 9.1 displays the hazard ratio of the lognormal distribution and its dependence on $\rho$ in a two-sample comparison with two competing causes. The mean log times to the event of interest and competing event in the Control group are $\beta_{11} = \beta_{12} = 0$; the mean log time to the event of interest on treatment is $\Delta = \beta_{21} = 0.5$. The mean log time to the competing event is unaffected by treatment ($\beta_{22} = 0$) and $\sigma_1 = \sigma_2 = 1$. In this case (a strongly competitive competing risk unaffected by treatment) the hazard ratio is time-dependent, and smaller for $\rho = 0.5$ than for the other two choices, meaning that treatment benefit is greater.

### 9.8.2 Estimation of $\rho$

The BNC linear model fit provides estimates of all regression parameters and of the covariance matrix $\mathbf{\Sigma}$. The estimation of $\rho$ is a well-posed problem. Hence MPL provides an estimate of $\rho$, the correlation between times to events of each cause.

For correlations between $-0.5$ and $0.5$, we generated 100 simulation samples of size $n = 1000$ and for each determined the ML estimate $\hat{\rho}$ using an EM algorithm from starting estimates $\boldsymbol{\mu}^0 = \mathbf{0}$, $\mathbf{\Sigma}^0 = \mathbf{I}$. Boxplots of ML estimates of $\rho$ returned in our simulation are provided in Figure 9.2. Density plots of the sampling distributions are shown in Figure 9.3. The boxplots demonstrate unbiased estimates of $\rho$ are obtained, despite considerable variability in ML estimation of this parameter when $\rho$ is large, or the fraction of data censored by end of follow-up is high. Density plots reveal that this variability extends to many ML estimates of $\rho$ approaching correlation 1, even in samples of n=1000.

### 9.8.3 2-sample estimation of treatment benefit

In a two-sample comparison, where subjects are randomized equally to treatment and control arms, we place focus on estimation of the treatment benefit for the event of interest. Varying the correlation allows us to assess its effect on estimation of treatment benefit using the BNC linear model or using standard approaches (Cox models and Fine-Gray estimation) assuming proportional hazards.

We simulated data from a clinical trial with two arms, imposing a balanced design with equal numbers $n/2$ of patients in treatment and control groups. The generated data follows an AFT model with expected log-survival-time $\mu_{ir} = \beta_{1r} + \beta_{2r} X_i$, for event cause $r = 1, 2$, subject $i = 1, \ldots, n$, with covariate $X_i$ the indicator of randomized allocation of subject $i$ to a treatment group. This is a particular case of the BNC linear model, where $B$ is a $2 \times 2$ matrix with $r$-th column the intercept and $\beta$-coefficient for cause $r$. A lognormal distribution has expected value $\exp(\mu + \sigma^2/2)$, so the AFT assumption implies the ratio of median survivals for cause 1, $\text{MR}_1 = E(\exp(Y_{i1}|X_i = 1)/E(\exp(Y_{i1}|X_i = 0) = \exp(\beta_{21})$ when variances $\sigma_r^2$ are independent of treatment allocation.

The estimation methods considered were:

1. HR estimates from Cox's partial likelihood estimator ("Cox model") treating competing events as independent censoring. We fitted hazard ratio estimates of event 1 with the function `survival::coxph()`.
2. regression coefficient estimates in the Generalized Estimating Equations (GEEs) fixing correlation: the Likelihood scores fix correlation $\rho = 0$. We assume the competing event is an (additional) independent censoring. We fitted regression coefficients of event 1 with the function `survival::survreg()` which implements MPL solution with $\rho$ fixed, in the particular case $\rho = 0$. To fix $\rho$ to other values, refer to Anderson and Olkin (1985).

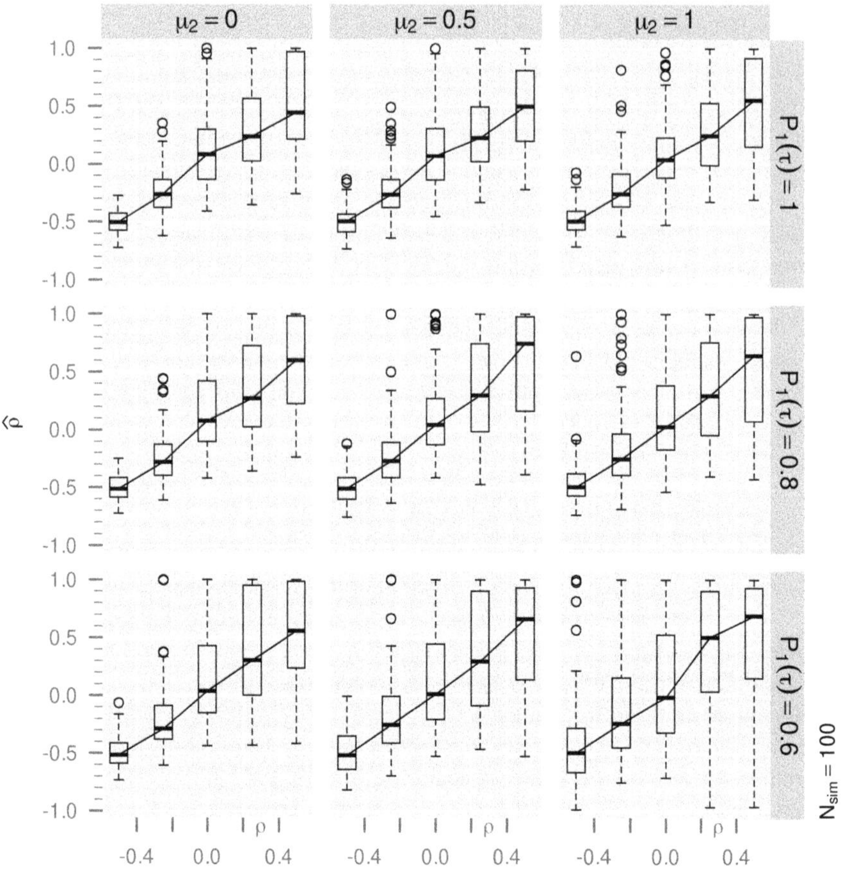


FIGURE 9.2: Boxplots of ML estimates $\hat{\rho}$ of correlation, for simulation sample size $n = 1000$. Parameters of the BNC linear model varied in the graphic are: mean difference in event time between competing event (with mean $\mu_2$ by column) and event of interest (with mean $\mu_1 = 0$ fixed), fraction of first events *not* subject to censoring by end of follow-up $P_1(\tau)$ in (1,0.8,0.6) (by row) and true correlation $\rho$ (x-axis). Fixed parameters are $\sigma_1 = \sigma_2 = 1$. The simulation comprised $N_{\text{sim}} = 100$ samples.

3. ML estimates of regression coefficient $\Delta = \beta_{21}$ in the BNC model obtained using a `squareEM` algorithm. Implementation within our `bnc` package.

A proportional Cox PH model estimates a time-averaged CS HR (Kalbfleisch & Prentice 1981).

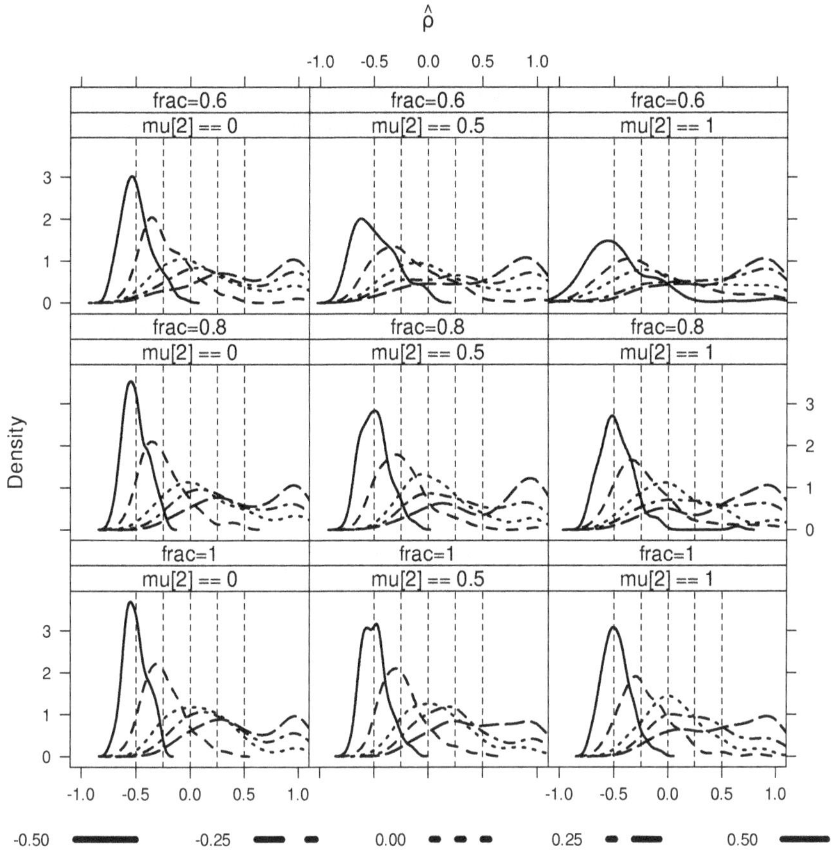


FIGURE 9.3: Density plots of correlation estimates $\hat{\rho}$ in samples of size $n = 1000$. Varying BNC parameters: Fraction (frac) of first events observed, $P_1(\tau) \in (0.6, 0.8, 1)$, by row; $\mu_2 \in (0, 0.5, 1)$, by column; $\rho \in (-0.5, -0.25, 0, 0.25, 0.5)$ coded by line type. Fixed BNC parameters, with cause of event indicated by subscript 1,2: $\mu_1 = 0$, $\sigma_1 = \sigma_2 = 1$, sample size $n = 1000$. The simulation comprised $N_{\text{sim}} = 100$ samples.

It is desirable for purposes of graphical display and comparison to present parameters of the proportional hazards model related to survival benefit $\Delta$. Were the hazard functions of time to event 1 in Treatment and Control arms each constant – $\lambda_1^T$ and $\lambda_1^C$ respectively – then median ratio is $\text{MR} = (1/\lambda_1^T)/(1/\lambda_1^C)$, the reciprocal of the HR for such data. As $1/\text{MR} = \exp(-\Delta)$ the corresponding estimate of $\Delta$ is the negative of coefficient for treatment group in the Cox competing risks model. For comparability of HR and median ratio estimates, we therefore display the estimates of the negative logarithm of the hazard ratio (HR) (for Cox PH models) and of

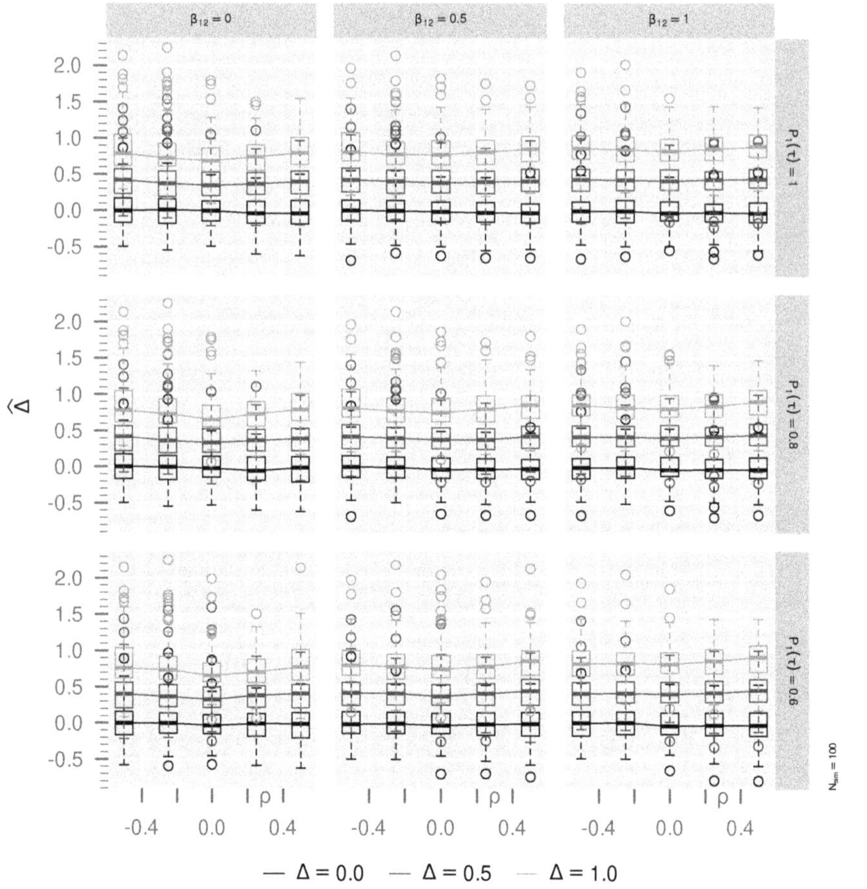


FIGURE 9.4: Boxplot of simulation ML estimates $\hat{\Delta}$ of Treatment Benefit, $n = 100$. Parameters of the BNC linear model varied in the graphic are: $P_1(\tau)$ (rows), mean delay $\beta_{12}$ (columns), correlation $\rho$ (x-axis). Fixed parameters are $\mu_1^C = 0$, $\beta_{22} = 0$ and $\sigma_1 = \sigma_2 = 1$. The simulation comprised $N_{\text{sim}} = 100$ samples.

treatment benefit $\Delta = E(Y_{ir}|X_i = 1) - E(Y_{ir}|X_i = 0) = \beta_1$ (for BNC models and `survfit` models fixing $\rho = 0$).

The matrix plot (mayplot) Figure 9.4 is a graphical display containing boxplots of ML estimates of $\Delta$ for simulated datasets generated under specified parameters. From it we can assess variability in individual simulated data sets, in this case for the smaller sample size $n = 100$. Observe that medians of treatment benefit estimates $\hat{\Delta}$ differ little from true values $\Delta$, though when treatment benefits are strong there is evidence of some bias for large positive $\rho$. The variability (measured by the inter-quartile range) is consistently small.

Figure 9.5 shows that estimates of $\Delta$ obtained under the assumption $\rho = 0$ from use of R package `survreg` are not robust for other $\rho$. The panels, for sample size $n = 100$, show increasing bias of this GEE-like estimator as $|\rho|$ grows.

Comparing this and the previous figure, we conclude that the BNC model provides better estimation of treatment benefits than a particular GEE solution fixing correlation.

Simulation estimates from fitting the Cox model provide Figure 9.6. The Figure displays Cox model estimates of $-\log(\text{HR})$ which increasingly depart from $\Delta$ as the treatment benefit grows, or censoring increases.

### 9.8.4 Findings:

These simulations demonstrate the influence of correlated competing risks on estimation of the treatment benefit in BVN data and show the superior robustness of BNC estimation of treatment benefit to correlation in comparison with Cox PH models.

The novel application of the EM algorithm provides reliable estimation.

In two-sample comparisons, regression coefficients provide estimates of treatment benefit. While even for large sample sizes ($n = 1000$) estimation of correlation is highly variable, the treatment benefit $\Delta$ is accurately estimable even in small sample sizes ($n = 100$).

We have demonstrated that the Cox model, in estimating treatment benefits with strong positive correlation between competing risks, is not robust to a violation of its PH assumption, unless the treatment benefit $\Delta = 0$.

The BNC linear model makes the assumption that the time-to-event distribution is log-Normal. While the parametric form of hazards is difficult to identify from data, simulations suggest the model is robust. (See vignettes of the R-package `bnc`). The model is appropriate for estimating treatment benefits under AFT assumptions.

Our simulations have shown the BNC linear model estimates regression coefficients well, but correlation only crudely. Regression coefficients estimate means and mean differences (relative differences when survival times are log-transformed). While the correlation estimation is unstable it exhibits little bias. The shrinkage of correlation estimates by penalized Likelihood offers potential benefits for improving stability. As asymptotic normality of correlation estimates is not present even in large samples, another approach is required for confidence intervals for $\rho$. Bootstrap methods using repeated samples and accelerated EM estimates provide a practical approach to obtaining a confidence interval for the correlation.

Because of the uncertain ability to estimate $\rho$, our simulations considered an alternative to ML estimates of treatment benefit in the BNC model, fixing $\rho$ (e.g. assuming $\rho = 0$) to provide restricted estimates of other BNC model parameters. This approach exhibited estimation bias, and so did not perform as well as BNC estimation. However, asymptotic properties of this GEE-like procedure may warrant further investigation.

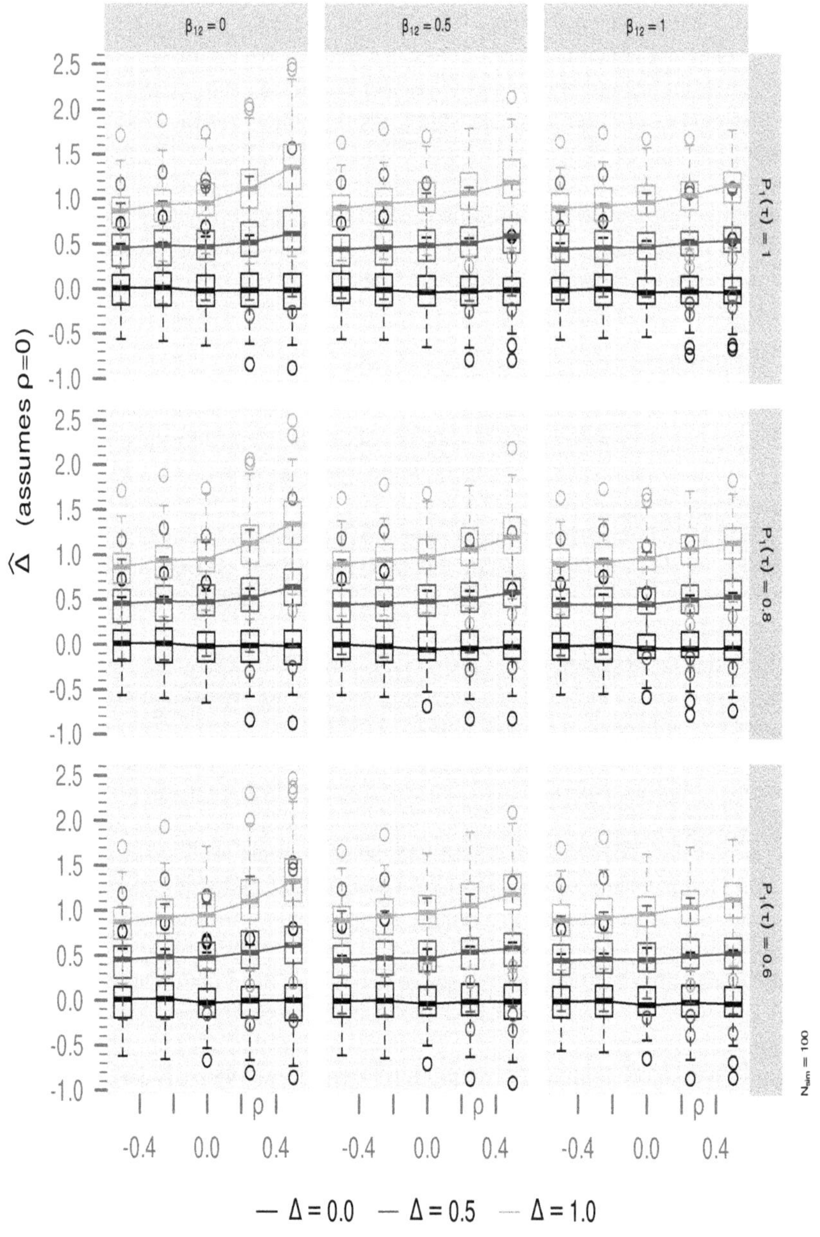


FIGURE 9.5: Boxplots of simulation restricted ML estimates $\hat{\Delta} = \hat{\beta}_{21}$ of Treatment Benefit for different expected benefits $\Delta \in \{0, 0.5, 1\}$ and sample size $n = 100$. Estimation for fixed $\rho = 0$ using R package `survreg`. BNC model parameters are : $P_1(\tau)$ (rows), mean difference $\mu_2$ (columns), correlation $\rho$ (x-axis), $\mu_1 = 0$, $\sigma_1 = \sigma_2 = 1$. The simulation comprised $N_{\text{sim}} = 100$ samples.

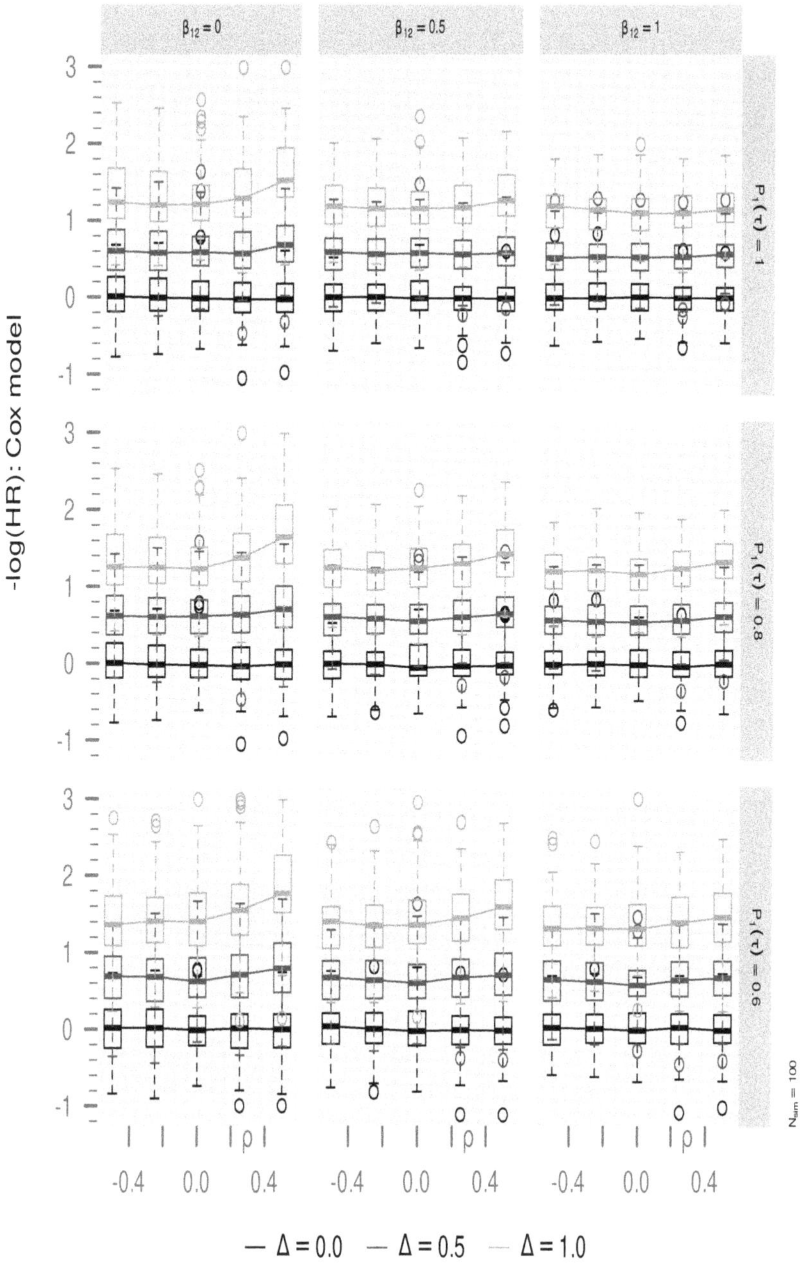


FIGURE 9.6: Boxplots of simulation estimates of the $\beta$-coefficient, i.e.negative log hazard ratio ($-\log \mathrm{HR}$) of the Cox model, $n = 100$, for different value of $\Delta \in \{0, 0.5, 1\}$. BNC model parameters are : $P_1(\tau)$ (rows), mean difference $\mu_2$ (columns), correlation $\rho$ (x-axis), $\mu_1 = 0$, $\sigma_1 = \sigma_2 = 1$. The simulation comprised $N_{\text{sim}} = 100$ samples.

# Appendices

# A

# *Proof of asymptotic results*

This chapter contains the sketched proofs of the asymptotic results presented in Chapter 2.

## A.1 Proof of Theorem 2.4

Our approach below closely follows the proofs given in Xue et al. (2004), Zhang et al. (2010) and Huang (1996).

Recall $\boldsymbol{\tau} = (\boldsymbol{\beta}, h_0(t)) \in B * A$ and $\boldsymbol{\tau}_n = (\boldsymbol{\beta}, h_n(t)) \in B * A_n \subset B * A$, where spaces $B$, $A$ and $A_n$ have already been defined in Section 2.6. Function $h_n(t)$ is an approximation to $h_0(t)$ using the basis functions. The MPL estimator is represented by $\widehat{\boldsymbol{\tau}}_n$. For $\boldsymbol{\tau}_1, \boldsymbol{\tau}_2 \in B * A$, recall that we defined a distance measure in Section 2.6, given by

$$\rho(\boldsymbol{\tau}_1, \boldsymbol{\tau}_2) = \{\|\boldsymbol{\tau}_1 - \boldsymbol{\tau}_2\|^2\}^{1/2} = \left\{ \|\boldsymbol{\beta}_1 - \boldsymbol{\beta}_2\|_2^2 + \sup_{t \in [a,b]} |h_{01}(t) - h_{02}(t)|^2 \right\}^{1/2},$$

and we noted this is a combination of $L_2$ norm and $L_\infty$ norm.

The proofs below require the concept of *covering number* of a space; its definition can be found in, for example, Pollard (1984). Briefly, this is the number of spherical balls of a given size required to cover a given space. For a space $A$ with measure $\kappa(A)$, we denote the covering number associated with a spherical radius $\varepsilon$ by $N(\varepsilon, A, \kappa(A))$.

Results 1 and 2 of Theorem 2.4 can be demonstrated if we can show that $\rho(\boldsymbol{\tau}_0, \widehat{\boldsymbol{\tau}}_n) \to 0$ (a.s.), where $\boldsymbol{\tau}_0 = (\boldsymbol{\beta}_0, h_0(t))$.

Since the smoothing parameter $\mu_n \to 0$ as $n \to \infty$ and the penalty function is bounded, we can ignore the penalty function and concentrate on the log-likelihood function only. The required result can be obtained through the following outcomes.

In the discussions below, to emphasize their dependence on the sample size $n$, we denote an approximate $h_0(t)$ by $h_n(t)$.

(1) Let $q(\boldsymbol{W}; \boldsymbol{\tau})$ denote the Fréchet derivative of the density functional $f(\boldsymbol{W}; \boldsymbol{\tau})$ with respect to $\boldsymbol{\tau}$. Let $\widetilde{\boldsymbol{\tau}}$ be a point in between $\widehat{\boldsymbol{\tau}}_n$ and $\boldsymbol{\tau}_0$. Since $\widetilde{\boldsymbol{\tau}}$ is not the maximum, the functional $q(\boldsymbol{W}; \widetilde{\boldsymbol{\tau}})$ is non-zero. Also,

DOI: 10.1201/9781351109710-A 

both $q(\boldsymbol{W};\widetilde{\boldsymbol{\tau}})$ and $f(\boldsymbol{W};\widetilde{\boldsymbol{\tau}})$ are bounded. According to the definition of $Pl(\boldsymbol{\tau})$ given in Section 2.6, and the fact that $\boldsymbol{\tau}_0$ maximizes $E_0 l(\boldsymbol{\tau};\boldsymbol{W})$, we have

$$\begin{aligned} &|Pl(\widehat{\boldsymbol{\tau}}_n;\boldsymbol{W}) - Pl(\boldsymbol{\tau}_0;\boldsymbol{W})| = E_0(l(\boldsymbol{\tau}_0;\boldsymbol{W}) - l(\widehat{\boldsymbol{\tau}}_n;\boldsymbol{W})) \\ &\geq \|f^{\frac{1}{2}}(\boldsymbol{\tau}_0;\boldsymbol{W}) - f^{\frac{1}{2}}(\widehat{\boldsymbol{\tau}}_n;\boldsymbol{W})\|_2^2 \\ &= \left\| \frac{q(\widetilde{\boldsymbol{\tau}};\boldsymbol{W})}{2f^{\frac{1}{2}}(\widetilde{\boldsymbol{\tau}};\boldsymbol{W})}(\boldsymbol{\tau}_0 - \widehat{\boldsymbol{\tau}}_n) \right\|_2^2 \\ &\geq C_4 \|\boldsymbol{\tau}_0 - \widehat{\boldsymbol{\tau}}_n\|_2^2, \end{aligned} \tag{A.1}$$

where the first inequality is established since the Kullback-Leibler distance is not less than the Hellinger distance (Wong & Shen 1995), the second equality comes from the mean value theorem and $C_4$ ($\neq 0$) is the lower bound of $|q(\widetilde{\boldsymbol{\tau}};\boldsymbol{W})/2f^{\frac{1}{2}}(\widetilde{\boldsymbol{\tau}};\boldsymbol{W})|$.

(2) It then suffices to show $Pl(\widehat{\boldsymbol{\tau}}_n) - Pl(\boldsymbol{\tau}_0) \to 0$ almost surely. However, since

$$|Pl(\widehat{\boldsymbol{\tau}}_n) - Pl(\boldsymbol{\tau}_0)| \leq |Pl(\widehat{\boldsymbol{\tau}}_n) - P_n l(\widehat{\boldsymbol{\tau}}_n)| + |P_n l(\widehat{\boldsymbol{\tau}}_n) - Pl(\boldsymbol{\tau}_0)|, \tag{A.2}$$

we wish to show that each term on the right-hand side of (A.2) converges to 0 almost surely.

For the first term of (A.2), we just need to implement the result from part (3) below, but the second term of (A.2) demands further analyzes. Define $\boldsymbol{\tau}_{0n} = (\boldsymbol{\beta}_0, h_{0n}(t)) \in B * A_n$, where $h_{0n}(t)$ is selected to satisfy $\rho(\boldsymbol{\tau}_{0n}, \boldsymbol{\tau}_0) \to 0$ (when $n \to \infty$), which is guaranteed by Assumption A2.4. Since $\boldsymbol{\tau}_0$ maximizes $Pl(\boldsymbol{\tau})$ for $\boldsymbol{\tau} \in B * A$ and $\widehat{\boldsymbol{\tau}}_n$ maximizes $P_n l(\boldsymbol{\tau})$ for $\boldsymbol{\tau} \in B * A_n$, we have

$$\begin{aligned} P_n l(\widehat{\boldsymbol{\tau}}_n) - Pl(\boldsymbol{\tau}_0) &\geq P_n l(\boldsymbol{\tau}_{0n}) - Pl(\boldsymbol{\tau}_0) \\ &= P_n l(\boldsymbol{\tau}_{0n}) - Pl(\boldsymbol{\tau}_{0n}) + Pl(\boldsymbol{\tau}_{0n}) - Pl(\boldsymbol{\tau}_0); \quad \text{and} \\ P_n l(\widehat{\boldsymbol{\tau}}_n) - Pl(\boldsymbol{\tau}_0) &\leq P_n l(\widehat{\boldsymbol{\tau}}_n) - Pl(\widehat{\boldsymbol{\tau}}_n). \end{aligned}$$

From the result of part (3) below, we have both $P_n l(\widehat{\boldsymbol{\tau}}_n) - Pl(\widehat{\boldsymbol{\tau}}_n)$ and $P_n l(\boldsymbol{\tau}_{0n}) - Pl(\boldsymbol{\tau}_{0n})$ converge to 0 almost surely; while $Pl(\boldsymbol{\tau}_{0n}) - Pl(\boldsymbol{\tau}_0)$ converges to 0 can be established from $\rho(\boldsymbol{\tau}_{0n}, \boldsymbol{\tau}_0) \to 0$ and the fact that $l(\cdot)$ is continuous and bounded.

(3) In this part, we will establish that

$$\sup_{\boldsymbol{\tau}_n \in B * A_n} |P_n l(\boldsymbol{\tau}_n) - Pl(\boldsymbol{\tau}_n)| \to 0$$

almost surely. This can be achieved through the following steps:

(i) Firstly, we show that

$$N(\varepsilon, A_n, L_\infty) \leq (6C_5C_6/\varepsilon)^m$$

where the constants $C_5$ and $C_6$ will be specified below. This is because for any $h_{n1}, h_{n2} \in A_n$ (hence each $h_{nv}(t) = \sum_u \theta_{un}^v \psi_u(t)$ with $v = 1, 2$),

$$\max_t |h_{n1}(t) - h_{n2}(t)| \le C_5 \max_u |\theta_{un}^1 - \theta_{un}^2|$$
$$\le C_5 \|\boldsymbol{\theta}^1 - \boldsymbol{\theta}^2\|_2$$

where $C_5$ is the upper bound of $\sum_u \psi_u(t)$ and $\boldsymbol{\theta}^v$ is an $m$-vector with elements $\theta_{un}^v$. Thus,

$$N(\varepsilon, A_n, L_\infty) \le N(\varepsilon/C_5, \{0 \le \theta_{un} \le C_6, 1 \le u \le m\}, L_2)$$
$$\le (6C_5C_6/\varepsilon)^m,$$

where the last inequality comes from Lemma 4.1 of Pollard (1984) and $C_6$ is the upper bound of all $\theta_{un}$.

(ii) Secondly, we wish to demonstrate that

$$N(\varepsilon, \mathcal{L}_n, L_\infty) \le K/\varepsilon^{p+m},$$

where the constant $K$ will be explicated below and

$$\mathcal{L}_n = \{l(\boldsymbol{\tau}_n), \boldsymbol{\tau}_n \in B * A_n\}.$$

In fact,

$$N(\varepsilon, \mathcal{L}_n, L_\infty) \le N(\varepsilon/2, B, L_2) \cdot N(\varepsilon/2, A_n, L_\infty)$$
$$\le (12C_1/\varepsilon)^p (12C_5C_6/\varepsilon)^m$$
$$= K/\varepsilon^{p+m},$$

where $C_1$ is the upper bound of $\{|\beta_j|, \forall j\}$ and $K = 12^{p+m} C_1^p (C_5C_6)^m$.

(iii) Select $\alpha_n = n^{-1/2+\phi_1}\sqrt{\log n}$ where $\phi_1 \in (\phi_0/2, 1/2)$ with $\phi_0 < 1$, and define $\varepsilon_n = \varepsilon\alpha_n$. Following Theorem 1 of Xue et al. (2004) we can show that

$$\text{Var}[P_n l(\boldsymbol{\tau}_n)]/(8\varepsilon_n^2) \to 0$$

when $n \to \infty$ and for any $\boldsymbol{\tau}_n \in B * A_n$.

(iv) Finally, from the result II.31 of Pollard (1984), we have

$$P[\sum_{B*A_n} |P_n l(\boldsymbol{\tau}_n) - P_n l(\boldsymbol{\tau})| > 8\varepsilon_n]$$
$$\le 8N(\varepsilon_n, \mathcal{L}_n, L_\infty) e^{-\frac{1}{128} n\varepsilon_n^2}$$
$$\le 8Ke^{-\left(\frac{1}{128}\varepsilon^2 + \frac{p+m}{2}\frac{\log(\varepsilon^2 n^{-1+2\phi_1}\log n)}{n^{2\phi_1}\log n}\right)n^{2\phi_1}\log n},$$

which converges to zero as the second term in bracket of the exponential function goes to zero when $n \to \infty$. Therefore, by the Borel-Cantelli lemma, we have

$$\sup_{B*A_n} |P_n l(\boldsymbol{\tau}_n) - Pl(\boldsymbol{\tau}_n)| \to 0$$

almost surely.

□

## A.2 Proof of Theorem 2.5

Let $\bar{l}(\boldsymbol{\eta}) = E_{\boldsymbol{\eta}_0}[n^{-1}l(\boldsymbol{\eta})]$. It follows from the strong law of large numbers that $n^{-1}l(\boldsymbol{\eta}) \to \bar{l}(\boldsymbol{\eta})$ almost surely and uniformly for $\boldsymbol{\eta} \in \Omega$.

This result, together with $\mu_n \to 0$ as $n \to \infty$ and $\boldsymbol{\eta}_0$ being the unique maximum of $\bar{l}(\boldsymbol{\eta})$ due to Assumption B2.2, implies that $\widehat{\boldsymbol{\eta}} \to \boldsymbol{\eta}_0$ almost surely by applying, for example, Corollary 1 of Honore & Powell (1994).

Next we prove the asymptotic normality result. From the KKT necessary conditions (2.27) and (2.28) we have that the constrained MPL estimate $\widehat{\boldsymbol{\eta}}$ satisfies

$$\mathbf{U}^{\mathsf{T}} \frac{\partial \Phi(\widehat{\boldsymbol{\eta}})}{\partial \boldsymbol{\eta}} = 0, \tag{A.3}$$

where matrix $\mathbf{U}$ is defined in equation (2.54) of Section 2.7.2. According to the Taylor expansion

$$\frac{\partial \Phi(\widehat{\boldsymbol{\eta}})}{\partial \boldsymbol{\eta}} = \frac{\partial \Phi(\boldsymbol{\eta}_0)}{\partial \boldsymbol{\eta}} + \frac{\partial^2 \Phi(\widetilde{\boldsymbol{\eta}})}{\partial \boldsymbol{\eta} \partial \boldsymbol{\eta}^{\mathsf{T}}}(\widehat{\boldsymbol{\eta}} - \boldsymbol{\eta}_0), \tag{A.4}$$

where $\widetilde{\boldsymbol{\eta}}$ is a vector between $\widehat{\boldsymbol{\eta}}$ and $\boldsymbol{\eta}_0$. Therefore,

$$0 = \mathbf{U}^{\mathsf{T}} \frac{\partial \Phi(\boldsymbol{\eta}_0)}{\partial \boldsymbol{\eta}} + \mathbf{U}^{\mathsf{T}} \frac{\partial^2 \Phi(\widetilde{\boldsymbol{\eta}})}{\partial \boldsymbol{\eta} \partial \boldsymbol{\eta}^{\mathsf{T}}}(\widehat{\boldsymbol{\eta}} - \boldsymbol{\eta}_0). \tag{A.5}$$

Next, let $\widehat{\boldsymbol{\eta}}^*$ be $\widehat{\boldsymbol{\eta}}$ after deleting the active constraints and $\boldsymbol{\eta}_0^*$ be similarly defined corresponding to $\boldsymbol{\eta}_0$, then

$$\widehat{\boldsymbol{\eta}} - \boldsymbol{\eta}_0 = \mathbf{U}(\widehat{\boldsymbol{\eta}}^* - \boldsymbol{\eta}_0^*). \tag{A.6}$$

Substituting (A.6) into (A.5), solving for $\widehat{\boldsymbol{\eta}}^* - \boldsymbol{\eta}_0^*$, and then using (A.6) to convert the result back to $\widehat{\boldsymbol{\eta}} - \boldsymbol{\eta}_0$ again, we eventually have

$$\sqrt{n}(\widehat{\boldsymbol{\eta}} - \boldsymbol{\eta}_0) = -\mathbf{U}\left(\mathbf{U}^{\mathsf{T}} \frac{1}{n} \frac{\partial^2 \Phi(\widetilde{\boldsymbol{\eta}})}{\partial \boldsymbol{\eta} \partial \boldsymbol{\eta}^{\mathsf{T}}} \mathbf{U}\right)^{-1} \mathbf{U}^{\mathsf{T}} \left(\frac{1}{\sqrt{n}} \frac{\partial l(\boldsymbol{\eta}_0)}{\partial \boldsymbol{\eta}} + o(1)\right). \tag{A.7}$$

In (A.7),

$$-\frac{1}{n}\frac{\partial^2 \Phi(\widetilde{\boldsymbol{\eta}})}{\partial \boldsymbol{\eta} \partial \boldsymbol{\eta}^{\mathsf{T}}} \to \mathbf{F}(\widetilde{\boldsymbol{\eta}}) \text{ (a.s.)}$$

by the law of large numbers. Also note that $\widetilde{\boldsymbol{\eta}} \to \boldsymbol{\eta}_0$ almost surely. Applying the central limit theorem to $n^{-1/2}\partial l(\boldsymbol{\eta}_0)/\partial \boldsymbol{\eta}$ we can show that this quantity converges in distribution to $N(\mathbf{0}, \mathbf{F}(\boldsymbol{\eta}_0))$. Then, the required asymptotic normality result follows from this result and equation (A.7). □

# B

# *Bivariate Normal identities*

## B.1 Conditional expectations of the standard BVN

The different conditional expectations used in the E-step of EM algorithm are given below.

Let $\Phi(z)$ denote the survival function (i.e. *upper-tail* cumulative distribution) of the univariate standard normal distribution. Its density is given by $\phi(z) = \exp(-z^2)/\sqrt{2\pi}$, for $-\infty < z < \infty$, and its hazard function $\Psi$ is $\Psi(z) = \phi(z)/\Phi(z)$.

Some properties of the bivariate Normal variates $(Z_1, Z_2)$ with mean vector $\mu = (0, 0)$, variance matrix $\mathbf{\Sigma}$, with $\sigma_{11} = \sigma_{22} = 1$ and $\sigma_{12} = \rho$ and density $p_{12}(z_1, z_2)$ are given below. In particular, we provide calculations for conditional expectations used in the E-step of EM algorithms for bivariate normal linear models.

Denote by $H$ the Heavyside function, the step function taking values 0 or 1 indicating whether its argument exceeds 0. Let $P(a, b) = P(Z_1 > a, Z_2 > b) = E[H(Z_1 - a)H(Z_2 - b)]$.

1. The conditional distribution of $(Z_1|Z_2 = b)$ is $\mathcal{N}(\rho b, 1-\rho^2)$ with density

$$p_{1|2}(z|b) = \frac{1}{\sqrt{1-\rho^2}}\,\phi\Big(\frac{z-\rho b}{\sqrt{1-\rho^2}}\Big). \tag{B.1}$$

2. Conditional expectations: moments of survival time to event given observed time to a competing event

$$E_{01}^{10}(a,b) = E(Z_1|Z_1 > a, Z_2 = b) \tag{B.2}$$

$$= \rho b + \sqrt{1-\rho^2}\,\Psi\Big(\frac{a-\rho b}{\sqrt{1-\rho^2}}\Big) \tag{B.3}$$

$$E_{01}^{20}(a,b) = E(Z_1^2|Z_1 > a, Z_2 = b) \tag{B.4}$$

$$= 1-\rho^2 + \sqrt{1-\rho^2}\,a\,\Psi\Big(\frac{a-\rho b}{\sqrt{1-\rho^2}}\Big) + \rho\, b\, E_{01}^{10} \tag{B.5}$$

DOI: 10.1201/9781351109710-B 

or,

$$\begin{aligned} E_{01}^{20}(a,b) &= E(Z_1^2|Z_1 > a, Z_2 = b) \\ &= 1-\rho^2+\sqrt{1-\rho^2}\,(a+\rho b)\,\Psi\Big(\frac{a-\rho\,b}{\sqrt{1-\rho^2}}\Big)+\rho^2\,b^2 \end{aligned} \tag{B.6}$$

3. Conditional expectations: moments given lower bounds

$$\begin{aligned} E_{00}^{11}(a,b) &= E(Z_1Z_2|Z_1 > a, Z_2 > b) \\ &= \rho+(1-\rho^2)\,\frac{p_{12}(a,b)}{P(a,b)} \\ &\quad + \rho\,\frac{a\,\phi(a)\,\Phi\left(\frac{b-\rho\,a}{\sqrt{1-\rho^2}}\right)+b\,\phi(b)\,\Phi\left(\frac{a-\rho\,b}{\sqrt{1-\rho^2}}\right)}{P(a,b)} \\ E_{00}^{10}(a,b) &= E(Z_1|Z_1 > a, Z_2 > b) \\ &= \frac{\phi(a)\Phi\left(\frac{b-\rho\,a}{\sqrt{1-\rho^2}}\right)+\rho\phi(b)\Phi\left(\frac{a-\rho\,b}{\sqrt{1-\rho^2}}\right)}{P(a,b)} \\ E_{00}^{20}(a,b) &= E(Z_1^2|Z_1 > a, Z_2 > b) \\ &= 1+\frac{a\phi(a)\Phi\left(\frac{b-\rho\,a}{\sqrt{1-\rho^2}}\right)+\rho^2b\phi(b)\Phi\left(\frac{a-\rho\,b}{\sqrt{1-\rho^2}}\right)+\rho(1-\rho^2)p_{12}(a,b)}{P(a,b)} \end{aligned} \tag{B.7}$$

These identities were developed by use of Stein identities (Liu, 1994). See also Tallis (1961). Proofs are available on request.

# *Bibliography*

Aalen, O. O. (1978), 'Non parametric inference for a family of counting processes', *Ann. Statist.* **6**, 701–726.

Aalen, O. O. (1989), 'A linear regression model for the analysis of life times', *Statistics in Medicine* **8**, 907–925.

Aitkin, M. (1981), 'A note on the regression analysis of censored data', *Technometrics* **23**(2), 161–163.
**URL:** *http://www.jstor.org/stable/1268032*

Allignol, A., Schumacher, M., Wanner, C., Drechsler, C. & Beyersmann, J. (2011), 'Understanding competing risks: a simulation point of view', *BMC Medical Research Methodology* **11**(1), 86.
**URL:** *https://doi.org/10.1186/1471-2288-11-86*

Amico, M. & van Keilegom, I. (2018), 'Cure models in survival analysis', *Annual Review of Statistics and its Application* **5**, 311–342.

Amico, M., Van Keilegom, I. & Han, B. (2021), 'Assessing cure status prediction from survival data using receiver operating characteristic curves', *Biometrika* **108**(3), 727–740.

Andersen, P. K. & Gill, R. D. (1982), 'Cox's regression model for counting processes: a large sample study', *Ann. Stats.* **10**, 1100–1120.

Andersen, P. K., Ørnulf, B., Richard, D. G. & Niels, K. (1993), *Statistical models based on counting processes*, Springer, New York.

Asano, J., Hirakawa, A. & Hamada, C. (2014), 'Assessing the prediction accuracy of cure in the cox proportional hazards cure model: an application to breast cancer data', *Pharmaceutical Statistics* **13**, 357–363.

Austin, P. C. (2012), 'Generating survival times to simulate cox proportional hazards models with time-varying covariates', *Statistics in Medicine* **31**, 3946–3958.

Barnard, J., McCulloch, R. & Meng, X. (2000), 'Modeling covariance matrices in terms of standard deviations and correlations, with application to shrinkage', *Statistica Sinica* **10**(4), 1281–1311.

Bertsekas, D. P. (1982), 'Projected newton methods for optimization problems with simple constraints', *SIAM J. Control and Optimization* **20**, 221–246.

Bobb, J. F. & Varadhan, R. (2018), *turboEM: A Suite of Convergence Acceleration Schemes for EM, MM and Other Fixed-Point Algorithms*. R package version 2018.1.
**URL:** *https://CRAN.R-project.org/package=turboEM*

Boyd, S. & Vandenberghe, L. (2004), *Convex Optimization*, Cambridge University Press, Cambridge.

Breslow, N. E. (1972), 'Discussion of the paper by D. R. Cox', *J. Roy. Statist. Soc. B* **34**, 216–217.

Brodaty, H., Woodward, M., Boundy, K., Ames, D. & Balshaw, R. (2011), 'Patients in Australian memory clinics: baseline characteristics and predictors of decline at six months', *Int Psychogeriatrics* **23**, 1086–1096.

Bryant, J. & Dignam, J. J. (2004), 'Semiparametric models for cumulative incidence functions', *Biometrics* **60**(1), 182–190.
**URL:** *https://EconPapers.repec.org/RePEc:bla:biomet:v:60:y:2004:i:1:p:182-190*

Cai, T. & Betensky, R. A. (2003), 'Hazard regression for interval-censored data with penalized spline', *Biometrics* **59**, 570–579.

Chan, R. H. & Ma, J. (2012), 'A multiplicative iterative algorithm for box-constrained penalized likelihood image restoration', *IEEE Trans. Image Processing* **21**, 3168–3181.

Chen, X., Hu, T. & Sun, J. (2017), 'Sieve maximum likelihood estimation for the proportional hazards model under informative censoring', *Computational Statistics and Data Analysis* **112**, 224–234.

Chen, Y. (2010), 'Semiparametric marginal regression analysis for dependent competing risks under an assumed copula', *Journal of the Royal Statistical Society. Series B* **72**, 235–251.

Chiou, S. H., Kang, S. & Yan, J. (2014), 'Fitting accelerated failure time models in routine survival analysis with r package aftgee', *Journal of Statistical Software* **61**(11), 1–23.
**URL:** *https://www.jstatsoft.org/index.php/jss/article/view/v061i11*

Chiou, S. H., Kang, S., Yan, J. et al. (2014), 'Fitting accelerated failure time models in routine survival analysis with R package aftgee', *Journal of Statistical Software* **61**(11), 1–23.

Choi, S. & Cho, H. (2019), 'Accelerated failure time models for the analysis of competing risks', *Journal of the Korean Statistical Society* **48**(3), 315 – 326.
**URL:** *http://www.sciencedirect.com/science/article/pii/S1226319218300723*

Clayton, D. G. (1978), 'A model for association in bivariate life tables and its application in epidemiological studies of familial tendency in chronic disease incidence', *Biomeirika* **65**, 141–151.

Corbière, F., Commenges, D., Taylor, J. M. & Joly, P. (2009), 'A penalized likelihood approach for mixture cure models', *Statistics in Medicine* **28**(3), 510–524.

Cox, D. R. (1972), 'Regression models and life tables', *J. Roy. Statist. Soc. B* **34**, 187–220.

Cox, D. R. (1975), 'Partial likelihood', *Biometrika* **62**, 269–276.

Cox, D. R. & Oakes, D. (1984), *Analysis of Survival Data*, Chapman Hall.

Crowder, M. (1984), 'On constrained maximum likelihood estimation with non-i.i.d. observations', *Ann. Inst. Stat. Math.* **36**, 239–249.

Crowder, M. (1991), 'On the identifiability crisis in competing risks analysis', *Scandinavian Journal of Statistics* **18**(3), 223–233.

Crowley, J. & Hu, M. (1977), 'Covariance analysis of heart transplant survival data', *J. Amer. Statist. Assoc.* **72**, 27–36.

DeBoor, C. & Daniel, J. W. (1974), 'Splines with nonnegative B-spline coefficients', *Mathematics of Computation* **126**, 565–568.

Dempster, A. P., Laird, N. M. & Rubin, D. B. (1977), 'Maximum likelihood from incomplete data via the EM algorithm', *Journal of the Royal Statistical Society. Series B (Methodological)* **39**(1), 1–38.
**URL:** *http://www.jstor.org/stable/2984875*

Deresa, N. & Van Keilegom, I. (2020), 'A multivariate normal regression model for survival data subject to different types of dependent censoring', *Computational Statistics & Data Analysis* **144**.

Dettoni, R., Marra, G. & Radice, R. (2020), 'Generalized link-based additive survival models with informative censoring', *J Comp. Grap. Stats.* **29**, 503–512.

Dignam, J. J. & Kocherginsky, M. N. (2008), 'Choice and interpretation of statistical tests used when competing risks are present', *Journal of Clinical Oncology* **26**(24), 4027–4034.

Dignam, J. J., Zhang, Q. & Kocherginsky, M. N. (2012), 'The use and interpretion of competing risks regression models', *Clinical Cancer Research* **18**(8), 2301–2308.

Dinse, G. & Larson, M. (1986), 'A note on semi-Markov models for partially censored data', *Biometrika* **73**, 379–86.

Dörre, A. & Emura, T. (2019), *Analysis of Doubly Truncated Data An Introduction*, Springer.

Efron, B. & Petrosian, R. (1999), 'Nonparametric methods for doubly truncated data', *J. Amer. Statist. Assoc.* **94**, 824–834.

Emura, T. & Chen, Y.-H. (2018), *Analysis of Survival Data with Dependent Censoring*, Springer.

Eubank, R. L. (1999), *Nonparametric Regression and Spline Smoothing*, Marcel Dekker.

Farewell, V. (1982), 'The use of mixture cure models for the analysis of survival data with long-term survivors', *Biometrics* **38**(4), 1041–1046.

Farrington, C. P. (1996), 'Interval censored survival data: a generalized linear modelling approach', *Statistics in Medicine* **15**(3), 283–292.

Fine, J. P., Jiang, H. & Chappell, R. (2001), 'On semi-competing risks data', *Biometrika* **88**, 907–919.

Finkelstein, D. M. (1986), 'A proportional hazard model for interval-censored failure time data', *Biometrics* **42**, 845–854.

Fix, E. & Neyman, J. (1951), 'A simple stochastic model of recovery, relapse, death and loss of patients', *Human Biology* **23**(3), 205–241.
**URL:** *http://www.jstor.org/stable/41448000*

Frank, M. J. (1979), 'On the simultaneous associativity of $f(x, y)$ and $x + y - f(x, y)$', *Aequationes Mathematicae* **19**, 194–226.

Gares, V., Hudson, M., Manuguerra, M. & Gebski, V. (2019), Bivariate normal censored linear models for competing risks, Technical report, Macquarie University. (submitted to SMMR).

Gaynor, J. J., Feuer, E. J., Tan, C. C., Wu, D. H., Little, C. R., Straus, D. J., Clarkson, B. D. & Brennan, M. F. (1993), 'On the use of cause-specific failure and conditional failure probabilities: Examples from clinical oncology data', *Journal of the American Statistical Association* **88**(422), 400–409.
**URL:** *http://www.jstor.org/stable/2290318*

Ghosh, D. (2001), 'Efficiency considerations in the additive hazards model with current status data', *Statistica Neerlandica* **55**(3), 367–376.

Grambsch, P. M. & Therneau, T. M. (1994), 'Proportional hazards tests and diagnostics based on weighted residuals', *Biometrika* **81**, 515–526.

Green, P. J. (1990), 'On use of the EM algorithm for penalized likelihood estimation', *Journal of the Royal Statistical Society. Series B (Methodological)* **52**(3), 443–452.
**URL:** *JSTOR, www.jstor.org/stable/2345668*

Green, P. J. & Silverman, B. W. (1994), *Nonparametric Regression and Generalized Linear Models–A roughness penealty approach*, Chapman and Hall, London.

Greenwood, M. (1926), 'A report on the natural duration of cancer', *Reports on Public Health and Medical Subjects* **33**, 1–26.

Grenander, U. (1981), *Abstract Inference*, J. Wiley, New York.

Groeneboom, P. & Wellner, J. A. (1992), *Information Bounds and Nonparametric Maximum Likelihood Estimation*, Birkhauser, Basel.

Gumbel, E. J. (1961), 'Distributions des valeurs extrême enplusiers dimensions', *Publications Institute Statistical University Pairs* **9**, 171–173.

Heinze, G. & Dunkler, D. (2008), 'Avoiding infinite estimates of time-dependent effects in small-sample survival studies', *Statistics in Medicine* **27**, 6455–6469.

Honore, B. E. & Powell, J. L. (1994), 'Pairwise difference estimators of censored and truncated regression models', *Journal of Econometrics* **64**, 241–278.

Hosmer, D., Lemeshow, S. & May, S. (2008), *Applied Survival Analysis: Regression Modeling of Time to Event Data, 2nd ed*, John Wiley.

Hougaard, P. (1986), 'A class of multivariate failure time distribution', *Biometrika* **73**, 671–678.

Hougaard, P. (1999), 'Fundamentals of survival data', *Biometrics* **55**(1), 13.

Hougaard, P. (2000), *Analysis of Multivariate Survival Data*, Springer.

Huang, J. (1996), 'Efficient estimation for the proportional hazard model with interval censoring', *The Annals of Statistics* **24**, 540–568.

Huang, J. & Rossini, A. J. (1997), 'Sieve estimation for the proportional-odds failure-time regression model with interval censoring', *J. Amer. Statist. Assoc.* **92**, 960–967.

Huang, X. & Wolfe, R. A. (2002), 'A frailty model for informative censoring', *Biometrics* **58**, 510–520.

Huang, X. & Zhang, N. (2008), 'Regression survival analysis with an assumed copula for dependent censoring: A sensitivity approach', *Biometrics* **64**, 1090–1099.

Hudson, H., Gares, V., Manuguerra, M. & Gebski, V. (2018), Correlated bivariate normal competing risks – simulation findings in an ill-posed problem. ISCB ASC Melbourne 2018. Presentation (slides) available on request.

Hudson, H. M., Manuguerra, M. & Gares, V. (2019), 'R package bnc: bivariate normal censored linear model for competing risks'. [Online; accessed 19-December-2019].
**URL:** *https://bitbucket.org/malcolm_hudson/bnc/src/master/*

Huffer, F. W. & McKeague, I. W. (1991), 'Weighted least squares estimation for aalen's additive risk model', *J. Amer. Statist. Assoc.* **86**, 114–129.

Hurvich, C. M., Simonoff, J. S. & Tsai, C.-L. (1998), 'Smoothing parameter selection in nonparametric regression using an improved Akaike information criterion', *J. Roy. Statist. Soc. B* **60**, 271–293.

Jamshidian, M. & Jennrich, R. I. (2000), 'Standard errors for EM estimation', *Journal of the Royal Statistical Society, Series B.* **62**, 257–270.

Jeong, J.-H. & Fine, J. P. (2006), 'Parametric regression on cumulative incidence function', *Biostatistics* **8**(2), 184–196.

Jeong, J.-H. & Fine, J. P. (2007), 'Parametric regression on cumulative incidence function', *Biostatistics* **8**(2), 184.

Jin, Y., Ton, T. G. N., Incerti, D. & Hu, S. (2022), 'Left truncation in linked data: a practical guide to understanding left truncation and applying it using sas and r', *Pharmaceutical Statistics* **22(1)**, 194–202. https://doi.org/10.1002/pst.2257.

Jin, Z., Lin, D. Y., Wei, L. J. & Ying, Z. (2003), 'Rank-based inference for the accelerated failure time model', *Biometrika* **90**, 341–353.

Joffe, M. M. (2001), 'Administrative and artificial censoring in censored regression models', *Statistics in Medicine* **20**, 2287–2304.

Joly, P., Commenges, D. & Letenneur, L. (1998), 'A penalized likelihood approach for arbitrarily censored and truncated data: application to age-specific incidence of dementia', *Biometrics* **54**, 185–194.

Kalbfleisch, J. D. & Prentice, R. (1981), 'Estimation of the average hazard ratio', *Biometrika* **68**(1), 105–112.

Kalbfleisch, J. D. & Prentice, R. L. (1973), 'Marginal likelihoods based on cox's regression and life model', *Biometrika* **60**, 267–278.

Kalbfleisch, J. D. & Prentice, R. L. (2002), *The Statistical Analysis of Failure Time Data*, Academic Press.

Kaplan, E. L. & Meier, P. (1958), 'Nonparametric estimation from incomplete observations', *J. Amer. Statist. Assoc.* **53**, 457–481.

Kendall, M. G. (1970), *Rank Correlation Method (4th ed.)*, Charles Griffin & Co., London.

Kim, J. S. (2003), 'Maximum likelihood estimation for the proportional hazard model with partly interval-censored data', *J. Roy. Statist. Soc. B* **65**, 489–502.

Klein, J. P. & Moeschberger, M. L. (2003), *Survival Analysis Techniques for Censored and Truncated Data*, Springer, New York.

Kleinbaum, D. G. & Klein, M. (2005), *Survival Analysis: A Self-Learning Text*, Springer, New York.

Komárek, A., Lesaffre, E. & Hilton, J. F. (2005), 'Accelerated failure time model for arbitrarily censored data with smoothed error distribution', *Journal of Computational and graphical Statistics* **14**, 726–745.

Korn, E. L. & Dorey, F. J. (1992), 'Applications of crude incidence curves.', *Statistics in Medicine* **11 6**, 813–29.
**URL:** *https://api.semanticscholar.org/CorpusID:8681892*

Lagakos, S., Sommer, C. & Zelen, M. (1978), 'Semi-Markov models for partially censored data', *Biometrika* **65**, 311–317.

Larson, M. G. & Dinse, G. E. (1985), 'A mixture model for the regression analysis of competing risks data', *Journal of the Royal Statistical Society: Series C (Applied Statistics)* **34**(3), 201–211.

Lawless, J. (2003), *Statistical Models and Methods for Lifetime Data*, 2nd ed. edn, John Wiley, Hoboken, N.J.

Lawless, J. F. (2002), *Statistical Models and Methods for Lifetime Data*, John Wiley.

LeCam, L. (1970), 'On the assumptions used to prove asymptotic normality of maximum likelihood estimates', *Ann. Math. Statist.* **41**, 802–828.

Li, J. & Ma, J. (2019), 'Maximum penalized likelihood estimation of additive hazards models with partly interval censoring', *Computational Statistics and Data Analysis* **137**, 170–180.

Li, J. & Ma, J. (2020), 'On hazard-based penalized likelihood estimation of accelerated failure time model with partly interval censoring', *Statistical Methods in Medical Research* **29**, 3804 – 3817.

Lin, D. (2007), 'On the breslow estimator', *Lifetime Data Anal* **13**, 471–480.

Lin, D., Oakes, D. & Ying, Z. (1998), 'Additive hazards regression with current status data', *Biometrika* **85**, 289–298.

Lin, D. Y. & Ying, Z. (1994), 'Semiparametric analysis of the additive risk model', *Biometrika* **81**, 61–71.

Lindsay, J. C. & Ryan, L. M. (1998), 'Methods for interval-censored data', *Statistics in Medicine* **17**, 219–238.

Liu, J. S. (1994), 'Siegel's formula via Stein's identities', *Statistics and Probability Letters* **21**(3), 247–251.

Liu, X. (2012), 'Planning of accelerated life tests with dependent failure modes based on a gamma frailty model', *Technometrics* **54**(4), 398–409.
**URL:** *http://www.tandfonline.com/doi/abs/10.1080/00401706.2012.707579*

Luenberger, D. (1984), *Linear and Nonlinear Programming (2nd edition)*, J. Wiley.

Luenberger, D. & Ye, Y. (2008), *Linear and Nonlinear Programming (3rd edition)*, Springer, New York.

Ma, D., Ma, J. & Graham, P. (2023), 'On semiparametric accelerated failure time models with time-varying covariates: a maximum penalised likelihood estimation', *Statistics in Medicine.* (doi: 10.1002/sim.9926) .

Ma, J., Couturier, D.-L., Heritier, S. & Marschner, I. (2021), 'Penalized likelihood estimation of the proportional hazards model for survival data with interval censoring', *International Journal of Biostatistics* **18(2)**, 553–575.

Ma, J., Heritier, S. & Lô, S. (2014), 'On the maximum penalized likelihood approach for proportional hazard models with right censored survival data', *Computational Statistics and Data Analysis* **74**, 142–156.

Ma, J. & Hudson, H. M. (1998), 'An augmented data scoring algorithm for maximum likelihood', *Communications in Statistics-Theory and Methods* **26**, 2761–2776.

Mandel, M., de Una-Alvarez, J., Simon, D. K. & Betensky, R. A. (2018), 'Inverse probability weighted cox regression for doubly-truncated data', *Biometrics* **74**(2), 481–487.

Martinussen, T. & Scheike, T. H. (2006), *Dynamic Regression Models for Survival Data*, Springer, New York.

Moore, T. J. & Sadler, B. M. (2006), 'Maximum-likelihood estimation and scoring under parametric constraints', *Army Research Laboratory* **ARL-TR-3805**.

Moore, T. J., Sadler, B. M. & Kozick, R. J. (2008), 'Maximum-likelihood estimation, the Cramér – Rao bound, and the method of scoring with parameter constraints', *IEEE Trans. Sig. Proc.* **56**, 895–908.

Morton, D., Thompson, J., Cochran, A., Mozzillo, N., Nieweg, O., Roses, D., Hoekstra, H., Karakousis, C., Puleo, C., Coventry, B., Kashani-Sabet, M., Smithers, B., Paul, E., Kraybill, W., McKinnon, J., Wang, H.-J., Elashoff, R., Faries, M. & for the MSLT Group (2014), 'Final trial report of sentinel-node biopsy versus nodal observation in melanoma', *The New England Journal of Medicine* **307**, 599–609.

Murphy, S. A. & van der Vaart, A. W. (2000), 'On profile likelihood', *J. Amer. Statist. Assoc.* **95**, 449–4651.

Nelsen, R. B. (2006), *An Introduction to copulas*, (2nd edition) edn, Springer, New York.

Nelson, W. (1972), 'Theory and applications of hazard plotting for censored failure data', *Technometrics* **42**, 12–25.

Nicolaie, M. A., van Houwelingen, H. C. & Putter, H. (2010), 'Vertical modeling: a pattern mixture approach for competing risks modeling', *Statist. Med.* **29**, 1190–1205.

Nocedal, J. & Wright, S. J. (2000), *Numerical Optimization*, Springer, New York.

O'Sullivan, F. (1986), 'A statistical perspective on ill-posed inverse problems', *Statistical Science* **1**(4), 502–518.
**URL:** *https://doi.org/10.1214/ss/1177013525*

Owen, A. B. (2001), *Empirical Likelihood*, Chapman and Hall, New York.

Pan, W. (1999), 'Extending the iterative convex minorant algorithm to the Cox model for interval-censored data', *Journal of Computational and Graphical Statistics* **8**, 109–120.

Pan, W. (2000), 'A multiple imputation approach to cox regression with interval-censored data', *Biometrics* **56**, 199–203.

Peng, Y. (2003), 'Estimating baseline distribution in proportional hazards cure model', *Computational Statistics  Data Analysis* **42**, 187–201.

Pepe, M. S. & Mori, M. (1993), 'Kaplan-Meier, marginal or conditional probability curves in summarizing competing risks failure time data?', *Statistics in Medicine* **12**, 737–751.

Pollard, D. (1984), *Convergence of Stochastic Processes*, Springer Verlag, New York.

Prentice, R. L. (1978), 'Linear rank tests with right censored data', *Biometrika* **65**, 167–179.

Prentice, R. L., Kalbfleisch, J. D., Peterson, A., Flournoy, N., Farewell, V. T. & Breslow, N. E. (1978), 'The analysis of failure times in the presence of competing risks', *Biometrics* **34**(4), 541–554.
**URL:** *http://www.jstor.org/stable/2530374*

Putter, H., Fiocco, M. & Geskus, R. B. (2007), 'Tutorial in biostatistics: competing risks and multi-state models', *Statistics in Medicine* **26**, 2389–2430.

Ramsay, J. O. (1988), 'Monotone regression splines in action', *Statistical Science* **3**, 425–441.

Rathnayake, K. (2019), *Parameter estimation for additive hazards model with partly interval-censored failure time data using a penalized likelihood approach* (PhD Thesis), Macquarie University, Sydney.

Rennert, L. & Xie, S. X. (2018), 'Cox regression model with doubly truncated data', *Biometrics* **74**(2), 725–733.

Rennert, L. & Xie, S. X. (2019), 'Bias induced by ignoring double truncation inherent in autopsy-confirmed survival studies of neurodegenerative diseases', *Statistics in Medicine* **38**, 3599–3613.

Rizopoulos, D. (2012), *Joint Models for Longitudinal and Time-to-Event Data With Applications in R*, Chapman and Hall, CRC.

Rodrıguez, G. (2005), 'Non-parametric estimation in survival models', online.
**URL:** *https://data.princeton.edu/pop509/NonParametricSurvival.pdf*

Rudin, L., Osher, S. & Fatemi, E. (1992), 'Nonlinear total variation based noise removal algorithms', *Physica D* **60**, 259–268.

Ruppert, D., Wand, M. P. & Carroll, R. J. (2003), *Semiparametric Regression*, Cambridge, New York.

Saha, A., Harowicz, M., Grimm, L., Kim, C., Ghate, S., Walsh, R. & Mazurowski, M. (2018), 'A machine learning approach to radiogenomics of breast cancer: a study of 922 subjects and 529 DCE-MRI features', *British Journal of Cancer* **119**(4), 508–516.

Schisterman, E. F., Cole, S. R., Ye, A. & Platt, R. W. (2013), 'Accuracy loss due to selection bias in cohort studies with left truncation', *Paediatric and Perinatal Epidemiology* **27**, 491–502.

Schmee, J. & Hahn, G. J. (1979), 'A simple method for regression analysis with censored data', *Technometrics* **21**(4), 417–432.
**URL:** *http://www.jstor.org/stable/1268280*

Schoenfeld, D. (1982), 'Partial residuals for the proportional hazards regression model', *Biometrika* **69**, 239–241.

Shao, L., Li, H., Li, S. & Sun, J. (2023), 'A pairwise pseudo-likelihood approach for regression analysis of left-truncated failure time data with various types of censoring', *BMC Medical Research Methodology* **23**, article number 82.

Shen, X. (1997), 'On methods of sieve and penalization', *Ann. Statist.* **25**, 2555–2591.

Shen, X. & Wong, W. H. (1994), 'Convergence rate of sieve estimates', *Ann. Statist.* **22**, 580–615.

Silverman, B. (1982), 'On the estimation of a probability density function by the maximum penalized likelihood method', *The Annals of Statistics* **10**, 795 – 810.

Speckman, P. (1988), 'Kernel smoothing in partial linear models', *J. Roy. Statist. Soc. B* **50**, 413–436.

Sun, J. (2006), *The Statistical Analysis of Interval-Censored Failure Time Data*, Springer.

Sun, W. & Yuan, Y. X. (2006), *Optimization Theory and Methods: nonlinear Programming*, Vol. 1, Springer.

Sy, J. P. & Taylor, J. M. (2000), 'Estimation in a cox proportional hazards cure model', *Biometrics* **56**(1), 227–236.

Thackham, M. & Ma, J. (2020), 'On maximum likelihood estimation of the semiparametric cox model with time-varying covariates', *Journal of Applied Statistics* **47**, 1511–1528.

Tikhonov, T. & Arsenin, V. (1977), *Solutions of Ill-Posed Problems*, Wiley, New York.

Tsiatis, A. (1975), 'A nonidentifiability aspect of the problem of competing risks', *Proceedings of the National Academy of Sciences of the United States of America* **72**(1), 20.
**URL:** *http://www.pnas.org/content/72/1/20.abstract*

Tsodikov, A. D. (1998), 'A proportional hazards model taking account of long-term survivors', *Biometrics* **54**, 1508–1516.

Tsodikov, A., Ibrahim, J. & Yakovlev, A. (2003), 'Estimating cure rates from survival data: an alternative to two-component mixture models', *Journal of the American Statistical Association* **98**(464), 1063–1078.

Wahba, G. (1985), 'A comparison of gcv and gml for choosing the smoothing parameter in the generalized spline smoothing problem', *The Annals of Statistics* **13**, 1378–1402.

Wang, L., McMahan, C., Hudgens, M. & Qureshi, Z. (2016), 'A flexible, computationally efficient method for fitting the proportional hazards model to interval-censored data', *Biometrics* **72**, 222–231.

Wang, L., Sun, J. & Tong, X. (2010), 'Regression analysis of case II interval-censored failure time data with the additive hazards model', *Statistica Sinica* **20**, 1709–1723.

Wang, P., Li, D. & Sun, J. (2021), 'A pairwise pseudo-likelihood approach for left-truncated and interval-censored data under the cox model', *Biometrics* **77**, 1303–1314.

Webb, A. & Ma, J. (2023), 'Cox models with time-varying covariates and partly-interval censoring – a maximum penalised likelihood approach', *Statistics in Medicine* **42**, 815–833.

Webb, A., Ma, J. & Lo, S. (2022), 'Penalized likelihood estimation of a mixture cure cox model with partly-interval censoring–an application to thin melanoma', *Statistics in Medicine* **41**(17), 3260–3280.

Wei, L. J. (1992), 'The accelerated failure time model: a useful alternative to the Cox regression model in survival analysis', *Statistics in Medicine* **11**, 1871–1879.

Wong, G. Y. C., Osborne, M. P., Diao, Q. & Yu, Q. (2017), 'Piecewise cox models with right-censored data', *Communications in Statistics - Simulation and Computation* **46**, 7894–7908.

Wong, W. H. & Severini, T. A. (1991), 'On maximum likelihood estimation in infinite dimensional parameter spaces', *Ann. Statist.* **19**, 603–632.

Wong, W. H. & Shen, X. (1995), 'Probability inequalities for likelihood ratios and convergene rates of sieve MLEs', *Ann. Statist.* **23**, 339–362.

Wood, S. (2011), 'Fast stable restricted maximum likelihood and marginal likelihood estimation of semiparametric generalized linear models', *J. Roy. Statist. Soc. B* **73**, 3–36.

Wright, S. J. (1997), *Primal-dual interior-point methods*, SIAM.

Wu, C. F. (1983), 'On the convergence properties of the EM algorithm', *The Annals of Statistics* **11**, 95–103.

Xu, J., Ma, J., Connors, M. H. & Brodaty, H. (2018), 'Proportional hazard model estimation under dependent censoring using copulas and penalized likelihood', *Statistics in Medicine* **37**, 2238–2251.

Xu, J., Ma, J. & Prvan, T. (2016), 'Nonparametric hazard estimation with dependent censoring using penalized likelihood and an assumed copula', *Communications in Statistics–Theory and Methods* **46**, 11383–11403.

Xue, H., Lam, K. F. & Li, G. (2004), 'Sieve maximum likelihood estimator for semi-parametric regression models with current status data', *J. Amer. Statist. Assoc.* **99**, 346–356.

Yakovlev, A., Tsodikov, A. & Asselain, B. (1996), *Stochastic Models of Tumor Latency and their Biostatistical Applications*, World Scientific, Singapore.

Yu, Y. & Ruppert, D. (2002), 'Penalized spline estimation for partially linear single index models', *J. Amer. Statist. Assoc.* **97**, 1042–1054.

Zeng, D., Cai, J. & Shen, Y. (2006), 'Semiparametric additive risks model for interval-censored data', *Statistica Sinica* **16**, 287–302.

Zhang, Y., Han, X. & Shao, Y. (2021), 'The roc of cox proportional hazard cure models with application in cancer studies', *Lifetime Data Analysis* **27**, 195–215.

Zhang, Y., Hua, L. & Huang, J. (2010), 'A spline-based semiparametric maximum likelihood esimation method for the Cox model with interval-censored data', *Scand. J. Statist.* **37**, 338–354.

# Index

Milton Keynes UK
Ingram Content Group UK Ltd.
UKHW031328071224
451979UK00004B/31

9 780815 362845